C Magno[?]
Dept of Zoology
University of Bristol.

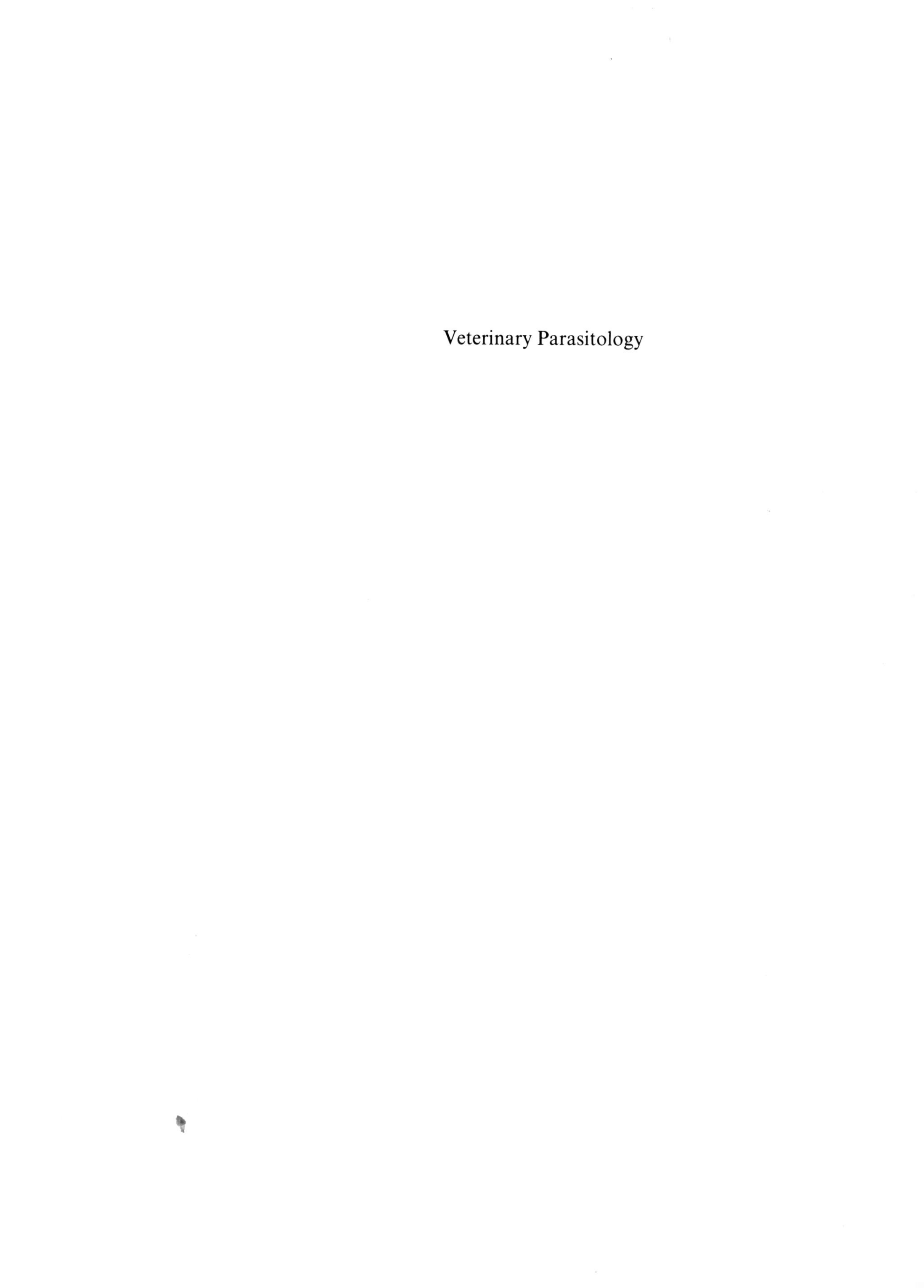

Veterinary Parasitology

Veterinary Parasitology

G. M. Urquhart
J. Armour
J. L. Duncan
A. M. Dunn
F. W. Jennings

Department of
Veterinary Parasitology,
The Faculty of Veterinary Medicine,
The University of Glasgow,
Scotland.

Longman Scientific & Technical,
Longman Group UK Limited,
Essex CM20 2JE, England
and Associated companies throughout the world.

Published in the United States of America
by Churchill Livingstone Inc., New York

First published 1987

British Library Cataloguing in Publication Data

Veterinary parasitology
1. Veterinary parasitology
I. Urquhart, G.M.
636.089'696 SF810.A3

ISBN 0-582-40906-3

Library of Congress Cataloging-in-Publication Data
Main entry under title:

Veterinary parasitology.
Bibliography: p.
Includes index.
1. Veterinary parasitology. I. Urquhart, G.M.
(George M.), 1925–
SF810.A3v425 1986 636.089'696 85–19789
ISBN 0–582–40906–3

Set in 9/10pt Lasercomp Times

Printed and Bound in
Great Britain at the Bath Press, Avon

CONTENTS

FOREWORD

This book is intended for students of veterinary parasitology, for practising veterinarians and for others requiring information on some aspect of parasitic disease.

Originally intended as a modestly expanded version of the printed notes issued to our students in the third and fourth years of the course, the text, perhaps inevitably, has expanded. This was due to three factors. First, a gradual realisation of the deficiencies in our notes; secondly, the necessity of including some of the comments normally imparted during the lecture course and thirdly, at the suggestion of the publishers, to the inclusion of certain aspects of parasitic infections not treated in any detail in our course.

We should perhaps repeat that the book is primarily intended for those who are directly involved in the diagnosis, treatment and control of parasitic diseases of domestic animals. The most important of these diseases have therefore been discussed in some detail, the less important dealt with more briefly and the uncommon either omitted or given a brief mention. Also, since details of classification are of limited value to the veterinarian we have deliberately kept these to the minimum sufficient to indicate the relationships between the various species. For a similar reason, taxonomic detail is only presented at the generic level and, occasionally, for certain parasites, at species level. We have also trod lightly on some other areas such as, for example, the identification of species of tropical ticks and the special significance and epidemiology of some parasites of regional importance. In these cases, we feel that instruction is best given by an expert aware of the significance of particular species in that region.

Throughout the text we have generally referred to drugs by their chemical, rather than proprietary, names because of the plethora of the latter throughout the world. Also, because formulations are often different, we have avoided stating doses; for these, reference should be made to the data sheets produced by the manufacturer. However, on occasions when a drug is recommended at an unusual dose, we have noted this in the text.

In the chapters at the end of the book we have attempted to review five aspects of veterinary parasitology, epidemiology, immunity, anthelmintics, ectoparasiticides and laboratory diagnosis. We hope that this broader perspective will be of value to students, and particularly to those dismayed by the many complexities of the subject.

There are no references in the text apart from those at the end of the chapter on diagnosis. This was decided with some regret and much relief on the grounds that it would have meant the inclusion, in a book primarily intended for undergraduates, of hundreds of references. We hope that those of our colleagues throughout the world who recognise the results of their work in the text will accept this by way of explanation and apology.

We would, however, like to acknowledge our indebtedness to the authors of several source books on veterinary parasitology whose work we have frequently consulted. These include *Medical and Veterinary Protozoology* by Adam, Paul and Zaman, *Veterinaermedizinische Parasitologie* by Boch and Supperer, Dunn's *Veterinary Helminthology*, Euzéby's *Les Maladies Vermineuses des Animaux Domestiques*, Georgi's *Parasitology for Veterinarians*, Reinecke's *Veterinary Helminthology*, Service's *A Guide to Medical Entomology* and Soulsby's *Helminths, Arthropods and Protozoa of Domesticated Animals*.

Any student seeking further information on specific topics should consult these or, alternatively, ask his tutor for a suitable review.

The ennui associated with repeated proof-reading may occasionally (we hope, rarely) have led to some errors in the text. Notification of these would be welcomed by the authors. Finally we hope that the stresses endured by each of us in this collaborative venture will be more than offset by its value to readers.

ACKNOWLEDGEMENTS

We would like to express our gratitude to the following individuals and organisations who assisted us in the preparation of this book.

First, to Drs R. Ashford and W. Beesley of Liverpool; Dr J. Bogan, Glasgow; Dr W. Campbell, Rahway, USA; Dr R. Dalgleish, Brisbane; Dr L. Joyner, Weybridge, England; Dr T. Miller, Florida; Dr M. Murray, Nairobi; Dr R. Purnell, Sandwich, England; Dr S. M. Taylor, Belfast; Professor K. Vickerman, Glasgow, Each of these read and commented on sections of the text in which they are expert. Any errors in these areas are, however, solely the responsibility of the authors.

Secondly, to the following individuals and companies who kindly allowed us to use their photographs or material as illustrations or plates:

Dr E. Allonby, Nairobi (Pl. I d, e, f); Dr K. Angus, Edinburgh (Fig. 167); Dr J. Arbuckle, Guildford, England (Fig. 61); Dr E. Batte, North Carolina, USA (Pl. IIIf); Dr I. Carmichael, Johannesburg, S. Africa (Fig. 142); Dr L. Cramer, Sao Paulo (Fig. 126b); Crown Copyright, UK (Pl. XIVb); Dr J. Dunsmore, Murdoch, W. Australia (Pl. IVd); Prof. J. Eckert, Zurich (Fig. 96); Glaxovet, Harefield, England (Pl. IIf); Dr I. Herbert, Bangor, Wales (Fig. 172); Dr A. Heydorn, W. Berlin (Figs 170, 171); Prof. F. Hörning, Berne (Fig. 82; Pl. Ve); Dr B. Iovanitti, Balcarce, Argentina (Figs 22, 23); Dr D. Jacobs, London (Fig. 38); Drs D. Kelly and A. Longstaffe, Bristol (Figs 156, 157); The late Dr I. Lauder, Glasgow (Fig. 65, Pl. XIc, e, XIIb); Drs B. Lindemann and J. McCall, Georgia, USA (Fig. 67); Dr N. McEwan, Glasgow (Pl. XId, XIIe); Dr G. Mitchell, Ayr, Scotland (Pl. VIe); Prof. M. Murray, Glasgow (Figs 68, 84, 152); Dr A. Nash, Glasgow (Fig. 138b, Pl. XIIc); Dr Julia Nicholls, Adelaide, Australia (Figs 6, 14c, d); Dr R. Purnell, Sandwich, England (Fig. 173, Pl. VIIId, e, f); Prof. H. Pirie, Glasgow (Fig. 40); Dr J. Reid, Brussels (Pl. XIIa); Dr Elaine Rose, Houghton Poultry Research Station, Huntingdon, England (Figs 160, 163b, 164a, b); Prof. I. Selman, Glasgow (Pl. XIf); Dr D. Taylor, Glasgow (Pl. XIVc); Dr M. Taylor, London (Fig. 85); Dr S. Taylor, Belfast (Pl. IIa); Dr H. Thompson, Glasgow (Fig. 92, Pl. IVb, c, VId); Dr R. Titchener, Ayr, Scotland (Fig. 113b, Pl. VIIIa); Dr A. Waddell, Brisbane, Australia (Fig. 66, Pl. IVe); Wellcome Research Laboratories, Berkhamsted, England (Pl. VIIIc); Dr A. Wright, Bristol (Pl. VIb, XIb, XIId, f).

In this context we are also extremely grateful to Miss E. Urquhart, Wrexham, Wales who prepared many of the line drawings.

Thirdly, to the pharmaceutical companies of Crown Chemical, Kent, England; Hoechst UK, Bucks; Merck Sharp & Dohme, Herts; Pfizer, Kent; Schering, New Jersey; Syntex Agribusiness, California. Their generosity enabled us to present many of the photographs in colour, thus enhancing their value.

Finally, to those members of the Faculty of Veterinary Medicine, Glasgow, whose cooperation was essential in the production of this book. We would especially like to thank Kenneth Bairden, our chief technician, who prepared much of the material for photography, often at inordinately short notice; Archie Finnie and Allan May, of the Photographic Unit, who, almost uncomplainingly, undertook the extra work of photographing many specimens; our two departmental secretaries, Elizabeth Millar and Julie Nybo without whose skill and attention to detail this book would certainly not have been written.

G. M. Urquhart
J. Armour
J. L. Duncan
A. M. Dunn
F. W. Jennings
September 1985

VETERINARY HELMINTHOLOGY

PRINCIPLES OF CLASSIFICATION

All animal organisms are related to one another, closely or remotely, and the study of the complex systems of inter-relationship is called **systematics**. It is essentially a study of the evolutionary process.

When organisms are examined it is seen that they form natural groups with features, usually morphological, in common. A group of this sort is called a **taxon**, and the study of this aspect of biology is called **taxonomy**.

The taxa in which organisms may be placed are recognised by international agreement, and the chief ones are: **Kingdom**, **Phylum**, **Class**, **Order**, **Family**, **Genus** and **Species**. The intervals between these are large, and some organisms cannot be allocated to them precisely, so that intermediate taxa, prefixed appropriately, have been formed; examples of these are the **Suborder** and the **Superfamily**. As an instance, the taxonomic status of one of the common abomasal parasites of ruminants may be expressed as shown in the next column.

The names of taxa must be adhered to according to the international rules, but it is permissible to anglicise the endings, so that members of the superfamily Trichostrongyloidea in the example above may also be termed trichostrongyloids.

The names of the genus and species are expressed in Latin form, the generic name having a capital letter, and

Kingdom	Animalia
Phylum	Nemathelminthes
Class	Nematoda
Order	Strongylida
Suborder	Strongylina
Superfamily	Trichostrongyloidea
Family	Trichostrongylidae
Subfamily	Haemonchinae
Genus	*Haemonchus*
Species	*contortus*

they must be in grammatical agreement. It is customary to print foreign words in italics, so that the name of an organism is usually underlined or italicised. Accents are not permitted, so that, if an organism is named after a person, amendment may be necessary; the name of Müller, for example, has been altered in the genus *Muellerius*.

The higher taxa containing helminths of veterinary importance are:

Major
Nemathelminthes (Roundworms)
Platyhelminthes (Flatworms)

Minor
Acanthocephala (Thornyheaded Worms)

Phylum NEMATHELMINTHES

Though the phylum Nemathelminthes has six classes only one of these, the **nematoda**, contains worms of parasitic significance. The nematodes are commonly called roundworms, from their appearance in cross-section.

Class NEMATODA

A system of classification of nematodes of veterinary importance is given in Table 1.

It must be emphasised that this is not an exact expression of the general system for parasitic nematodes, but is a simplified presentation intended for use in the study of veterinary parasitology. It is based on the ten superfamilies in which nematodes of veterinary importance occur, and which are conveniently divided into **bursate** and **non-bursate** groups as shown in Table 1.

Table 1 Parasitic Nematoda of veterinary importance Simplified Classification

Superfamily	Typical features
Bursate nematodes	
Trichostrongyloidea *Trichostrongylus, Ostertagia, Dictyocaulus, Haemonchus*, etc.	Buccal capsule small Life cycle **direct;** infection by L_3.
Strongyloidea *Strongylus, Ancylostoma, Syngamus*, etc.	Buccal capsule well developed; leaf crowns and teeth usually present. Life cycle **direct;** infection by L_3.
Metastrongyloidea *Metastrongylus, Muellerius, Protostrongylus*, etc.	Buccal capsule small. Life cycle **indirect**; infection by L_3 in intermediate host.
Non-bursate nematodes	
Rhabditoidea *Strongyloides, Rhabditis*, etc.	Very small worms; buccal capsule small. Free-living and parasitic generations. Life cycle **direct**; infection by L_3.
Ascaridoidea *Ascaris, Toxocara, Parascaris*, etc.	Large white worms. Life cycle **direct**; infection by L_2 in egg.
Oxyuroidea *Oxyuris, Skrjabinema*, etc.	Female has long, pointed tail. Life cycle **direct**; infection by L_3 in egg.
Spiruroidea *Spirocerca, Habronema, Thelazia*, etc.	Spiral tail in male. Life cycle **indirect**; infection by L_3 from insect.
Filarioidea *Dirofilaria, Onchocerca, Parafilaria*, etc.	Long thin worms. Life cycle **indirect**; infection by L_3 from insect.
Trichuroidea *Trichuris, Capillaria, Trichinella*, etc.	Whip-like or hair-like worms. Life cycle **direct** or **indirect**; infection by L_1.
Dioctophymatoidea *Dioctophyma*, etc.	Very large worms. Life cycle **indirect**; infection by L_3 in aquatic annelids.

STRUCTURE AND FUNCTION

Most nematodes have a cylindrical form, tapering at either end, and the body is covered by a colourless, somewhat translucent, layer, the cuticle.

The cuticle is secreted by the underlying hypodermis, which projects into the body cavity forming two lateral cords, which carry the excretory canals, and a dorsal and ventral cord carrying the nerves (Fig. 1). The muscle cells, arranged longitudinally, lie between the hypodermis and the body cavity. The latter contains fluid at a high pressure which maintains the turgidity and shape of the body. Locomotion is effected by undulating waves of muscle contraction and relaxation which alternate on the dorsal and ventral aspects of the worm.

Most of the internal organs are filamentous and suspended in the fluid-filled body cavity (Fig. 2).

The **digestive system** is tubular. The mouth of many nematodes is a simple opening, which may be surrounded by two or three lips, and leads directly into the oesophagus. In others, such as the strongyloids, it is large,

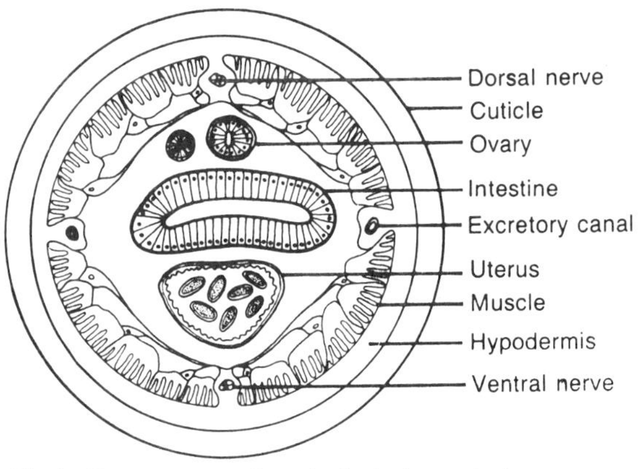

Fig 1 Transverse section of a typical nematode.

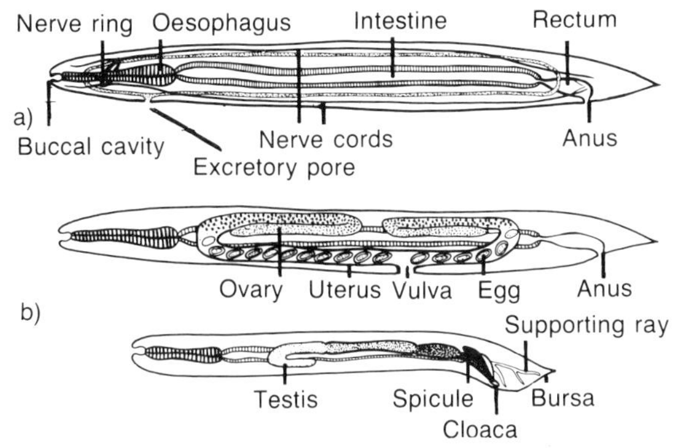

Fig 2 Longitudinal sections of a nematode illustrating:
(a) Digestive, excretory and nervous system.
(b) Reproductive system of female and male nematodes.

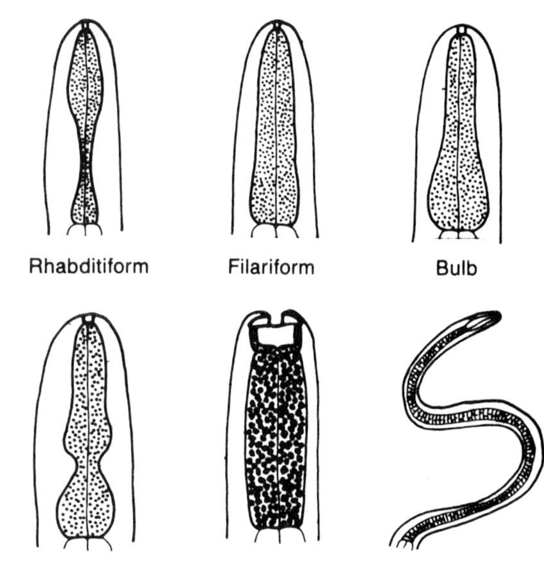

Fig 4 The basic forms of oesophagus found in nematodes.

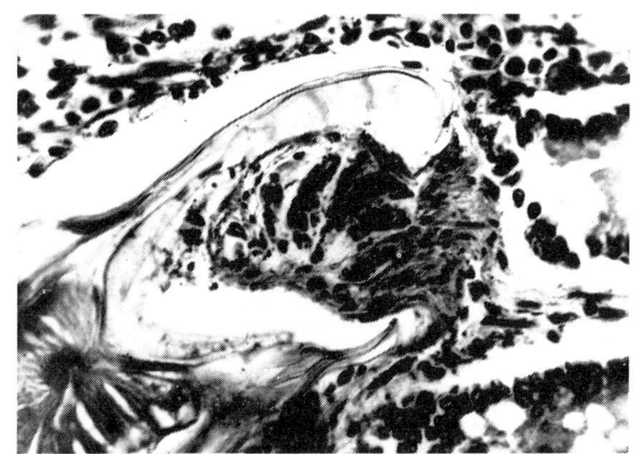

Fig 3 Large buccal capsule of strongyloid nematode ingesting plug of mucosa.

and opens into a **buccal capsule**, which may contain teeth; such parasites, when feeding, draw a plug of mucosa into the buccal capsule (Fig. 3), where it is broken down by the action of enzymes which are secreted into the capsule from adjacent glands. Some of these worms may also secrete anticoagulant, and small vessels, ruptured in the digestion of the mucosal plug, may continue to bleed for some minutes after the worm has moved to a fresh site.

Those with very small buccal capsules, like the trichostrongyloids, or simple oral openings, like the ascaridoids, generally feed on mucosal fluid and cell debris, while others, such as the oxyuroids, appear to scavenge on the contents of the lower gut. Worms living in the bloodstream or tissue spaces, such as the filarioids, feed exclusively on fluids.

The **oesophagus** is usually muscular and pumps food into the intestine. It is of variable form (Fig. 4), and is a useful preliminary identification character for groups of worms. It may be **filariform**, simple and slightly thickened posteriorly, as in the bursate nematodes; **bulb-shaped**, with a large posterior swelling, as in the ascaridoids; or **double bulb-shaped**, as in the oxyuroids. In some groups this wholly muscular form does not occur: the filarioids and spiruroids have a **muscular-glandular** oesophagus which is muscular anteriorly, the posterior part being glandular; the **trichuroid** oesophagus has a capillary form, passing through a single column of cells, the whole being known as a stichosome. A **rhabditiform** oesophagus, with slight anterior and posterior swellings, is present in the preparasitic larvae of many nematodes, and in adult free-living nematodes.

The **intestine** is a tube whose lumen is enclosed by a single layer of cells or by a syncytium. Their luminal surfaces possess microvilli which increase the absorptive capacity of the cells. In female worms the intestine terminates in an anus while in males there is a cloaca which functions as an anus, and into which opens the vas deferens and through which the copulatory spicules may be extruded.

The so-called **'excretory system'** is very primitive, consisting of a canal within each lateral cord joining at the excretory pore in the oesophageal region.

The **reproductive systems** consist of filamentous tubes. The **female organs** comprise ovary, oviduct and uterus, which may be paired, ending in a common short vagina which opens at the vulva. At the junction of uterus and vagina in some species there is a short muscular organ, the ovejector, which assists in egg-laying. A vulval flap may also be present (Fig. 5).

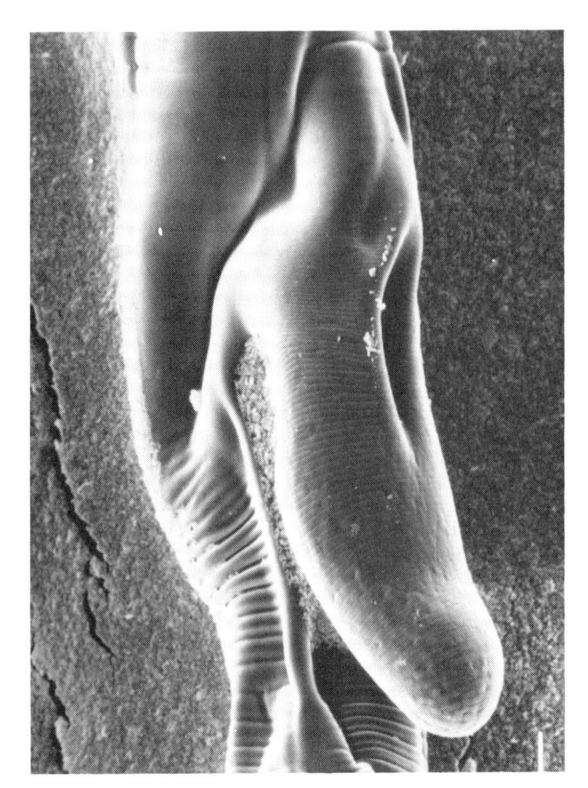

Fig 5 Scanning electron micrograph of a vulval flap of a trichostrongyloid nematode.

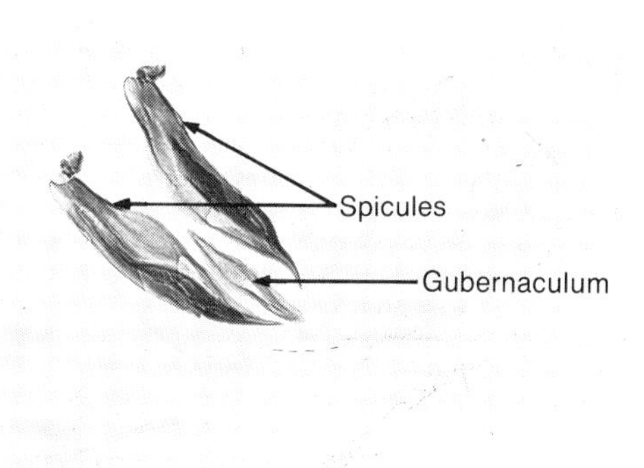

Fig 6 Spicules and gubernaculum of a trichostrongyloid nematode.

The **male organs** consist of a single continuous testis and a vas deferens terminating in an ejaculatory duct into the cloaca. Accessory male organs are sometimes important in identification, especially of the trichostrongyloids, the two most important being the spicules and gubernaculum (Fig. 6). The **spicules** are chitinous organs, usually paired, which are inserted in the female genital opening during copulation. The **gubernaculum**, also chitinous, is a small structure which acts as a guide for the spicules. With the two sexes in close apposition the amoeboid sperm are transferred from the cloaca of the male into the uterus of the female.

The **cuticle** may be modified to form various structures, the more important (Fig. 7) of which are:

Leaf crowns consisting of rows of papillae occurring as fringes round the rim of the buccal capsule (external leaf crowns) or just inside the rim (internal leaf crowns).

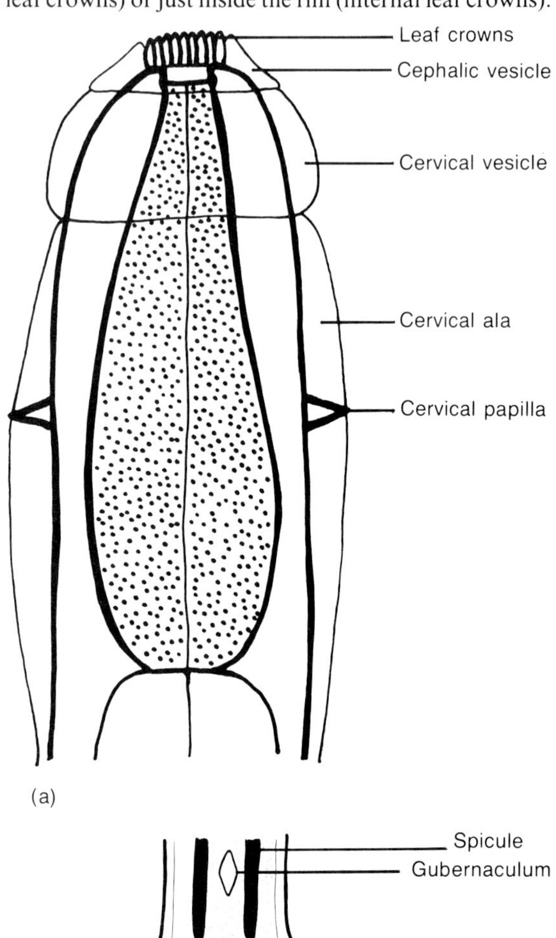

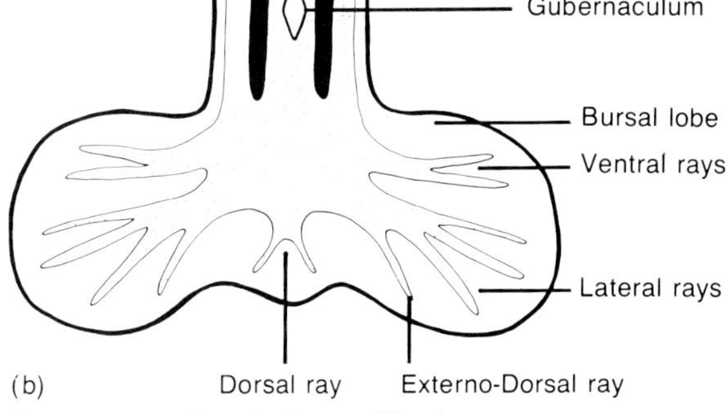

Fig 7 Nematode cuticular modifications.
(a) Anterior.
(b) Posterior of male.

They are especially prominent in certain nematodes of horses. Their function is not known, but it is suggested that they may be used to pin a patch of mucosa in position during feeding, or that they may prevent the entry of foreign matter into the buccal capsule when the worm has detached from the mucosa.

Cervical papillae occur anteriorly in the oesophageal region, and **caudal papillae** posteriorly at the tail. They are spine-like or finger-like processes, and are usually diametrically placed. Their function may be sensory or supportive.

Cervical and **caudal alae** are flattened wing-like expansions of the cuticle in the oesophageal and tail regions.

Cephalic and **cervical vesicles** are inflations of the cuticle around the mouth opening and in the oesophageal region.

The **copulatory bursa**, which embraces the female during copulation, is important in the identification of certain male nematodes and is derived from much expanded caudal alae which are supported by elongated caudal papillae called **bursal rays**. It consists of two lateral lobes and a single small dorsal lobe.

Plaques and **cordons** are plate-like and cord-like ornamentations present on the cuticle of many nematodes of the superfamily Spiruroidea.

BASIC LIFE CYCLE

In the Nematoda, the sexes are separate and the males are generally smaller than the females which lay eggs or larvae. During development, a nematode moults at invervals shedding its cuticle. In the complete life cycle there are four moults, the successive larval stages being designated L_1, L_2, L_3, L_4 and finally L_5, which is the immature adult.

One feature of the basic nematode life cycle is that immediate transfer of infection from one **final host** to another rarely occurs. Some development usually takes place either in the faecal pat or in a different species of animal, the **intermediate host**, before infection can take place.

In the common form of **direct** life cycle, the free-living larvae undergo two moults after hatching and infection is by ingestion of the free L_3. There are some important exceptions however, infection sometimes being by larval penetration of the skin or by ingestion of the egg containing a larva.

In **indirect** life cycles, the first two moults usually take place in an intermediate host and infection of the final host is either by ingestion of the intermediate host or by inoculation of the L_3 when the intermediate host, such as a blood sucking insect, feeds.

After infection, two further moults take place to produce the L_5 or immature adult parasite. Following copulation a further life cycle is initiated.

In the case of gastrointestinal parasites, development may take place entirely in the gut lumen or with only limited movement into the mucosa.

However, in many species, the larvae travel considerable distances through the body before settling in their final (predilection) site and this is the migratory form of life cycle. One of the most common routes is the **hepatic-tracheal**. This takes developing stages from the **gut** via the portal system to the **liver** then via the hepatic vein and posterior vena cava to the **heart** and from there via the pulmonary artery to the **lungs**. Larvae then travel via the bronchi, trachea and oesophagus to the **gut**. It should be emphasised that the above is a basic description of nematode life cycles and that there are many variations.

DEVELOPMENT OF THE PARASITE

EGG

Nematode eggs differ greatly in size and shape, and the shell is of variable thickness usually consisting of three layers.

The inner membrane, which is thin, has lipid characteristics and is impermeable. A middle layer which is tough and chitinous gives rigidity and, when thick, imparts a yellowish colour to the egg. In many species this layer is interrupted at one or both ends with an operculum (lid) or plug. The third outer layer consists of protein which is very thick and sticky in the ascaridoids and is important in the epidemiology of this superfamily.

In contrast, in some species the egg shell is very thin and may be merely present as a sheath around the larva.

The survival potential of the egg outside the body varies, but appears to be connected with the thickness of the shell, which protects the larva from desiccation. Thus parasites whose infective form is the larvated egg usually have very thick-shelled eggs which can survive for years on the ground.

HATCHING

Depending on the species, eggs may hatch outside the body or after ingestion.

Outside the body, hatching is controlled partly by factors such as temperature and moisture and partly by the larva itself. In the process of hatching, the inner impermeable shell membrane is broken down by enzymes secreted by the larva and by its own movement. The larva is then able to take up water from the environment and enlarges to rupture the remaining layers and escape.

When the larvated egg is the infective form, the host initiates hatching after ingestion by providing stimuli for the larva which then completes the process. It is important for each nematode species that hatching should occur in appropriate regions of the gut and hence the stimuli will differ, although it appears that dissolved carbon dioxide is a constant essential.

LARVAL DEVELOPMENT AND SURVIVAL

Three of the important superfamilies, the trichostrongyloids, the strongyloids and the rhabditoids, have a completely free-living preparasitic phase. The first two larval stages usually feed on bacteria, but the L_3, sealed off from the environment by the retained cuticle of the L_2, cannot feed and must survive on the stored nutrients acquired in the early stages. Growth of the larva is interrupted during moulting by periods of lethargus in which it neither feeds nor moves.

The cuticle of the L_2 is retained as a sheath around the L_3; this is important in larval survival with a protective role analogous to that of the egg shell in egg-infective groups.

The two most important components of the external environment are temperature and humidity.

The optimal temperature for the development of the maximum number of larvae in the shortest feasible time is generally in the range 18–26 °C. At higher temperatures, development is faster and the larvae are hyperactive, thus depleting their lipid reserves. The mortality rate then rises, so that few will survive to L_3. As the temperature falls the process slows, and below 10 °C the development from egg to L_3 usually cannot take place. Below 5 °C movement and metabolism of L_3 is minimal, which in many species favours survival.

The optimal humidity is 100%, although some development can occur down to 80% relative humidity. It should be noted that even in dry weather where the ambient humidity is low, the microclimate in faeces or at the soil surface may be sufficiently humid to permit continuing larval development.

In the trichostrongyloids and strongyloids, the embryonated egg and the ensheathed L_3 are best equipped to survive in adverse conditions such as freezing or desiccation; in contrast, the L_1 and L_2 are particularly vulnerable. Although desiccation is generally considered to be the most lethal influence in larval survival, there is increasing evidence that by entering a state of anhydrobiosis, certain larvae can survive severe desiccation.

On the ground most larvae are active; although they require a film of water for movement and are stimulated by light and temperature, it is now thought that larval movement is mostly random and encounter with grass blades accidental.

INFECTION

As noted previously, infection may be by ingestion of the free-living L_3, and this occurs in the majority of trichostrongyloid and strongyloid nematodes. In these, the L_3 sheds the retained sheath of the L_2 within the alimentary tract of the host, the stimulus for exsheathment being provided by the host in a manner similar to the hatching stimulus required by egg-infective nematodes. In response to this stimulus the larva releases its own exsheathing fluid, containing an enzyme leucine aminopeptidase, which dissolves the sheath from within, either at a narrow collar anteriorly so that a cap detaches, or by splitting the sheath longitudinally. The larva can then wriggle free of the sheath.

As in the preparasitic stage, growth of the larva during parasitic development is interrupted by two moults, each of these occurring during a short period of lethargus.

The time taken for development from infection until mature adult parasites are producing eggs or larvae is known as the **prepatent period** and this is of known duration for each nematode species.

METABOLISM

The main food reserve of preparasitic nematode larvae, whether inside the egg shell or free-living, is lipid which may be seen as droplets in the lumen of the intestine; the infectivity of these stages is often related to the amount present, in that larvae which have depleted their reserves are not as infective as those which still retain quantities of lipid.

Apart from these reserves the free-living first and second stage larvae of most nematodes feed on bacteria. However, once they reach the infective third stage, they are sealed in the retained cuticle of the second stage, cannot feed and are completely dependent on their stored reserves.

In contrast, the adult parasite stores its energy as glycogen, mainly in the lateral cords and muscles, and this may constitute 20% of the dry weight of the worm.

Free-living and developing stages of nematodes usually have an aerobic metabolism whereas adult nematodes can metabolise carbohydrate by both glycolysis (anaerobic) and oxidative decarboxylation (aerobic). However, in the latter, pathways may operate which are not present in the host and it is at this level that some antiparasitic drugs operate.

The oxidation of carbohydrates requires the presence of an electron transport system which in most nematodes can operate aerobically down to oxygen tensions of 5.0 mm Hg or less. Since the oxygen tension at the mucosal surface of the intestine is around 20 mm Hg, nematodes in close proximity to the mucosa normally have sufficient oxygen for aerobic metabolism. Otherwise, if the nematode is temporarily or permanently some distance from the mucosal surface, energy metabolism is probably largely anaerobic.

As well as the conventional cytochrome and flavoprotein electron transport system, many nematodes have 'haemoglobin' in their body fluids which gives them a red pigmentation. This nematode haemoglobin is chemically similar to myoglobin and has the highest affinity for oxygen of any known animal haemoglobin. The main function of nematode haemoglobin is thought to be to transport oxygen, acquired by diffusion through the cuticle or gut, into the tissues; blood-sucking worms presumably ingest a considerable amount of oxygenated nutrients in their diet.

The end products of the metabolism of carbohydrates, fats or proteins are excreted through the anus or cloaca, or by diffusion through the body wall. Ammonia, the terminal product of protein metabolism, must be excreted rapidly and diluted to non-toxic levels in the surrounding fluids. During periods of anaerobic carbohydrate metabolism, the worms may also excrete pyruvic acid rather than retaining it for future oxidation when aerobic metabolism is possible.

The 'excretory system' terminating in the excretory pore is almost certainly not concerned with excretion, but rather with osmoregulation and salt balance.

Two phenomena which affect the normal parasitic life cycle of nematodes and which are of considerable biological and epidemiological importance are **arrested larval development** and the **periparturient rise** in faecal egg counts.

ARRESTED LARVAL DEVELOPMENT

(Synonyms: inhibited larval development, **hypobiosis**).

This phenomenon may be defined as the temporary cessation in development of a nematode at a precise point in its parasitic development. It is usually a facultative characteristic and affects only a proportion of the worm population. Some strains of nematodes have a high propensity for arrested development while in others this is low.

Conclusive evidence for the occurrence of arrested larval development can only be obtained by examination of the worm population in the host. It is usually recognised by the presence of large numbers of larvae at the same stage of development in animals withheld from infection for a period longer than that required to reach that particular larval stage.

The nature of the stimulus for arrested development and for the subsequent maturation of the larvae is still a matter of debate. Although there are apparently different circumstances which initiate arrested larval development, most commonly the stimulus is an environmental one received by the free-living infective stages prior to ingestion by the host. It may be seen as a ruse by the parasite to avoid adverse climatic conditions for its progeny by remaining sexually immature in the host until more favourable conditions return. The name commonly applied to this seasonal arrestment is **hypobiosis**. Thus the accumulation of arrested larvae often coincides with the onset of cold autumn/winter conditions in the northern hemisphere, or very dry conditions in the subtropics or tropics. In contrast, the maturation of these larvae coincides with the return of environmental conditions suitable to their free-living development, although it is not clear what triggers the signal to mature and how it is transmitted.

The degree of adaptation to these seasonal stimuli and therefore the proportion of larvae which do become arrested seems to be a heritable trait and is affected by various factors including grazing systems and the degree of adversity in the environment. For example, in Canada where the winters are severe, most Trichostrongyloid larvae ingested in late autumn or winter become arrested, whereas in southern Britain with moderate winters, about 50–60% are arrested. In the humid tropics where free-living larval development is possible all the year round, relatively few become arrested.

However, arrested development may also occur as a result of both acquired and age immunity in the host and although the proportions of larvae arrested are not usually so high as in hypobiosis they can play an important part in the epidemiology of nematode infections. Maturation of these arrested larvae seems to be linked with the breeding cycle of the host and occurs at or around parturition.

The epidemiological importance of arrested larval development from whatever cause is that, first, it ensures the survival of the nematode during periods of adversity; secondly, the subsequent maturation of arrested larvae increases the contamination of the environment and can sometimes result in clinical disease.

PERIPARTURIENT RISE (PPR) IN FAECAL EGG COUNTS

(Synonyms: Post-parturient rise, Spring rise).

This refers to an increase in the numbers of nematode eggs in the faeces of animals around parturition. The phenomenon is most marked in ewes, sows and goats.

The etiology of this phenomenon has been principally studied in sheep and seems to result from a temporary relaxation in immunity associated with changes in the circulating levels of the lactogenic hormone, prolactin. It appears that a decrease in parasite-specific immune responses occurs following elevation of serum prolactin levels. These are rapidly restored when prolactin levels drop at the end of lactation or more abruptly if lambs are weaned early and the suckling stimulus removed.

The source of the periparturient rise (PPR) is threefold:

(i) Maturation of larvae arrested due to host immunity.
(ii) An increased establishment of infections acquired from the pastures and a reduced turnover of existing adult infections.
(iii) An increased fecundity of existing adult worm populations.

Contemporaneously, but not associated with the relaxation of host immunity, the PPR may be augmented by the maturation of hypobiotic larvae.

The importance of the PPR is that it occurs at a time when the numbers of new susceptible hosts are increasing and so ensures the survival and propagation of the worm species. Depending on the magnitude of infection, it may also cause a loss of production in lactating animals and by contamination of the environment lead to clinical disease in susceptible young stock.

Superfamily TRICHOSTRONGYLOIDEA

The trichostrongyloids are small, often hair-like, worms in the bursate group which, with the exception of the lungworm *Dictyocaulus*, parasitise the alimentary tract of animals and birds.

Structurally they have few cuticular appendages and the buccal capsule is vestigial. The males have a well developed bursa and two spicules, the configuration of which is used for species differentiation. The life cycle is direct and usually non-migratory and the ensheathed L_3 is the infective stage.

The trichostrongyloids, including *Dictyocaulus*, are responsible for considerable mortality and widespread morbidity, especially in ruminants. The most important alimentary genera are *Ostertagia*, *Haemonchus*, *Trichostrongylus*, *Cooperia*, *Nematodirus*, *Hyostrongylus*, *Marshallagia* and *Mecistocirrus*.

Ostertagia

This genus is the major cause of parasitic gastritis in ruminants in temperate areas of the world.

Hosts: Ruminants

Site: Abomasum

Species:

Ostertagia ostertagi	cattle
O. circumcincta	sheep and goats
O. trifurcata	sheep and goats

Minor species are *O.* (syn. *Skrjabinagia*) *lyrata* and *kolchida*, in cattle and *O. leptospicularis* in cattle, sheep and goats

Distribution:
Worldwide; *Ostertagia* is especially important in temperate climates and in subtropical regions with winter rainfall

IDENTIFICATION

The adults are slender reddish-brown worms up to 1.0 cm long, occurring on the surface of the abomasal mucosa and are only visible on close inspection. The larval stages occur in the gastric glands and can only be seen microscopically following processing of the gastric mucosa.

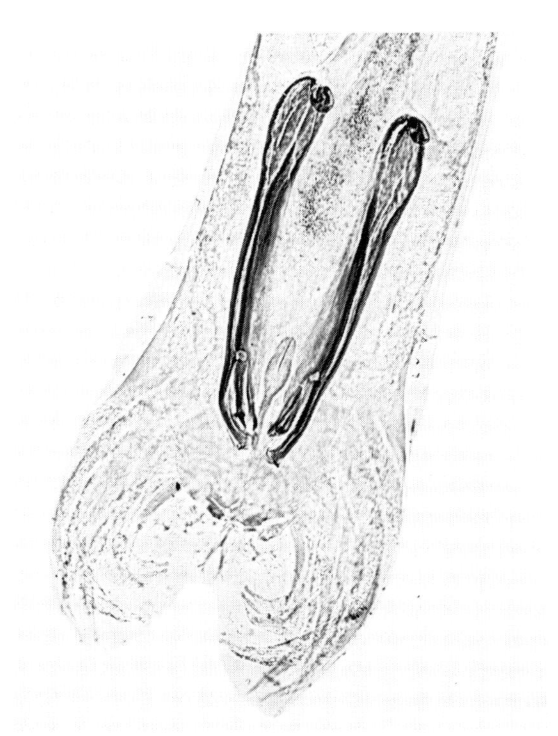

Fig 8 Structure of spicules from five *Ostertagia* species.
(a) *O. ostertagi.*

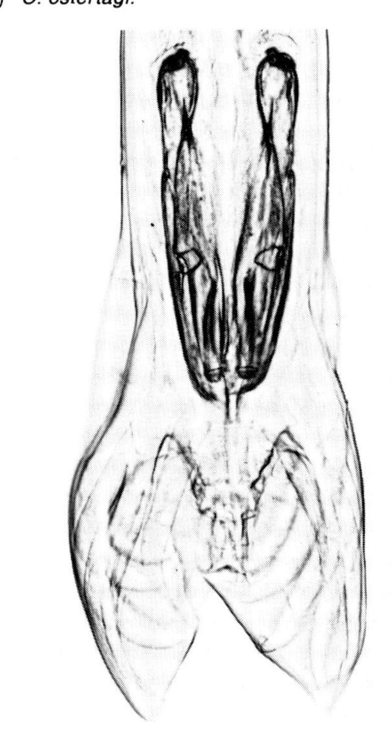

(b) *O. lyrata.*

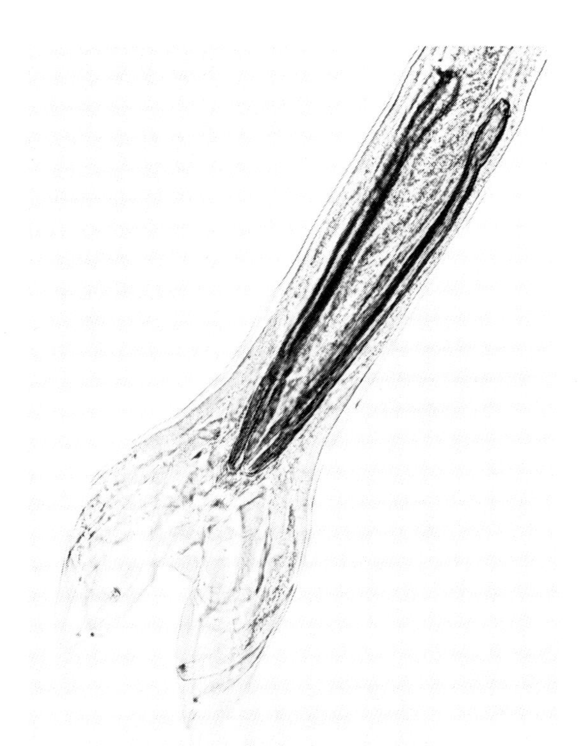

(c) *O. circumcincta.*

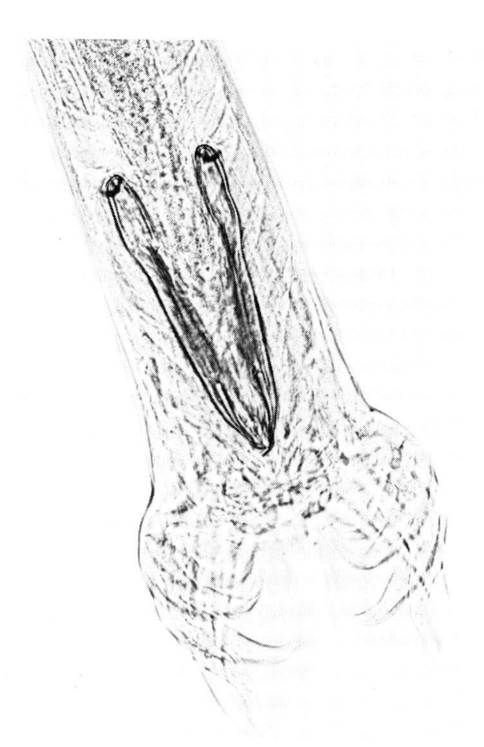

(e) *O. leptospicularis.*

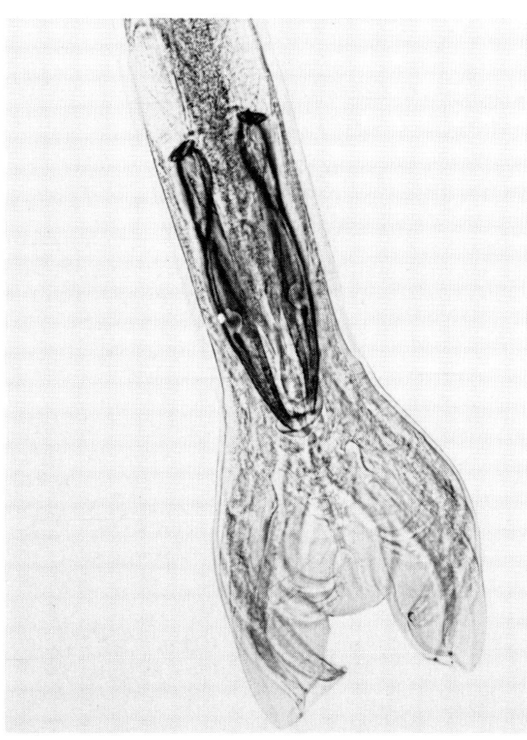

(d) *O. trifurcata.*

Species differentiation is based on the structure of the spicules which usually have three distal branches (Fig. 8).

BOVINE OSTERTAGIASIS

Since *O. ostertagi* is the most prevalent of the species in cattle it is considered in detail.

Ostertagia ostertagi

O. ostertagi is perhaps the most common cause of parasitic gastritis in cattle. The disease, often simply known as ostertagiasis, is characterised by weight loss and diarrhoea and typically affects young cattle during their first grazing season, although herd outbreaks and sporadic individual cases have also been reported in adult cattle. cattle.

LIFE CYCLE

O. ostertagi has a direct life cycle. The eggs (Fig. 9), which are typical of the Trichostrongyloidea, are passed in the faeces and under optimal conditions develop within the faecal pat to the infective third stage within two weeks. When moist conditions prevail, the L_3 migrate from the faeces on to the herbage.

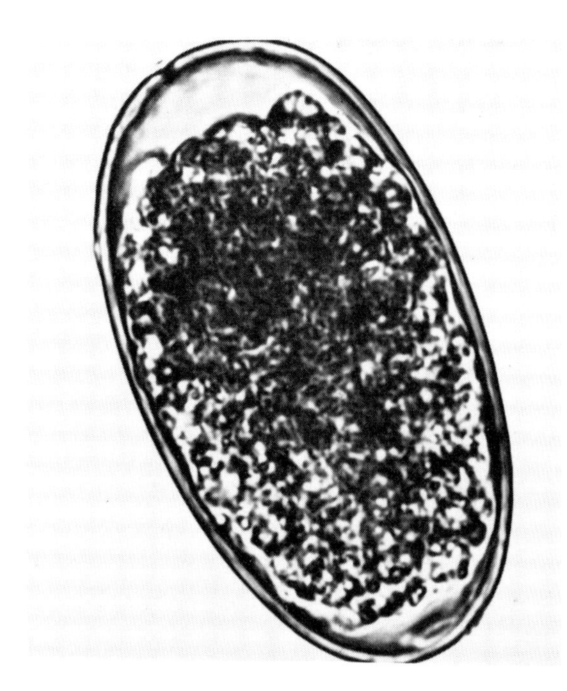

Fig 9 Typical trichostrongyloid egg.

Fig 10 *Ostertagia ostertagi* infection showing larva in gastric gland.

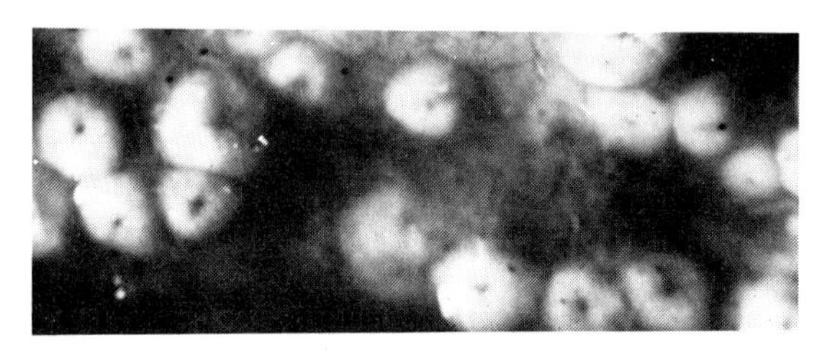

Fig 11 Umbilicated nodules on mucosal surface after emergence of *Ostertagia* larvae.

After ingestion, the L_3 exsheaths in the rumen and further development takes place in the lumen of an abomasal gland. Two parasitic moults occur before the L_5 emerges from the gland around 18 days after infection to become sexually mature on the mucosal surface.

The entire parasitic life cycle usually takes three weeks, but under certain circumstances many of the ingested L_3 become arrested in development at the early fourth larval stage (EL_4) for periods of up to six months.

PATHOGENESIS

The presence of *O. ostertagi* in the abomasum in sufficient numbers gives rise to extensive pathological and biochemical changes and severe clinical signs. These changes are maximal when the parasites are emerging from the gastric glands (Pl. I). This is usually about 18 days after infection, but it may be delayed for several months when arrested larval development occurs.

The developing parasites cause a reduction in the functional gastric gland mass responsible for the production of the highly acidic proteolytic gastric juice; in particular, the parietal cells, which produce hydrochloric acid, are replaced by rapidly dividing, undifferentiated non-acid secreting cells. Initially, these cellular changes occur in the parasitised gland (Fig. 10), but as it becomes distended by the growing worm which increases from 1.3–8.0 mm in length, these changes spread to the surrounding non-parasitised glands, the end result being a thickened hyperplastic gastric mucosa (Pl. I).

Macroscopically, the lesion is a raised nodule with a visible central orifice (Fig. 11); in heavy infections these nodules coalesce to produce an effect reminiscent of morocco leather. The abomasal folds are often very oedematous and hyperaemic and sometimes necrosis and sloughing of the mucosal surface occurs (Pl. I); the regional lymph nodes are enlarged and reactive.

In heavy infections of 40,000 or more adult worms the principal effects of these changes are, first, a reduction in the acidity of the abomasal fluid, the pH increasing from 2.0 up to 7.0. This results in a failure to activate pepsinogen to pepsin and so denature proteins. There is also a loss of bacteriostatic effect in the abomasum. Secondly, there is an enhanced permeability of the abomasal epithelium to macromolecules such as pepsinogen and plasma proteins. One explanation is that the cell junctions between the rapidly dividing and undifferentiated cells which come to line the parasitised mucosa appear to be incompletely formed, and as a result, macromolecules may pass into and out of the epithelial sheet.

The results of these changes are a leakage of pepsinogen into the circulation leading to elevated plasma pepsinogen levels and the loss of plasma proteins into the gut lumen eventually leading to hypoalbuminaemia. Another more recent theory is that, in response to the presence of the adult parasites, the zymogen cells secrete increased amounts of pepsin directly into the circulation. Clinically the consequences are reflected as inappetence, weight loss and diarrhoea, the precise cause of the diarrhoea being unknown.

In lighter infections the main effects are sub-optimal weight gains.

Although reduced feed consumption and diarrhoea affect liveweight gain they do not wholly account for the loss in production. Current evidence suggests that this is primarily because of substantial leakage of endogenous protein into the gastrointestinal tract. Despite some reabsorption, this leads to a disturbance in post-absorptive nitrogen and energy metabolism due to the increased demands for the synthesis of vital proteins, such as albumin and the immunoglobulins, which occur at the expense of muscle protein and fat deposition.

These disturbances are of course influenced by the level of nutrition, being exacerbated by a low protein intake and alleviated by a high protein diet.

CLINICAL SIGNS

Bovine ostertagiasis is known to occur in two clinical forms. In temperate climates with cold winters the seasonal occurence of these is as follows:

The **Type I** disease is usually seen in calves grazed intensively during their first grazing season, as the result of larvae ingested 3–4 weeks previously; in the northern hemisphere this normally occurs from mid-July onwards.

The **Type II** disease occurs in yearlings, usually in late winter or spring following their first grazing season and results from the maturation of larvae ingested during the previous autumn and subsequently arrested in their development at the early fourth larval stage.

The main clinical sign in both Type I and Type II disease is a profuse watery diarrhoea and in Type I, where calves are at grass, this is usually persistent and has a characteristic bright green colour. In contrast, in the majority of animals with Type II, the diarrhoea is often intermittent and anorexia and thirst are usually present. The coats of affected animals in both syndromes are dull and the hind quarters heavily soiled with faeces.

In Type II ostertagiasis, hypoalbuminaemia is more marked and there is a moderate anaemia of unknown etiology. As a result of the hypoalbuminaemia, submandibular oedema is often present. In both forms of the disease, the loss of body weight is considerable during the clinical phase and may reach 20% in 7–10 days. Carcass quality may also be affected since there is a reduction in total body solids relative to total body water.

In Type I disease, the morbidity is usually high, often exceeding 75%, but mortality is rare provided treatment is instituted within 2–3 days. In Type II the prevalence of clinical disease is comparatively low and often only a proportion of animals in the group are affected; mortality in such animals is very high unless early treatment with an anthelmintic effective against both arrested and developing larval stages is instituted.

EPIDEMIOLOGY

The epidemiology of ostertagiasis in temperate countries of the northern hemisphere can be conveniently considered under the headings of dairy herds and beef herds; important differences in subtropical climates are summarised later.

Dairy herds

From epidemiological studies the following important facts have emerged (Fig. 12):

(i) A considerable number of L_3 can survive the winter on pasture and in soil. Sometimes the numbers are sufficient to precipitate Type I disease in calves 3–4 weeks after they are turned out to graze in the spring. However, this is unusual and the role of the surviving L_3 is rather to infect calves at a level which produces patent subclinical infection and ensures contamination of the pasture for the rest of the grazing season.

(ii) A high mortality of overwintered L_3 on the pasture occurs in spring and only negligible numbers can usually be detected by June. This mortality combined with the dilution effect of the rapidly growing herbage renders most pastures, not grazed in the spring, safe for grazing after mid-summer.

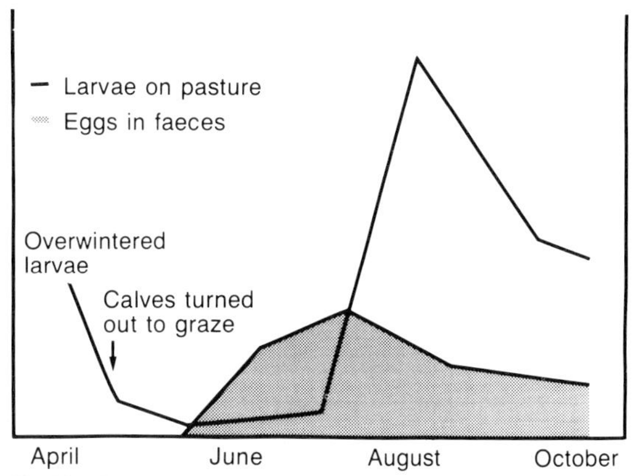

Fig 12 Epidemiology of bovine ostertagiasis in temperate zones of the northern hemisphere showing mid-summer rise of infective larvae on pasture.

However, despite the mortality of L_3 on the pasture it now seems that many survive in the soil for at least another year and on occasion appear to migrate on to the herbage. Whether this is a common occurence and whether the larvae migrate or are transported by terrestrial populations of earthworms or beetles is not definitely known, but the occurrence of this apparent reservoir of larvae in soil may be important in relation to certain systems of control based on grazing management.

(iii) The eggs deposited in the spring develop slowly to L_3; this rate of development becomes more rapid towards mid-summer as temperatures increase, and as a result, the majority of eggs deposited during April, May and June all reach the infective stage from mid-July onwards. If sufficient numbers of these L_3 are ingested, the Type I disease occurs any time from July until October. Development from egg to L_3 slows during the autumn and it is doubtful if many of the eggs deposited after September ever develop to L_3.

(iv) As autumn progresses and temperatures fall an increasing proportion (up to 80%) of the L_3 ingested do not mature but become inhibited at the early fourth larval stage (EL_4). In late autumn, calves can therefore harbour many thousands of these EL_4 but few developing forms or adults. These infections are generally asymptomatic until maturation of the EL_4 takes place during winter and early spring and if large numbers of these develop synchronously, Type II disease materialises. Where maturation is not synchronous, clinical signs may not occur but the adult worm burdens which develop can play a significant epidemiological role by contributing to pasture contamination in the spring.

Two factors, one management and one climatic, appear to increase the prevalence of Type II ostertagiasis.

First, the practice of grazing calves from May until late July on permanent pasture, then moving these to hay or silage aftermath before returning them to the original grazing in late autumn. In this system the accumulation of L_3 on the original pasture will occur from mid-July, ie. after the calves have moved to aftermath. These L_3 are still present on the pastures when the calves return in the late autumn and, when ingested, the majority will become arrested.

Secondly, in dry summers the L_3 are retained within the crusted faecal pat and cannot migrate on to the pasture until sufficient rainfall occurs to moisten the pat. If rainfall is delayed until late autumn many larvae liberated on to pasture will become arrested following ingestion and so increase the chance of Type II disease. Indeed, epidemics of Type II ostertagiasis are typically preceded by dry summers.

Although primarily a disease of young dairy cattle, ostertagiasis can nevertheless affect groups of older cattle in the herd, particularly if these have had little previous exposure to the parasite, since there is no significant age immunity to infection.

Acquired immunity in ostertagiasis is slow to develop and calves do not achieve a significant level of immunity until the end of their first grazing season. If they are then housed for the winter the immunity acquired by the end of the grazing season has waned by the following spring and yearlings turned out at that time are partially susceptible to reinfection and so contaminate the pasture with small numbers of eggs. However, immunity is rapidly re-established and any clinical signs which occur are usually of a transient nature. During the second and third year of grazing, a strong acquired immunity develops and adult stock in endemic areas are highly immune to reinfection and of little significance in the epidemiology. An exception to this rule occurs around the periparturient period when immunity wanes, particularly in heifers, and there are reports of clinical disease following calving. The reason is unknown but may be due to the development of larvae which were arrested in their development as a result of host immunity.

Beef herds

Although the basic epidemiology in beef herds is similar to dairy herds, the influence of immune adult animals grazing alongside susceptible calves has to be considered. Thus, in beef herds where calving takes place in the spring, ostertagiasis is uncommon since egg production by immune adults is low, and the spring mortality of the overwintered L_3 occurs prior to the suckling calves ingesting significant quantities of grass. Consequently only low numbers of L_3 become available on the pasture later in the year.

However, where calving takes place in the autumn or winter, ostertagiasis can be a problem in calves during the following grazing season once they are weaned, the epidemiology then being similar to dairy calves. Whether Type I or Type II disease subsequently occurs depends on the grazing management of the calves following weaning.

In countries in the southern hemisphere with temperate climates, such as New Zealand, the seasonal pattern is similar to that reported for Europe with Type I disease occurring in the summer and burdens of arrested larvae accumulating in the autumn.

In those countries with subtropical climates and winter rainfall such as parts of southern Australia, South West Africa and some regions of Argentina, Chile and Brazil, the increase in L_3 population occurs during the winter and outbreaks of Type I disease are seen towards the end of the winter period. Arrested larvae accumulate during the spring and where Type II disease has been reported it has occurred in late summer or early autumn.

A basically similar pattern of infection is seen in some southern parts of the U.S.A. with non-seasonal rainfall, such as Louisiana and Texas. There, larvae accumulate on pasture during winter and arrested development occurs in late winter and early spring with outbreaks of Type II disease occurring in late summer or early autumn.

The environmental factors which produce arrested larvae in subtropical zones are not yet known.

THE EFFECT OF OSTERTAGIA INFECTION ON LACTATION YIELDS OF GRAZING COWS

Although burdens of adult *Ostertagia* spp. in dairy cows are usually low there is some evidence that a single anthelmintic treatment of such cows at, or soon after, calving can improve milk yields. However, the economic benefit gained from such treatment varies considerably from farm to farm and also apparently from country to country and there are as yet insufficient grounds for advocating routine treatment of herds at calving.

It has also been suggested that during lactation a reduction in milk yield might result from oedema and increased permeability of the abomasal mucosa, possibly due to hypersensitivity reaction associated with the continued ingestion and destruction of large numbers of L_3.

DIAGNOSIS

In young animals this is based on:

(i) **The clinical signs** of inappetence, weight loss and diarrhoea.
(ii) **The season.** For example, in Europe Type I occurs from July until September and Type II from March to May.
(iii) **The grazing history.** In Type I disease, the calves have usually been set-stocked in one area for several months; in contrast, Type II disease often has a typical history of calves being grazed on a field from spring to mid-summer, then moved and brought back to the original field in the autumn.

Affected farms usually also have a history of ostertagiasis in previous years.
(iv) **Faecal egg counts.** In Type I disease these are usually more than 1,000 eggs per gram (epg) and are a useful aid to diagnosis; in Type II the count is highly variable, may even be negative and is of limited value.
(v) **Plasma pepsinogen levels.** In clinically affected animals up to two years old these are usually in excess of 3.0 i.u. tyrosine (normal levels are 1.0 i.u. in non-parasitised calves). The test is less reliable in older cattle where high values are not necessarily correlated with large adult worm burdens but, instead, may reflect plasma leakage from a hypersensitive mucosa under heavy larval challenge.
(vi) **Post-mortem examination.** If this is available, the appearance of the abomasal mucosa is characteristic. There is a putrid smell from the abomasal contents due to the accumulation of bacteria and the high pH. The adult worms, reddish in colour and 1.0 cm in length, can be seen on close inspection of the mucosal surface. Adult worm burdens are typically in excess of 40,000, although lower numbers are often found in animals which have been diarrhoeic for several days prior to necropsy.

In older animals the clinical signs and history are similar but laboratory diagnosis is more difficult since faecal egg counts and plasma pepsinogen levels are less reliable. A useful technique to employ in such situations is to carry out a pasture larval count on the field on which the animals had been grazing. Where the level of infection is more than 1,000 larvae per kg of dried herbage, the daily larval intake of grazing cows is in excess of 10,000 larvae. This level is probably sufficient to cause clinical disease in susceptible adult animals or to upset the normal functioning of the gastric mucosa in immune cows.

TREATMENT

Type I disease responds well to treatment at the standard dosage rates with any of the modern benzimidazoles (albendazole, fenbendazole or oxfendazole), the pro-benzimidazoles (febantel and thiophanate), levamisole, or ivermectin. All of these drugs are effective against developing larvae and adult stages. Following treatment, calves should be moved to pasture which has not been grazed by cattle in the same year.

For the successful treatment of Type II disease it is necessary to use drugs which are effective against arrested larvae as well as developing larvae and adult stages. Only the modern benzimidazoles listed above or ivermectin are effective in the treatment of Type II disease when used at standard dosage levels, although the pro-benzimidazoles are also effective at higher dose rates. Sometimes with the orally administered benzimidazoles the drug by-passes the rumen and enters the abomasum directly and this appears to lower efficacy because of its more rapid absorption and excretion.

The field where the outbreak has originated may be grazed by sheep or rested until the following June.

CONTROL

Traditionally, ostertagiasis has been prevented by routinely treating young cattle with anthelmintics over the period when pasture larval levels are increasing. For example, in Europe this involved one or two treatments usually in July and September and on many farms this prevented disease and produced acceptable growth rates. However, it has the disadvantage that since the calves are under continuous larval challenge their performance may be impaired. With this system, effective anthelmintic treatment at housing is also necessary using a drug effective against hypobiotic larvae in order to prevent Type II disease.

Today, it is accepted that the prevention of ostertagiasis by limiting exposure to infection is a more efficient method of control.

This may be done by grazing calves on new grass leys, although it is doubtful if this should be recommended for replacement dairy heifers, as it would result in a pool of susceptible adult animals. A better policy is to permit young cattle sufficient exposure to larval infection to stimulate immunity but not sufficient to cause a loss in production. The provision of this 'safe pasture' may be achieved in two ways:

First, by using anthelmintics to limit pasture contamination with eggs during periods when the climate is optimal for development of the free-living larval stages, ie. spring and summer in temperate climates, or autumn and winter in the sub-tropics.

Alternatively, by resting pasture or grazing it with another host, such as sheep, which are not susceptible to *O. ostertagi*, until most of the existing L_3 on the pasture have died out.

Sometimes a combination of these methods is employed. The timing of events in the systems described below is applicable to the calendar of the northern hemisphere.

Prophylactic anthelmintic medication

Since the crucial period of pasture contamination with *O. ostertagi* eggs is the period up to mid-July, one of the efficient modern anthelmintics may be given on two or three occasions between turn-out in the spring and July to minimise the numbers of eggs deposited on the pasture. For calves going to pasture in early May two treatments, three and six weeks later, after first treatment, are used, whereas calves turned out in April require three treatments at intervals of three weeks. Where parenteral ivermectin is used the interval after first treatment may be extended to five weeks due to its residual activity.
oped which releases the anthelmintic morantel over a 90 day period. When this sustained-release bolus is administered to calves just prior to turn-out, it prevents the development of infections acquired from overwintered larvae and so prevents the deposition of eggs during the spring. In extensive trials throughout Europe this method of control has given good results and may well be the forerunner of other similar devices.

Anthelmintic prophylaxis has the advantage that animals can be grazed throughout the year on the same pasture and is particularly advantageous for the small heavily stocked farm where grazing is limited.

Anthelmintic treatment and move to safe pasture in mid-July

This system, usually referred to as 'dose and move', is based on the knowledge that the annual increase of L_3 occurs after mid-July. Therefore if calves grazed from early spring are given an anthelmintic treatment in early July and moved immediately to a second pasture such as silage or hay aftermath, the level of infection which develops on the second pasture will be low.

The one reservation with this technique is that in certain years the numbers of L_3 which overwinter are sufficient to cause heavy infections in the spring and clinical ostertagiasis can occur in calves in April and May. However, once the 'dose and move' system has operated for a few years this problem is unlikely to arise.

In some European countries such as the Netherlands, the same effect has been obtained by delaying the turn-out of calves until mid-summer. This method has given good control of ostertagiasis, but many farmers are unwilling to continue housing and feeding calves when there is ample grazing available.

Alternate grazing of cattle and sheep

This system ideally utilises a three-year rotation of cattle, sheep and crops. Since the effective life-span of most *O. ostertagi* L_3 is under one year and cross-infection between cattle and sheep in temperate areas is largely limited to *O. leptospicularis*, *Trichostrongylus axei*, and occasionally *C. oncophora* good control of bovine ostertagiasis should, in theory, be achieved. It is particularly applicable to farms with a high proportion of land suitable for cropping or grassland conservation and less so for marginal or upland areas. However, in the latter, reasonable control has been reported using an annual rotation of beef cattle and sheep.

The drawback of alternate grazing systems is that they impose a rigorous and inflexible regimen on the use of land which the farmer may find impractical. Furthermore, in warmer climates where *Haemonchus* spp. are prevalent, this system can prove dangerous since this very pathogenic genus establishes in both sheep and cattle.

Rotational grazing of adult and young stock

This system involves a continuous rotation of paddocks in which the susceptible younger calves graze ahead of the immune adults and remain long enough in each paddock to remove only the leafy upper herbage before being moved on to the next paddock. The incoming immune adults then graze the lower more fibrous echelons of the herbage which contain the majority of the L_3. Since the faeces produced by the immune adults contains few if any *O. ostertagi* eggs the pasture contamination is greatly reduced. The success of this method depends on having sufficient fenced paddocks available to prevent over-grazing and the adults must have a good acquired immunity.

While this system has many attractions, its main disadvantage is that it is costly in terms of fencing and again requires careful supervision. Its main attractions are the optimal utilisation of permanent grassland and the control of internal parasitism without resort to therapy.

OVINE OSTERTAGIASIS

In sheep *O. circumcincta* and *O. trifurcata* are responsible for outbreaks of clinical ostertagiasis, particularly in lambs. In Europe a clinical syndrome analogous to Type I bovine ostertagiasis occurs from August to October; thereafter arrested development of many ingested larvae occurs and a Type II syndrome has been occasionally reported in late winter and early spring, especially in young adults.

In subtropical areas with winter rainfall ostertagiasis occurs primarily in late winter.

LIFE CYCLE

Both the free-living and parasitic phases of the life cycle are similar to those of the bovine species.

PATHOGENESIS

In clinical infections, this is similar to the situation in cattle and the same lesions are present at necropsy. In subclinical infections, it has been shown under both experimental and natural conditions that *O. circumcincta* causes a marked depression in appetite and this, together with losses of plasma protein into the gastrointestinal tract, results in interference with the post-absorptive metabolism of protein and to a lesser extent the utilisation of metabolisable energy. In lambs with moderate infections of *Ostertagia* spp., carcass evaluation shows poor protein, fat and calcium deposition.

CLINICAL SIGNS

The most frequent clinical sign is a marked loss of weight. Diarrhoea is intermittent and although stained hindquarters are common, the fluid faeces which characterise bovine ostertagiasis are less frequently seen.

EPIDEMIOLOGY

In Europe the herbage numbers of *Ostertagia* spp. L_3 increase markedly from mid-summer onwards and this is when disease appears.

These larvae are derived mainly from eggs passed in the faeces of ewes during the periparturient period, from about two weeks prior to lambing until six weeks post-lambing (Fig. 13). Eggs passed by lambs, from worm burdens which have accrued from the ingestion of overwintered larvae, also contribute to the pasture contamination.

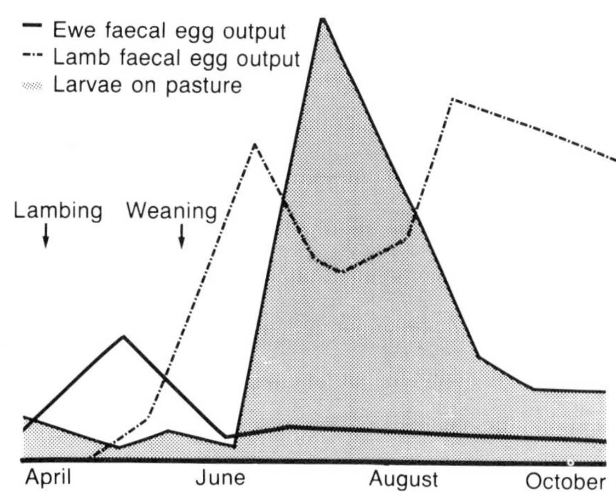

Fig 13 Epidemiology of ovine parasitic gastroenteritis in temperate zones of the northern hemisphere showing periparturient rise in the faecal egg counts of ewes and the mid-summer rise of infective larvae on pasture.

It is important to realise that it is these eggs deposited in the first half of the grazing season from April to June, which give rise to the potentially dangerous populations of L_3 from July to October.

If ingested prior to October, the majority of these larvae mature in three weeks; thereafter, many become arrested in development for several months and may precipitate Type II disease when they mature.

Immunity is acquired slowly and requires exposure over two grazing seasons before a significant resistance to infection develops. Subsequently, adult ewes harbour only very low populations of *Ostertagia* spp. except during the annual periparturient rise (PPR).

The epidemiology in subtropical areas is basically similar to that in temperate zones, except that the seasonal timing of events is different. In many of these areas lambing is geared to an increase in the growth of pasture which occurs with the onset of rain in late autumn or winter. This coincides with conditions which are favourable to the development of the free-living stages of *Ostertagia* spp. and so infective larvae accumulate during the winter to cause clinical problems or production loss in the second half of the winter; arrested larval development occurs at the end of the winter or early spring. The sources of pasture contamination are again the ewes during the PPR and the lambs following ingestion of larvae which have survived the summer.

The relative importance of these sources in any country varies according to the conditions during the adverse period for larval survival. Where the summer is very dry and hot, the longevity of L_3 is reduced except in areas with shade and these can act as reservoirs of infection until the following winter. Although L_3 can persist in sheep faeces during adverse weather conditions the protection is probably less than that afforded by the more abundant bovine faecal pat.

DIAGNOSIS

This is based on clinical signs, seasonality of infection and faecal egg counts, and if possible, post-mortem examination, when the characteristic lesions can be seen in the abomasum. Plasma pepsinogen levels are above the normal of 1.0 i.u. tyrosine and usually exceed 2.0 i.u. in sheep with heavy infections.

TREATMENT

Ovine ostertagiasis responds well to treatment with any of the benzimidazoles or pro-benzimidazoles, levamisole, which in sheep is effective against arrested larvae, or ivermectin. Treated lambs should preferably be moved to safe pasture and if this is not possible, treatment may have to be repeated at monthly intervals until the pasture larval levels decrease in early winter.

CONTROL

See the 'Treatment and control of parasitic gastroenteritis (PGE) in sheep', p. 32.

CAPRINE OSTERTAGIASIS

Increasing numbers of goats are being kept worldwide and generally these run on permanent grazing. It has been shown that goats are very susceptible to the *Ostertagia* spp. which predominate in sheep, *O. circumcincta* and *O. trifurcata*, and also to *O. leptospicularis* which establishes equally well in sheep and cattle. There is also some evidence that *O. ostertagi* can establish in goats.

As in sheep there is a marked PPR in female goats; these eggs are the main source of pasture contamination and, eventually, the L_3 which may then infect grazing kids.

The pathogenesis, diagnosis, treatment and control measures are as for the other ruminants, but care must be taken in choosing the anthelmintic since many of those recommended for sheep and cattle are not registered for use in goats. Where goat milk or milk products are used for human consumption, milk-withholding periods for different drugs should be observed. Thiabendazole has anti-fungal properties and should not be used when milk is processed for cheese.

Marshallagia marshalli

Found in the abomasum of small ruminants in the tropics and subtropics including southern Europe, U.S.A., South America, India and Russia. It is similar to *Ostertagia* spp. and can be differentiated by its greater length (up to 2.0 cm). The eggs are much larger and resemble those of *Nematodirus battus*.

The life cycle is similar to *Ostertagia* and there is penetration of the gastric glands with resultant nodule formation. Each nodule contains three or four developing parasites and measures 2.0–4.0 mm in diameter.

The pathogenicity of *M. marshalli* is not known.

Haemonchus

This blood-sucking abomasal nematode may be responsible for extensive losses in sheep and cattle, especially in tropical areas.

Hosts:
Cattle, sheep and goats

Site:
Abomasum

Species:
Haemonchus contortus
H. placei
H. similis

Until recently the sheep species was called *H. contortus* and the cattle species *H. placei.* However there is now increasing evidence that these are the single species *H. contortus* with only strain adaptations for cattle and sheep

Distribution:
Worldwide. Most important in tropical and subtropical areas

IDENTIFICATION

Gross: The adults are easily identified because of their specific location in the abomasum and their large size (2.0–3.0 cm). In fresh specimens, the white ovaries winding spirally around the blood-filled intestine produce a 'barber's pole' appearance.

Microscopic: The male has an asymmetrical dorsal lobe and barbed spicules; the female usually has a vulval flap. In both sexes there are cervical papillae and a tiny lancet inside the buccal capsule (Fig. 14).

LIFE CYCLE

This is direct and the preparasitic phase is typically trichostrongyloid. The females are prolific egg layers. The eggs hatch to L_1 on the pasture and may develop to L_3 in as short a period as five days but development may be delayed for weeks or months under cool conditions. After ingestion, and exsheathment in the rumen, the larvae moult twice in close apposition to the gastric glands. Just before the final moult they develop the piercing lancet which enables them to obtain blood from the mucosal vessels. As adults they move freely on the sur-

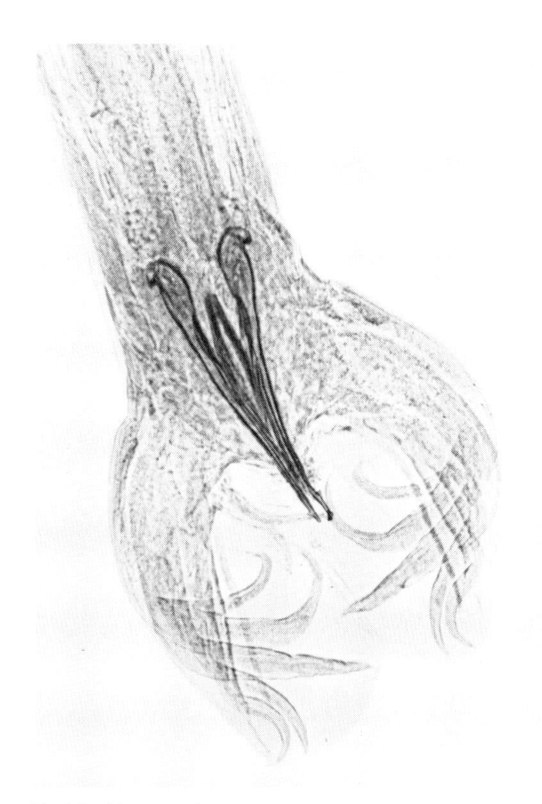

Fig 14 *Haemonchus contortus.*
(a) Male-bursa and spicules.

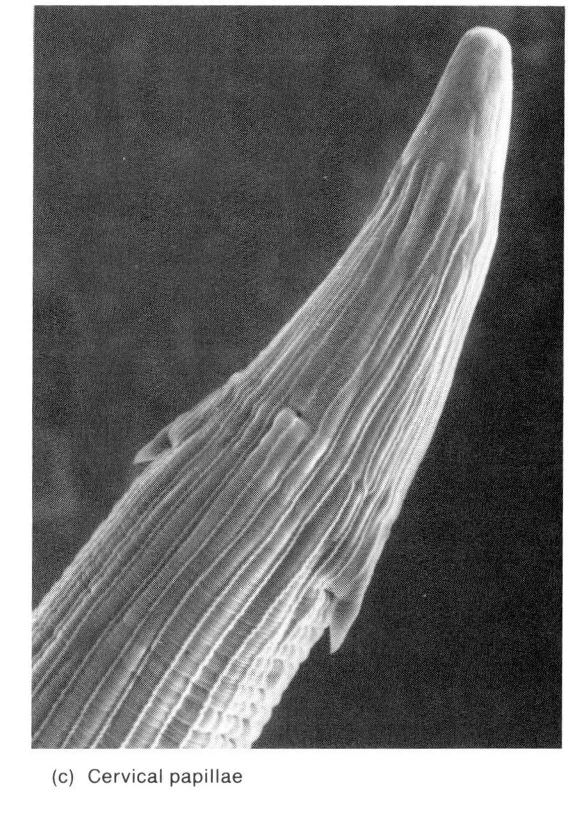

(c) Cervical papillae

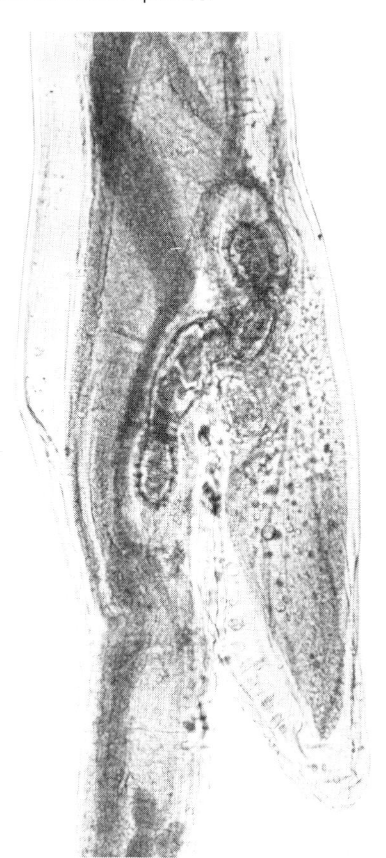

(b) Female-vulval flap.

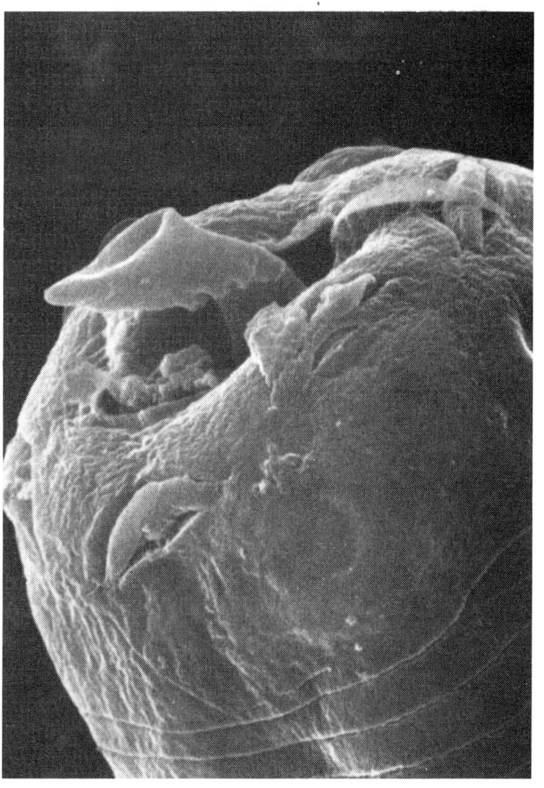

(d) Buccal lancet.

face of the mucosa. The prepatent period is 2–3 weeks in sheep and four weeks in cattle.

OVINE HAEMONCHOSIS

PATHOGENESIS

Essentially the pathogenesis of haemonchosis is that of an acute haemorrhagic anaemia due to the blood-sucking habits of the worms. Each worm removes about 0.05 ml of blood per day by ingestion and seepage from the lesions so that a sheep with 5000 *H. contortus* may lose about 250 ml daily.

In **acute haemonchosis** anaemia becomes apparent about two weeks after infection and is characterised by a progressive and dramatic fall in the packed red cell volume. During the subsequent weeks the haematocrit usually stabilises at a low level, but only at the expense of a two- to three-fold compensatory expansion of erythropoiesis. However due to the continual loss of iron and protein into the gastrointestinal tract and increasing inappetence, the marrow eventually becomes exhausted and the haematocrit falls still further before death occurs.

When ewes are affected, the consequent agalactia may result in the death of the suckling lambs.

At necropsy, between 2,000 and 20,000 worms may be present on the abomasal mucosa which shows numerous small haemorrhagic lesions (Pl. I). The abomasal contents are fluid and dark brown due to the presence of altered blood. The carcass is pale and oedematous and the red marrow has expanded from the epiphyses into the medullary cavity (Pl. I).

Less commonly, in heavier infections of up to 30,000 worms, apparently healthy sheep may die suddenly from severe haemorrhagic gastritis. This is termed **hyperacute haemonchosis**.

Perhaps as important as acute haemonchosis in tropical areas is the lesser known syndrome of **chronic haemonchosis**. This develops during a prolonged dry season when reinfection is negligible, but the pasture becomes deficient in nutrients. Over such a period the continual loss of blood from small persisting burdens of several hundred worms are sufficient to produce clinical signs associated primarily with loss of weight, weakness and inappetence rather than marked anaemia.

CLINICAL SIGNS

In hyperacute cases, sheep die suddenly from haemorrhagic gastritis.

Acute haemonchosis is characterised by anaemia, variable degrees of oedema, of which the submandibular form and ascites are most easily recognised (Pl. I), lethargy, dark coloured faeces and falling wool. Diarrhoea is not generally a feature.

Chronic haemonchosis is associated with progressive weight loss and weakness, neither severe anaemia nor gross oedema being present.

EPIDEMIOLOGY

The epidemiology of *H. contortus* is best considered separately depending on whether it occurs in tropical and subtropical or in temperate areas.

Tropical and subtropical areas

Because larval development of *H. contortus* occurs optimally at relatively high temperatures, haemonchosis is primarily a disease of sheep in warm climates. However, since high humidity, at least in the microclimate of the faeces and the herbage, is also essential for larval development and survival, the frequency and severity of outbreaks of disease is largely dependent on the rainfall in any particular area.

Given these climatic conditions, the sudden occurrence of acute clinical haemonchosis appears to depend on two further factors. First, the high faecal worm egg output of between 2,000 and 20,000 epg, even in moderate infections, means that massive pasture populations of L_3 may appear very quickly. Second, in contrast to many other helminth infections, there is little evidence that sheep in endemic areas develop an effective acquired immunity to *Haemonchus*, so that there is continuous contamination of the pasture.

In certain areas of the tropics and subtropics such as Australia, Brazil, the Middle East and Nigeria, the survival of the parasite is also associated with the ability of *H. contortus* larvae to undergo hypobiosis. Although the trigger for this phenomenon is unknown, hypobiosis occurs at the start of a prolonged dry season and permits the parasite to survive in the host as arrested L_4 instead of maturing and producing eggs which would inevitably fail to develop on the arid pasture. Resumption of development occurs just before the onset of seasonal rains. In other tropical areas such as East Africa, no significant degree of hypobiosis has been observed and this may be due to more frequent rainfall in these areas making such an evolutionary development unnecessary.

The survival of *H. contortus* infection on tropical pastures is variable depending on the climate and degree of shade, but the infective larvae are relatively resistant to desiccation and some may survive for 1–3 months on pasture or in faeces.

Temperate areas

In the British Isles, the Netherlands and presumably in other parts of northern Europe and in Canada, which are among the least favourable areas for the survival of *H. contortus*, the epidemiology is different from that of tropical zones. From the information available, infections seem to develop in two ways. Perhaps most common is the single annual cycle. Infective larvae which have developed from eggs deposited by ewes in the spring are ingested by ewes and lambs in early summer. The majority of these become arrested in the abomasum as EL_4 and do not complete development until the following spring. During the period of maturation of these hypobiotic larvae, clinical signs of acute haemonchosis may occur and in the ewes this often coincides with lambing.

In some years however clinical haemonchosis is seen in grazing lambs in late summer. The underlying epidemiology is unknown, but is perhaps associated with pasture contamination by that proportion of ingested larvae which did not undergo hypobiosis in early summer.

DIAGNOSIS

The history and clinical signs are often sufficient for the diagnosis of the acute syndrome especially if supported by faecal worm egg counts.

Necropsy, paying attention to both the abomasum and the marrow changes in the long bones, is also useful. Changes are usually evident in both, although in sheep which have just undergone 'self cure' (see below) or are in a terminal stage of the disease, the bulk of the worm burden may have been lost from the abomasum.

In hyperacute haemonchosis, only the abomasum may show changes since death may have occurred so rapidly that marrow changes are minimal.

The diagnosis of chronic haemonchosis is more difficult because of the concurrent presence of poor nutrition and confirmation may have to depend on the gradual disappearance of the syndrome after anthelmintic treatment.

TREATMENT

When an acute outbreak has occurred the sheep should be treated with one of the benzimidazoles, levamisole or ivermectin and immediately moved to pasture not recently grazed by sheep. When the original pasture is grazed again, prophylactic measures should be undertaken, as enough larvae may have survived to institute a fresh cycle of infection. Chronic haemonchosis is dealt with in a similar fashion. If possible the new pasture should have a good nutritional value; alternatively some supplementary feeding may be given.

CONTROL

In the tropics and subtropics this varies depending on the duration and number of periods in the year when rainfall and temperature permit high pasture levels of *H. contortus* larvae to develop. At such times it may be necessary to use an anthelmintic at intervals of 2–4 weeks depending on the degree of challenge. Sheep should also be treated at least once at the start of the dry season and preferably also before the start of prolonged rain to remove persisting hypobiotic larvae whose development could pose a future threat. For this purpose, one of the modern benzimidazoles or ivermectin is recommended.

In some wool producing areas where *Haemonchus* is endemic, disophenol, which has a residual prophylactic effect of up to 6 weeks, may be used. Because of its long withdrawal period it is of limited use in meat producing animals.

Apart from anthelmintic prophylaxis some studies, especially in Kenya, have indicated the potential value of some indigenous breeds of sheep which seem to be naturally highly resistant to *H. contortus* infection. Presumably such breeds could be of value in developing areas of the world where veterinary surveillance is poor.

In temperate areas, the measures outlined for the control of parasitic gastroenteritis in sheep are usually sufficient to pre-empt outbreaks of haemonchosis.

CAPRINE HAEMONCHOSIS

Goats are highly susceptible to *Haemonchus contortus* particularly when they are precluded from browsing and derive all their food intake from pasture.

BOVINE HAEMONCHOSIS

The disease caused by *H. placei* or *H. similis*, the latter possessing a characteristic vulval flap (Fig. 15), is similar in most respects to haemonchosis in sheep and is important in the tropics and subtropics during seasonal rains when severe outbreaks may occur. However the disease has also been recorded at the end of a long dry season due to the maturation of hypobiotic larvae.

Unlike haemonchosis in sheep, grazing cattle over two years old are relatively immune although this may be broken down by drought conditions which lead to poor nutrition and heavy challenge from congregation of animals around watering points. Treatment and control are similar to that described for *H. contortus* in sheep.

THE SELF-CURE PHENOMENON

In areas of endemic haemonchosis it has often been observed that after the advent of a period of heavy rain the faecal worm egg counts of sheep infected with *H. contortus* drop sharply to near zero levels due to the expulsion of the major part of the adult worm burden. This event is commonly termed the **self-cure** phenomenon, and has been reproduced experimentally by superimposing an infection of *H. contortus* larvae on an existing

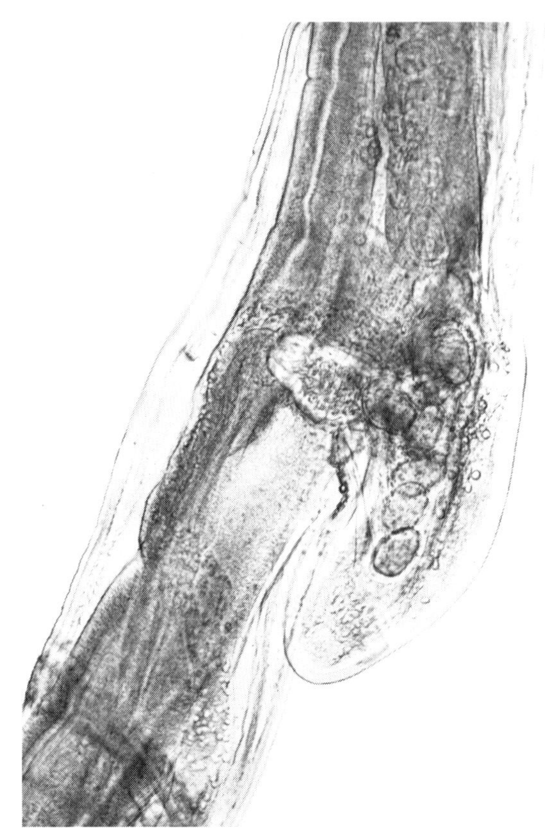

Fig 15 *Haemonchus similis* showing the vulva situated at the tip of the vulval flap.

adult infection in the abomasum. The expulsion of the adult worm population is considered to be the consequence of an immediate-type hypersensitivity reaction to antigens derived from the developing larvae. It is thought that a similar mechanism operates in the naturally occurring self-cure when large numbers of larvae mature to the infective stage on pasture after rain.

Although this phenomenon has an immunological mechanism it is not necessarily associated with protection against reinfection since the larval challenge often develops to maturity.

Another explanation of the self-cure phenomenon as it occurs in the field is based on the observation that it may happen in lambs and adults contemporaneously and on pasture with insignificant numbers of infective larvae. This suggests that the phenomenon may also be caused, in some non-specific way, by the ingestion of fresh growing grass.

Whatever the cause, self-cure is probably of mutual benefit to both host and parasite. The former gains a temporary respite from persistent blood loss while the ageing parasite population is eventually replaced by a vigorous young generation.

Mecistocirrus digitatus

This blood-sucking abomasal parasite, which to the naked eye is indistinguishable from *H. contortus*, is common in buffalo and cattle in certain areas of Asia. Microscopically it is most readily distinguished from the latter by having long narrow spicules. The prepatent period is also longer, being 60–80 days.

The pathogenesis is similar to that of *H. contortus* in sheep and it is of similar economic importance.

Trichostrongylus

Trichostrongylus is rarely a primary pathogen in temperate areas, but is usually a component of parasitic gastroenteritis in ruminants. By contrast, in the subtropics it is one of the most important causes of parasitic gastroenteritis. One species, *T. axei*, is also responsible for gastritis in horses while *T. tenuis* has been implicated in outbreaks of severe enteritis in game birds.

Hosts:
Ruminants, horses, pigs, rabbits and fowl

Site:
Small intestine except *T. axei* and *T. tenuis*

Species:

Trichostrongylus axei	abomasum of ruminants and stomach of horses and pigs
T. colubriformis	ruminants
T. vitrinus } *T. capricola* }	sheep and goats
T. retortaeformis	rabbits
T. tenuis	small intestine and caeca of game birds

There are a number of other species of ruminants with more local distribution and importance such as *T. rugatus*, *T. falculatus*, *T. probolurus* and *T. longispicularis*

Distribution:
Worldwide

IDENTIFICATION

Gross: The adults are small and hair-like, usually less than 7.0 mm long and difficult to see with the naked eye.
Microscopic: The words have no obvious buccal capsule. A most useful generic character is the distinct excretory notch in the oesophageal region (Fig. 16). The spicules are thick and unbranched and in the case of *T. axei* are also unequal in length; in the female the tail is bluntly tapered (Fig. 16) and there is no vulval flap. In *T. axei* the eggs are arranged pole to pole longitudinally. There is no vulval flap.

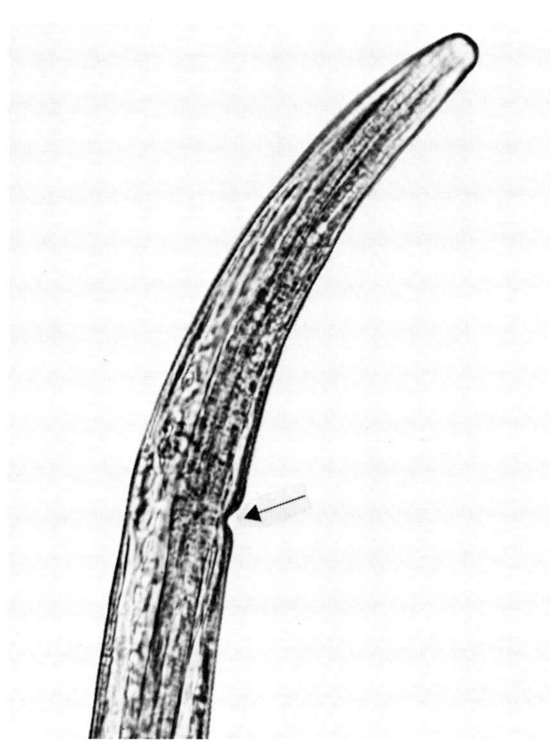

Fig 16 Some characteristics of *Trichostrongylus* species.
(a) Excretory notch in oesophageal region (↑).

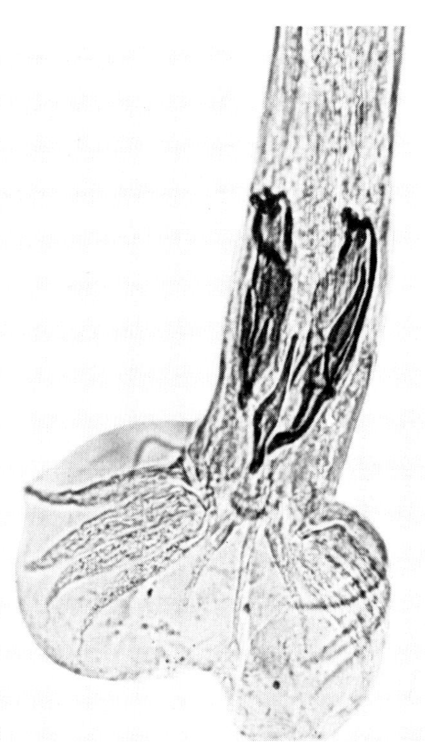

(c) Unequal spicules of *T. axei*.

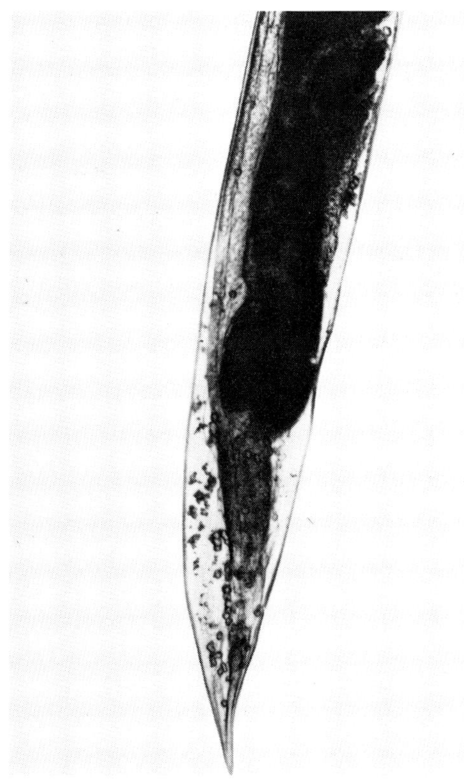

(b) Appearance of female tail.

(d) Leaf-like spicules of *T. vitrinus*.

Fig 16 (continued)

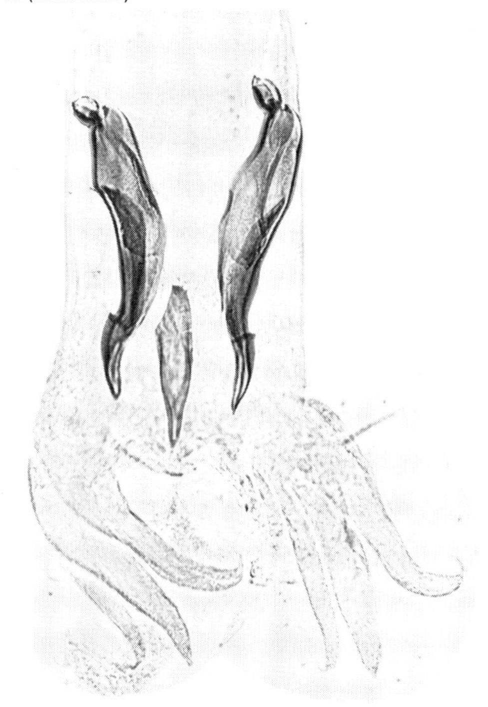

(e) 'Stepped' tip of spicules of *T. colubriformis*.

LIFE CYCLE

This is direct and the preparasitic phase is typically trichostrongyloid, except that exsheathment of the L_3 of intestinal species occurs in the abomasum. Under optimal conditions, development from the egg to infective stage occurs in 1–2 weeks.

The parasitic phase is non-migratory and the prepatent period in ruminants is 2–3 weeks. In the horse, *T. axei* has a prepatent period of 25 days while in game birds infected with *T. tenuis* it is only 10 days.

PATHOGENESIS

Following ingestion, the L_3 of the intestinal species penetrate between the epithelial glands of the mucosa with formation of tunnels beneath the epithelium, but above the lamina propria (Pl. II). When the sub-epithelial tunnels containing the developing worms rupture to liberate the young worms about 10–12 days after infection, there is considerable haemorrhage and oedema and plasma proteins are lost into the lumen of the gut. Grossly, there is an enteritis, particularly in the duodenum; the villi become distorted and flattened reducing the area available for absorption of nutrients and fluids. However many such areas appear normal. Where parasites are congregated within a small area, erosion of the mucosal surface is apparent (Pl. II). In heavy infections diarrhoea occurs, and this, together with the loss of plasma protein into the lumen of the intestine, leads to weight loss. A reduced deposition of protein, calcium and phosphorus has also been recorded.

In the case of *T. axei* the changes induced in the gastric mucosa are similar to those of *Ostertagia* with an alteration in pH and an increased permeability of the mucosa. One difference is that the worms penetrate between the glands rather than into the glands as in *Ostertagia*. Coalescence of the subsequent nodular lesions often results in plaques or ring-like lesions (Pl. II).

CLINICAL SIGNS

The principal clinical signs in heavy infections are rapid weight loss and diarrhoea. At lower levels of infection, inappetence and poor growth rates, sometimes accompanied by soft faeces, are the common signs. It is often difficult to distinguish the effects of low infections from malnutrition.

EPIDEMIOLOGY

The embryonated eggs and infective L_3 of *Trichostrongylus* have a high capacity for survival under adverse conditions whether these are extreme cold or desiccation. In temperate areas such as in Britain the L_3 survive the winter well, sometimes in sufficient numbers to precipitate clinical problems in the spring. More commonly, larval numbers increase on pasture in summer and autumn giving rise to clinical problems during these seasons.

In the southern hemisphere larvae accumulate in late winter and outbreaks are usually seen in spring. In Australia, following a period of drought the advent of rain has been shown to rehydrate large numbers of apparently desiccated L_3 (anhydrobiosis) which then become active and rapidly available to grazing animals. Under similar circumstances in southern Brazil in 1958, 20% of a total sheep population of 12 million died from intestinal trichostrongylosis.

Until recently hypobiosis was not considered to be a feature of this genus. However, there is now ample evidence in temperate areas that hypobiosis plays an important part in the epidemiology, the seasonal occurrence being similar to that of *Ostertagia* spp. In contrast to other trichostrongyles hypobiosis occurs at the L_3 stage although their role in outbreaks of disease has not been established.

Immunity to *Trichostrongylus* as in *Ostertagia* is slowly acquired and in sheep and probably goats it wanes during the periparturient period.

DIAGNOSIS

This is based on clinical signs, seasonal occurrence of

disease and, if possible, lesions at post-mortem examination. Faecal egg counts are a useful aid to diagnosis, although faecal cultures are necessary for generic identification of larvae.

TREATMENT AND CONTROL

Depending on the host, this is as described for bovine ostertagiasis, parasitic gastroenteritis in sheep and strongylosis in the horse.

T. TENUIS INFECTION

In game birds, heavy infections produce an acute and fatal haemorrhagic typhlitis. Lighter infections result in a chronic syndrome characterised by anaemia and emaciation. On game farms, therapy with levamisole in the drinking water has proved useful, but it is more important that pens should be moved regularly to prevent the accumulation of larvae. If possible the runs should not be placed in the same areas in successive years.

Cooperia

In temperate areas, members of the genus *Cooperia* usually play a secondary role in the pathogenesis of parasitic gastroenteritis of ruminants although they may be the most numerous trichostrongyle present. However, in some tropical and subtropical areas, some species are responsible for severe enteritis in calves.

Hosts:
Ruminants

Site:
Small intestine

Species:

Cooperia oncophora *C. punctata* *C. pectinata*	cattle
C. surnabada (syn. *C. mcmasteri*)	cattle and sheep
C. curticei	sheep and goats

Distribution:
Worldwide

IDENTIFICATION

Gross: In size *Cooperia* are similar to *Ostertagia*. The most notable features are the 'watch spring-like' posture of *C. curticei* and the very large bursa in all species.

Microsopic: The main generic features (Fig. 17) are the small cephalic vesicle and the transverse cuticular striations in the oesophageal region. The spicules usually have a distinct wing-like expansion in the middle region and often bear ridges; there is no gubernaculum.

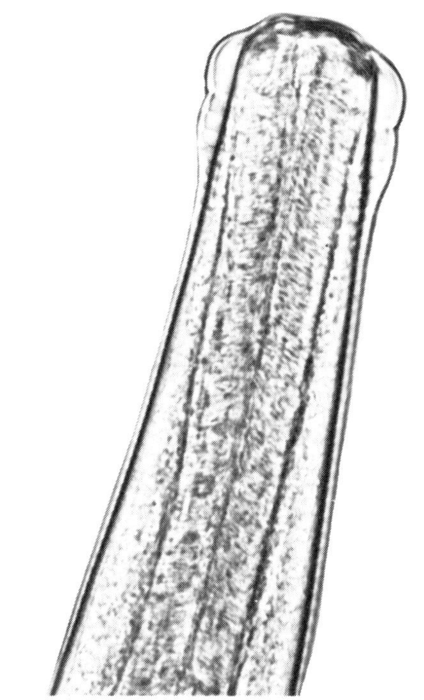

Fig 17 Some characteristics of *Cooperia* species.
(a) Cephalic vesicle and cuticular striations.

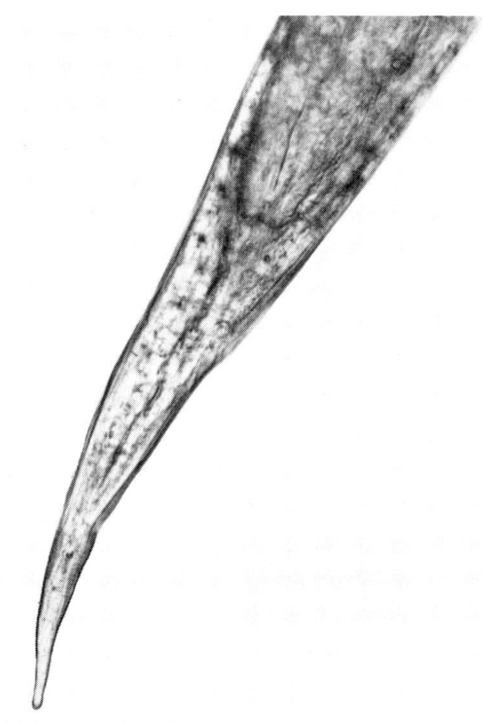

(b) Elongated female tail.

Fig 17 (continued)

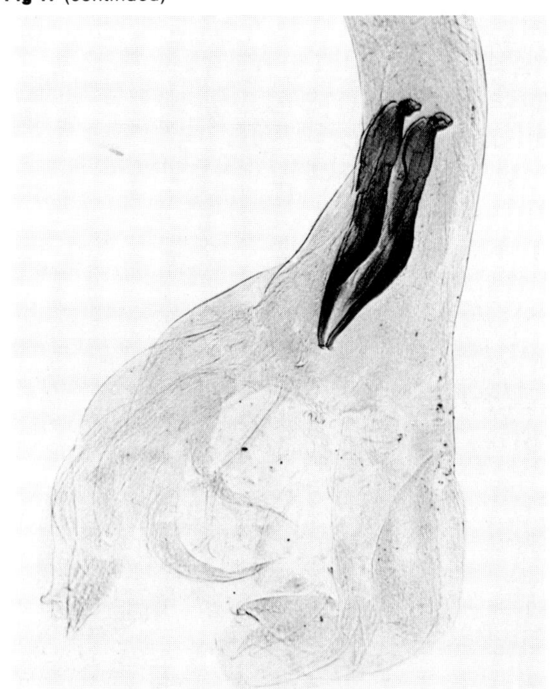

(c) Spicules of *C. oncophora.*

(e) Spicules of *C. pectinata.*

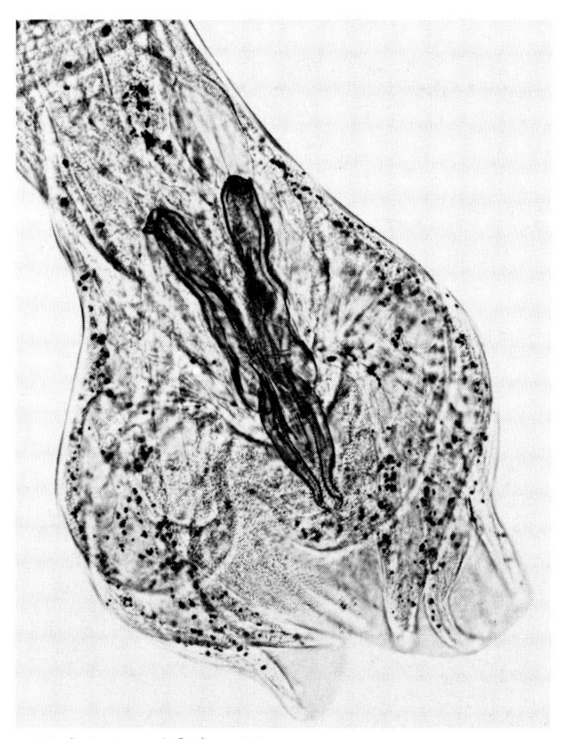

(d) Spicules of *C. punctata.*

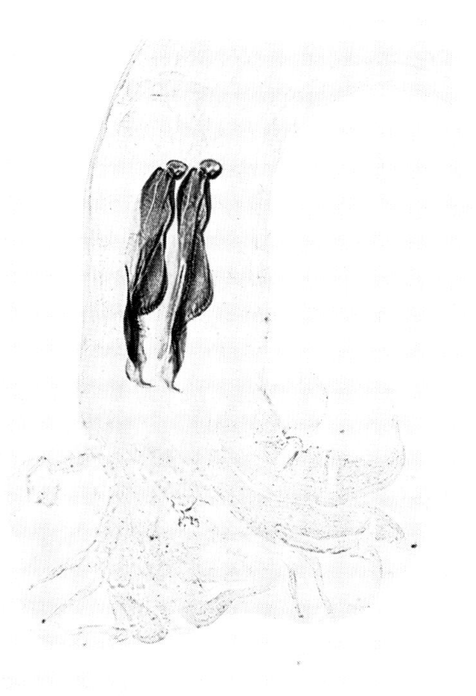

(f) Spicules of *C. curticei.*

The females usually have a small vulval flap and a long tapering tail.

LIFE CYCLE

This is direct and typical of the superfamily. The bionomic requirements of the free-living stages vary according to species. Thus, for example, those of *C. oncophora* and *C. curticei* which are primarily found in temperate areas are similar to those of *Ostertagia*; *C. punctata* and *C. pectinata* which are more common in warmer areas have similar requirements to *Haemonchus*. In the parasitic phase the two common temperate zone species develop on the surface of the intestinal mucosa whereas with the others some penetration of the epithelium takes place. The prepatent period varies from 15–18 days.

PATHOGENESIS

C. oncophora and *C. curticei* are generally considered to be mild pathogens in calves and lambs respectively although in some studies they have been associated with inappetence and poor weight gains. A strong immunity to reinfection develops after one year.

C. punctata, *C. pectinata* and probably *C. surnabada* are more pathogenic since they penetrate the epithelial surface of the small intestine and cause a disruption similar to that of intestinal trichostrongylosis which leads to villous atrophy and a reduction in the area available for absorption. In heavy infections diarrhoea has been reported.

CLINICAL SIGNS

These are loss of appetite, poor weight gains and with *C. punctata* and *C. pectinata*, diarrhoea, severe weight loss and submandibular oedema.

EPIDEMIOLOGY

In temperate areas this is identical to that of *Ostertagia*. Hypobiosis at the EL_4 is a regular feature during late autumn and winter in the northern hemisphere, and spring and summer in the southern hemisphere.

In the subtropics the epidemiology is similar to that of *Haemonchus* though *Cooperia* does not have the same high biotic potential and the L_3 survive rather better under arid conditions. Hypobiosis is a feature during prolonged dry seasons.

A closely related parasite *Paracooperia nodulosa* is responsible for a severe nodular enteritis in buffaloes in Asia, Africa and South America.

DIAGNOSIS, TREATMENT AND CONTROL

The principles are similar to those applied in bovine ostertagiasis and PGE in sheep.

Hyostrongylus

This parasite is responsible for a chronic gastritis in pigs, particularly gilts and sows.

Host:
Pig

Site:
Stomach

Species:
Hyostrongylus rubidus

Distribution:
Worldwide

IDENTIFICATION

Gross: Slender reddish worms 5–8 mm in length.

Microscopic: A small cephalic vesicle is present and the spicules resemble *Ostertagia*, but have only two distal branches.

LIFE CYCLE

Typical of the superfamily. The free-living and parasitic stages are similar to those of *Ostertagia*. The prepatent period is three weeks.

PATHOGENESIS

Similar to ostertagiasis, with penetration of the gastric glands by the L_3 and replacement of the parietal cells by rapidly dividing undifferentiated cells which proliferate to give rise to nodules on the mucosal surface. The pH becomes elevated in heavy infections. Sometimes there is ulceration and haemorrhage of the nodular lesions, but more commonly light infections occur and these are associated with decreased appetite and poor feed conversion rates.

CLINICAL SIGNS

These are inappetence, anaemia and loss of condition.

EPIDEMIOLOGY

Because of the preparasitic larval requirements infection is confined to pigs with access to pasture or those kept in straw yards. It is therefore more common in breeding stock, particularly gilts. The epidemiology, at least in temperate zones, is similar to that of *Ostertagia* in ruminants with seasonal hypobiosis a feature.

DIAGNOSIS

This is based on a history of access to permanent pig pastures and the clinical signs. Confirmatory diagnosis is by examination of faeces for eggs; for differentiation from other nematodes, larval identification following faecal culture may be necessary.

TREATMENT

When *Hyostrongylus* infection is diagnosed, particularly in breeding stock, it is important to use a drug such as a modern benzimidazole or ivermectin which will remove hypobiotic larvae.

CONTROL

The same principles apply as for the control of parasitic gastroenteritis in ruminants. For example, in temperate areas there should be an annual rotation of pasture with other livestock or crops. The timing of the move to other pastures may be dependent on other farming activities, but if it can be delayed until October or later and accompanied by an anthelmintic treatment then eggs from any worms which survive the treatment are unlikely to develop due to the unfavourable winter temperatures. A second treatment, again using a modern benzimidazole or ivermectin, is recommended 3–4 weeks later to remove any residual infection.

Nematodirus

Nematodirus is of special importance as a parasite of lambs in temperate regions.

Hosts:
Ruminants

Site:
Small intestine

Species:

Nematodirus battus	sheep, occasionally calves
N. filicollis	sheep and goats
N. spathiger	sheep and goats, occasionally cattle
N. helvetianus	cattle

N. abnormalis and *N. oiratianus* have also been recorded from sheep and goats in southern Europe, Asia, America and Australia

Distribution:
Worldwide, more commonly in temperate zones. *N. battus* is most important in the British Isles, but also occurs in Norway and the Netherlands

IDENTIFICATION

Gross: The adults are slender worms about 2.0 cm long. The intertwining of the thin, twisted worms produces an appearance similar to that of cotton wool.
Microscopic: A small but distinct cephalic vesicle is

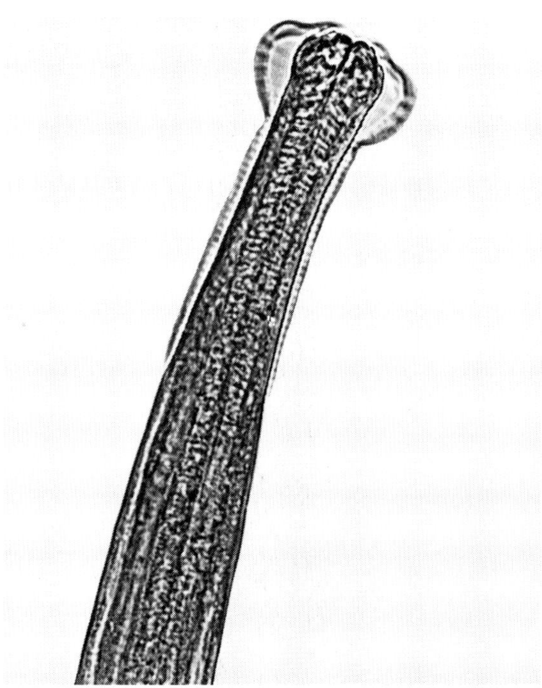

Fig 18 Characteristics of most *Nematodirus* species. (a–d)
(a) Cephalic vesicle.

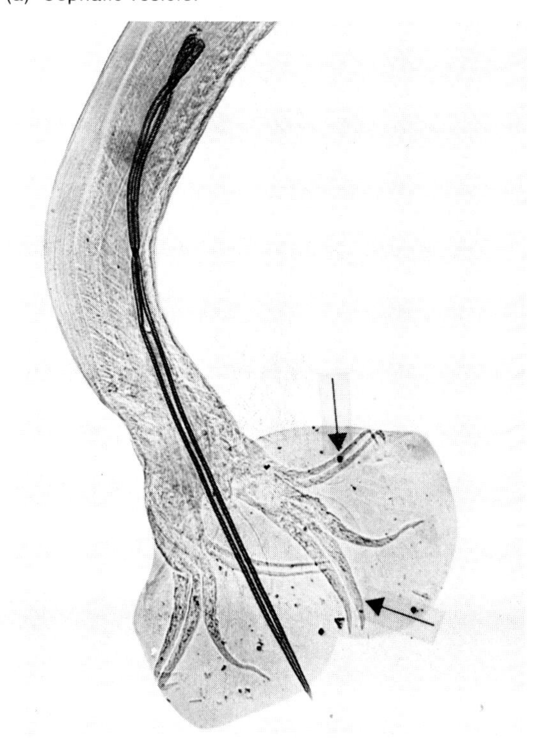

(b) Long slender spicules and 2 sets of parallel rays in each bursal lobe (↑).

(c) Tail of female showing spine.

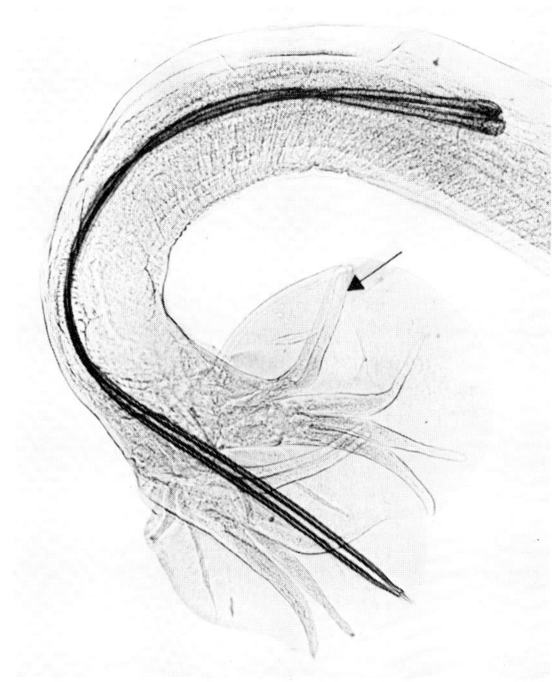

Exceptional features of *N. battus.* (e–g)
(e) 1 set of parallel rays in each bursal lobe (↑).

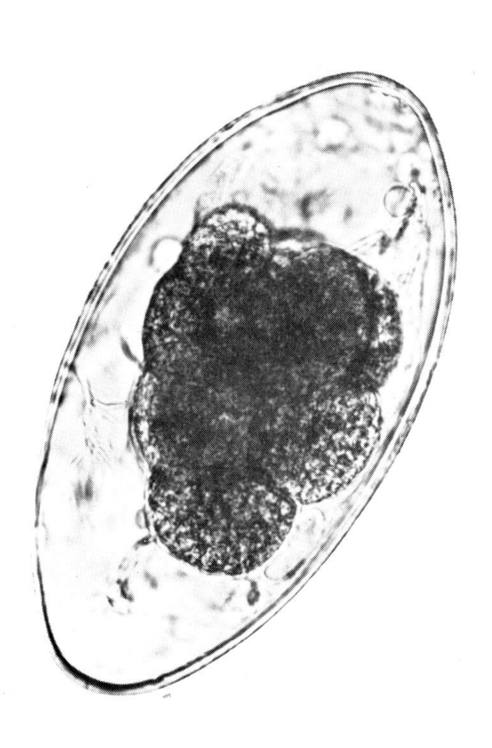

(d) Large ovoid egg.

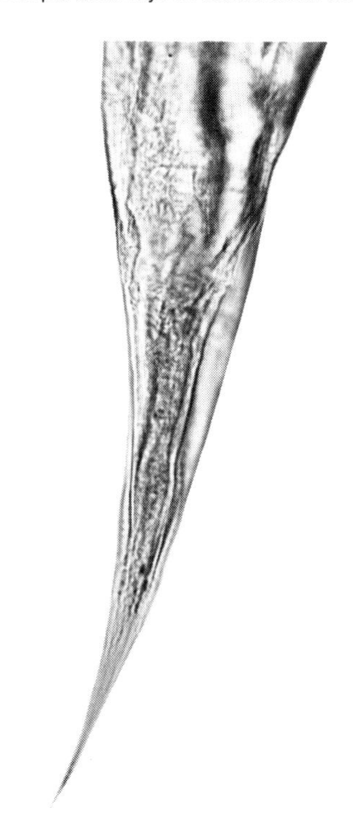

(f) Long pointed female tail.

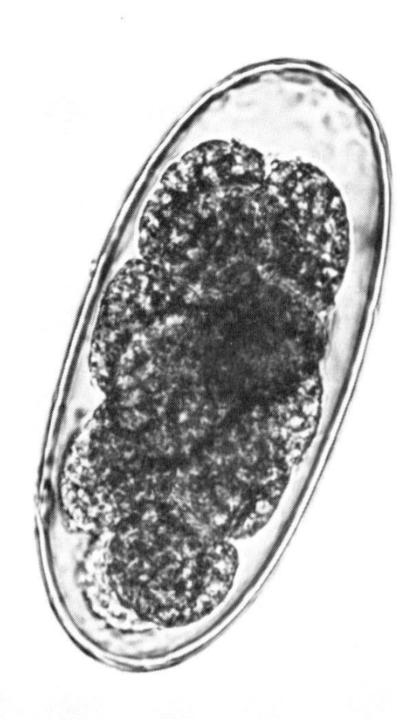

(g) Large egg with parallel sides.

present (Fig. 18). The spicules are long and slender with fused tips. In all except *N. battus* the male has two sets of parallel rays in each of the main bursal lobes; the female has a truncate tail with a small spine, and the egg is large, ovoid and colourless (Fig. 18) and twice the size of the typical trichostrongyle egg.

N. battus is characterised by having only one set of parallel rays in each bursal lobe while the female worm has a long pointed tail and the large egg is brownish with parallel sides (Fig. 18).

LIFE CYCLE

The preparasitic phase is almost unique in the trichostrongyloids in that development to the L_3 takes place within the egg shell. This development is generally very slow and in temperate climates takes at least two months. Once the L_3 is present there is often a lag period before hatching occurs, the duration varying according to the species.

Nematodirus battus

In the case of *N. battus*, the most important species in Britain, hatching of most eggs requires a prolonged period of chill followed by a mean day/night temperature of more than 10 °C, conditions which occur in late spring. Hence most of the eggs from one season's grazing must remain unhatched on the ground during the winter and only one generation is possible each year for the bulk of this species. However, some *N. battus* eggs deposited in the spring are capable of hatching in the autumn of the same year resulting in significant numbers of L_3 on the pasture at this time.

The parasitic phase is non-migratory and the prepatent period is 15 days.

Other Nematodirus species

The other species do not have the same critical hatching requirements as *N. battus* and so the L_3 appear on the pasture within 2–3 months of the eggs being excreted in the faeces. More than one annual generation is therefore possible.

PATHOGENESIS

Nematodiriasis, due to *N. battus* infection, is an example of a parasitic disease where the principal pathogenic effect is attributable to the larval stages. Following ingestion of large numbers of L_3 there is disruption of the intestinal mucosa, particularly in the ileum, although the majority of developing stages are found on the mucosal surface. Development through L_4 to L_5 is complete by 10–12 days from infection and this coincides with severe damage to the villi and erosion of the mucosa leading to villous atrophy (Fig. 19). The ability of the intestine to exchange fluids and nutrients is grossly reduced, and with the onset of diarrhoea the lamb rapidly becomes dehydrated. At necropsy the carcass has a dehydrated appearance and there is an enteritis in the ileum.

While the pathogenesis of infections with the other *Nematodirus* spp. is probably similar there is some controversy on the extent of their pathogenic effect. For example, though *N. helvetianus* has been incriminated in outbreaks of bovine parasitic gastroenteritis, experimental attempts to reproduce the disease have been unsuccessful.

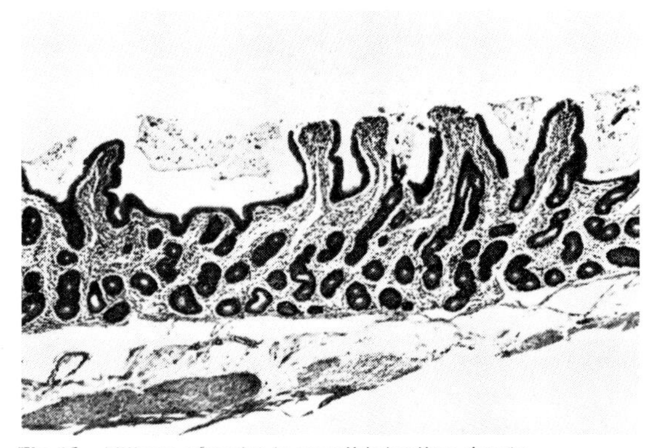

Fig 19 Villous atrophy in small intestine due to *Nematodirus battus* infection.

CLINICAL SIGNS

In severe infections, diarrhoea is the most prominent clinical sign. As dehydration proceeds the affected animals become thirsty and in infected flocks the ewes continue to graze, apparently unaffected by the larval challenge, while their inappetent and diarrhoeic lambs congregate round drinking places.

EPIDEMIOLOGY

This is best considered separately for *N. battus* and other *Nematodirus* spp.

The three most important features of the epidemiology of *N. battus* infections are:

(i) The capacity of the free-living stages, particularly the egg containing the L_3, to survive on pasture, some for up to two years.

(ii) The critical hatching requirements of most eggs which ensure the appearance of a large number of L_3 on the pasture simultaneously, usually in May and June. Though the flush of larvae on the pasture may be an annual event, the appearance of clinical nematodiriasis is not; thus if the flush of L_3 is early the suckling lambs may not be consuming sufficient grass to acquire large numbers of L_3, and if it is late the lambs may be old enough to resist the larval challenge. There is some evidence that there is an age resistance to *N. battus* which commences when lambs are about three months old. However, lambs of 6–7 months can have considerable *N. battus* burdens and it is therefore doubtful if this age immunity is absolute.

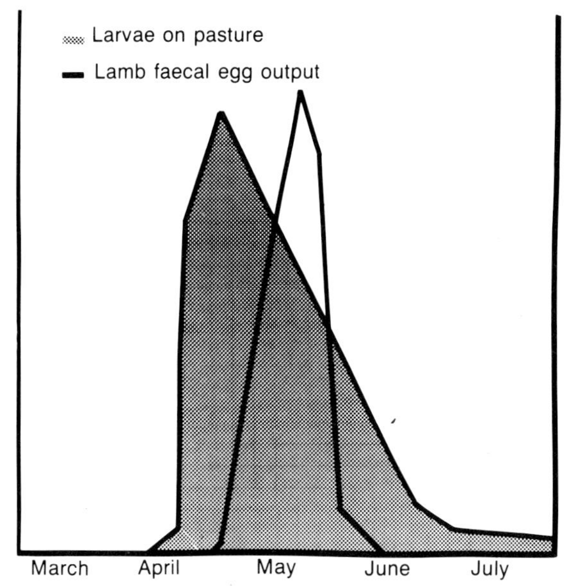

Fig 20 Epidemiological features of *Nematodirus battus* infection illustrating the sudden appearance of infective larvae in spring following mass hatching of overwintered eggs.

(iii) The negligible role played by the ewe in the annual cycling of *N. battus* which can thus be considered as a lamb-to-lamb disease. Adult sheep often have a few *N. battus* eggs in their faeces, but these are insufficient to precipitate a larval flush although enough to ensure the persistence of infection on the pastures.

The epidemiology is illustrated in Fig. 20.

Since the other *Nematodirus* species do not have such critical hatching requirements, a sudden flush of L_3 does not occur and although *N. filicollis*, *N. spathiger* and *N. helvetianus* have all been associated with outbreaks of nematodiriasis in sheep and cattle, it is more common to find them in conjunction with the other trichostrongyles.

Though L_4 of *Nematodirus* spp. apparently arrested in their development have been recorded at necropsy, there is no obvious seasonal pattern to their occurrence and it seems more likely that they have accumulated as a consequence of host resistance rather than hypobiosis.

DIAGNOSIS

Because the clinical signs appear during the prepatent period faecal egg counts are of little value in diagnosis which is best made on grazing history, clinical signs, and if possible, a post-mortem examination.

TREATMENT

Several drugs are highly effective against *Nematodirus* infections especially levamisole, ivermectin or one of the modern benzimidazoles, fenbendazole, oxfendazole or albendazole. The response to treatment is usually rapid and if diarrhoea persists coccidiosis should be considered as a complicating factor.

CONTROL

With the exception of *N. battus*, which requires special consideration, disease due to monospecific *Nematodirus* infections is rarely seen. Instead, they are usually part of the worm burden of trichostrongyloid species which are responsible for the syndrome of parasitic gastroenteritis in sheep and as such may be controlled by the measures outlined below.

Since *N. battus* infection of lambs has a unique epidemiology, its control is best considered separately. Due to the annual hatching of *N. battus* eggs in spring, the disease can be controlled by avoiding the grazing of successive lamb crops on the same pasture. Where such alternative grazing is not available each year, control can be achieved by anthelmintic prophylaxis, the timing of treatments being based on the knowledge that the peak time for the appearance of *N. battus* L_3 is May to early June. Ideally, dosing should be at three week intervals over May and June and it is unwise to await the appearance of clinical signs of diarrhoea before administering the drugs.

The Ministry of Agriculture in Britain has developed a forecasting system based primarily on soil temperature in the early spring which can predict the likely severity of nematodiriasis. In years when the forecast predicts severe disease, three treatments are recommended during May and June; in other years two treatments in May should suffice. Several drugs are recommended, including levamisole, or any of the modern benzimidazoles or ivermectin.

THE TREATMENT AND CONTROL OF PARASITIC GASTROENTERITIS (PGE) IN SHEEP

The recommendations outlined below are applicable to temperate areas of the northern hemisphere, but the principles can be adapted to local conditions elsewhere.

TREATMENT

Because of the short period between birth and marketing, the treatment of PGE in lambs is an inferior policy compared with the preventive measures discussed below. However, when necessary, treatment with any of the benzimidazoles, levamisole or ivermectin will remove adult worms and developing stages. Following treatment, lambs should be moved to pasture not grazed by sheep that year, otherwise they will immediately become reinfected.

The occasional outbreaks of Type II ostertagiasis in young adult sheep in the spring may be treated with the same anthelmintics. Unlike *O. ostertagi* in calves the arrested stages of the common sheep nematodes are susceptible to thiabendazole and levamisole.

CONTROL

Although the control of PGE in sheep is based on the same principles as that described for *O. ostertagi* in cattle, its practice is somewhat different for the following reasons:

(i) The PPR (periparturient rise in faecal egg counts) is very marked in ewes and is the most important cause of pasture contamination with nematode eggs in the spring.

(ii) PGE in sheep is generally associated with a variety of nematode genera with differing epidemiological characteristics.

(iii) Most sheep graze throughout their lives so that pasture contamination with nematode eggs and the intake of infective larvae is almost continuous and modified only by climatic restrictions.

In selecting the best method of prophylaxis much depends on whether the farm consists primarily of permanent pasture or has pastures which are rotated with crops so that new leys or hay and silage aftermaths are available each year.

FARMS CONSISTING OF MAINLY PERMANENT PASTURE

On such farms control may be obtained either by anthelmintic prophylaxis or by alternate grazing on an annual basis with cattle and sheep. The former is the only feasible method where the farm stock is primarily sheep while the latter can be used where cattle and sheep are both present in reasonable proportions.

Prophylaxis by anthelmintics

Adult sheep

The most important source of infection for the lamb crop is undoubtedly the increase in nematode eggs in ewe faeces during the PPR and prophylaxis will only be efficient if this is kept to a minimum. Effective anthelmintic therapy of ewes during the fourth month of pregnancy should eliminate most of the worm burdens present at this time including arrested larval stages and in the case of ewes on extensive grazing, where nutritional status is frequently low, this treatment often results in improved general body condition. However, during late pregnancy and early lactation, such treated ewes will soon become reinfected from the ingestion of overwintered larvae on the pasture. It is therefore recommended that for optimal prophylaxis, a further treatment be given at 4–6 weeks post-lambing.

Young adults and rams should also be treated at these times.

An alternative to the gathering of ewes for these treatments is to provide anthelmintic incorporated in a feed or energy block during the periparturient period. The results obtained with the latter system appear to be best when the ewes are contained in small paddocks or fields as the uptake of drug is less consistent under extensive grazing systems.

Rumen boluses designed for the slow release of anthelmintics over a prolonged period are under development for sheep and seem an ideal way of minimising worm egg output.

Lambs

Apart from specific treatment for *N. battus* infection, lambs should be treated at weaning, and if possible moved to safe pastures ie. those not grazed by sheep since the previous year. Where such grazing is not available, prophylactic treatments should be repeated until autumn or marketing. The number of treatments will vary depending on the stocking rate, one treatment in September sufficing for lambs under extensive grazing and two between weaning and marketing for those under more intensive conditions.

For prophylactic treatments levamisole, the benzimidazoles, pro-benzimidazoles and ivermectin may be used.

For low level administration in feed blocks thiophanate or fenbendazole have proved useful.

The prophylactic programmes outlined above are relatively costly in terms of drugs and labour but are currently the only methods available on farms where the enterprise is heavily dependent on one animal species.

Prophylaxis by alternate grazing of sheep and cattle

On farms where sheep and cattle are both present in significant numbers, good control is theoretically possible by alternating the grazing of fields on an annual basis with each host, due to the relative insusceptibility of cattle to sheep nematodes and vice versa.

In practice, control is best achieved by exchanging, in the spring, pastures grazed by sheep and beef cattle over the previous year, preferably combined with anthelmintic treatment at the time of exchange.

FARMS WITH ALTERNATIVE GRAZING

In these mostly intensive farms, rotation of crops and grass is often a feature, and therefore new leys and hay and silage aftermaths are available as safe pastures each year and can be reserved for susceptible stock. In such a situation, control should be based on a combination of grazing management and anthelmintic prophylaxis.

Prophylaxis by grazing management and anthelmintics

Good control is possible with only one annual anthelmintic treatment of ewes when they leave the lambing field. This will terminate the PPR in faecal egg counts prior to moving the ewes and lambs to a safe pasture. At weaning, the lambs should be moved to another safe pasture and an anthelmintic treatment of the lambs at this time is good policy.

A second system, not costly in labour or drugs, has been devised for farms where arable crops, sheep and cattle are major components and involves a three year rotation of cattle, sheep and crops.

With this system the aftermath grazing available after cropping may be used for weaned calves and weaned lambs.

It has been suggested that anthelmintic prophylaxis can be disposed of completely under this system but clinical PGE has sometimes occurred when treatment has been omitted. It is worth remembering that even the highly efficient drugs currently available will not remove all the worms present; that some cattle nematodes can infect sheep and vice versa; and that a few infective larvae on the pasture can survive for beyond two years. It is therefore advisable to give at least one annual spring treatment to all stock prior to moving to new pastures.

Prophylaxis by grazing management

Many schemes were devised to control the acquisition of L_3 based solely on grazing management. One recommendation was to rotate sheep through paddocks, but since it involved the return of the sheep to their previously grazed paddocks in the same season, this was of little value. Two other methods, which did not involve a return to the original pasture, were strip grazing, in which sheep were confined to a narrow strip across the field by fences which were moved every few days, and creep grazing, in which a single fence confined the ewes, but since it possessed a 'creep' or hole, allowed the lambs to graze forward. The systems may have been highly effective in preventing PGE, but were costly in fencing and labour and are little used nowadays.

Dictyocaulus

This genus living in the bronchi of cattle, sheep, horses and donkeys is the major cause of parasitic bronchitis in these hosts.

Hosts:
Ruminants, horses and donkeys

Site:
Trachea and bronchi, particularly of the diaphragmatic lobes

Species:

Dictyocaulus viviparus	cattle and deer
D. filaria	sheep and goats
D. arnfieldi	donkeys and horses

Distribution:
Worldwide, but especially important in temperate climates.

IDENTIFICATION

The adults are slender thread-like worms up to 8.0 cm in length. Their location in the trachea and bronchi and their size are diagnostic.

Since *D. viviparus* is the most pathogenic of the three species it is presented in detail. For the other two species, only those features which are different to *D. viviparus* are discussed.

Dictyocaulus viviparus

Dictyocaulus viviparus is the cause of parasitic bronchitis in cattle, also known as husk, hoose, verminous pneumonia or dictyocauliasis. The disease is characterised by bronchitis and pneumonia and typically affects young cattle during their first grazing season on permanent or semi-permanent pastures. The disease is prevalent in temperate areas with high rainfall.

LIFE CYCLE

The female worms are ovo-viviparous, producing eggs containing fully developed larvae which hatch almost immediately. The L_1 migrate up the trachea, are swallowed and pass out in the faeces. The larvae are unique in that they are present in fresh faeces, are characteristically sluggish, and their intestinal cells are filled with dark brown food granules (Fig. 21). In consequence the preparasitic stages do not require to feed. Under optimal conditions the L_3 stage is reached within five days, but usually takes longer in the field. The L_3 leave the faecal pat to reach the herbage either by their own motility or through the agency of the fungus *Pilobolus*.

After ingestion, the L_3 penetrate the intestinal mucosa and pass to the mesenteric lymph nodes where they moult. Thence the L_4 travel via the lymph and blood to the lungs, and break out of the capillaries into the alveoli about one week after infection. The final moult occurs in the bronchioles a few days later and the young adults then move up the bronchi and mature. The prepatent period is around 3–4 weeks.

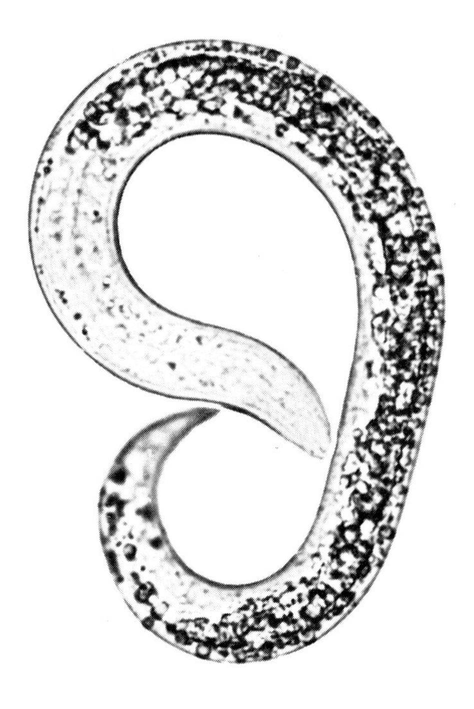

Fig 21 First stage larvae of *Dictyocaulus vivparus.*

PATHOGENESIS

This may be divided into four phases:

(i) Penetration phase. Days 1–7.

During this period the larvae are making their way to the lungs and pulmonary lesions are not yet apparent.

(ii) Prepatent phase. Days 8–25.

This phase starts with the appearance of larvae within the alveoli where they cause alveolitis. This is followed by bronchiolitis and finally bronchitis as the larvae become immature adults and move up the bronchi. Cellular infiltrates of neutrophils, eosinophils and macrophages temporarily plug the lumina of the bronchioles and cause collapse of other groups of alveoli. This lesion (Fig. 22) is largely responsible for the first clinical signs.

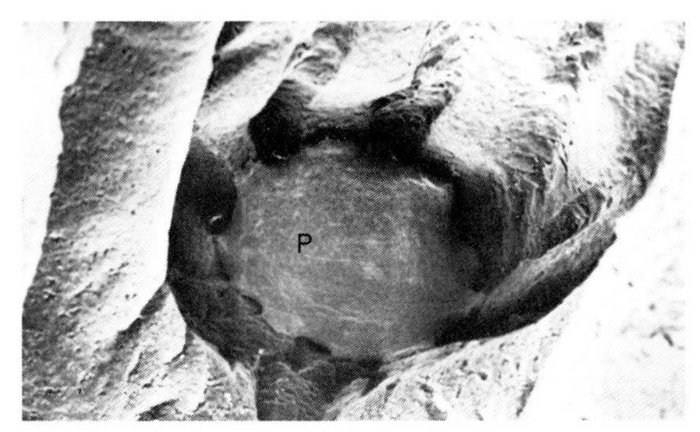

Fig 22 Scanning electron micrograph of mucus and cellular plug (p) in small bronchiole following infection with *Dictyocaulus viviparus.*

Towards the end of this phase bronchitis develops, characterised by mucus containing immature lungworms in the airways, which may only be seen with the aid of a low-power microscope, and by cellular infiltration of the epithelium.

Heavily infected animals, whose lungs contain several thousand developing worms, may die from day 15 onwards due to respiratory failure following the development of severe interstitial emphysema and pulmonary oedema.

(iii) Patent phase. Days 26–60.

This is associated with two main lesions.

First, a parasitic bronchitis characterised by the presence of hundreds or even thousands of adult worms in the frothy white mucus in the lumina of the bronchi (Pl. II). The bronchial epithelium is hyperplastic and heavily infiltrated by inflammatory cells, particularly eosinophils.

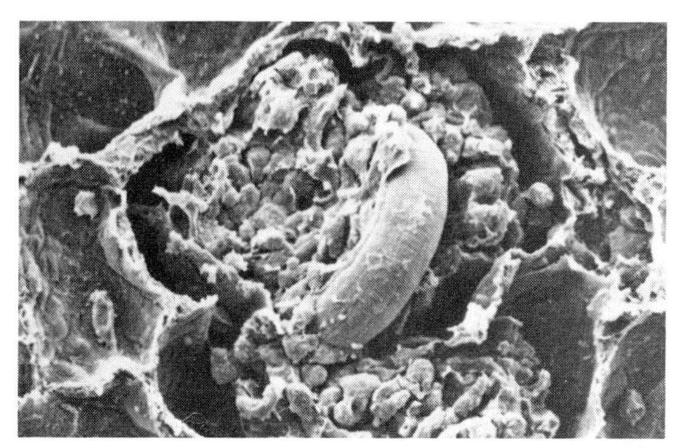

Fig 23 Aspirated *Dictyocaulus viviparus* L_1 surrounded by inflammatory cells.

Secondly, the presence of dark-red collapsed areas around infected bronchi. This is a parasitic pneumonia (Pl. II) caused by the aspiration of eggs and L_1 into the alveoli. These 'foreign bodies' quickly provoke dense infiltrates of polymorphs, macrophages and multinucleated giant cells around them (Fig. 23).

Depending on the extent of the infection there may be varying degrees of interstitial emphysema and oedema.

(iv) Postpatent phase. Days 61–90.

In untreated calves, this is normally the recovery phase after the adult lungworms have been expelled. Although the clinical signs are abating the bronchi are still inflamed and residual lesions such as bronchial and peribronchial fibrosis may persist for several weeks or months. Eventually the broncho-pulmonary system becomes completely normal and coughing ceases. However, in about 25% of animals which have been heavily infected, there is a flare-up of clinical signs during this phase which is frequently fatal. This is caused by one of two entities.

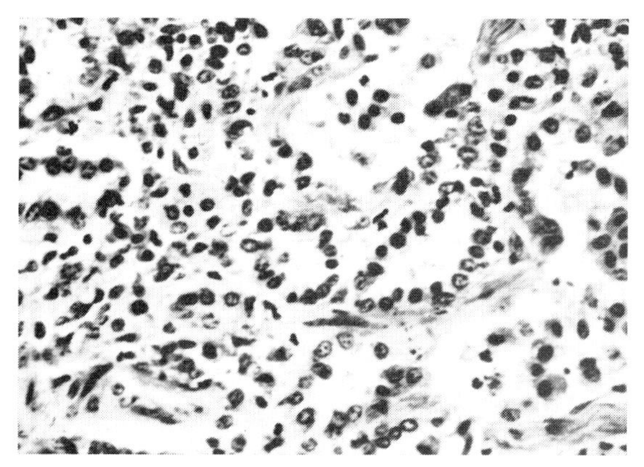

Fig 24 'Alveolar Epithelialisation' which may occur during postpatent parasitic bronchitis.

Most commonly, there is a proliferative lesion so that much of the lung is pink and rubbery and does not collapse when the chest is opened. This, often described as 'epithelialisation', is due to the proliferation of Type 2 pneumocytes on the alveoli giving the appearance of a gland-like organ (Fig. 24). Gaseous exchange at the alveolar surface is gravely hindered and the lesion is often accompanied by interstitial emphysema and pulmonary oedema. The etiology is unknown, but is thought to be due to the dissolution and aspiration of dead or dying worm material into the alveoli. The clinical syndrome is often termed **postpatent parasitic bronchitis**.

The other cause, usually in animals convalescing indoors, is a superimposed bacterial infection of the imperfectly healed lungs leading to acute interstitial pneumonia.

CLINICAL SIGNS

Within any affected group, differing degrees of clinical severity are usually apparent; typically a few animals are mildly affected, most are moderately affected and a few are severely affected.

Mildly affected animals cough intermittently, particularly when exercised.

Moderately affected animals have frequent bouts of coughing at rest, tachypnoea (>60 respirations per minute) and hyperpnoea. Frequently, squeaks and crackles over the posterior lung lobes are heard on auscultation.

Severely affected animals show severe tachypnoea (>80 respirations per minute) and dyspnoea and frequently adopt the classic 'air-hunger' position of mouth-breathing with the head and neck outstretched. There is usually a deep harsh cough, squeaks and crackles over the posterior lung lobes, salivation, anorexia and sometimes mild pyrexia. Often the smallest calves are most severely affected.

Calves may show clinical signs during the prepatent period and occasionally a massive infection can cause severe dyspnoea of sudden onset often followed by death in 24–48 hours.

Most animals gradually recover although complete return to normality may take weeks or months. However, a proportion of convalescing calves suddenly develop severe respiratory signs, unassociated with pyrexia, which usually terminates fatally 1–4 days later. This syndrome of postpatent parasitic bronchitis has been described above.

EPIDEMIOLOGY

Generally only calves during their first grazing season are clinically affected, since on farms where the disease is endemic older animals have a strong acquired immunity.

In endemic areas in the northern hemisphere infection may persist from year to year in two ways:

(i) Overwintered larvae

L_3 may survive on pasture from autumn until late spring in sufficient numbers to initiate infection or occasionally to cause disease. A similar effect may result when infected slurry or manure is spread on pastures in the spring.

(ii) Carrier animals

Small numbers of adult worms can survive in the bronchi of infected animals, particularly yearlings, until the next grazing season. Until recently it was assumed that they all persisted as adults, but it has now been shown that the chilling of infective larvae before administration to calves will produce arrested L_5; hypobiosis at this stage has also been observed in naturally infected calves in Switzerland, Austria and Canada, although the extent to which this occurs naturally after ingestion of larvae in late autumn and its significance in the transmission of the infection has not yet been fully established.

The dispersal of larvae from the faecal pat during the grazing season appears to be effected by a fungus rather than by simple migration. This fungus, *Pilobolus*, is commonly found growing on the surface of bovine faecal pats about one week after these have been deposited. The larvae of *D. viviparus*, crawling on the surface of the pats, migrate in large numbers up the stalks of the fungi on to, and even inside, the sporangium or seed-capsule (Pl.II). When the sporangium is discharged it is projected a distance of up to three metres in still air to land on the surrounding herbage. It is possible that this very effective method of dissemination of larvae may be further enhanced given a moderate wind. Perhaps also the rare outbreaks of parasitic bronchitis in housed calves are associated with the dispersal of L_3 by *Pilobolus* growing in the dung of adjacent infected cattle.

Parasitic bronchitis is predominantly a problem in areas such as northern Europe which have a mild climate, a high rainfall and abundant permanent grass. Outbreaks of disease occur from June until November, but are most common from July until September. It is not clear why the disease is usually not apparent until calves, turned out to graze in the spring, have been at grass for 2–5 months. One explanation is that the initial infection, acquired from the ingestion of overwintered larvae in May, involves so few worms that neither clinical signs nor immunity is produced; however, sufficient numbers of larvae are seeded on to the pasture so that by July the numbers of L_3 on pasture are sufficient to produce clinical disease. Young calves, added to such a grazing herd in July, may develop clinical disease within 2–3 weeks.

An alternative explanation is that L_3 overwinter in the soil rather than on the grass and only migrate on to pasture at some point between June and October as a result of some factor, as yet unknown, perhaps involving earthworms or coprophagic beetles. At present there is only circumstantial evidence to support this theory.

Although dairy or dairy-cross calves are most commonly affected it should be recognised that autumn-born single-suckled beef calves are just as susceptible when turned out to grass in early summer. Spring-born suckled beef calves grazed with their dams until housed or sold do not usually develop clinical signs, although coughing due to a mild infection is common. However, the typical disease may occur in weaned calves grazed until late autumn.

This epidemiological picture, typical of temperate countries, may be modified in some areas by factors such as climate or husbandry. In tropical countries, where disease due to *D. viviparus* may occur intermittently, the epidemiology is presumably quite different and probably depends more on pasture contamination by carrier animals such as may occur during flooding when cattle congregate on damp, high areas, rather than on the prolonged survival of infective larvae.

DIAGNOSIS

Usually the clinical signs, the time of the year and a history of grazing on permanent or semi-permanent pastures are sufficient to enable a diagnosis to be made.

Larvae are found (50–1,000/g) only in the faeces of patent cases so that faecal samples should be obtained from a number of affected individuals. To avoid contamination with soil nematodes, samples should be obtained from the rectum.

TREATMENT

The anthelmintics available for the treatment of bovine parasitic bronchitis are the modern benzimidazoles, levamisole or ivermectin.

These drugs have been shown to be effective against all stages of lungworms with a consequent amelioration of clinical signs. In the past diethylcarbamazine was widely used, but it has been largely superseded by the drugs mentioned above.

For maximum efficiency all of these drugs should be used as early as possible in the treatment of the disease since clinical signs associated with pulmonary pathology are not rapidly resolved by mere removal of adult lungworms.

Where the disease is severe and well established in a number of calves, the stockowner should be warned that anthelmintic treatment, while being the only course available, may exacerbate the clinical signs in one or more animals with a possible fatal termination. The reasons underlying this are still under study, but are probably similar to those which produce postpatent parasitic bronchitis.

Whatever treatment is selected, it is advisable to divide affected calves into two groups as the prognosis will vary according to the severity of the disease. Those calves which are only coughing and/or tachypnoeic are usually in the prepatent stage of the disease or have a small adult worm burden and treatment of these animals should result in rapid recovery. Calves in this category may not have developed a strong immunity and after treatment should not be returned to the field which was the source of infection; if this is impossible, parenteral ivermectin is the drug of choice since its residual effect prevents reinfection for a further three weeks.

Any calves which are dyspnoeic, anorexic and possibly pyrexic should be kept indoors for treatment and further observation. The prognosis must be guarded and the owner informed that a proportion of these animals may not recover while others may remain permanently stunted. As well as being treated with an anthelmintic, severely affected animals may require antibiotics if pyrexic and may be in need of hydration if they are not drinking. In the case of valuable animals, treatment with oxygen may be merited.

CONTROL

The only reliable method of preventing parasitic bronchitis is to immunise all young calves with lungworm vaccine. This live vaccine, consisting of larvae attenuated by irradiation, is currently only available in Europe and is given orally to calves aged eight weeks or more. Two doses of vaccine are given at an interval of four weeks and, in order to allow a high level of immunity to develop, vaccinated calves should be protected from challenge until two weeks after their second dose.

Although vaccination is effective in preventing clinical disease, it does not completely prevent the establishment of small numbers of lungworms. Consequently, pastures may remain contaminated, albeit at a very low level. For this reason it is important that all of the calves on any farm should be vaccinated whether they go to pasture in the spring or later in the year. Also once a vaccination programme has been undertaken it must be continued annually for each calf crop. Although the limited pasture larval contamination will serve to boost the immunity of vaccinated calves it can lead to clinical disease in susceptible animals.

The vaccination programme for dairy calves should, if possible, be completed before they go to grass in the spring or early summer. However, these or suckled calves can be vaccinated successfully at grass provided the vaccine is given in spring or early summer, that is, prior to encountering a significant larval challenge.

Attempts have been made to control parasitic bronchitis by grazing systems designed to ensure that young susceptible cattle are not exposed to large numbers of infective larvae. This approach is rarely successful, particularly if the farm is heavily stocked.

The introduction of calves to infected pasture followed by anthelmintic treatment after an arbitrary period has also been suggested as a means of 'immunisation'. However this can be hazardous since the initial larval challenge may not be sufficient to stimulate immunity and disease may subsequently occur.

It is also worth noting that, because of the unpredictable epidemiology, the technique commonly used in ostertagiasis of 'dose and move' in midsummer does not prevent parasitic bronchitis.

PARASITIC BRONCHITIS IN ADULT CATTLE

Parasitic bronchitis is only seen in adult cattle under two circumstances. First as a herd phenomenon, or in a particular age group within a herd, if they have failed to acquire immunity through natural challenge in earlier years. Such animals may develop the disease if exposed to heavy larval challenge as might occur on pasture recently vacated by calves suffering from clinical husk. Secondly, disease is occasionally seen in an individual adult penned in a heavily contaminated calf paddock because it requires daily attention for some other reason.

The disease is most commonly encountered in the patent phase although the other forms have been recognised. In addition to coughing and tachypnoea, a reduction in milk yield in cows is a common presenting sign.

Treatment is similar to that discussed for calves but in selecting a drug one should consider the withdrawal period of milk for human consumption. If possible, an annual programme of calf vaccination should be started.

THE REINFECTION SYNDROME IN PARASITIC BRONCHITIS

Normally the natural challenge of adult cattle, yearlings or calves which have acquired immunity to *D. viviparus*, whether by natural exposure or by vaccination, is not associated with clinical signs.

Occasionally, however, clinical signs do occur to produce the 'reinfection syndrome' which is usually mild, but occasionally severe. It arises when an immune animal is suddenly exposed to a massive larval challenge, usually from a heavily contaminated field. Significant numbers of larvae reach the lungs and migrate to the bronchioles where they are killed by the animal's immune response. The resulting proliferation of lympho-reticular cells around each dead larva causes bronchiolar obstruction and ultimately the formation of a macroscopically visible greyish-green, lymphoid nodule about 5.0 mm in diameter (Fig. 25). In addition,

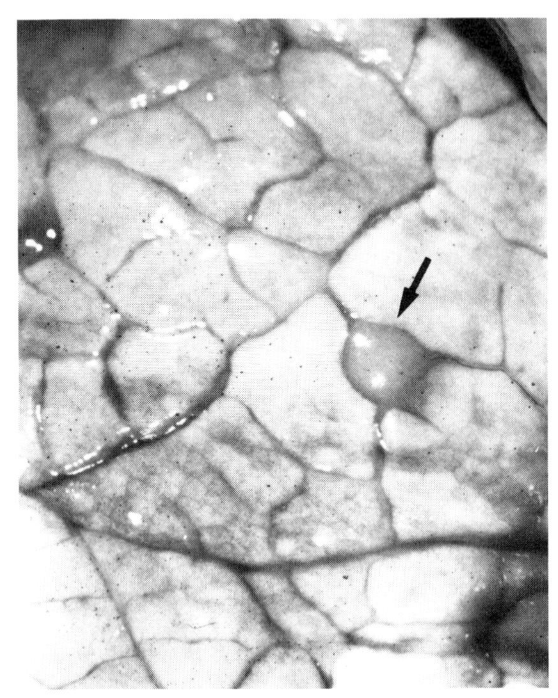

Fig 25 Small lymphoid nodules on lung surface, a sequel of reinfection with *Dictyocaulus viviparus* (↑).

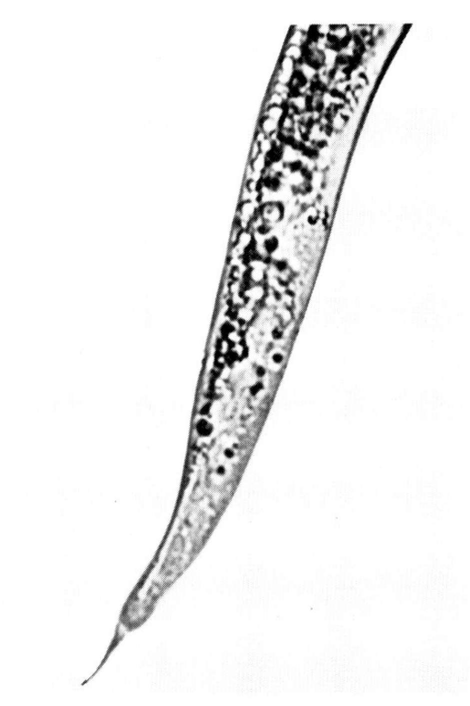

Fig 26 The first stage larva of *Dictyocaulus arnfieldi* resembles that of *D. viviparus*, but possesses a small terminal lancet.

dense infiltration of the lung by eosinophils occurs, and where these accumulate in the small bronchi in large numbers, they may be seen as greenish plugs at necropsy.

Usually the syndrome is associated with frequent coughing and slight tachypnoea over a period of a few days; less frequently there is marked tachypnoea, hyperpnoea and in dairy cows a reduction in milk yield. Deaths rarely occur.

In the absence of a good history it may be impossible to differentiate this syndrome from the early stages of a severe primary infection. The only course of action in these instances is treatment with one of the anthelmintics described above and a change of pasture.

Dictyocaulus arnfieldi

This ubiquitous parasite of donkeys is rarely associated with signs of clinical disease. In horses, its prevalence is difficult to establish since infections rarely become patent although it is frequently incriminated as a cause of chronic coughing.

LIFE CYCLE

The detailed life cycle is not fully known, but is considered to be similar to that of the bovine lungworm, *D. viviparus*, except in the following respects.

The adult worms are most often found in the small bronchi and their eggs, containing the first stage larvae, hatch soon after being passed in the faeces (Fig. 26).

The prepatent period is between 2–4 months. Patent infections are common in donkeys of all ages, but in horses generally only occur in foals and yearlings. In older horses the adult lungworms rarely attain sexual maturity.

PATHOGENESIS

The characteristic lesion is similar in both horses and donkeys and is somewhat different from bovine parasitic bronchitis.

In the caudal lung lobes particularly, there are raised circumscribed areas of over-inflated pulmonary tissue 3.0–5.0 cm in diameter (Fig. 27). On section, at the centre of each lesion is a small bronchus containing lungworms and mucopurulent exudate. Microscopically, the epithelium is hyperplastic with an increase in the size and number of mucus-secreting cells while the lamina propria is heavily infiltrated and often surrounded by inflammatory cells, predominantly lymphocytes.

CLINICAL SIGNS

Despite the prevalence of patent *D. arnfieldi* infection in

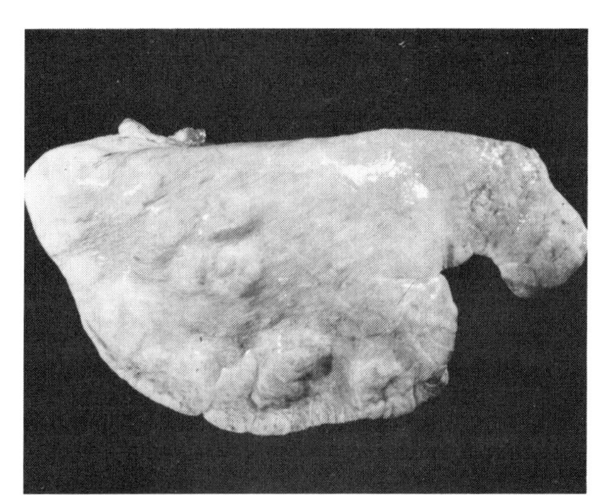

Fig 27 Circumscribed lung lesions typical of *Dictyocaulus arnfieldi* infection.

donkeys, overt clinical signs are rarely seen; however, on close examination slight hyperpnoea and harsh lung sounds may be detected. This absence of significant clinical abnormality may be partly a reflection of the fact that donkeys are rarely required to perform sustained exercise.

Infection is much less prevalent in horses. However, patent infections may develop in foals and these are not usually associated with clinical signs. In older horses infections rarely become patent but are often associated with persistent coughing and an increased respiratory rate.

EPIDEMIOLOGY

Donkeys acquire infection as foals and yearlings and tend to remain infected, presumably through re-exposure, all their lives. Horses are thought to acquire infection mainly from pastures contaminated by donkeys during the summer months. Most commonly this occurs when donkeys are grazed as companion animals with horses. *Pilobolus* fungi may play a role in the dissemination of *D. arnfieldi* larvae from faeces, as in *D. viviparus*.

DIAGNOSIS

In donkeys, patent infections are common and L_1 are readily recovered from fresh faeces. In horses, although a history of donkey contact and clinical signs may be suggestive of *D. arnfieldi* infection, it is often not possible to confirm a diagnosis by demonstrating larvae in the faeces. When attempted, a modified Baerman technique is employed using 50 g of faeces from the rectum. Recently the detection of eosinophils in tracheal mucus has been described as an ancillary aid to diagnosis.

In practice, a presumptive diagnosis of lungworm infection in horses is often only possible in retrospect, when resolution of the clinical signs occurs after treatment.

TREATMENT

Successful treatment of both horses and donkeys has been reported using various regimens of diethylcarbamazine, levamisole, febendazole or mebendazole. High efficacy has been reported in trials with oral ivermectin at normal dose rates.

CONTROL

Ideally, horses and donkeys should not be grazed together, but if they are, it is advisable to treat the donkeys, preferably in the spring, with a suitable anthelmintic. A similar regimen should be practised in donkey studs and visiting animals should be isolated in separate paddocks.

Dictyocaulus filaria

This species, the most important lungworm of sheep and goats, is commonly associated with a chronic syndrome of coughing and unthriftiness which usually affects lambs and kids.

LIFE CYCLE

Similar to that of *D. viviparus* except that the prepatent period is five weeks.

PATHOGENESIS

Similar to that of *D. viviparus* infection. However, since the number of lungworms in individual animals is generally low, the widespread lesions associated with the bovine infection are not common.

Nevertheless, in severe cases, pulmonary oedema and emphysema may occur and the lung surface may be studded with purulent areas of secondary infection.

CLINICAL SIGNS

The most common signs are coughing and unthriftiness which, in endemic areas, is usually confined to young animals. In more severe cases dyspnoea and a tenacious nasal discharge are also present.

These signs may be accompanied by diarrhoea or anaemia due to concurrent gastrointestinal trichostrongylosis or fascioliasis.

EPIDEMIOLOGY

Although this parasite is prevalent throughout the world, it is only responsible for sporadic outbreaks of disease in temperate countries such as Britain and North America. However, it occurs more frequently as a clinical problem in some warmer areas such as Mediterranean countries, the Middle East and India.

In temperate areas the epidemiology is somewhat similar to that of *D. viviparus* in that both the survival of overwintered larvae on pasture and the role of the ewe as a carrier are significant factors in the persistence of infection on pasture from year to year in endemic areas. In ewes it seems likely that the parasites are present largely as hypobiotic larvae in the lungs during each winter and mature in the spring.

Development to the L_3 only occurs during the period from spring to autumn. In lambs, patent infections first occur in early summer, but the heaviest infections are usually seen in autumn. In ewes the prevalence of infection is lower and their larval output smaller. As with the other trichostrongyloids it seems likely that only two cycles of the parasite occur during each grazing season.

In warmer climates, where conditions are often unsuitable for larval survival, the carrier animal is probably a more important source of pasture contamination and outbreaks of disease in lambs and kids are most likely to occur after a period of prolonged rain around the time of weaning. Goats appear to be more susceptible to infection than sheep and are thought to play a prominent role in the dissemination of infection where both are grazed together.

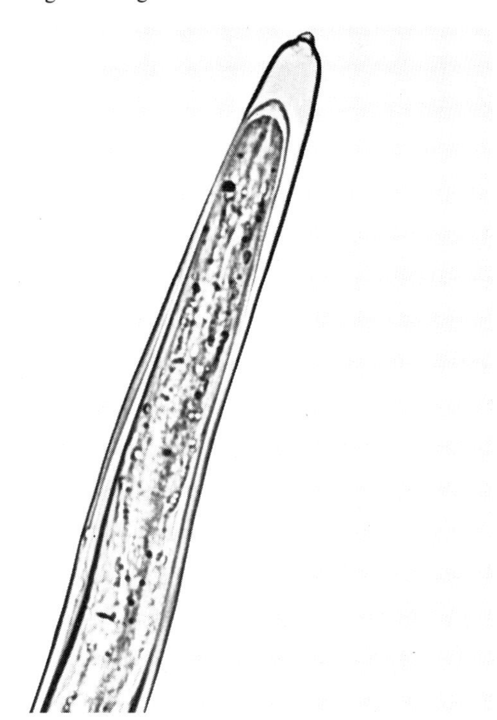

Fig 28 Head of infective larva of *Dictyocaulus filaria* showing cuticular knob on retained first larval sheath.

DIAGNOSIS

This is based on history and clinical signs, but should be confirmed by faecal examinations from a large sample of the flock. The L_1 resembles that of *D. viviparus*, but has a characteristic cuticular knob at the anterior extremity (Fig. 28). It is differentiated from other ovine lungworms by its larger size and straight tail.

TREATMENT AND CONTROL

Where sporadic outbreaks occur, the affected animals, or preferably the whole flock, should be gathered, treated with a suitable anthelmintic and then, if possible, moved to fresh pasture. It is probable that the prophylactic regimens of control currently recommended for the control of gastrointestinal nematodes in sheep will, in normal years, be effective to a large extent in suppressing *D. filaria* infection.

Where it is necessary to apply specific control measures, it is suggested that the flock should be annually treated with a suitable anthelmintic in late pregnancy. The ewes and lambs should then be grazed on pasture which, in temperate areas at least, should not have been used by sheep during the previous year.

Three trichostrongyloid genera of minor importance are *Amidostomum*, *Ollulanus* and *Ornithostrongylus*.

Amidostomum anseris

This parasite found in the upper alimentary tract, particularly the gizzard, may cause heavy mortality in goslings, ducklings and other young aquatic fowls.

The adult worms, bright red in colour and up to 2.5 cm in length, are easily recognised at necropsy where they predominate in the horny lining of the gizzard. Microscopically they are characterised by a shallow buccal capsule with three teeth.

Eggs passed in the faeces are already embryonated and, like *Nematodirus*, develop to the L_3 in the egg.

Treatment with one of the benzimidazoles or levamisole is effective and the condition may be prevented by ensuring that birds do not run on the same ground each year.

Ollulanus tricuspis

This very small trichostrongyle (0.7–1.0 mm long) occurs in the stomach of cats, pigs, wild felids, foxes and occasionally domestic dogs. It is identified microscopically by the spiral coil of the head and the fact that the viviparous female has a tail with three or four short points.

The whole life cycle may be completed endogenously and transmission, at least in the cat, is thought to be via ingestion of vomit containing the L_3. The worms live under a layer of mucus in the stomach wall.

Little is known of its pathogenicity although a chronic gastritis has been reported in the pig. Treatment with benzimidazoles is effective.

Ornithostrongylus quadriradiatus

This trichostrongyle, found in the small intestine and crop of pigeons, causes an enteritis and anaemia which, in heavy infections, may result in severe mortality in domestic pigeons.

The adult worms which measure up to 2.5 cm are bloodsuckers, have a reddish colour and can be seen by the naked eye.

Superfamily STRONGYLOIDEA

There are several important parasites of domestic mammals and birds in this superfamily of bursate nematodes.

Most are characterised by a large buccal capsule which often contains teeth or cutting plates and in some there are prominent leaf crowns surrounding the mouth opening. The adults occur on mucosal surfaces of the gastrointestinal and respiratory tracts and feeding is generally by the ingestion of plugs of mucosa.

With the exception of three genera, *Syngamus* and *Mammomonogamus*, which are parasitic in the trachea and major bronchi, and *Stephanurus* found in the perirenal area, all other genera of veterinary importance in this superfamily are found in the intestine and can be conveniently divided into two groups, the **strongyles** and **hookworms**.

The strongyles are parasitic in the large intestine and the important genera are *Strongylus*, *Triodontophorus*, *Trichonema*, *Chabertia* and *Oesophagostomum*.

Hookworms are parasites of the small intestine and the three genera of veterinary importance are *Ancylostoma*, *Uncinaria*, and *Bunostomum*.

STRONGYLES OF HORSES

Strongylus

Members of this genus live in the large intestine of horses and donkeys and, with *Triodontophorus*, are commonly known as the **large strongyles**.

Hosts:
Horses and donkeys

Site:
Caecum and colon

Species:
Strongylus vulgaris

S. edentatus

S. equinus

Distribution:
Worldwide

IDENTIFICATION

Gross: Robust dark-red worms which are easily seen against the intestinal mucosa. The well developed buccal capsule of the adult parasite is prominent as is the bursa of the male.

Microscopic: Species differentiation is based on size and the presence and shape of the teeth in the base of the buccal capsule (Fig. 29).

S. vulgaris	1.5–2.5 cm	Two ear-shaped rounded teeth
S. edentatus	2.5–4.5 cm	No teeth
S. equinus	2.5–5.0 cm	Three conical teeth. One is situated dorsally and is larger than the others and bifid.

Fig 29 Buccal capsules of *Strongylus* species
(a) *S. vulgaris*.

Fig 29 (continued)

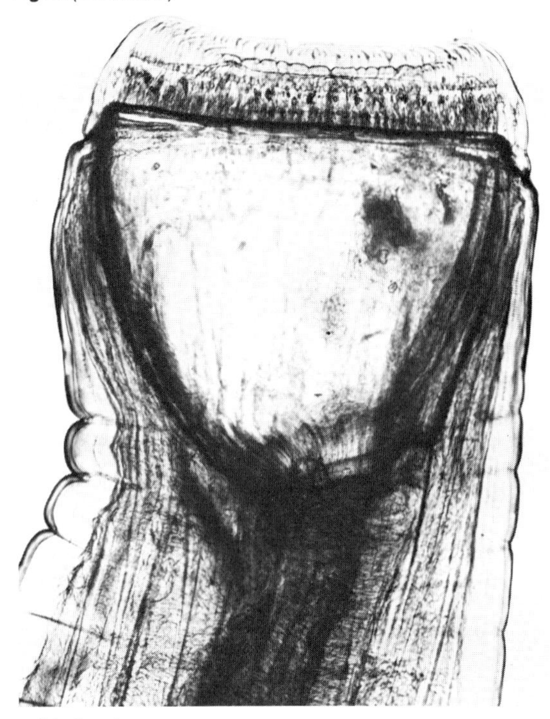

(b) *S. edentatus.*

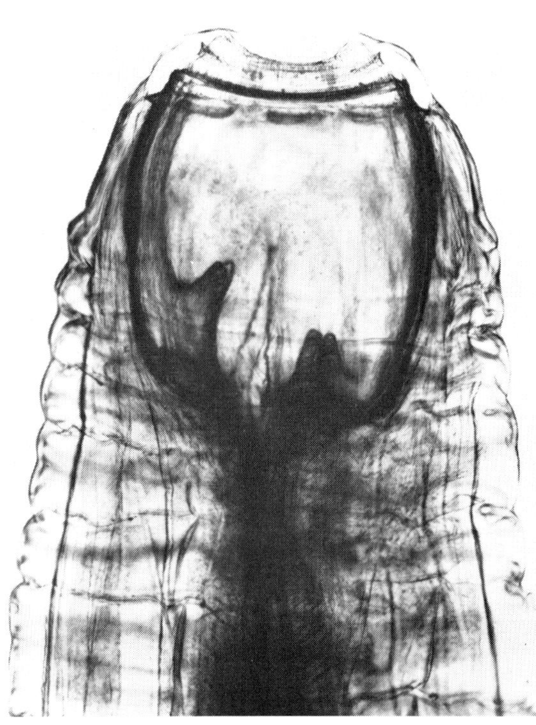

(c) *S. equinus.*

LIFE CYCLE

The adult parasites live in the caecum and colon. Eggs which resemble those of the trichostrongyles (Fig. 30) are passed in the faeces and development from egg to the L_3 under summer conditions in temperate climates requires approximately two weeks. Infection is by ingestion of the L_3. Subsequently, parasitic larval development of the three species differs and will be dealt with separately.

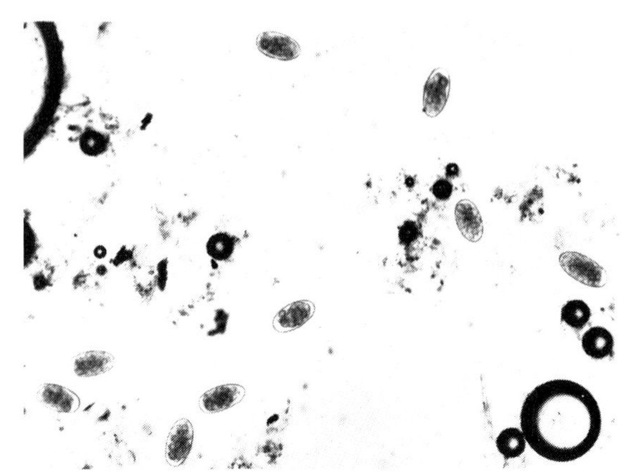

Fig 30 Strongyle eggs in horse faeces.

S. vulgaris

The L_3 penetrate the intestinal mucosa and moult to L_4 in the submucosa. These then enter small arteries and migrate on the endothelium to their predilection site in the **cranial mesenteric artery and its main branches**. After a period of development of several months the larvae moult to L_5 and return to the intestinal wall via the arterial lumina. Nodules are formed around the larvae mainly in the wall of the caecum and colon when, due to their size, they can travel no further within the arteries and subsequent rupture of these nodules releases the young adult parasites into the lumen of the intestine.

The prepatent period is 6–7 months.

S. edentatus

After penetration of the intestinal mucosa L_3 travel via the portal system, and reach the liver parenchyma within a few days. About two weeks later the moult to L_4 takes place, further migration then occurs in the liver and, by 6–8 weeks post-infection, larvae can be found subperitoneally around the hepatorenal ligament. The larvae then travel under the peritoneum to many sites with a predilection for the **flanks and hepatic ligaments** (Fig. 31). The final moult occurs after four months and each L_5 then migrates, still subperitoneally, to the wall of the large intestine where a large purulent nodule is

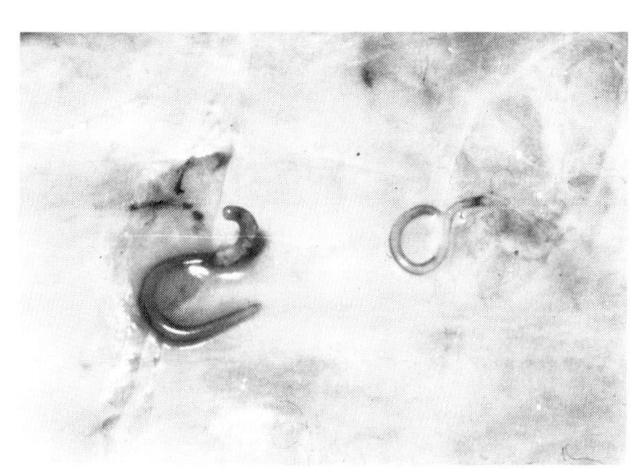

Fig 31 *Strongylus edentatus* larvae, 2–3 cm, from the parietal peritoneum of the flank.

formed, which subsequently ruptures with release of the young adult parasite into the lumen.

The prepatent period is 10–12 months.

S. equinus

Of the three *Strongylus* species, least is known of the larval migration of *S. equinus*. It appears that the L_3 lose their sheaths while penetrating the wall of the caecum and ventral colon and within one week provoke the formation of nodules in the muscular and subserosal layers of the intestine. The moult to L_4 occurs within these nodules and the larvae then travel across the peritoneal cavity to the liver where they migrate within the parenchyma for six weeks or more. After this time L_4 and L_5 have been found in and around the **pancreas** before their appearance in the large intestinal lumen.

The prepatent period is 8–9 months.

PATHOGENESIS

Larvae

Despite the invasive behaviour of the parasitic larval stages, little specific pathogenic effect can be attributed to them; the exception is *S. vulgaris*, 90% of horses in Britain having lesions in the arterial system of the intestine caused by this species. Lesions are most common in the cranial mesenteric artery and its main branches, and consist of thrombus formation provoked by larval damage to the endothelium together with a marked inflammation and thickening of the arterial wall (Pl. III). True aneurysms with dilatation and thinning of the arterial wall, although uncommon, may be found especially in animals which have experienced repeated infection.

Much of the information concerning *S. vulgaris* has been derived from experimental infection of foals. A few weeks after infection with several hundred L_3 a clinical syndrome of fever, inappetence and dullness occurs sometimes accompanied by colic. At necropsy, these signs are associated with arteritis and thrombosis of intestinal blood vessels with subsequent infarction and necrosis of areas of bowel. However, a syndrome of this severity is not commonly reported in foals under natural conditions, probably because larval intake is continuous during grazing; it has been shown experimentally that foals may tolerate large numbers of larvae administered in small doses over a long period.

In *S. edentatus* infection there are gross changes in the liver associated with early larval migration, but these rarely result in clinical signs. Similarly, the haemorrhages and fluid-filled nodules which accompany later larval development in subperitoneal tissues, rarely result in clinical signs.

There has been little work on the pathogenesis of migrating larvae of *S. equinus*.

Adults

The pathogenesis of infection with adult *Strongylus* spp. is associated with damage to the large intestinal mucosa due to the feeding habits of the worms (Pl. III) and, to some extent, to the disruption caused by emergence of young adults into the intestine following completion of their parasitic larval development.

These worms have large buccal capsules and feed by ingestion of plugs of mucosa as they move over the surface of the intestine. Although the worms appear to feed entirely on mucosal material the incidental damage to blood vessels can cause considerable haemorrhage. Ulcers which result from these bites eventually heal leaving small circular scars. The effects of infection with the adult worms have not been quantified, but the gross damage and subsequent loss of blood and tissue fluids is certainly partly responsible for the unthriftiness and anaemia associated with intestinal helminthiasis in the horse.

Since members of this genus form only one component of the total parasitic burdens of the large intestine, the other aspects of infection will be dealt with after a description of the life cycles and pathogenesis of the other genera.

Triodontophorus

Members of this common genus of non-migratory large strongyles frequently occur in large numbers in the colon and contribute to the deleterious effects of mixed strongyle infection.

Hosts:
Horses and donkeys

Site:
Colon and caecum

Species:
Triodontophorus serratus
T. tenuicollis
T. brevicauda
T. minor

Distribution:
Worldwide

IDENTIFICATION

Gross: Robust, reddish worms 1.0–2.5 cm in length readily visible on the colonic mucosa. In one species, *T. tenuicollis*, groups of adult worms are characteristically found feeding in groups.
Microscopic: Species differentiation is based on buccal capsule characteristics, especially the number and shape of the teeth present in all species.

LIFE CYCLE

Little information is available on the developmental cycle of this genus, but it is thought to be similar to that of the genus *Trichonema*.

PATHOGENESIS

Like the other horse strongyles, the pathogenic effect of these worms is damage to the large intestinal mucosa from the feeding habits of the adult parasites; in particular, *T. tenuicollis*, whose adults feed in groups and cause the formation of large deep ulcers which may be several centimetres across (Pl. III).

Trichonema

This genus embraces over 40 species, popularly known as **trichonemes**, **cyathostomes** or **small strongyles**. These parasites are found in the large intestine of horses and their effects on the host range from poor performance to clinical signs of severe enteritis.

Hósts:
Horses and donkeys

Site:
Caecum and colon

Species:
For many years there has been a great deal of confusion in the classification of this group of parasites and in a recent revision it has been proposed that the genus *Trichonema* be discarded and replaced by four genera, namely *Cyathostomum*, *Cylicocyclus*, *Cylicodontophorus* and *Cylicostephanus*, these being collectively referred to as cyathostomes. Since the majority of species involved are similar both morphologically and in behaviour they will be referred to in this text as trichonemes or small strongyles

Fifteen species of these are commonly present in large numbers

Distribution:
Worldwide

IDENTIFICATION

Gross: Small to medium sized (<1.5 cm in length) bursate nematodes ranging in colour from white to dark red, the majority being visible on close inspection of the large intestinal mucosa or contents.

Microscopic: The well developed buccal capsule is cylindrical and species differentiation is based on characteristics of the buccal capsule, and the internal and external leaf crowns.

LIFE CYCLE

Hatching of eggs and development to L_3 is complete within two weeks during the summer in temperate areas, after which the larvae migrate from the faeces on to the surrounding herbage. After ingestion, the L_3 exsheath and invade the wall of the large intestine where they develop to L_4 (Pl. III) before emerging into the gut lumen and moulting to become young adult worms.

The prepatent periods of members of this genus are generally between 2–3 months although this may be extended, due to hypobiosis in some species.

PATHOGENESIS

Parasitic larval development of most species takes place entirely in the mucosa of the caecum and colon, but a few penetrate the muscularis and develop in the submucosa. The entry of larval trichonemes into the lumina of the tubular glands generally provokes an inflammatory response together with marked goblet cell hypertrophy. Emergence of the bright red L_4 into the gut lumen appears to be associated with a massive infiltration of the gut mucosa with eosinophils. Many thousand L_4 may be present, but their pathogenic significance has been little studied. There are, however, reports of heavy natural infections of adult worms and larvae associated with catarrhal and haemorrhagic enteritis, with thickening and oedema of the mucosa, especially in animals of six months to three years of age.

Mature parasites are frequently present in large numbers in the lumen of the large intestine and during feeding those species with small buccal capsules take in only glandular epithelium while large species may damage deeper layers of the mucosa. Although the erosions caused by individual parasites may be slight, when large numbers are present a desquamative enteritis may result.

THE CLINICAL SIGNS, EPIDEMIOLOGY, DIAGNOSIS, TREATMENT AND CONTROL OF EQUINE STRONGYLOSIS

CLINICAL SIGNS

Grazing horses usually carry a mixed burden of large and small strongyles and the major clinical signs associated with heavy infections in animals up to 2–3 years of age are unthriftiness, anaemia and sometimes diarrhoea. Marked clinical signs are less common in older animals, although general performance may be impaired.

In temperate countries an acute syndrome of severe diarrhoea and death in young ponies in the spring has been reported which is associated with the simultaneous mass emergence of trichoneme L_4 from the intestinal mucosa and submucosa. This may have etiological and epidemiological similarities to Type II ostertagiasis in young cattle.

The significance of migrating larvae of *S. vulgaris* in natural cases of colic is difficult to assess, but it is generally recognised that where strongyle infections of horses are efficiently controlled the incidence of colic is markedly decreased.

EPIDEMIOLOGY

Strongylosis is most frequently a problem in young horses reared on permanent horse pastures, although cases of severe disease may occur in adult animals kept in suburban paddocks and subjected to overcrowding and poor management.

Although the preparasitic larval requirements of the horse strongyles are similar to those of the trichostrongyles of ruminants, adult horses, unlike cattle, may carry substantial worm burdens and therefore have a considerable influence on the epidemiology of infection. Thus there are two sources of infection during the grazing season in temperate areas. First there are infective larvae which developed during the previous grazing season and have survived on pasture over winter. The second and probably more important source of infective larvae are the eggs passed in the current grazing season by horses, including nursing mares, sharing the same grazing area. Pasture larval levels increase markedly during the summer months when conditions are optimal for rapid development of eggs to L_3 (Fig. 32).

At present there is little evidence for a periparturient rise in faecal egg output in breeding mares due to a relaxation of immunity since the egg rise in the spring occurs in both breeding and non-breeding animals and is often unrelated to parturition.

There is increasing evidence that many trichoneme L_3 ingested during the autumn show a degree of hypobiosis and remain in the large intestinal mucosa until the following spring. Mass emergence of these larvae results in the severe clinical signs described previously.

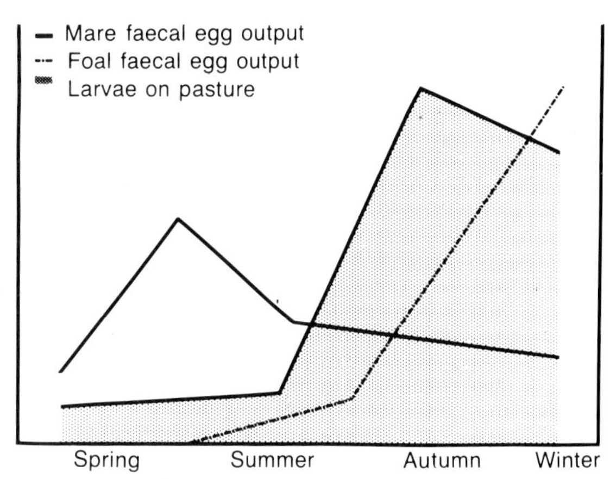

Fig 32 The pattern of strongyle infection in mares and foals at pasture in temperate areas.

DIAGNOSIS

This is based on the grazing history and clinical signs of loss of condition and anaemia. Although the finding of typical, oval, thin-shelled strongyle eggs on faecal examination may be a useful aid to diagnosis, it is important to remember that substantial worm burdens may be associated with faecal egg counts of only a few hundred epg, due either to low fecundity of adult worms or to the presence of many immature parasites. On some occasions when heavy trichoneme infections in the spring cause severe diarrhoea, thousands of bright red trichoneme L_4, apparently unable to establish, may be present in the faeces.

TREATMENT

Treatment for clinical strongylosis should not be necessary if prophylactic measures are adequate.

There are a number of broad spectrum anthelmintics including the benzimidazoles, pyrantel, dichlorvos and ivermectin, which are effective in removing lumen-dwelling adult and larval strongyles and these are usually marketed as in-feed or oral preparations. Two of these compounds, dichlorvos and ivermectin, have the additional advantage of activity against larvae of horse bot flies (*Gasterophilus* spp.) which develop in the stomach.

Some modern benzimidazoles and ivermectin are also efficient against both developing trichoneme larvae in the gut wall and some migrating stages of the large strongyles.

CONTROL

Since horses of any age can become infected and excrete eggs, all grazing animals over two months of age should be treated every 4–6 weeks with an effective broad spectrum anthelmintic. This regimen will also control infections with other intestinal parasites such as *Parascaris equorum* and *Oxyuris equi*.

Any new animals joining a treated group should receive an anthelmintic and be isolated for 48–72 hours before being introduced.

If possible, a paddock rotation system should be adopted so that nursing mares and their foals do not graze the same area in successive years.

If horses are housed in the winter, treatment at that time with an anthelmintic effective against larval trichonemes will reduce the risk of disease due to their mass emergence in the spring.

There is evidence that some species of trichonemes may become resistant to benzimidazole compounds and to avoid this it is suggested that these should be alternated with chemically unrelated anthelmintics on an annual or a six-monthly basis. Faecal samples from groups of horses should be examined at regular intervals to monitor drug efficiency.

STRONGYLES OF OTHER ANIMALS

Chabertia

Chabertia ovina is present, usually in low numbers, in the majority of sheep and goats. It contributes to the syndrome of parasitic gastroenteritis and only occasionally occurs in sufficient numbers to cause clinical disease on its own.

Hosts:
Sheep, goats and occasionally cattle

Site:
Colon

Species:
Chabertia ovina

Distribution:
Worldwide

IDENTIFICATION

Gross: The adults are 1.5–2.0 cm in length and are the largest nematodes found in the colon of ruminants. They are white with a markedly truncated and enlarged anterior end due to the presence of the very large buccal capsule.

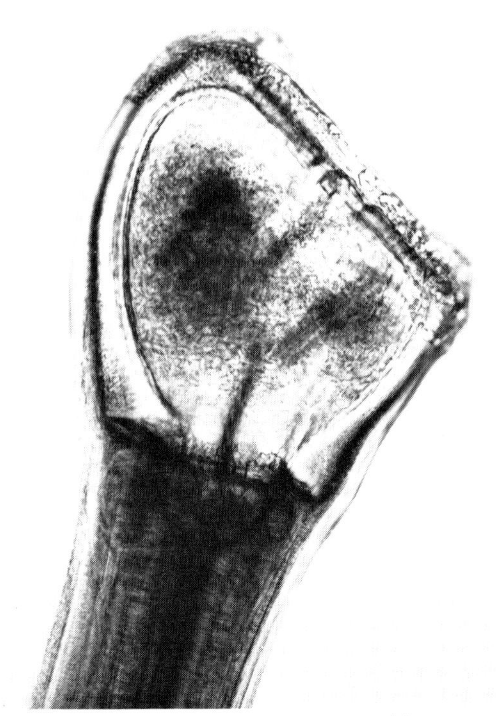

Fig 33 Large bell-shaped head of *Chabertia ovina*.

Microscopic: The huge buccal capsule, which is bell shaped, has a double row of small papillae around the rim. There are no teeth (Fig. 33).

LIFE CYCLE

This is direct and the preparasitic phase is similar to that of the trichostrongyles of ruminants.

In the parasitic phase the L_3 enter the mucosa of the small intestine and occasionally that of the caecum and colon; after a week they moult, the L_4 emerge on to the mucosal surface and migrate to congregate in the caecum where development to the L_5 is completed about 25 days after infection. The young adults then travel to the colon. The prepatent period is 42 days.

PATHOGENESIS

The major pathogenic effect is caused by the L_5 and by mature adults which feed by ingesting large plugs of mucosa resulting in local haemorrhage and loss of protein through the damaged mucosa.

A burden of 250–300 worms is considered pathogenic and in severe outbreaks the effects become evident during the late prepatent period. The wall of the colon becomes oedematous, congested and thickened with small haemorrhages at the sites of worm attachment (Fig. 34).

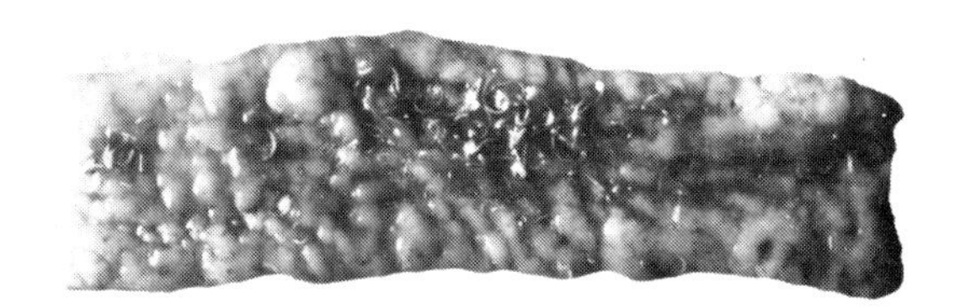

Fig 34 Oedema and thickening of colon associated with *Chabertia ovina* infection.

CLINICAL SIGNS

In severe infections, diarrhoea, which may contain blood and in which worms may be found, is the most common clinical sign. The sheep become anaemic and hypoalbuminaemic and can suffer severe weight loss.

EPIDEMIOLOGY

In temperate areas, L_3 are capable of surviving the winter. The parasite may also overwinter in the host as hypobiotic L_4 in the wall of the intestine emerging in the late winter and early spring.

Although outbreaks of chabertiasis have been recorded in goats and sheep in Europe, the disease is more important in the winter rainfall areas of Australasia and South Africa.

DIAGNOSIS

Since much of the pathogenic effect occurs within the prepatent period, the faecal egg count may be very low. However, during the diarrhoeic phase, the worms may be expelled and they are easily recognised. At necropsy, diagnosis is generally based on the lesions since the worm burden may be negligible following the expulsion of worms in the faeces.

TREATMENT AND CONTROL

Since the epidemiology of *C. ovina* is the same as that of the trichostrongyles the control and treatment is similar.

Oesophagostomum

Oesophagostomum species are responsible for an enteritis in ruminants and pigs. The more pathogenic species in ruminants occur in the subtropics and tropics and are associated with nodule formation in the intestine.

Hosts:
Ruminants, pigs

Site:
Caecum and colon

Species:

Oesophagostomum columbianum	sheep and goat
Oe. venulosum	sheep and goat
Oe. radiatum	cattle and buffalo
Oe. dentatum	pig
Oe. quadrispinulatum	pig

Other species found in the pig are *Oe. longicaudatum*, *Oe. granatensis* and *Oe. brevicaudum* and in sheep and goats *Oe. asperum*.

Distribution:
Worldwide; more important in tropical and subtropical areas

IDENTIFICATION

Gross: Stout white worm 1.0–2.0 cm long. Readily differentiated by its tapered head from *Chabertia*.

Microscope: The buccal capsule is small. In many species it is surrounded by leaf crowns. The external crown, if present, is compressed and so there is only a narrow opening into the buccal capsule. Around the

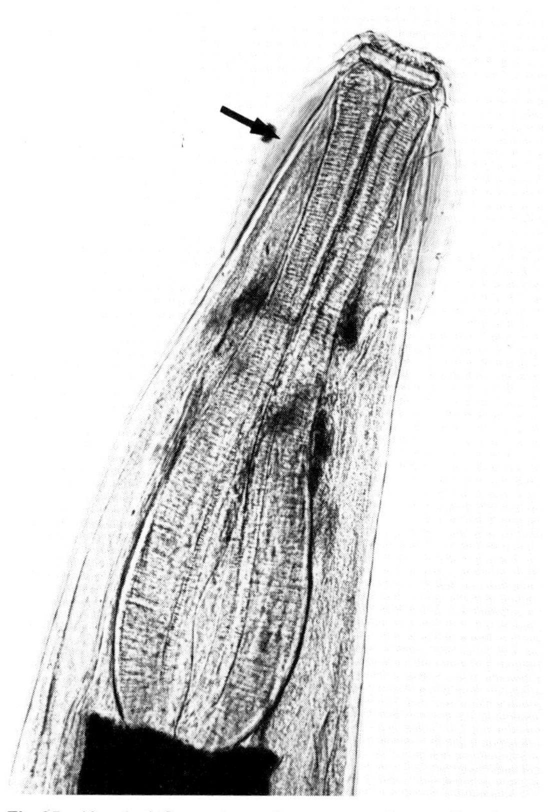

Fig 35 Head of *Oesophagostomum venulosum* showing cervical vesicle (↑) characteristic of the genus.

capsule there is a small cephalic vesicle behind which is a cervical vesicle (Fig. 35). This terminates in a cervical groove which is followed in some species by broad cervical alae. The position of cervical papillae and the leaf crown arrangements are used to identify species.

LIFE CYCLE

The preparasitic phase is typically strongyloid and infection is by ingestion of L_3 although there is limited evidence that skin penetration is possible, at least in pigs. The L_3 enter the mucosa of any part of the small or large intestine and in some species (*Oe. columbianum*, *Oe. radiatum*, *Oe. quadrispinulatum*) become enclosed in obvious nodules in which the moult to L_4 takes place (Pl. III). These L_4 then emerge on to the mucosal surface, migrate to the colon, and develop to the adult stage. The prepatent period is about 45 days.

On reinfection with most species the larvae may remain arrested as L_4 in nodules for up to one year; however with *Oe. venulosum* nodules are absent while in *Oe. dentatum* they are barely visible.

PATHOGENESIS

All species are capable of causing a severe enteritis including *Oe. venulosum*, which does not provoke nodule formation.

In the intestine *Oe. columbianum* L_3 migrate deep into the mucosa, provoking an inflammatory response with the formation of nodules which are visible to the naked eye. On reinfection, this response is more marked, the nodules reaching 2.0 cm in diameter and containing greenish eosinophilic pus and an L_4. When the L_4 emerge there may be ulceration of the mucosa. Diarrhoea occurs coincident with emegence about a week after primary infection and up to one year after reinfection. In heavy infections, there may be ulcerative colitis and the disease runs a chronic debilitating course with effects on the production of wool and mutton. The nodules in the gut wall also render the intestines useless for processing as sausage skins and surgical suture material.

In *Oe. radiatum* infections in cattle, the pathogenic effect is also attributed to the nodules (up to 5.0 mm in diameter) in the intestine and it appears that as few as 500 larvae are sufficient to produce clinical signs. Necropsy reveals a severely inflamed mucosa studded with yellowish-green purulent nodules. In the later stages of the disease, anaemia and hypoalbuminaemia develop due to the combined effects of protein loss and leakage of blood through the damaged mucosa.

Oesophagostomum infections in the pig are less often associated with clinical disease, but are responsible for poor productivity.

CLINICAL SIGNS

In acute infections of ruminants, severe dark green diarrhoea is the main clinical sign and there is usually a rapid loss of weight and sometimes submandibular oedema. In chronic infections, which occur primarily in sheep, inappetence and emaciation with intermittent diarrhoea and anaemia are the main signs of oesophagostomiasis.

Pregnant sows show inappetence, become very thin, and following farrowing, milk production is reduced with effects on litter performance.

EPIDEMIOLOGY

In temperate areas, there is evidence that *Oe. venulosum* undergoes hypobiosis at the L_4 stage in sheep during autumn and winter, and that this is the principal manner in which this species survives until the next spring. It is not yet known if hypobiosis occurs in *Oe. radiatum*. Both species are also capable of overwintering on pasture as L_3.

In the pig species, *Oe. dentatum* and *Oe. quadrispinulatum*, survival of both free-living L_3 on the pasture and hypobiotic L_4 in the host occur during autumn and winter; the hypobiotic larvae complete their development in the spring often coincident with farrowing. There is also some evidence that larvae develop in faeces on the skin of pigs and it seems likely that, in housed animals, transmission is by contact between sows and their litters, infection occurring either orally or percutaneously. Pen to pen transmission may also occur via dipteran flies which can carry L_3 on their legs.

In tropical and subtropical areas, *Oe. columbianum* and *Oe. radiatum*, in sheep and cattle respectively, are especially important. In *Oe. columbianum* infections, the prolonged survival of the L_4 within the nodules in the gut wall and the lack of an effective immunity made control difficult until the advent of effective anthelmintics. In contrast, cattle develop a good immunity to *Oe. radiatum*, partly due to age and partly to previous exposure so that it is primarily a problem in weaned calves.

DIAGNOSIS

This is based on clinical signs and post-mortem examination. Since the acute disease occurs within the prepatent period, eggs of *Oesophagostomum* spp. are not usually present in the faeces. In the chronic disease eggs are present and L_3 can be identified following faecal culture.

TREATMENT AND CONTROL

The treatment and control of ruminant infections with *Oesophagostomum* spp. is similar to that of the trichostrongyles while infections in pigs can be controlled by the methods described for *Hyostrongylus*.

Stephanurus

This is the 'kidney worm' of swine which is of economic importance in its endemic areas.

Host:
Pig

Site:
Kidneys and perirenal tissues

Species:
Stephanurus dentatus

Distribution:
Mainly warm to tropical regions of all continents. It does not occur in western Europe

IDENTIFICATION

A large stout worm up to 4.5 cm long, with a prominent buccal capsule and transparent cuticle through which the internal organs may be seen (Pl. III). The colour is usually pinkish. The size and site are diagnostic.

LIFE CYCLE

Preparasitic development from egg to L_3 is typically strongyloid, though earthworms may intervene as transport hosts. There are three modes of infection: by ingestion of the free L_3, ingestion of earthworms carrying the L_3, and percutaneously. After entering the body, there is an immediate moult, and the L_4 travel to the liver in the bloodstream, either from the intestine by the portal stream, or from the skin by the lungs and systemic circulation. In the liver the final moult takes place, and the young adults wander in the parenchyma for three months or more before piercing the capsule and migrating in the peritoneal cavity to the perirenal region. There they are enclosed in a cyst by host reaction, and complete their development. The cyst communicates with the ureter either directly or, if it is more distant, by a fine connecting canal, allowing the worm eggs to be excreted in the urine.

Though the favoured site is in the perirenal fat, some worms occur in the kidney itself, in the calyces and pelvis.

Prenatal infection has been reported.

Erratic migration is common in *Stephanurus* infection, and larvae have been found in most organs and in muscle. In these sites they are trapped by encapsulation and never reach the perirenal area.

The prepatent period ranges from 6–19 months and the worms have a longevity of about two years.

PATHOGENESIS

The main pathogenic effect is due to the larvae which, by the late L_4 stage, have heavily sclerotised buccal capsules capable of tearing tissue and they cause much damage to the liver and occasionally other organs in their wanderings. In heavy infections there may be severe cirrhosis and ascites and, in rare cases, liver failure and death. In most infections, however, the effects are seen only after slaughter as patchy cirrhosis, and the main importance of the worm is economic, from liver condemnation.

Usually the adult worms, soon after arrival at the perirenal site, are encapsulated in cysts, which may contain greenish pus. In rare cases the ureters may be thickened and stenosed, with consequent hydronephrosis.

Stephanurus may occasionally cause severe liver damage in calves grazing on contaminated ground.

CLINICAL SIGNS

In most infections the only sign is failure to gain weight or, in more severe cases, weight loss. Where there is more extensive liver damage there may be ascites, but it is only when there is massive invasion, comparable to acute fascioliasis in sheep, that death occurs.

EPIDEMIOLOGY

Though the adult worms are never numerous, they are very fecund, and an infected pig may pass a million eggs per day.

The L_3 is susceptible to desiccation, so that stephanuriasis is mainly associated with damp ground. Since it infects readily by skin penetration, the pigs' habit of lying around the feeding area when kept outside presents a risk, as does damp, unhygienic accommodation for housed animals. Such conditions, coupled with prenatal infection and the longevity of the worm, ensure continuity of infection through many generations of pigs.

DIAGNOSIS

The clinical signs are likely to be few, and since most of the damage occurs during the prepatent phase, eggs may not be found in the urine. However, in endemic areas, where pigs are failing to thrive and where local abattoirs record appreciable numbers of cirrhotic livers, a presumptive diagnosis can be made.

TREATMENT

Levamisole, the modern benzimidazoles and ivermectin are effective.

CONTROL

One approach to control is based on the susceptibility of the L_3 to desiccation and on the fact that a major route of infection is percutaneous. It follows that the provision of impervious surfaces around the feeding areas for outdoor-reared pigs, and simple hygiene, ensuring clean dry flooring, in pig houses will help to limit infection. This approach may be supplemented by segregating young pigs from those of more than nine months of age which will be excreting eggs.

The 'gilt only' scheme, which was advocated by workers in the United States, consists essentially of using only gilts for breeding. The gilts are reared on land which is dry and exposed to the sun. A single litter is taken from them, and as soon as the piglets can be weaned the gilts are marketed. The scheme takes advantage of the extremely long prepatent period which allows a single breeding cycle by the gilts to be completed before egg-laying begins and so progressively eliminates infection. The boars used in the scheme are housed on concrete.

Regimes incorporating anthelmintic control recommend treatment of sows and gilts 1–2 weeks before putting to the boar, and again 1–2 weeks before farrowing.

It should be remembered in designing a control system that the earthworm transport hosts present a continuous reservoir of infection.

Syngamus

Only one member of this genus, *Syngamus trachea*, is of veterinary significance and parasitises the upper respiratory tract of non-aquatic birds; it is commonly known as the 'gapeworm' and may be responsible for respiratory distress and death.

Hosts:
Domestic fowl and game birds such as pheasants and partridges

Site:
Trachea

Species:
Syngamus trachea

Distribution:
Worldwide

IDENTIFICATION

Gross: The reddish, large female (up to 2.0 cm) and small male (up to 0.5 cm) worms, are permanently *in copula* forming a Y shape: they are the only parasites found in the trachea of domestic birds (Pl. IV).
Microscopic: The worms have large shallow buccal capsules which have up to 10 teeth at their base.

The ellipsoidal egg of *S. trachea* has an operculum at both ends (Fig. 36).

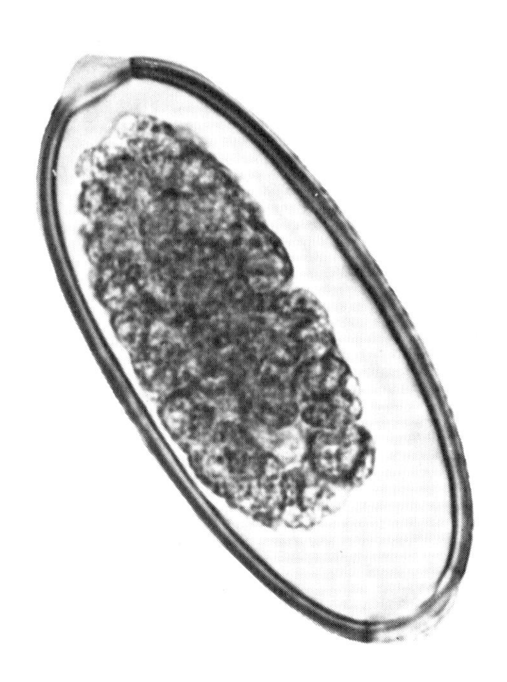

Fig 36 Egg of *Syngamus trachea* showing an operculum at each end.

LIFE CYCLE

Eggs escape under the bursa of the male and are carried up the trachea in the excess mucus produced in response to infection: they are then swallowed and passed in the faeces. Unlike other strongyloids the L_3 develops within the egg.

Infection may occur by one of three ways, firstly by ingestion of the L_3 in the egg, secondly by ingestion of the hatched L_3 or thirdly by ingestion of a transport host containing the L_3.

The most common transport host is the common earthworm, but a variety of other invertebrates including slugs, snails and beetles, may act as transport hosts. After penetrating the intestine of the final host the L_3 travel to the lungs, probably in the blood since they are found in the alveoli 4–6 hours after experimental infection. The two parasitic moults take place in the lungs within five days by which time the parasites are 1.0–2.0 mm long. Copulation occurs around day seven in the trachea or bronchi after which the female grows rapidly. The prepatent period is 18–20 days.

PATHOGENESIS

The effects of *S. trachea* are most severe in young birds especially game chicks and turkey poults. In these, migration through the lungs in heavy infections may result in pneumonia and death. In less severe infections the adult worms cause a haemorrhagic tracheitis with excess mucus production which leads to partial occlusion of the airways and difficulty in breathing.

CLINICAL SIGNS

Pneumonia during the prepatent phase may cause signs of dyspnoea and depression, whereas the presence of adult worms and excess mucus in the trachea lead to signs of asphyxia or suffocation with the bird gasping for air; often there is a great deal of head shaking and coughing as it tries to rid itself of the obstruction. The clinical picture of 'gapes' may thus range from gasping, dyspnoea and death to, in less severely affected animals, weakness, anaemia and emaciation.

EPIDEMIOLOGY

Gapeworm infection primarily affects young birds, but turkeys of all ages are susceptible, the adults often acting as carriers. Eggs may survive for up to nine months in soil and L_3 for years within the earthworm or other transport hosts. Disease is seen most frequently in breeding and rearing establishments where outdoor pens, such as are used for breeding pheasants, are in use. Infection may be initiated by eggs, passed by wild birds such as rooks and blackbirds; these may also infect earthworms.

DIAGNOSIS

This is based on clinical signs and the finding of eggs in the faeces. Disease is probably best confirmed by post-mortem examination of selected cases when worms will be found attached to the tracheal mucosa.

TREATMENT

In-feed thiabendazole and fenbendazole are effective, administered usually over periods of 3–14 days. Nitroxynil and levamisole are also very efficacious when given in the water.

CONTROL

Young birds should not be reared with adults, especially turkeys, and to prevent infection becoming established runs or yards should be kept dry and contact with wild birds prevented.

Drug prophylaxis may be practised over the period when outbreaks are normally expected.

Mammomonogamus

This genus, closely related to *Syngamus*, is parasitic in the respiratory passages of mammals. Two species, *M. larnygeus* and *M. nasicola*, are parasites of cattle, buffalo and goats in the Far East, Africa and Central and South America. Another species, *M. ierei*, in the nasal cavities of cats has been reported from the Caribbean. Little is known of the life cycle or effects of members of this genus, but they are not considered serious pathogens.

HOOKWORMS OF DOGS AND CATS

The family Ancylostomidae, whose members are commonly called hookworms because of the characteristic hook posture of their anterior ends, are responsible for widespread morbidity and mortality in animals primarily due to their blood-sucking activities in the intestine.

Ancylostoma

Hosts:
Dog, cat and fox

Site:
Small intestine

Species:

Ancylostoma caninum	dog and fox
A. tubaeforme	cat
A. braziliense	dog and cat

Distribution:
Worldwide in the tropics and warm temperate areas. In other countries it is sometimes seen in dogs imported from endemic regions

IDENTIFICATION

Gross: They are readily recognised on the basis of size (1.0–2.0 cm), being much smaller than the common ascarid nematodes which are also found in the small intestine, and by their characteristic 'hook' posture.
Microscopic: The buccal capsule is large with marginal teeth (Fig. 37), there being three pairs in *A. caninum* and *A. tubaeforme* and two pairs in *A. braziliense*.

Since the most important species is *A. caninum* this is discussed in detail.

Ancylostoma caninum

LIFE CYCLE

The life cycle is direct and given optimal conditions the eggs may hatch and develop to L_3 in as little as five days.

Infection is by skin penetration or by ingestion, both methods being equally successful. In percutaneous infection, larvae migrate via the blood stream to the lungs

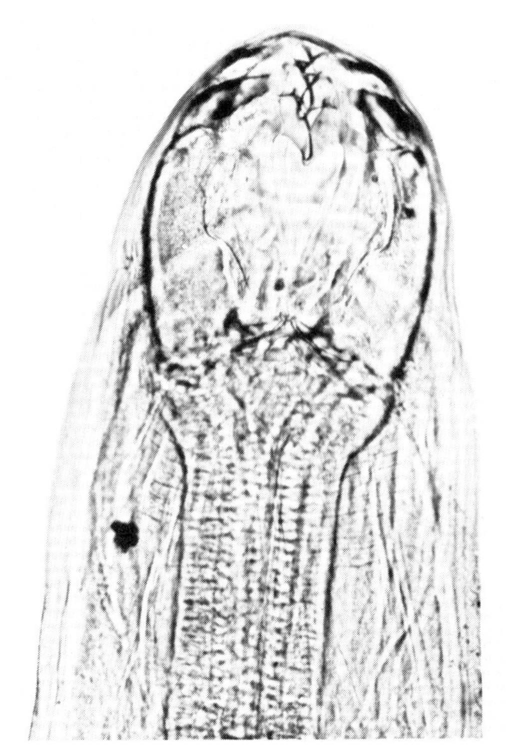

Fig 37 Buccal capsule of *Ancylostoma caninum* showing three pairs of marginal teeth.

where they moult to L_4 in the bronchi and trachea, and are then swallowed and pass to the small intestine where the final moult occurs. If infection is by ingestion the larvae may either penetrate the buccal mucosa and undergo the pulmonary migration described above or pass direct to the intestine and develop to patency. Whichever route is taken the prepatent period is 14–21 days. The worms are prolific egg layers and an infected dog may pass millions of eggs daily for several weeks.

An important feature of *A. caninum* infection is that, in susceptible bitches, a proportion of the L_3 which reach the lungs migrate to the skeletal muscles where they remain dormant until the bitch is pregnant. They are then reactivated and, still as L_3, are passed in the milk of the bitch for a period of about three weeks after whelping. This transmammary infection is often responsible for severe anaemia in litters of young pups in their second or third week of life. Infection of the bitch on a single occasion has been shown to produce transmammary infections in at least three consecutive litters.

It also appears that dormant L_3 in the muscles of both bitches and dogs can recommence migration months or years later to mature in the host's intestine. Stress, severe illness or repeated large doses of corticosteroids can all precipitate these apparently new infections in dogs, which may perhaps now be resident in a hookworm-free environment.

A final point is that, experimentally, L_3 of some strains of *A. caninum* exposed to chilling before oral administration have been shown to remain in arrested development in the intestinal mucosa for weeks or months. The significance of this observation is still unknown, but it is thought that such larvae may resume development if the adult hookworm population is removed by an anthelmintic or at times of stress such as lactation.

PATHOGENESIS

This is essentially that of an acute or chronic haemorrhagic anaemia. The disease is most commonly seen in dogs under one year old and young pups, infected by the transmammary route, are particularly susceptible due to their low iron reserves. Blood loss starts about the eighth day of infection when the immature adult has developed the toothed buccal capsule which enables it to grasp plugs of mucosa containing arterioles. Each worm removes about 0.1 ml of blood daily and in heavy infections of several hundred worms, pups quickly become profoundly anaemic.

In lighter infections, common in older dogs, the anaemia is not so severe, as the marrow response is able to compensate for a variable period. Ultimately however, the dog may become iron deficient and develop a microcytic hypochromic anaemia.

In previously sensitised dogs, skin reactions such as moist eczema and ulceration at the sites of percutaneous infection occur especially affecting the inter-digital skin.

CLINICAL SIGNS

In acute infections, there is anaemia and lassitude and occasionally respiratory embarrassment. In suckled pups the anaemia is often severe and is accompanied by diarrhoea which may contain blood and mucus. Respiratory signs may be due to larval damage in the lungs or to the anoxic effects of anaemia.

In more chronic infections, the animal is usually underweight, the coat is poor, and there is loss of appetite and perhaps pica. Inconsistently there are signs of respiratory embarrassment, skin lesions and lameness.

EPIDEMIOLOGY

In endemic areas the disease is most common in dogs under one year old. In older animals, the gradual development of age resistance makes clinical disease less likely, particularly in dogs reared in endemic areas whose age resistance is reinforced by acquired immunity.

The epidemiology is primarily associated with the two main sources of infection, transmammary in suckled pups and percutaneous or oral from the environment.

An important aspect of transmammary infection is that disease may occur in suckled pups reared in a clean environment and nursed by a bitch which may have been recently treated with an anthelmintic and has a negative faecal egg count.

Contamination of the environment is most likely when dogs are exercised on grass or earth runs which retain moisture and also protect larvae from sunlight. On such surfaces larvae may survive for some weeks. In contrast, dry impervious surfaces, particularly if exposed to sunlight, are lethal to larvae within a day or so. Housing is also important and failure to remove soiled bedding, especially if the kennels are damp or have porous or cracked floors, can lead to a massive build-up of infection.

DIAGNOSIS

This depends on the clinical signs and history supplemented by haematological and faecal examination. High faecal worm egg counts are valuable confirmation of diagnosis, but it should be noted that suckled pups may show severe clinical signs before eggs are detected in the faeces. Also, a few hookworm eggs in the faeces, although confirmatory evidence of infection, do not necessarily indicate that an ailing dog is suffering from hookworm disease.

TREATMENT

Affected dogs should be treated with an anthelmintic, such as thenium, mebendazole, fenbendazole, dichlorvos and nitroscanate, all of which will kill both adult and developing intestinal stages. If the disease is severe, it is advisable to give parenteral iron and to ensure that the dog has a protein-rich diet. Young pups may require a blood transfusion.

CONTROL

A system of regular anthelmintic therapy and hygiene should be adopted. Weaned pups and adult dogs should be treated every three months.

Pregnant bitches should be dosed at least once during pregnancy and the nursing litters dosed at least twice, at 1–2 weeks of age and again 2 weeks later with a drug specifically recommended for use in pups. This will also control ascariasis.

A new but expensive regimen which has been shown to reduce the perinatal transfer of both *Ancylostoma* and *Toxocara* larvae is the oral administration of large doses of fenbendazole daily from three weeks before to three weeks after whelping.

Kennel floors should be free of crevices and dry and the bedding should be disposed of daily. Runs should preferably be of tarmac or concrete and kept as clean and dry as possible; faeces should be removed with a shovel before hosing. If an outbreak has occurred, earth runs may be treated with sodium borate which is lethal to hookworm larvae, but this also kills grass. A second possibility which is often used in fox farms is the provision of wire-mesh flooring in the runs.

A. tubaeforme

The life cycle and treatment of this hookworm of cats are similar to that of *A. caninum* in the dog but there is no evidence of transmammary infection.

A. braziliense

This hookworm occurs in both dogs and cats. Its life cycle is similar to that of *A. caninum* although evidence of transmammary infection is lacking. While it may cause a degree of hypoalbuminaemia through an intestinal leak of plasma, it is not a blood-sucker and consequently is of little pathogenic significance in dogs, causing only mild digestive upsets and occasional diarrhoea. Treatment is similar to that for *A. caninum*.

The main importance of *A. braziliense* is that it is regarded as the primary cause of **cutaneous larva migrans** in man. This lesion, characterised by tortuous erythematous inflammatory tracts within the dermis and by severe pruritus, is caused by infective larvae of *A. braziliense*, and less frequently *Uncinaria*, penetrating the skin and wandering in the dermis. These larvae do not develop, but the skin lesions usually persist for weeks.

Similar lesions, although only transient and pinpoint, may be caused by *A. caninum* larvae.

Uncinaria

Hosts:
Dog, cat and fox

Site:
Small intestine

Species:
Uncinaria stenocephala

Distribution:
Temperate and sub-arctic areas; the 'northern hookworm'

IDENTIFICATION

A small worm, up to 1.0 cm long, it possesses two cutting plates on the border of the buccal capsule (Fig. 38) and at the base a small pair of teeth.

LIFE CYCLE

Similar to *A. caninum* except that oral infection, without pulmonary migration, is the usual route. Although the infective larvae can penetrate the skin, the infection

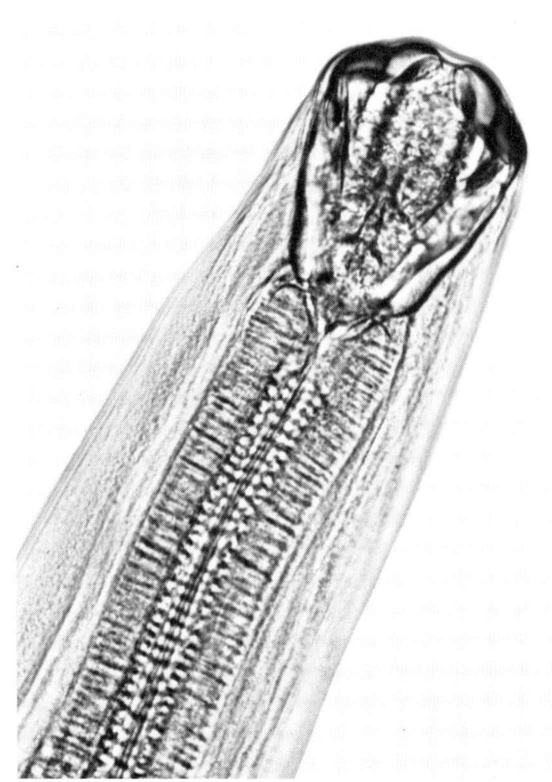

Fig 38 Buccal capsule of *Uncinaria stenocephala* showing cutting plates on border.

rarely matures and there is no evidence as yet of transmammary infection. The prepatent period is about 15 days.

PATHOGENESIS AND CLINICAL SIGNS

The infection is not uncommon in groups of sporting and working dogs. The worm is not a voracious blood-sucker like *A. caninum*, but hypoalbuminaemia and low-grade anaemia, accompanied by diarrhoea, anorexia and lethargy, have been recorded in heavily infected pups. Probably the most common lesion in dogs made hypersensitive by previous exposure is pedal dermatitis, affecting particularly the inter-digital skin.

EPIDEMIOLOGY

In England, in a paddock used continuously throughout the year by greyhounds, the seasonal pattern of infective larvae on the pasture followed closely that described for gastrointestinal trichostrongyloids in ruminants with a sharp rise in July and a peak in September; this suggests that development to the L_3 is heavily dependent on temperature.

DIAGNOSIS

In areas where *A. caninum* is absent, the clinical signs of the patent infection together with the demonstration of strongyle eggs in the faeces is indicative of uncinariasis. Where *Ancylostoma* is also endemic, differential diagnosis may require larval culture although the treatment is similar.

TREATMENT AND CONTROL

Regular anthelmintic treatment and good hygiene as outlined for *Ancylostoma* will control *Uncinaria* infection. The pedal dermatitis responds poorly to symptomatic treatment, but regresses gradually in the absence of reinfection.

HOOKWORMS OF RUMINANTS

Bunostomum

Hosts:
Ruminants

Site:
Small intestine

Species:

Bunostomum trigonocephalum	sheep and goats
B. phlebotomum	cattle

Distribution:
Worldwide

IDENTIFICATION

Gross: *Bunostomum* is one of the larger nematodes of the small intestine of ruminants, being 1.0–3.0 cm long and characteristically hooked at the anterior end.
Microscopic: The large buccal capsule bears on the margin a pair of cutting plates and internally a large dorsal cone. (Fig. 39)

LIFE CYCLE

Infection with the L_3 may be percutaneous or oral, only the former being followed by pulmonary migration. The prepatent period ranges from 1–2 months.

PATHOGENESIS AND CLINICAL SIGNS

The adult worms are blood-suckers and infections of 100–500 worms will produce anaemia, hypoalbuminaemia, loss of weight and occasionally diarrhoea. In calves, skin penetration of the larvae may be accompanied by foot-stamping and signs of itching.

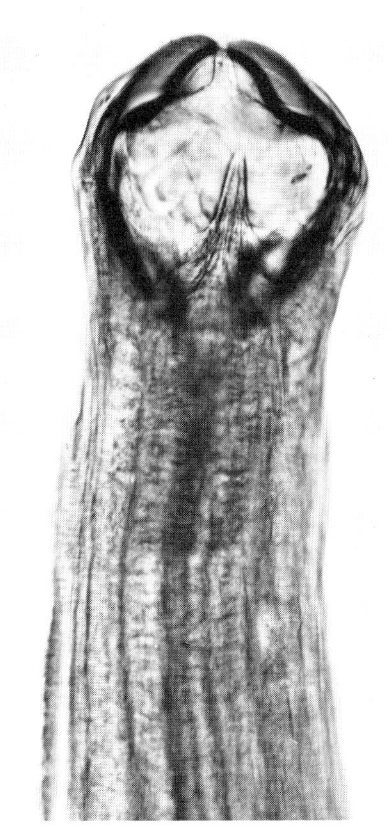

Fig 39 Head of *Bunostomum* showing the marginal cutting plates. The cone-like structure in the buccal capsule is not a tooth, but carries the duct of the oesophageal gland.

EPIDEMIOLOGY

In temperate countries, high worm burdens are uncommon and in Britain, for example, only one outbreak has been recorded, occurring in young cattle housed in a damp yard; in sheep it is unusual to find more than 50 adult worms. In contrast, pathogenic infections are more common in the tropics and in some areas, such as Nigeria, the highest worm burdens are found at the end of the dry season apparently due to the maturation of hypobiotic larvae.

DIAGNOSIS

The clinical signs of anaemia and perhaps diarrhoea in calves or young sheep are not in themselves pathognomonic of bunostomiasis. However, in temperate areas, the epidemiological background may be useful in eliminating the possibility of *Fasciola hepatica* infection. In the tropics, haemonchosis must be considered, possibly originating from hypobiotic larvae.

Faecal worm egg counts are useful in that these are lower than in *Haemonchus* infection while the eggs are more bluntly rounded, with relatively thick sticky shells to which debris is often adherent. For accurate differentiation, larval cultures should be prepared.

TREATMENT AND CONTROL

The prophylactic anthelmintic regimens practised for *Ostertagia* or *Haemonchus* are usually sufficient to control this parasite. Otherwise treatment of outbreaks should be accompanied by measures to improve hygiene, particularly with regard to the disposal of manure, and by the provision of dry bedding for housed or yarded animals.

Gaigeria

Gaigeria pachyscelis, which closely resembles *Bunostomum* in most respects, is found in sheep and goats in South America, Africa and Asia. It is a voracious bloodsucker and even 100–200 worms are sufficient to produce death in sheep within a few weeks of infection.

Agriostomum

Agriostomum vryburgi is a common hookworm of the large intestine of cattle and buffaloes in Asia and South America.

Its life cycle is probably direct and its pathogenicity, although unknown, presumably depends on its haematophagic habits.

[**Hookworms in man:** Two hookworms, *Ancylostoma duodenale* and *Necator americanus*, occur in man in the tropics. Their pathogenesis is similar to that of *A. caninum*, but transmammary infection does not occur.]

Superfamily METASTRONGYLOIDEA

Most worms in this superfamily inhabit the lungs or the blood vessels adjacent to the lungs. The typical life cycle is indirect, and the intermediate host is usually a mollusc.

They may be conveniently divided into three groups according to host; those occurring in pigs, in sheep and goats, and in the domestic carnivores.

METASTRONGYLES OF PIGS

Only one genus occurs in pigs, *Metastrongylus*, and it is exceptional in having earthworms, rather than molluscs, as intermediate hosts.

Metastrongylus

Host:
Pig

Intermediate hosts:
Earthworms

Site:
Small bronchi and bronchioles, especially those of the posterior lobes of the lungs

Species:
Metastrongylus apri (syn. *elongatus*)

M. salmi

M. pudendotectus

Distribution:
Worldwide

IDENTIFICATION

Slender white worms, up to 6.0 cm in length; the host, site and long slender form are sufficient for generic identification.

The eggs have rough, thick shells, and are larvated when laid.

LIFE CYCLE

In cold temperatures the eggs are very resistant and can survive for over a year in soil. Normally, however, they hatch almost immediately, the intermediate host ingesting the L_1. In the earthworm, development to L_3 takes about ten days at optimal temperatures of 22–26 °C. The longevity of the L_3 in the earthworm is similar to that of the intermediate host itself, and may be up to seven years.

The pig is infected by ingestion of earthworms and the L_3, released by digestion, travel to the mesenteric lymph nodes, moult and the L_4 then reach the lungs by the lymphatic-vascular route, the final moult occuring after arrival in the air passages.

The prepatent period is about 4 weeks.

PATHOGENESIS

During the prepatent period areas of pulmonary consolidation, bronchial muscular hypertrophy, and peribronchial lymphoid hyperplasia develop (Fig. 40), often accompanied by areas of overinflation.

When the worms are mature, and eggs are aspirated into the smaller air passages and parenchyma, consolidation increases and emphysema is more marked. Hypersecretion of bronchiolar mucus also occurs during this stage.

About six weeks after infection, chronic bronchitis and emphysema are established and small greyish nodules may be found in the posterior part of the diaphragmatic lobes; these may aggregate to form larger areas and are slow to resolve. Purulent staphylococcal infection in the lungs has been noted in many cases of metastrongylosis.

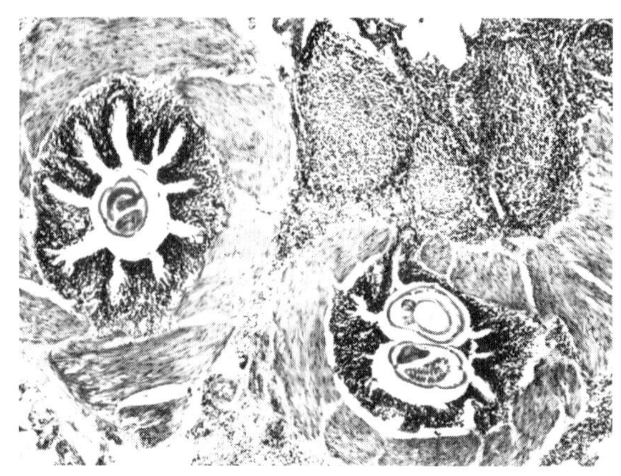

Fig 40 Bronchial muscular hypertrophy and peribronchial lymphoid hyperplasia associated with *Metastrongylus* infection of pig lung.

CLINICAL SIGNS

Most infections are light and asymptomatic. However in heavy infections coughing is marked, and is accompanied by dyspnoea and nasal discharge. Secondary bacterial infection may complicate the signs.

EPIDEMIOLOGY

Metastrongylosis shows a characteristic age distribution, being most prevalent in pigs of 4–6 months old. The parasite is common in most countries although outbreaks of disease do not often occur, probably due to the fact that most systems of pig husbandry do not allow ready access to earthworms by pigs. Though it is often suggested that *Metastrongylus* may transmit some of the porcine viruses, and may enhance the effect of viruses already present in the lungs, the role of the worm is not conclusively proven.

DIAGNOSIS

For faecal examination saturated magnesium sulphate should be used as the flotation solution because of the heavy density of the eggs. The small rough-shelled larvated eggs are characteristic, but it should be recollected that *Metastrongylus* is often present in normal pigs, and pulmonary signs may be referable to virus infection rather than lungworms. The disease is most often encountered in pigs on pasture, though an occasional outbreak has occurred in yarded pigs.

TREATMENT

Many anthelmintics including the modern benzimidazoles, levamisole and ivermectin are highly effective.

CONTROL

When pig husbandry is based on pasture, control is extremely difficult because of the ubiquity and longevity of the earthworm intermediate host. On farms where severe outbreaks have occurred pigs should be housed, dosed, and the infected pasture cultivated or grazed with other stock.

METASTRONGYLES OF SHEEP AND GOATS

These worms all inhabit the lungs, but none is a major pathogen and, though common, they are of little economic importance compared with the other helminth parasites of sheep and goats. Although there are several different genera they are sufficiently similar in behaviour to be considered together.

Hosts:
Sheep and goats

Intermediate hosts:
Molluscs: *Muellerius* in snails and slugs; *Protostrongylus* in snails

Genera and sites:
Muellerius capillaris found in alveoli
Protostrongylus, many species, found in small bronchioles

Minor related genera are *Cystocaulus*, *Spiculocaulus* and *Neostrongylus*

Distribution:
Worldwide except for arctic and subarctic regions

IDENTIFICATION

These are brown hair-like worms 1.0–3.0 cm long which are difficult to discern with the naked eye as they are embedded in lung tissue.

LIFE CYCLE

The worms are ovo-viviparous, the L_1 being passed in the faeces; these penetrate the foot of the molluscan intermediate host, and develop to L_3 in a minimum period of 2–3 weeks. The sheep is infected by ingesting the mollusc and the L_3, freed by digestion, travel to the lungs by the lymphatic-vascular route, the parasitic moults occurring in the mesenteric lymph nodes and lungs.

The prepatent period of *Muellerius* is 6–10 weeks and that of *Protostrongylus* is 5–6 weeks. The period of patency is very long, exceeding two years in all the genera examined.

PATHOGENESIS

Muellerius is associated with small, spherical, nodular lesions, which occur most commonly near, or on, the lung surface, and on palpation have the feel and size of lead shot (Fig. 41). Nodules containing single worms are almost imperceptible, and the visible ones enclose several of the tiny worms as well as eggs and larvae.

In *Protostrongylus* infection there is a somewhat larger area of lung involvement, the occlusion of a small bronchus by worms resulting in its lesser branches which occur toward the lung surface being filled with eggs, larvae, and cellular debris; the whole lesion has a roughly conical form, with the base on the surface of the lung.

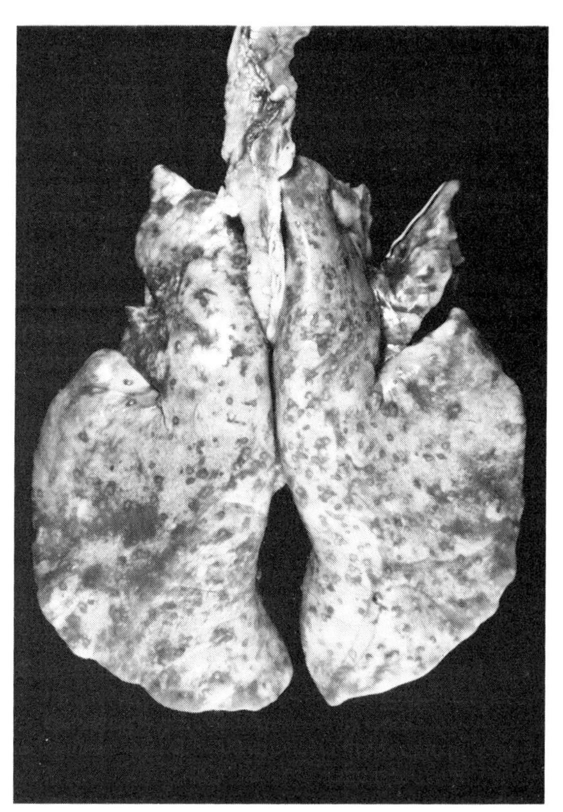

Fig 41 Characteristic nodular lesions associated with *Muellerius* infection of sheep lung.

CLINICAL SIGNS

Pneumonic signs have rarely been observed, and infections are almost always inapparent, being identified only at necropsy.

EPIDEMIOLOGY

Muellerius is by far the commonest genus, and in many temperate areas such as Britain, the eastern states of the U.S.A. and the winter rainfall regions of Australia, almost all sheep carry the infection; the extensive distribution and high prevalence are partly attributable to its wide range of intermediate hosts.

Protostrongylus, whose intermediate host range is restricted to certain species of snail, has a lower prevalence, though its geographic range is just as wide.

Additional factors which play a part in ensuring the endemicity of these worms are, first, the ability of the L_1 to survive for months in the faecal pellet, and secondly, the persistence of the L_3 in the intermediate host for the lifetime of the mollusc. Also important in this respect are the long periods of patency and the apparent inability of the final host to develop acquired immunity, so that adult sheep have the heaviest infections and the high prevalence.

DIAGNOSIS

The presence of infection is usually noted only during routine faecal examination. The L_1 are first differentiated from those of *Dictyocaulus filaria* by the absence of an anterior protoplasmic knob, and then on the individual characters of the larval tail (Fig. 42).

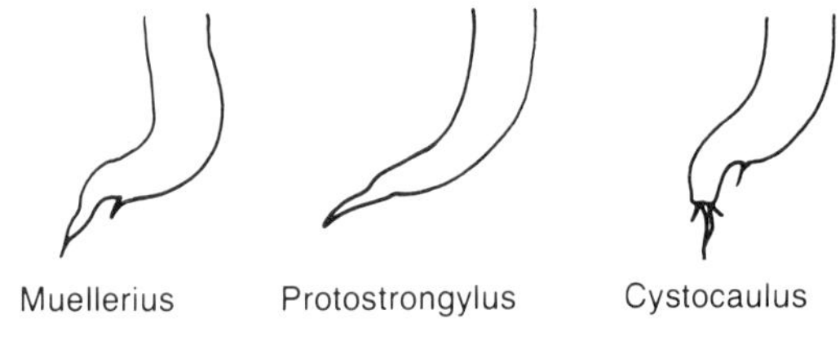

Fig 42 Differential features of larvae of 3 genera of metastrongyloids of sheep.

TREATMENT

The modern benzimidazoles, levamisole and ivermectin have been shown to be effective.

CONTROL

Because of the ubiquity of the molluscan intermediate hosts, and the fact that the L_3 can survive as long as the molluscs, specific control is difficult, but fortunately rarely necessary.

Elaphostrongylus cervi is a metastrongyloid which occurs in farmed deer, including reindeer, in Europe, Asia and New Zealand. It is found in the intermuscular connective tissue of the thorax and back, the eggs being carried in the bloodstream to the lungs, where they hatch and are coughed up and swallowed. However, sometimes the worms invade the CNS to cause a meningoencephalitis with paralysis and occasional death. In the muscles the parasites are harmless, but their presence may necessitate trimming of the carcass.

A related species, *Parelaphostrongylus tenuis*, whose natural host is the white-tailed deer in N. America occasionally invade the CNS of sheep causing paralysis.

METASTRONGYLES OF DOGS AND CATS

Like most members of the superfamily, these worms live in, or adjacent to, the lungs. The few genera of veterinary interest are considered in order of importance.

Oslerus (syn. *Filaroides*)

This genus was, until recently, part of the larger genus *Filaroides*, but has now been separated on morphological grounds from the other members. Though distinction has been made on morphology it is also useful from the veterinary standpoint, for it separates the single harmful species, *Oslerus osleri*, living in the upper air passages, from the relatively harmless species which are retained in the genus *Filaroides*, and which live in the lung parenchyma.

Hosts:
Domestic and wild dogs

Site:
The worms are embedded in fibrous nodules in the trachea at the region of bifurcation, and in the adjacent bronchi

Species:
Oslerus osleri

Distribution:
Worldwide

IDENTIFICATION

Small, pale, slender worms, up to 1.5 cm long; the site and lesions are diagnostic.

LIFE CYCLE

Oslerus, and its closely related genus, *Filaroides*, are exceptional in the superfamily in having direct life cycles. The females are ovo-viviparous, and most eggs hatch in the trachea. Many larvae are coughed up and swallowed, and passed in the faeces and infection may occur by ingestion of these; more commonly, transmission occurs when an infected bitch licks the pup and transfers the newly hatched L_1 which are present in her sputum.

After ingestion, the first moult occurs in the small intestine and the L_2 travel to the lungs by the lymphatic-vascular route. Development through to L_5 takes place in the alveoli and bronchi, and the adults migrate to the trachea.

The prepatent period varies from 10–18 weeks.

PATHOGENESIS

The nodules in which the worms live first appear at about two months from infection. They are pinkish grey, and the small worms may be seen partly protruding from their surfaces. These nodules are fibrous in character and are very firmly applied to the mucosa; they may be up to 2.0 cm in diameter. Though the majority occur near the tracheal bifurcation a few may be found several centimetres from this area (Fig. 43).

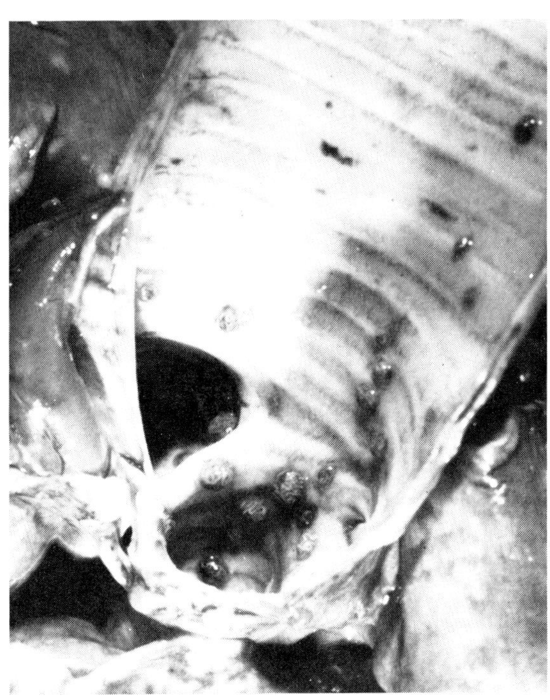

Fig 43 Nodules at tracheal bifurcation caused by infection with *Oslerus osleri*.

CLINICAL SIGNS

Many infections are clinically inapparent, and the characteristic nodules are only discovered incidentally at necropsy.

The major signs of *Oslerus* infection are respiratory distress and a rasping cough, especially after exercise. The most severe cases have usually been seen in dogs of 6–12 months old, and obviously the infection is of greater importance in working dogs. In household pets whose exercise is limited the presence of the tracheal nodules is well tolerated, and animals can survive for long periods with little distress.

EPIDEMIOLOGY

Though *Oslerus* has been recorded from many countries there is little data on its local prevalence. In the south of England one survey has given a figure of 6% for all types of dog. In further surveys in that area, greyhounds have shown a prevalence rate of 18%, but there is no evidence of breed susceptibility. In general the focus of infection appears to be the nursing bitch.

DIAGNOSIS

Swabs of pharyngeal mucus give variable results and repeated sampling may be necessary. However, in paroxysmal coughing, large amounts of bronchial mucus are often expelled, containing large numbers of larvae. Less rewarding techniques are those based on faecal examination, either by flotation or by the Baerman method.

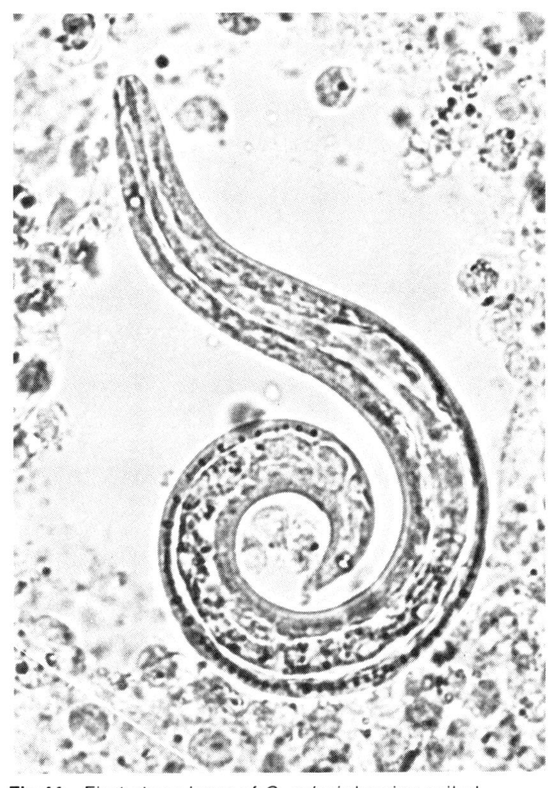

Fig 44 First-stage larva of *O. osleri* showing coiled appearance and S-shaped tip of tail.

Although requiring general anaesthesia, bronchoscopy is the most reliable method, as it will indicate not only the presence, size and location of many of the nodules, but will also allow the collection of tracheal mucus for confirmatory examination for eggs and larvae; the latter are invariably coiled, sluggish and have an S-shaped tail (Fig. 44).

Large nodules may be detected by lateral thoracic radiography.

TREATMENT

Evidence for successful treatment is conflicting. There are reports of amelioration of clinical signs, apparently due to a reduction in the size of the nodules, after prolonged treatment with some benzimidazoles, such as fenbendazole and oxfendazole at increased dosage rates. However, consistent evidence of therapeutic success is lacking, although total suppression of larval output has been demonstrated during the period of drug administration.

CONTROL

This is difficult unless infected bitches can be identified and treated before whelping and during lactation, but even in this event, until drugs are available which will give total clearance, the only certain measure is the removal of pups from infected dams at birth, and hand-rearing or fostering on uninfected bitches.

Filaroides

Hosts:
Domestic dog and wild carnivores

Site:
Lung parenchyma

Species:
Filaroides milksi

F. hirthi

Distribution:
North America, Europe and Japan

IDENTIFICATION

The worms are very small, slender, hair-like and greyish, and are not only difficult to see with the naked eye in the lung parenchyma, but are unlikely to be recovered intact from the tissue. A squeeze preparation from a cut surface of the lung will show worm fragments, eggs and larvae, and this, with the host and site, is sufficient for generic diagnosis.

LIFE CYCLE

The worms are ovo-viviparous and the hatched L_1 are passed in faeces or expelled in sputum. Though infection may be acquired by ingestion of faecal larvae, the important route, as in *Oslerus* infection, is thought to be by transfer of L_1 in the bitch's saliva when the pup is licked. The prepatent period of *F. hirthi* is five weeks; that of *F. milksi* is unknown.

PATHOGENESIS

The chief lesions are the small, soft, greyish miliary nodules which are associated with the presence of worms and which are distributed subpleurally and throughout the lung parenchyma; in heavy infections, sometimes observed in experimental dogs subjected to immunosuppressive drugs, the nodules may coalesce into greyish masses.

CLINICAL SIGNS

Infection is almost invariably asymptomatic, and is discovered only at post mortem examination. However, in the rare heavy infection, hyperpnoea may occur.

EPIDEMIOLOGY

Little is known of the epidemiology. *F. hirthi* was first observed in a breeding colony of experimental beagles, and it would be fair to suggest, in view of its mode of transmission, that a high prevalence could be expected in dogs from breeding kennels.

DIAGNOSIS

Only *F. hirthi* has been diagnosed in the live animal and this was in experimental dogs. The L_1, present in the faeces and sputum, is coiled, and the tail has a notch, followed by a constriction, and has a terminal lance-like point.

TREATMENT

Albendazole has been reported to be effective.

CONTROL

Unlikely to be required.

Aelurostrongylus

One species, *Aelurostrongylus abstrusus*, is common in the lungs of the domestic cat.

Hosts:
Cats

Intermediate hosts:
Many molluscs

Site:
Lung parenchyma and small bronchioles

Species:
Aelurostrongylus abstrusus

Distribution:
Worldwide

IDENTIFICATION

Aggregations of worms, eggs and larvae are present throughout the lung tissue. The worms, about 1.0 cm long, are very slender and delicate, and are difficult to recover intact for examination; a squeeze preparation from a cut surface of the lung will show the worm material including the characteristic L_1.

LIFE CYCLE

The worms are ovo-viviparous, and the L_1 are passed in the faeces. These penetrate the foot of the molluscan intermediate host and develop to the infective L_3, and during this phase the mollusc may be eaten by paratenic hosts such as birds and rodents. The cat is infected by ingestion of these hosts and the L_3, released in the alimentary tract, travel to the lungs by the lymphatic or blood stream.

The prepatent period is between 4–6 weeks, and the duration of patency is about four months, though some worms may survive in the lungs for several years despite the absence of larvae in the faeces.

PATHOGENESIS

The worm generally has a low pathogenicity, and the majority of infections are discovered only incidentally at post mortem examination. In most cases the lungs show only multiple small foci with greyish centres containing the worms and tissue debris, but in the rare severe infections larger nodules are present, up to 1.0 cm in diameter with caseous centres, projecting from the lung surface; these nodules may coalesce to form areas of consolidation. Microscopically the alveoli are seen to be blocked with worms, eggs, larvae, and cellular aggregations which may progress to granuloma formation (Fig. 45). A characteristic change is muscular hypertrophy and hyperplasia, which affects not only the bronchioles and alveolar ducts, but also the media of the pulmonary arteries.

With the exception of the muscular changes, which appear to be irreversible, resolution is rapid, and the lungs appear almost completely normal within six months of experimental infection, though a few worms may still be present.

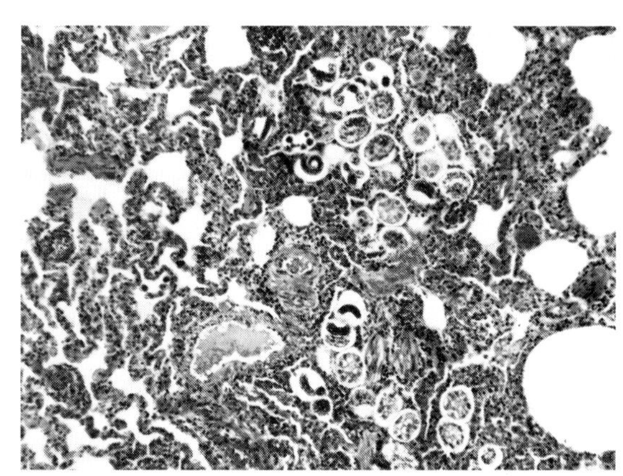

Fig 45 Section of cat lung infected with *Aelurostrongylus abstrusus*.

CLINICAL SIGNS

The clinical effects are slight, and in the resting cat are limited to a chronic mild cough; following exercise or handling, there may be coughing and sneezing with slight dyspnoea and production of mucoid sputum. In heavy experimental infections the most severe signs have appeared at 6–12 weeks after infection when egg-laying is maximal.

EPIDEMIOLOGY

Aelurostrongylus infection is widespread partly because it is almost indiscriminate in its ability to develop in slugs and snails, and partly because of its wide range of paratenic hosts. So far all surveys have shown prevalences greater than 5%.

DIAGNOSIS

Repeated faecal examination by smear, flotation, or Baerman technique may be necessary to find the characteristic L_1, which bears a subterminal spine on its S-shaped tail. Examination of pharyngeal swabs may be a useful additional procedure. Radiography has revealed the increased vascular and focal parenchymal densities which would be expected from the changes described above.

TREATMENT

Fenbendazole has proved effective.

CONTROL

In household pets, and especially in those of a nomadic disposition, access to the intermediate and paratenic hosts is difficult to prevent.

Angiostrongylus

The single species of veterinary importance is not found in lung tissue, but in the heart and associated pulmonary vessels.

Host:
Dog

Intermediate hosts:
Land snails and slugs

Site:
Right ventricle and pulmonary artery

Species:
Angiostrongylus vasorum

Distribution:
Worldwide except in the Americas

IDENTIFICATION

Slender worms, up to 2.5 cm long. In the female the white ovaries are coiled round the red intestine as in *Haemonchus* spp.

LIFE CYCLE

The genus is ovo-viviparous. The adult worms in the larger pulmonary vessels lay eggs which are carried to the capillaries, where they hatch. The L_1 break into the alveoli, migrate to the trachea and thence to the alimentary tract to be passed in the faeces. Further development takes place after entry into the intermediate host, the infective stage being reached in 17 days.

After the mollusc has been ingested by the dog the infective L_3, freed by digestion, travel to the lymph nodes adjacent to the alimentary tract, where both parasitic moults take place, and then to the vascular predilection site; L_5 have also been found in the liver.

The prepatent period is seven weeks, and the worms can live in the dog for more than two years.

PATHOGENESIS

Canine angiostrongylosis is usually a chronic condition, extending over months or even years.

Much of the pathogenic effect is attributable to the presence of the adult worms in the larger vessels and eggs and larvae in the pulmonary arterioles and capillaries. Blockage of these results in circulatory impediment which may lead eventually to congestive cardiac failure.

In the larger vessels, there is endarteritis and periarteritis which progresses to fibrosis, and at necropsy the vessels have a pipe-stem feel on palpation. The vascular change may extend to the right ventricle, with endocarditis involving the tricuspid valve.

The cut surface of the lung is mottled and reddish-purple. One reported systemic effect which is unusual in helminth infections is interference with the blood-clotting mechanism, so that subcutaneous haematomata may be present.

CLINICAL SIGNS

In recently established infections the resting dog usually shows no clinical signs, but if a substantial number of worms is present the active animal will show tachypnoea, with a heavy productive cough, the sputum sometimes showing blood.

In longer established severe infections signs are present even in the resting dog. There may be recurrent syncope. As a consequence of reduced blood-clotting capacity, slowly developing painless swellings may appear in dependent areas such as the lower abdomen and intermandibular space, and on the limbs where bruising has occurred.

The rare acute infection shows dyspnoea and violent cough, with white-yellow, occasionally bloody, sputum.

EPIDEMIOLOGY

Though worldwide in general distribution, *A. vasorum* is only prevalent in certain localities, and these are invariably rural. In Europe, endemic foci have been recognised in France, Spain, Eire and England.

DIAGNOSIS

The L_1, which may be present in faeces and sputum, has a small cephalic button, and a wavy tail with a subterminal notch, and its presence in association with respiratory and circulatory signs is accepted as confirmatory.

TREATMENT

Mebendazole is effective if given daily for five days.

CONTROL

Control is impractical in most cases, due to the ubiquity of the molluscan intermediate hosts.

[***Angiostrongylus cantonensis***, which is normally parasitic in the pulmonary artery of rats in the Far East, may cause disease in humans if the infected molluscs or crustacean paratenic hosts are ingested. The L_3 travel to the brain, where they cause an eosinophilic meningoencephalitis, which may prove fatal.]

Crenosoma

This genus contains several species parasitic in

carnivores and insectivores, but only one is of veterinary interest.

Hosts:
Dogs and farmed foxes

Intermediate hosts:
Mainly land snails

Site:
Trachea, bronchi, and bronchioles

Species:
Crenosoma vulpis

Distribution:
Worldwide

IDENTIFICATION

Slender white worms, up to 1.5 cm long. The host and site are sufficient for generic diagnosis. Microscopic confirmation is based on the presence of annular folds of the cuticle which bear small backwardly-directed spines on their margins.

LIFE CYCLE

C. vulpis is ovo-viviparous and L_1 are passed in the faeces. After ingestion of the molluscan intermediate host by the final host the L_3 are released by digestion, and travel to the lungs where both parasitic moults take place. The prepatent period is 19 days.

PATHOGENESIS

The spiny cuticular folds abrade the mucosa of the air passages with resulting bronchopneumonia and occlusion of the smaller bronchi and bronchioles.

CLINICAL SIGNS

The symptons are those of a chronic respiratory infection, with coughing, sneezing, and nasal discharge associated with tachypnoea. Foxes may become emaciated, with fur of poor quality.

In the infrequent acute infections there may be high mortality.

EPIDEMIOLOGY

C. vulpis is commoner in the fox than in the dog, and can be a problem in farmed foxes. The infection has a seasonality corresponding to fluctuations in population of its snail vectors so that, though cubs may begin to acquire L_3 in early summer, the highest incidence of clinical crenosomiasis is seen in autumn.

DIAGNOSIS

Examination of faeces by smear, flotation, or Baerman technique will reveal the L_1 with a straight tail which differentiates it in fresh canine faeces from those of *Oslerus, Filaroides* and *Angiostrongylus.* The L_1 resembles somewhat that of *Strongyloides spp.*

TREATMENT

Diethycarbamazine has been reported to be effective.

CONTROL

The snail vectors may be eliminated by spraying fox runs with molluscicide and painting woodwork with creosote up to 20 cm from the ground. Faeces should be disposed of in a manner which will avoid access by molluscs.

OTHER METASTRONGYLOIDS

Several other metastrongyloid genera occur in the domestic carnivores, but they are limited in distribution. They include *Metathelazia,* found in domestic cats in Russia, and in wild cats in the U.S.A., *Anafilaroides* in the domestic cat in U.S.A., Sri Lanka, and Israel, and *Gurltia* in cats in South America. *Metathelazia* and *Anafilaroides* inhabit the lung parenchyma, whilst *Gurltia* is found in the veins of the upper hind limb and is an occasional cause of paralysis.

Superfamily RHABDITOIDEA

This is a primitive group of nematodes which are mostly free-living, or parasitic in lower vertebrates and invertebrates.

Although a few normally free-living genera such as *Micronema* and *Rhabditis* occasionally cause problems in animals, the only important genus from the veterinary point of view is *Strongyloides.*

Strongyloides

Members of this genus are common parasites of the small intestine in very young animals and, although generally of little pathogenic significance, under certain circumstances may give rise to a severe enteritis.

Hosts:
Most animals

Site:
Small intestine. Also caecum in poultry

Species:

Strongyloides westeri	Horses and donkeys
S. papillosus	Ruminants
S. ransomi	Pigs
S. stercoralis	Dogs and cats; man
S. avium	Poultry

Distribution:
Worldwide

IDENTIFICATION

Gross: Slender, hair-like worms generally less than 1.0 cm long.
Microscopic: Only females are parasitic. The long oesophagus may occupy up to one-third of the body length and the uterus is intertwined with the intestine giving the appearance of twisted thread. Unlike other intestinal parasites of similar size the tail has a blunt point (Fig. 46).

Strongyloides eggs are oval, thin-shelled and small, being half the size of typical strongyle eggs (Fig. 47). In herbivores it is the larvated egg which is passed out in the faeces but in other animals it is the hatched L_1.

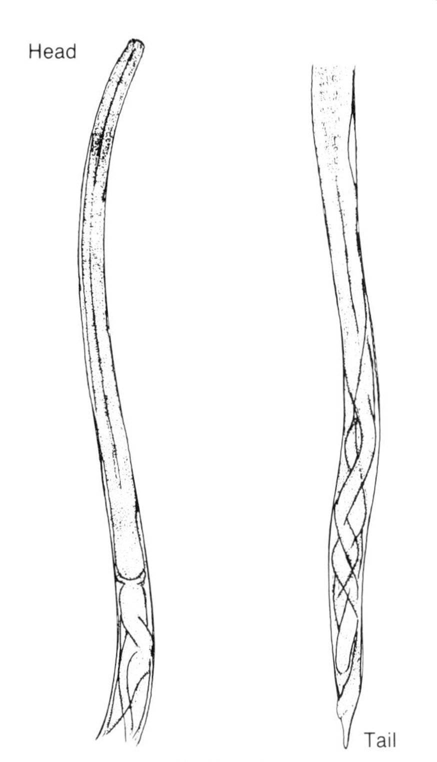

Fig 46 Adult *Strongyloides* female.

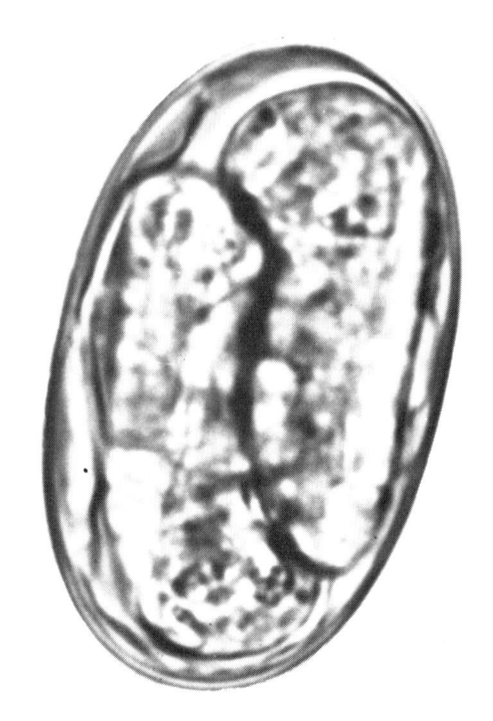

Fig 47 Egg of *Strongyloides westeri*.

LIFE CYCLE

Strongyloides is unique among the nematodes of veterinary importance being capable of both parasitic and free-living reproductive cycles. The parasitic phase is composed entirely of female worms in the small intestine and these produce larvated eggs by parthenogenesis, ie. development from an unfertilised egg. After hatching, larvae may develop through four larval stages into free-living adult male and female worms and this can be followed by a succession of free-living generations. However under certain conditions, possibly related to temperature and moisture, the L_3 can become parasitic infecting the host by skin penetration or ingestion and migrating via the venous system, the lungs and trachea to develop into adult female worms in the small intestine.

Foals, lambs and piglets may acquire infection immediately after birth from the mobilisation of arrested larvae in the tissues of the ventral abdominal wall of the dam which are subsequently excreted in the milk. In addition, prenatal infection has been demonstrated experimentally in pigs and cattle.

The prepatent period is from 8–14 days.

PATHOGENESIS

Skin penetration by infective larvae may cause an erythematous reaction which in sheep can allow the entry of *Bacteroides nodosus*, the causative organism of foot-rot. Passage of larvae through the lungs has been shown experimentally to result in multiple small haemorrhages visible over most of the lung surfaces.

Mature parasites are found in the duodenum and proximal jejunum and if present in large numbers may cause inflammation with oedema and erosion of the epithelium. This results in a catarrhal enteritis with impairment of digestion and absorption.

CLINICAL SIGNS

The common clinical signs usually seen only in very young animals, are diarrhoea, anorexia, dullness, loss of weight or reduced growth rate.

EPIDEMIOLOGY

Strongyloides infective larvae are not ensheathed and are susceptible to extreme climatic conditions. However warmth and moisture favour development and allow the accumulation of large numbers of infective stages. For this reason it can be a major problem in housed calves up to six months of age in some Mediterranean countries.

A second major source of infection for the very young animal is the reservoir of larvae in the tissues of their dams and this may lead to clinical strongyloidiasis in foals and piglets in the first few weeks of life. Successive progeny from the same dam often show heavy infections.

DIAGNOSIS

The clinical signs in very young animals, usually within the first few weeks of life, together with the finding of large numbers of the characteristic eggs or larvae in the faeces are suggestive of strongyloidiasis. It should be emphasised however that high faecal egg counts may be found in apparently healthy animals.

TREATMENT AND CONTROL

Specific control measures for *Strongyloides* infection are rarely called for. The benzimidazoles and ivermectin may be used for the treatment of clinical cases and a single dose of ivermectin 4–16 days prior to farrowing has been shown to suppress larval excretion in the milk of sows.

Rhabditis

Several members of this free-living genus of nematodes may become casual parasites, invading the skin and causing an intense pruritus. Cases have been most frequently reported in dogs housed in kennels with damp hay or straw bedding and the lesions, usually confined to areas of the body in contact with the ground, show hair loss, erythema, and pustule formation if infected with bacteria. The very small worms 1.0–2.8 mm in length with a rhabditiform oesophagus may be recovered from skin scrapings. Treatment is symptomatic and the condition can be prevented by housing animals on clean, dry bedding.

In East African cattle, otitis externa associated with *Rhabditis* infection has been reported.

Micronema

Occasional cases of infection of horses with the saprophytic, free-living nematode *Micronema deletrix* have been described from various parts of the world. In affected animals the very small worms, less than 0.5 mm in length, have been found in nasal and maxillary granulomata and in the brain and kidney. Severe nervous signs and death appear to accompany infection of the central nervous system.

[**Strongyloidiasis in man:** *S. stercoralis* occurs in man in warm climates. It produces diarrhoea, especially in young children, and in immunologically compromised adults may multiply within the host with fatal consequences. The dog may act as a natural host for this species.]

Superfamily ASCARIDOIDEA

The ascaridoids are among the largest nematodes and occur in most domestic animals, both larval and adult stages being of veterinary importance. While the adults in the intestine may cause unthriftiness in young animals, and occasional obstruction, an important feature of the group is the pathological consequences of the migratory behaviour of the larval stages.

With a few exceptions the genera have the following characters in common:

They are large, white opaque worms which inhabit the small intestine. There is no buccal capsule, the mouth consisting simply of a small opening surrounded by three lips. The common mode of infection is by ingestion of the thick-shelled egg containing the L_2. However, the cycle may involve transport and paratenic hosts.

Ascaris

Host:
Pig

Site:
Small intestine

Species:
Ascaris suum

Distribution:
Worldwide

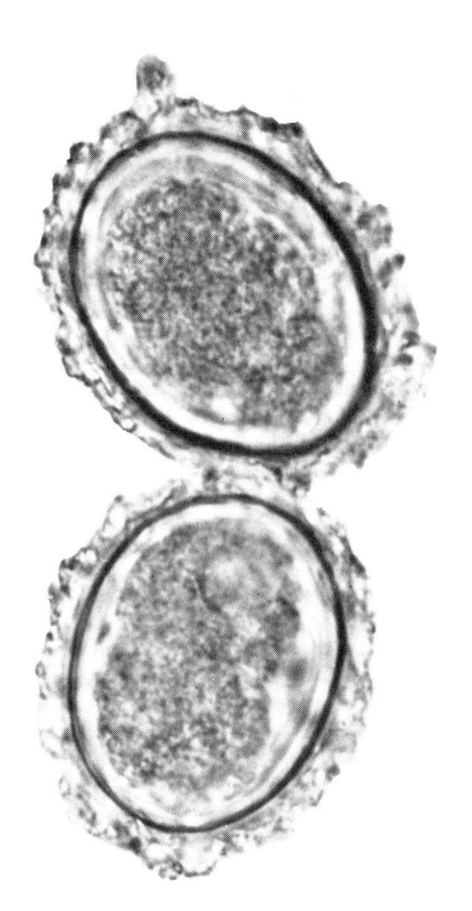

Fig 48 Eggs of *Ascaris suum.*

IDENTIFICATION

A. suum is by far the largest nematode of the pig; the females are up to 40.0 cm long, and there is no possibility of confusion with any other pig parasite.

The egg is ovoid and yellowish, with a thick shell, the outer layer of which is irregularly mamillated (Fig. 48).

LIFE CYCLE

The life cycle is direct. Though the single preparasitic moult occurs at about three weeks after the egg is passed, a period of maturation is necessary, and it is not infective until a minimum of four weeks after being passed, even in the optimal temperature range of 22–26 °C.

The egg is very resistant to temperature extremes, and is viable for more than four years.

After infection, the egg hatches in the small intestine and the L_2 travels to the liver, where the first parasitic moult takes place. The L_3 then passes in the bloodstream to the lungs and thence to the small intestine via the trachea. In the intestine the final two parasitic moults occur.

If the eggs are ingested by an earthworm or dung beetle they will hatch, and the L_2 travel to the tissues of these paratenic hosts, where they can remain, fully infective for pigs, for a long period.

The prepatent period is between 6–8 weeks, and each female worm is capable of producing more than 200,000 eggs per day.

PATHOGENESIS

The migrating larval stages in large numbers may cause a transient pneumonia, but it is now recognised that many cases of so-called '*Ascaris* pneumonia' may be attributable to other infections, or to piglet anaemia.

In the liver, the migrating L_2 and L_3 can cause 'milk spot' which appears as cloudy whitish spots of up to 1.0 cm in diameter, and represents the fibrous repair of inflammatory reactions to the passage of larvae in the livers of previously sensitised pigs (Pl. IV).

The adult worms in the intestine cause little apparent damage to the mucosa, but occasionally, if large numbers are present, there may be obstruction, and rarely a worm may migrate into the bile duct, causing obstructive jaundice and carcass condemnation.

Experimental infections have shown that in young pigs the important effect of alimentary ascariasis is economic, with poor feed conversion and slower weight gains, leading to an extension of the fattening period by 6–8 weeks.

CLINICAL SIGNS

The main effect of the adult worms is to cause production loss in terms of diminished weight gain. Otherwise, clinical signs are absent except in the occasional case of intestinal or biliary obstruction. In piglets under four months old, larval activity during the pulmonary phase of migration may cause a clinically evident pneumonia which is usually transient and rapidly resolving.

EPIDEMIOLOGY

A partial age immunity operates in pigs from about four months of age onwards, and this, coupled with the fact that the worms themselves have a limited life-span of several months, would suggest that the main source of infection is the highly resistant egg on the ground, a common characteristic of the ascaridoids. Hence 'milk spot', which is economically very important, since it is a cause of much liver condemnation, presents a continuous problem in some pig establishments. This condition has been widely noted to have a distinct seasonality of occurrence, appearing in greatest incidence in temperate areas during the warm summer months, and almost disappearing when the temperatures of autumn, winter and spring are too low to allow development of eggs to the infective stage.

A. suum may occasionally infect cattle, causing an acute, atypical, interstitial pneumonia, which may prove fatal. In most cases reported the cattle have had access to housing previously occupied by pigs, sometimes several years before, or to land fertilised with pig manure. In lambs, *A. suum* may also be a cause of clinical pneumonia as well as 'milk spot' lesions, resulting in condemnation of livers. In most cases lambs have been grazed on land fertilised with pig manure or slurry, such pasture remaining infective for lambs, even after ploughing and cropping. Young adults of *A. suum* are occasionally found in the small intestine of sheep.

There are a few recorded cases of patent *A. suum* infection in man.

DIAGNOSIS

Diagnosis is based on clinical signs, and in infections with the adult worm, on the presence in faeces of the yellow-brown ovoid eggs, with thick mamillated shells. Being dense, the eggs float more readily in saturated solutions of zinc sulphate or magnesium sulphate than in the saturated sodium chloride solution which is used in most faecal examination techniques.

TREATMENT

The intestinal stages are susceptible to most of the anthelmintics in current use in pigs, and the majority of these, such as the benzimidazoles, dichlorvos, or tetramisole are given in the feed. In cases of suspected *Ascaris* pneumonia injectable levamisole and ivermectin may be more convenient.

CONTROL

In the past, elaborate control systems have been designed for ascariasis in pigs, but with the appearance of highly effective anthelmintics these labour-intensive systems are falling into disuse.

The chief problem in control is the great survival capacity of the eggs, but in housed pigs, strict hygiene in feeding and bedding, with frequent hosing of walls and floors, will limit the risk of infection. In pigs on free range the problem is greater, and where there is serious ascariasis it may be necessary to discontinue the use of paddocks for several years, since the eggs can survive cultivation.

It is good practice to treat in-pig sows at entry to the farrowing pen, and on farms where ascariasis is prevalent young pigs should receive anthelmintic treatment at 5–6 weeks of age, and again 4 weeks later.

[**Ascariasis in man:** The type species, *Ascaris lumbricoides*, occurs in man, and at one time it was not differentiated from *A. suum*, so that the pig was thought to present a zoonotic risk for man. With morphological distinction now possible, *A lumbricoides* is accepted as specific for man, and is irrelevant to veterinary medicine.]

Toxocara

Though the members of this genus are in many respects typical ascaridoids, their biology is sufficiently varied for it to be necessary to consider each species separately.

Toxocara canis

Apart from its veterinary importance, this species is responsible for the most widely recognised form of visceral larva migrans in man.

Host:
Dog

Site:
Small intestine

Distribution:
Worldwide

IDENTIFICATION

Toxocara canis is a large white worm up to 10.0 cm in length, and in the dog can be confused only with *Toxascaris leonina*. Differentiation of these two species is difficult, as the only useful character, visible with a hand lens, is the presence of a small finger-like process on the tail of the male *T. canis*.

The egg is dark brown and subglobular, with a thick, pitted shell (Fig. 49).

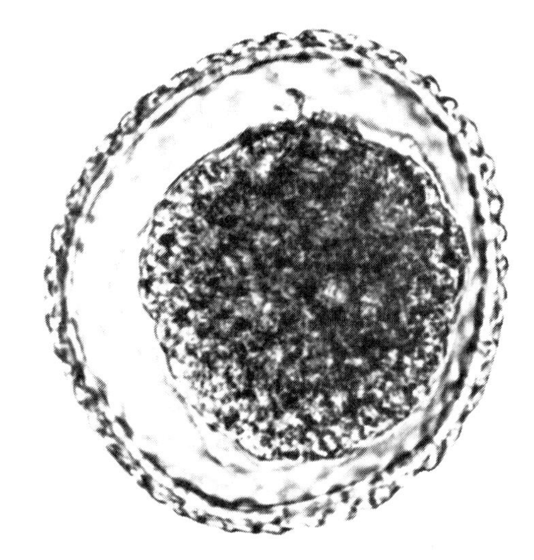

Fig 49 Egg of *Toxocara canis* showing thick pitted shell. Compare with that of *Toxascaris* in Fig. 51.

LIFE CYCLE

This species has the most complex life cycle in the superfamily, with four possible modes of infection:

The basic form is typically ascaridoid, the **egg containing the L_2** being infective, at optimal temperatures, four weeks after being passed. After ingestion, and hatching in the small intestine, the L_2 travel by the bloodstream via the liver to the lungs, where the second moult occurs, the L_3 returning via the trachea to the intestine where the final two moults take place. This form of infection occurs regularly only in dogs of up to three months old.

In dogs over three months of age, the hepatic-tracheal migration occurs less frequently, and at six months it has almost ceased. Instead, the L_2 travel to a wide range of tissues including the liver, lungs, brain, heart and skeletal muscle, and the walls of the alimentary tract. In the pregnant bitch, **prenatal infection** occurs, larvae becoming mobilised at about three weeks prior to parturition and migrating to the lungs of the foetus where they moult to L_3 just before birth. In the newborn pup the cycle is completed when the larvae travel to the intestine via the trachea, and the final moults occur. **A bitch, once infected, will usually harbour sufficient larvae to infect all her subsequent litters, even if she never again encounters the infection. A few of these mobilised larvae, instead of going to the uterus, complete the normal migration in the bitch, and the resulting adult worms produce a transient but marked increase in faecal *Toxocara* egg output in the weeks following parturition.**

The suckling pup may also be infected by ingestion of **L_3 in the milk** during the first three weeks of lactation. There is no migration in the pup following infection by this route.

Paratenic hosts such as rodents or birds may ingest the infective eggs, and the L_2 travel to their tissues where they remain until eaten by a dog when subsequent development is apparently confined to the gastrointestinal tract.

A final complication is recent evidence that bitches may be reinfected during late pregnancy or lactation, leading directly to transmammary infection of the suckling pups, and, once patency is established in the bitch, to contamination of the environment with eggs.

The known minimum prepatent periods are:

Direct infection following ingestion of eggs or larvae in a paratenic host:	4–5 weeks
Prenatal infection:	3 weeks.

PATHOGENESIS

In moderate infections, the larval migratory phase is accomplished without any apparent damage to the tissues, and the adult worms provoke little reaction in the intestine.

In heavy infections the pulmonary phase of larval migration is associated with pneumonia, which is sometimes accompanied by pulmonary oedema; the adult worms cause a mucoid enteritis, there may be partial or complete occlusion of the gut (Pl. IV) and, in rare cases, perforation with peritonitis or in some instances blockage of the bile duct.

CLINICAL SIGNS

In mild to moderate infections, there are no clinical signs during the pulmonary phase of larval migration. The adults in the intestine may cause pot-belly, with failure to thrive, and occasional diarrhoea. Entire worms are sometimes vomited or passed in the faeces.

The signs in heavy infections during larval migration result from pulmonary damage and include coughing, increased respiratory rate, and a frothy nasal discharge. Most fatalities from *T. canis* infection occur during the pulmonary phase, and pups which have been heavily infected transplacentally may die within a few days of birth.

Nervous convulsions have been attributed by some clinicians to toxocariasis, but there is still some disagreement on whether the parasite can be implicated as a cause of these signs.

EPIDEMIOLOGY

Surveys of *T. canis* prevalence in dogs have been carried out in most countries and have shown a wide range of infection rates, from 5% to over 80%. The highest prevalences have been recorded in dogs of less than six months of age, with the fewest worms in adult animals.

The widespread distribution and high intensity of infection with *T. canis* depend essentially on three factors:

First, the females are extremely fecund, one worm being able to contribute about 700 eggs to each gram of faeces per day, and egg counts of 15,000 epg are not uncommon in pups.

Second, the eggs are highly resistant to climatic extremes, and can survive for years on the ground.

Third, there is a constant reservoir of infection in the somatic tissues of the bitch, and larvae in these sites are insusceptible to most anthelmintics.

DIAGNOSIS

Only a tentative diagnosis is possible during the pulmonary phase of heavy infections when the larvae are migrating, and is based on the simultaneous appearance of pneumonic signs in a litter, often within two weeks of birth.

The eggs in faeces, subglobular and brown with thick pitted shells, are species-diagnostic. The egg production of the worms is so high that there is no need to use flotation methods, and they are readily found in simple faecal smears to which a drop of water has been added.

TREATMENT AND CONTROL

The adult worms are easily removed by anthelmintic treatment. The most popular drug used has been piperazine although this is being superseded by the benzimidazoles, fenbendazole and mebendazole and by nitroscanate.

A simple and frequently recommended regime for control of toxocariasis in young dogs is as follows:

All pups should be dosed at 2 weeks of age, and again 2 weeks later, to eliminate prenatally-acquired infection. It is also recommended that the bitch should also be treated at the same time as the pups.

A further dose should be given to the pups at two months old, to eliminate any infection acquired from the milk of the dam or from any increase in faecal egg output by the dam in the weeks following whelping.

Newly purchased pups should be dosed twice at an interval of 14 days.

Since there are likely to be a few worms present, even in adult dogs, in spite of the diversion of the majority of larvae to the somatic tissues, it is recommended that adult dogs should be treated every six months throughout their lives.

It has recently been shown that daily administration of high doses of fenbendazole to the bitch from three weeks pre-partum to three weeks post-partum has largely eliminated transmammary and prenatal infection of the pups, although residual infection in the brain may persist. This regimen may be useful in breeding kennels.

VISCERAL LARVA MIGRANS

Though this term was originally applied to invasion of the visceral tissues of an animal by parasites whose natural hosts were other animals, it has now, in common usage, come to represent this type of invasion in humans alone and, in particular, by the larvae of *Toxocara canis*. Its complementary term is cutaneous larva migrans, for infections by 'foreign' larvae which are limited to the skin.

The condition occurs most commonly in children who have had close contact with household pets, or who have frequented areas such as public parks where there is contamination of the ground by dog faeces. Surveys of such areas in many countries have almost invariably shown the presence of viable eggs of *T. canis* in around 10% of soil samples.

Despite this high risk of exposure to infection, the reported incidence of clinical cases is small. For example, in 1979 a French survey of the world literature reported that only 430 cases of ocular, and 350 cases of visceral, larva migrans had been recorded. However, it has been suggested that 50–60 clinical cases occur in Britain each year, since many are not recorded.

In many cases larval invasion is limited to the liver, and may give rise to hepatomegaly and eosinophilia, but on some occasions a larva escapes into the general circulation and arrives in another organ, the most frequently noted being the eye. Here, a granuloma forms around the larva on the retina, often resembling a retinoblastoma, and there have been cases of precipitate removal of the eye in children following misdiagnosis. Only in rare cases does the granuloma involve the optic disc, with total loss of vision, and most reports are of partial impairment of vision, with endophthalmitis or granulomatous retinitis. Such cases are currently treated using laser therapy. In a few cases of epilepsy, *T. canis* infection has been identified serologically, but the significance of the association has yet to be established.

Control of visceral larva migrans is based on the anthelmintic regimen described above, on the safe disposal of dog faeces in houses and gardens, and on the limitation of access by dogs to areas where children play, such as public parks.

Toxocara cati

Host:
Cat

Site:
Small intestine

Distribution:
Worldwide

IDENTIFICATION

Typically of the superfamily, *Toxocara cati* is a large white worm, often occurring as a mixed infection with the other ascaridoid of carnivores, *Toxascaris leonina*.

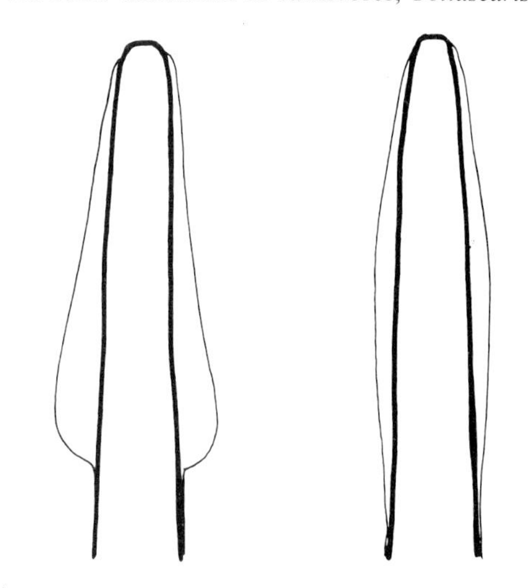

Fig 50 Differentiation of *Toxocara cati* and *Toxascaris leonina* may be made on the shape of the cervical alae.

Differentiation (Fig. 50) is readily made between the two on gross examination or with a hand lens, when the cervical alae of *T. cati* are seen to have an arrow-head form, with the posterior margins almost at a right angle to the body, whereas those of *Toxascaris* taper gradually into the body. The male, like that of *T. canis*, has a small finger-like process at the tip of the tail.

The egg, subglobular, with a thick, pitted shell and almost colourless, is characteristic in cat faeces.

LIFE CYCLE

Like *T. canis*, the life cycle of *T. cati* is migratory when infection occurs by ingestion of the L_2 in the egg and non-migratory after transmammary infection with L_3 or after ingestion of a paratenic host. However, unlike *T. canis* prenatal infection does not occur.

The prepatent period from egg infection is about eight weeks.

PATHOGENESIS AND CLINICAL SIGNS

Because the majority of infections are acquired either in the milk of the dam or by ingestion of paratenic hosts, there is no migratory phase and any changes are usually confined to the intestine, showing as pot-belly, diarrhoea, poor coat and failure to thrive.

EPIDEMIOLOGY

The epidemiology of *T. cati* depends largely on a reservoir of larvae in the tissues of the dam which are mobilised late in pregnancy and excreted in the milk throughout lactation. The paratenic host is also of considerable significance because of the strong hunting instinct in cats. Exposure to the latter route of infection does not occur until kittens begin to hunt for themselves or to share the prey of their dams.

DIAGNOSIS

The subglobular eggs, with thick, pitted shells, are easily recognised in faeces.

TREATMENT

This is similar to that described for *T. canis* in dogs.

CONTROL

Since infection is first acquired during suckling, complete control would be based on removal of kittens from the dam and artificial rearing. In most cases, adequate control is achieved by early and repeated administration of anthelmintics to kittens along the lines recommended for *T. canis* in pups.

T. cati has been reported as a rare cause of visceral larva migrans in man.

Toxocara vitulorum (syn. Neoascaris vitulorum)

Hosts:
Cattle and buffalo

Site:
Small intestine

Distribution:
Mainly in tropical and warm regions

IDENTIFICATION

T. vitulorum is the largest intestinal parasite of cattle, the females being up to 30.0 cm long. It is a thick worm, pinkish when fresh, and the cuticle is rather transparent so that the internal organs can be seen.

The egg is subglobular, with a thick pitted shell, and is almost colourless.

LIFE CYCLE

The life cycle of this species resembles that of *T. cati*, in that the most important source of infection is the milk of the dam in which larvae are present for up to 30 days after parturition. There is no tissue migration in the calf following infection and the prepatent period is 3–4 weeks.

The ingestion of larvated eggs by calves over 6 months old seldom results in patency, the larvae migrating to the tissues where they are stored; in female animals, resumption of development in late pregnancy allows further transmammary transmission.

PATHOGENESIS AND CLINICAL SIGNS

The main effects of this infection appear to be caused by the adult worms in the intestines of calves up to six months old. Heavy infections are associated with poor thriving and intermittent diarrhoea, and in buffalo calves particularly, fatalities may occur.

EPIDEMIOLOGY

The most important feature is the reservoir of larvae in the tissues of the cow, with subsequent milk-borne transmission ensuring that calves are exposed to infection from the first day of life.

DIAGNOSIS

The subglobular eggs, with thick, pitted shells, are characteristic in bovine faeces.

TREATMENT

The adult worms are susceptible to a wide range of

anthelmintics including piperazine, levamisole and the benzimidazoles. All these drugs are also effective against developing stages in the intestine.

CONTROL

The prevalence of infection can be dramatically reduced by treatment of calves at three and six weeks of age preventing developing worms reaching patency.

Toxascaris

This genus occurs in domestic carnivores, and though common, is of less significance than *Toxocara* because its parasitic phase is non-migratory.

Hosts:
Dog and cat

Site:
Small intestine

Species:
Toxascaris leonina

Distribution:
Worldwide

IDENTIFICATION

Toxascaris is almost indistinguishable grossly from *Toxocara canis*, the only point of difference being the absence of a finger-like process at the tip of the male tail of the latter. In the cat, differentiation from *T. cati* is based on the shape of the cervical alae, which are lanceolate in *Toxascaris* but arrow-head shaped in *T. cati* (Fig. 50).

The egg is slightly ovoid, with a smooth thick shell, and is characteristic in dog and cat faeces (Fig. 51).

LIFE CYCLE

Infection is by ingestion of the L_2 in the egg or as larvae in the tissues of mice and subsequent development takes place entirely in the wall and lumen of the intestine, there being no migratory phase. The prepatent period is around 11 weeks.

TREATMENT AND CONTROL

Since ascarid infections in the domestic carnivores invarably include *Toxocara*, the measures recommended for control of the latter will also have an effect on *Toxascaris*.

Since the two main reservoirs of infection are larvae in the prey or eggs on the ground, control has to be based on treatment of worm infection in the host animals, and on adequate hygiene to limit the possibility of acquisition of infection by ingestion of eggs.

Parascaris

Infection with *Parascaris equorum* is common throughout the world and is a major cause of unthriftiness in young foals.

Hosts:
Horses and donkeys

Site:
Small intestine

Species:
Parascaris equorum

Distribution:
Worldwide

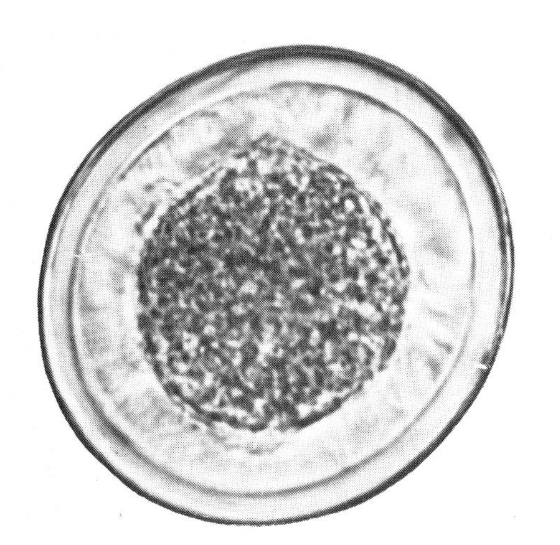

Fig 51 Egg of *Toxascaris leonina* showing smooth shell.

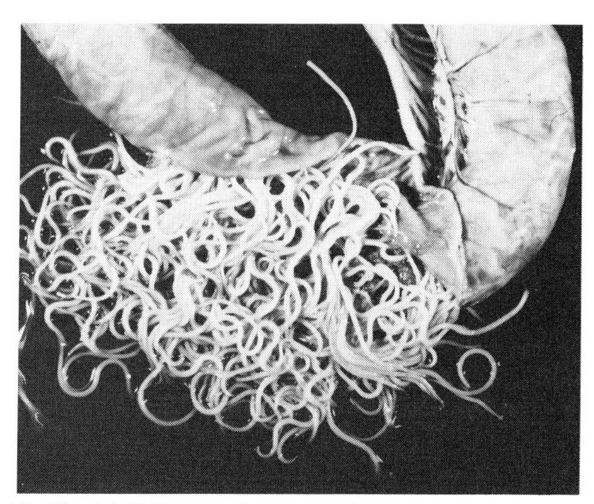

Fig 52 Adult *Parascaris equorum* in small intestinal contents.

IDENTIFICATION

Gross: This very large whitish nematode, up to 40 cms in length, cannot be confused with any other intestinal parasite of equines (Fig. 52).

Microscopic: The adult parasites have a simple mouth opening surrounded by three large lips and in the male the tail has small caudal alae.

The egg of *P. equorum* is almost spherical, brownish and thick-shelled with an outer pitted coat.

LIFE CYCLE

The life cycle is direct. Eggs produced by the adult female worms are passed in the faeces and can reach the infective stage containing the L_2 (Fig. 53) in as little as 10–14 days, although development may be delayed at low temperatures. After ingestion and hatching the larvae penetrate the intestinal wall and within 48 hours have reached the liver. By two weeks they have arrived in the lungs where they migrate up the bronchi and trachea, are swallowed, and return to the small intestine. The site of occurrence and timing of the parasitic larval moults of *P. equorum* are not precisely known, but it would appear that the moult from L_2 to L_3 occurs between the intestinal mucosa and the liver and the two subsequent moults in the small intestine.

The minimum prepatent period of *P. equorum* is 10 weeks. There is no evidence of prenatal infection.

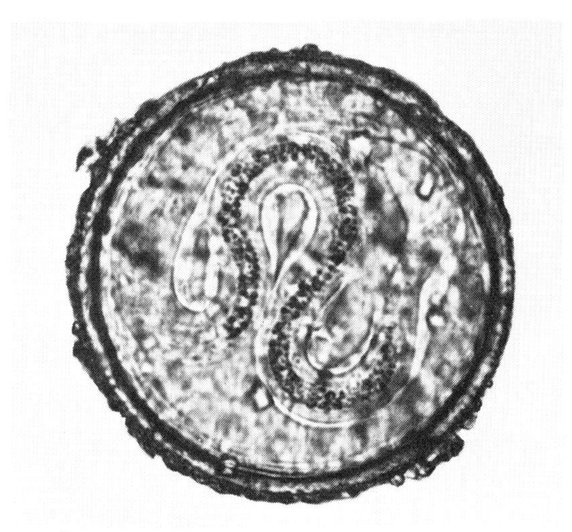

Fig 53 Egg of *P. equorum* which has developed to the infective stage.

PATHOGENESIS

Gross changes are provoked in the liver and lungs by migrating *P. equorum* larvae. In the liver, larvae cause focal haemorrhages and eosinophilic tracts which resolve leaving whitish areas of fibrosis. Larval migration in the lungs also leads to haemorrhage and infiltration by eosinophils which are later replaced by accumulations of lymphocytes, while sub-pleural greyish-green lymphocytic nodules develop around dead or dying larvae; these nodules are more numerous following reinfection.

Although the presence of worms in the small intestine is not associated with any specific lesions, occasionally, heavy infections have been reported as a cause of impaction and perforation leading to peritonitis. However, under experimental conditions, unthriftiness is a major sign and despite maintaining a good appetite infected foals lose weight and may become emaciated. Competition between a large mass of parasites and the host for nutrients may be the underlying cause of this weight loss.

CLINICAL SIGNS

During the migratory phase of experimental infections, up to four weeks following infection, the major signs are frequent coughing accompanied in some cases by a greyish nasal discharge although the foals remain bright and alert. Light intestinal infections are well tolerated, but moderate to heavy infections will cause unthriftiness in young animals with poor growth rates, dull coats and lassitude.

A wide variety of other clinical signs including fever, nervous disturbances and colic have been attributed to field cases of parascariasis, but these have not been observed in experimental studies.

EPIDEMIOLOGY

There are two important factors. First, the high fecundity of the adult female parasite, some infected foals passing millions of eggs in the faeces each day. Secondly, the extreme resistance of the egg in the environment ensures its persistence for several years. The sticky nature of the outer shell may also facilitate passive spread of eggs.

In the northern hemisphere, summer temperatures are such that many eggs become infective at a time when a population of susceptible foals is present. The infections acquired by these result in further contamination of pasture with eggs which may survive during several subsequent grazing seasons. Although mature horses may harbour a few adult worms, heavy burdens are usually confined to yearlings and to foals, which become infected from the first month or so of life, and infection is maintained largely by seasonal transmission between these groups of young animals.

DIAGNOSIS

This depends on clinical signs and the presence of spherical thick-shelled eggs on faecal examination. If disease due to prepatent infection is suspected, faecal examination having proved negative, diagnosis may be confirmed by administration of an anthelmintic when large numbers of immature worms may be observed in the faeces.

CONTROL

Anthelmintic prophylaxis for the horse strongyles will effectively control *P. equorum* infection. Since transmission is largely on a foal-to-foal basis it is good policy to avoid using the same paddocks for nursing mares and their foals in successive years.

Ascaridia

This is a non-migratory ascaridoid, and its appearance and biology are typical of the Superfamily.

Hosts:
Domestic and wild birds

Site:
Small intestine

Species:
Ascaridia galli

Two other species are *A. dissimilis* in turkeys and *A. columbae* in pigeons

Distribution:
Worldwide

IDENTIFICATION

The worms are stout and densely white, the females measuring up to 12.0 cm in length. *Ascaridia* is by far the largest nematode of poultry.

The egg is distinctly oval, with a smooth shell (Fig. 54), and cannot easily be distinguished from that of the other common poultry ascaridoid, *Heterakis*.

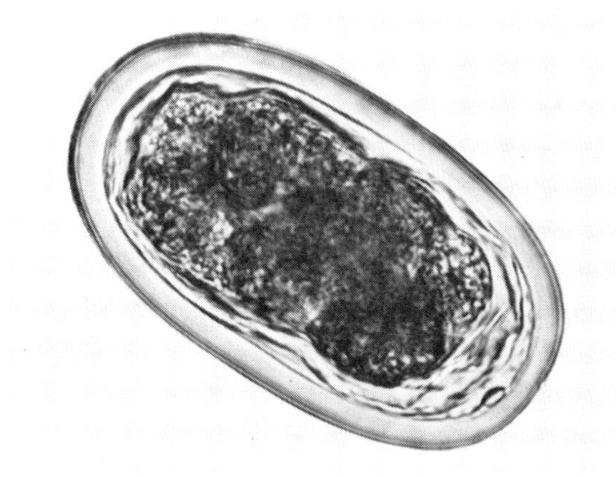

Fig 54 Smooth-shelled egg of *Ascaridia galli*.

LIFE CYCLE

The egg becomes infective at optimal temperatures in a minimum of three weeks and the parasitic phase is non-migratory. The egg is sometimes ingested by earthworms, which may act as transport hosts.

The prepatent period ranges from 5–6 weeks in chicks to eight weeks or more in adult birds. The worms live for about one year.

PATHOGENESIS AND CLINICAL SIGNS

Ascaridia is not a highly pathogenic worm, and any effects are seen in young birds, adults appearing relatively unaffected. The main effect is seen during the prepatent phase, when the larvae are in the mucosa. There they cause an enteritis which is usually catarrhal, but in very heavy infections may be haemorrhagic. In moderate infections the adult worms are tolerated without clinical signs, but when considerable numbers are present the large size of these worms may cause intestinal occlusion and death.

EPIDEMIOLOGY

Adult birds are symptomless carriers, and the reservoir of infection is on the ground, either as free eggs or in earthworm transport hosts.

DIAGNOSIS

In infections with adult worms, the eggs will be found in faeces, but since it is difficult to distinguish these from *Heterakis* eggs, confirmation must be made by postmortem examination of a casualty when the large white worms will be found. In the prepatent period, larvae will be found in the intestinal contents and in scrapings of the mucosa.

TREATMENT AND CONTROL

When birds are reared on a free-range system, and ascariasis is a problem, the young birds should, if possible, be segregated and reared on ground previously unused by poultry.

Since the nematode may also be a problem in deep litter houses, feeding and watering systems which will limit the contamination of food and water by faeces should be used.

In either case treatment with piperazine salts or levamisole is best administered in the drinking water, uptake being facilitated by withdrawal of water overnight before its substitution with medicated water in the morning.

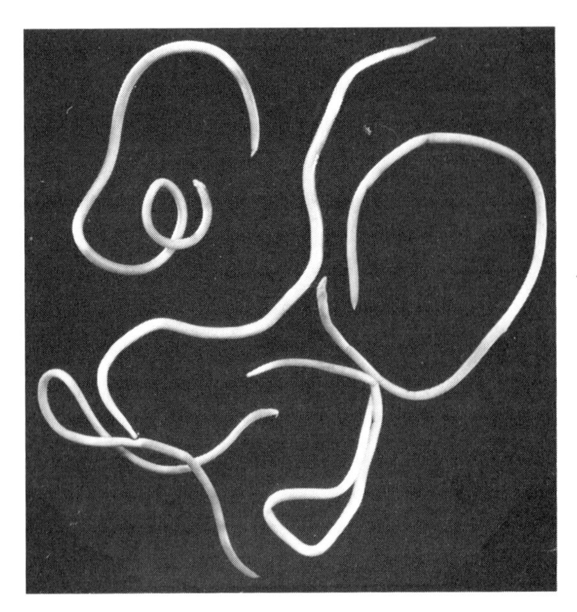

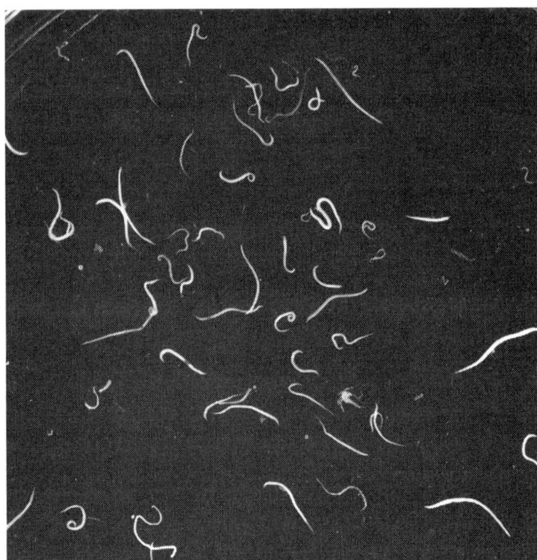

Fig 55 The two common poultry ascarids are *Ascaridia galli* (up to 12 cm long) and *Heterakis gallinarum* (up to 1.5 cm long).

Heterakis

This genus is exceptional in its small size and in its location in the large intestine, in contrast to *Ascaridia* which is large and inhabits the small intestine (Fig. 55).

Hosts:
Domestic and wild birds

Site:
Caeca

Species:
Heterakis gallinarum

Another species, *H. isolonche*, occurs in game birds, notably pheasants

Distribution:
Worldwide

IDENTIFICATION

Whitish worms up to 1.5 cm long, with elongated pointed tails. Gross examination readily indicates the genus, but for specific identification microscopic examination is necessary to demonstrate the spicules, which are unequal in length in *H. gallinarum*, but of equal length in *H. isolonche*. Microscopically, also, generic identity may be confirmed by the presence of a large precloacal sucker in the male, and prominent caudal alae supported by large caudal papillae.

The egg is ovoid and smooth-shelled, and is difficult to distinguish from that of *Ascaridia*.

LIFE CYCLE

The egg is infective on the ground in about two weeks at optimal temperatures. Earthworms may be transport hosts, the eggs simply passing through the gut, or paratenic hosts in which the egg hatches and the L_2 travels to the tissues to await ingestion by the fowl. In *H. gallinarum* all three parasitic moults appear to occur in the caecal lumen, but in *H. isolonche* infection the hatched larvae enter the caecal mucosa, and develop to maturity in nodules. Each nodule has an opening into the gut through which the eggs reach the lumen.

The prepatent period of the genus is about four weeks.

PATHOGENESIS AND CLINICAL SIGNS

H. gallinarum is the commonest nematode parasite of poultry, and is usually regarded as being non-pathogenic. Its chief pathogenic importance is as a vector of the protozoan, *Histomonas meleagridis*, the causal agent of 'blackhead' in turkeys. The organism can be transmitted from fowl to fowl in the egg of *Heterakis* and in earthworms containing hatched larvae of the worm.

H. isolonche of game birds is in itself pathogenic, causing a severe inflammation of the caeca with nodules projecting from both peritoneal and mucosal surfaces. These cause diarrhoea with progressive emaciation and there may be high mortality in heavily infected flocks.

EPIDEMIOLOGY

H. gallinarum is widespread in most poultry flocks and is of little pathogenic significance in itself, but is of great importance in the epidemiology of *Histomonas*. In contrast *H. isolonche* in game birds occurs as a clinical entity.

DIAGNOSIS

H. gallinarum infection is usually only diagnosed accidentally, by the finding of eggs in faeces or the presence of worms at necropsy.

H. isolonche infection is diagnosed at necropsy by the finding of caecal nodules containing adult worms, and if necessary, confirmed microscopically by examination of the spicules.

TREATMENT

Like *Ascaridia*, *Heterakis* is susceptible to piperazine and levamisole.

CONTROL

Control of *H. gallinarum* is only necessary when histomoniasis is a problem in turkeys. It is largely based on hygiene, and in backyard flocks two main points are: the segregation of turkeys from other domestic poultry, and the removal and disposal of litter from poultry houses. Where the problem is serious and continuous, it may be advisable to administer either piperazine or levamisole intermittently in the feed or water in addition to continuous *Histomonas* chemoprophylaxis.

Where *H. isolonche* infection is endemic in pheasantries, the runs should be abandoned and pheasant chicks reared on fresh ground.

Anisakid infection

The Anisakidae are ascaridoids whose adults are parasitic in a wide range of animals, including marine mammals and birds. The larvae occur in many fish which have ingested either the eggs or crustacean paratenic hosts carrying larvae. If the fish are eaten by humans, the larvae will migrate from the alimentary tract into other tissues, causing a form of visceral larva migrans which can be fatal. One out-break in the Netherlands involed the consumption of raw herring which harboured the larvae in their muscles, but the most widespread endemic cycle is usually recognised as being between seals and cod, and for this reason there is pressure in some fishing communities for the reduction of seal populations in order to diminish the economic loss resulting from rejection of fish at inspection.

Superfamily OXYUROIDEA

Adult oxyuroids of animals inhabit the large intestine and are commonly called pinworms because of the pointed tail of the female parasite. They have a double bulb oesophagus and a direct life cycle. The only genera of veterinary interest are *Oxyuris* and *Probstmayria*, both parasitic in the horse, and *Skrjabinema* which is a parasite of ruminants.

Oxyuris

Infection with the horse pinworm, *Oxyuris equi*, is extremely common and, although of limited pathogenic significance in the intestine, the female parasites may cause an intense anal pruritis during the process of egg-laying.

Hosts:
Horses and donkeys

Site:
Caecum, colon and rectum

Species:
Oxyuris equi

Distribution:
Worldwide

IDENTIFICATION

Gross: The mature females are large white worms with pointed tails which may reach 10.0 cms in length (Fig. 56) whereas the mature males are generally less than 1.0 cm long. *O. equi* L_4 vary from 5–10 mm in length, have tapering tails (Fig. 57) and are often attached orally to the intestinal mucosa.

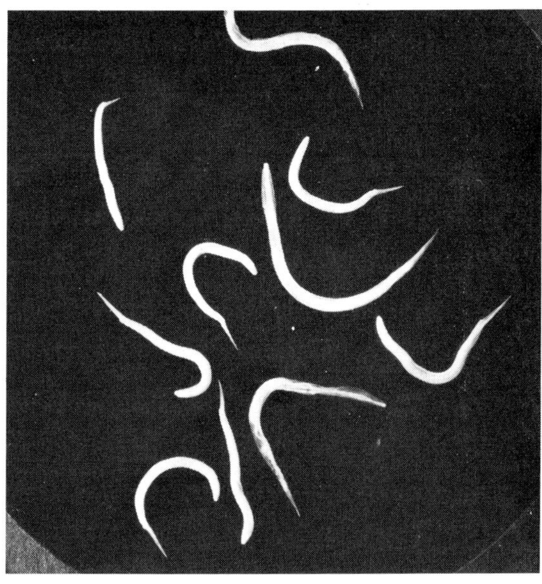

Fig 56 Adult *Oxyuris equi* females.

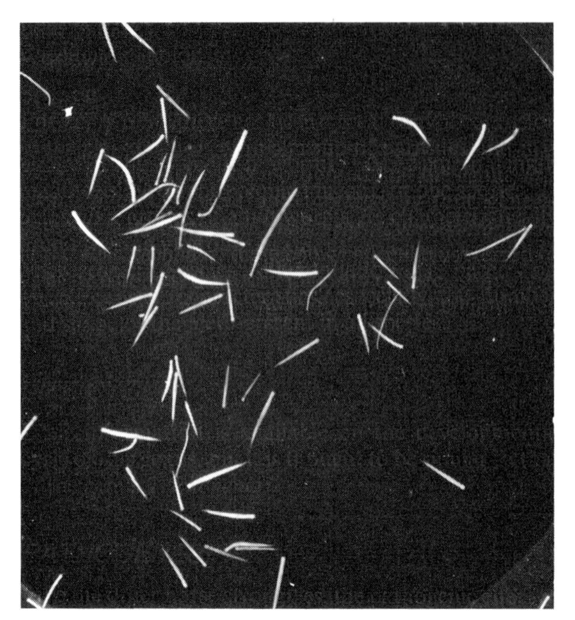

Fig 57 Fourth-stage larvae of *Oxyuris equi* which resemble small carpet tacks.

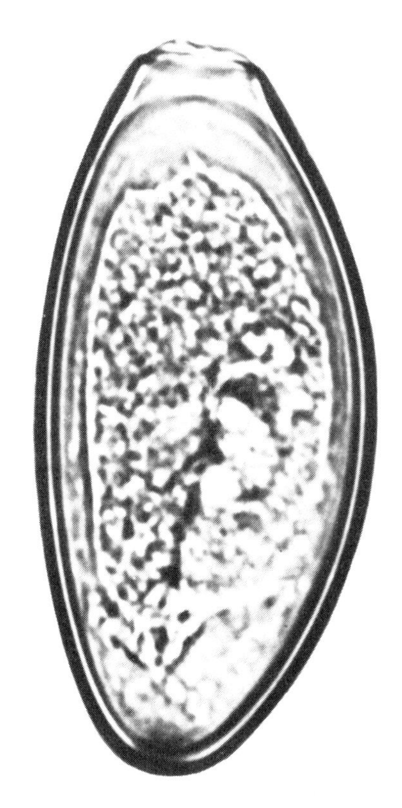

Fig 58 *Oxyuris equi* egg.

Microscopic: There is a double oesophageal bulb and the tiny males have caudal alae and a single spicule. In the female the vulva is situated anteriorly.

O. equi eggs are ovoid, yellow and slightly flattened on one side with a mucoid plug at one end (Fig. 58).

LIFE CYCLE

The adult worms are found in the lumen of the colon. After fertilisation the gravid female migrates to the anus, extrudes her anterior end and lays her eggs in clumps, seen grossly as yellowish white gelatinous streaks on the perineal skin. Development is rapid and within 4–5 days the egg contains the infective L_3. Infection is by ingestion of the eggs and the larvae are released in the small intestine, move into the large intestine and migrate into the mucosal crypts of the caecum and colon where development to L_4 takes place within 10 days. The L_4 then emerge and feed on the mucosa before maturing to adult stages which feed on intestinal contents.

The prepatent period of *O. equi* is 5 months.

PATHOGENESIS

Most of the pathogenic effects of *O. equi* in the intestine are due to the feeding habits of the L_4 which result in small erosions of the mucosa and, in heavy infections, these may be widespread and accompanied by an inflammatory response. Normally, a more important effect is the perineal irritation caused by the adult females during egg-laying.

CLINICAL SIGNS

The presence of parasites in the intestine rarely cause any clinical signs. However, intense pruritis around the anus causes the animal to rub, resulting in broken hairs, bare patches and inflammation of the skin over the rump and tail head.

EPIDEMIOLOGY

Although the infective stage may be reached on the skin, more often flakes of material containing eggs are dispersed in the environment by the animal rubbing on stable fittings, fencing posts or other solid objects. Heavy burdens may build up in horses in infected stables and there appears to be little immunity to reinfection.

DIAGNOSIS

This is based on signs of anal pruritis and the finding of greyish-yellow egg masses on the perineal skin. The large white long-tailed female worms are often seen in the faeces having been dislodged while laying their eggs.

O. equi eggs are rarely found on faecal examination of samples taken from the rectum, but may be observed in material from the perineum or in faecal material taken from the ground.

TREATMENT AND CONTROL

O. equi is susceptible to many broad spectrum anthelmintics and should be controlled by routine chemotherapy for the more important horse parasites such as the strongyles and *P. equorum*.

Where animals are showing clinical signs, the perineal skin and underside of the tail should be frequently cleaned using a disposable cloth, in addition to anthelmintic treatment. A high standard of stable hygiene should be observed.

Probstmayria

Probstmayria vivipara is a 2.0–3.0 mm long oxyuroid parasite which is unusual in that it is a perpetual parasite and lives from generation to generation in the equine caecum and colon. The females are viviparous and adults and larvae may be passed in the faeces; transmission is therefore probably via coprophagia. Although millions of these pinworms may be present they have never given rise to any clinical signs. Like *Oxyuris*, this parasite is susceptible to most modern anthelmintics.

Skrjabinema

This genus of small pinworms contains several species parasitic in the caecum and colon of domestic and wild ruminants. *Skrjabinema ovis*, for example, is a pinworm, up to 1.0 cm long, of goats and less commonly of sheep. It has rarely been incriminated as a cause of disease and is usually recognised only at necropsy.

[**Pinworms in man:** *Enterobius vermicularis*, the human pinworm, is common throughout the world and although it occurs in all age groups it is most prevalent in children. It is important to note that *E. vermicularis*, like other oxyuroids, is highly host-specific, so that there is no possibility of household pets becoming infected and acting as reservoirs of human pinworm infection.]

Superfamily SPIRUROIDEA

The precise classification of a number of genera currently assigned to this superfamily is controversial, but there are five of significance in veterinary medicine, *Spirocerca*, *Habronema*, *Draschia*, *Thelazia* and *Gnathostoma*. A major characteristic of this group is the tight spirally coiled tail of the male. The life cycles are indirect involving arthropod intermediate hosts.

Spirocerca

The adult nematodes are found in large granulomatous nodules in the wall of the oesophagus. These may cause a variety of clinical signs including, infrequently, those of oesophageal osteosarcoma.

Host:
Dog and occasionally cat

Intermediate hosts:
Coprophagous beetles

Site:
The migrating larvae produce characteristic lesions in the wall of the aorta while the adults are found in granulomatous lesions in the wall of the oesophagus and stomach

Species:
Spirocerca lupi

Distribution:
Tropical and subtropical areas

Fig 59 Large oesophageal granuloma due to *Spirocerca lupi* infection.

IDENTIFICATION

The appearance of the granulomatous lesions, up to golf-ball size, is usually sufficient for identification (Fig. 59). Numerous pink worms may be seen on section of the granulomas (Fig. 60), but these are difficult to extricate intact since they are coiled and up to 8.0 cm long.

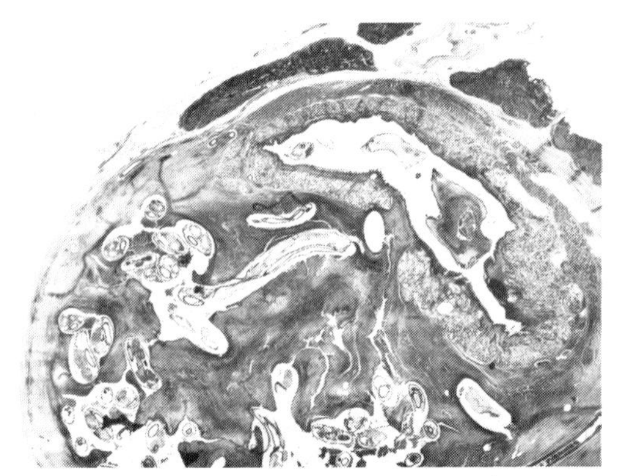

Fig 60 Transverse section showing *Spirocerca lupi* within a granuloma and in the oesophageal lumen.

LIFE CYCLE

The thick-shelled elongate egg, containing a larva, is passed in the faeces or vomit and does not hatch until ingested by a dung-beetle. In this, the intermediate host, the larva develops to the L_3 and encysts. Paratenic hosts may also be involved if the dung-beetle, in turn, is ingested by any of a variety of other animals including the domestic chicken, wild birds, mice and lizards. In these the L_3 becomes encysted in the viscera.

On ingestion of the intermediate or paratenic host by the final host the L_3 are liberated, penetrate the stomach wall and migrate via the coeliac artery to the thoracic aorta. About three months later they cross to the adjacent oesophagus where they provoke the development of granulomas as they develop to the adult stage in a further three months. The prepatent period is therefore six months. Eggs, however, may not be found in the faeces of a proportion of animals with adult infections where the granulomas have no openings into the oesophageal lumen.

PATHOGENESIS AND CLINICAL SIGNS

The migrating larvae produce scarring of the internal wall of the aorta which, if particularly severe, may cause stenosis or even rupture.

The oesophageal granulomas, up to 4.0 cm in size, associated with the adult worms may be responsible for a variety of clinical signs including dysphagia and vomiting arising from obstruction and inflammation.

Two further complications are, first, the development of oesophageal osteosarcoma in a small proportion of infected dogs. These may be highly invasive and produce metastases. Secondly, also relatively rare, is the occurrence of spondylosis of the thoracic vertebrae or of hypertrophic pulmonary osteoarthropathy of the long bones. The etiology of these lesions is unknown.

However, despite the potential pathogenicity of this parasite, many infected dogs do not exhibit clinical signs even when extensive aortic lesions and large, often purulent, oesophageal granulomas are present.

EPIDEMIOLOGY

In endemic areas the incidence of infection in dogs is often extremely high, sometimes approaching 100%. Probably this is associated with the many opportunities of acquiring infection from the variety of paratenic hosts.

DIAGNOSIS

Eggs may be found in the faeces or vomit if there are fistulae in the oesophageal granulomas. Otherwise diagnosis may depend on endoscopy or radiography.

TREATMENT

Diethylcarbamazine and disophenol have both been reported to be of value. Disophenol is given subcutaneously at 7 mg/kg body weight and the treatment is repeated after seven days. Diethylcarbamazine is administered orally at 20 mg/kg daily for 10 days. Neither drug is likely to be effective against immature worms.

CONTROL

This is difficult because of the ubiquity of the intermediate and paratenic hosts. Dogs should not be fed uncooked viscera from wild birds or from free-range domestic chickens.

Habronema

Members of this and the closely related genus *Draschia* are parasitic in the stomach of the horse: *Habronema* may cause a catarrhal gastritis, but is not considered an important pathogen while *Draschia* provokes the formation of large fibrous nodules which are occasionally significant. The chief importance of these parasites is as a cause of cutaneous habronemiasis or 'summer sores' in warm countries.

Hosts:
Horses and donkeys

Intermediate hosts:
Muscid flies

Site:
Stomach

Species:
Habronema muscae

H. microstoma (syn. *H. majus*)

Distribution:
Worldwide

IDENTIFICATION

Slender white worms 1.0–2.5 cm long. In the male the tail has a spiral twist. It is unlikely to be confused with other nematodes in the stomach since *Draschia* is associated with characteristic lesions and *T. axei* is less than 1.0 cm in length.

The elongated eggs are thin shelled and larvated when laid.

LIFE CYCLE

Eggs or L_1 are passed in the faeces and the L_1 are ingested by the larval stages of various muscid flies including *Musca*, *Stomoxys*, and *Haematobia* which are often present in faeces. Development to L_3 occurs synchronously with the development to maturity of the fly intermediate host. When the fly feeds around the mouth of the horse the larvae pass from its mouthparts on to the skin and are swallowed. Alternatively infected flies may be swallowed whole. Development to adult takes place in the glandular area of the stomach in approximately two months.

When *Habronema* larvae are deposited on a skin wound or around the eyes they invade the tissues, but do not complete their development.

PATHOGENESIS

The adults in the stomach may cause a mild catarrhal gastritis with excess mucus production. More important are the granulomatous lesions of cutaneous habronemiasis, commonly known as 'summer sores' (Pl. IV), and the persistent conjunctivitis with nodular thickening and ulceration of the eyelids associated with invasion of the eyes. Larvae have also been found associated with small lung abscesses.

CLINICAL SIGNS

These are usually absent in gastric habronemiasis. Lesions of cutaneous habronemiasis are most common in areas of the body liable to injury and occur during the fly season in warm countries. During the early stages, there is intense itching of the infected wound or abrasion which may cause further self-inflicted damage. Subsequently a reddish-brown non-healing granuloma develops which protrudes above the level of the surrounding skin and may be up to 8.0 cm in diameter. Later the lesion may become more fibrous and inactive, but will not heal until the advent of cooler weather when fly activity ceases. Invasion of the eye produces a persistent conjunctivitis with nodular ulcers especially at the medial canthus.

EPIDEMIOLOGY

The seasonality of cutaneous lesions is related to the activity of the fly vectors.

DIAGNOSIS

This is based on the finding of non-healing, reddish, cutaneous granulomas. The larvae, recognised by spiny knobs on their tails, may be found in material from these lesions. Gastric infection is not easily diagnosed since *Habronema* eggs and larvae are not readily demonstrable in the faeces by routine techniques.

TREATMENT AND CONTROL

A number of modern broad spectrum anthelmintics have been shown to have activity against the adult parasites in the stomach. Cutaneous lesions are difficult to treat, but ivermectin may prove effective. Symptomatic treatment and the use of insect repellents have some benefit and radiation therapy and cryosurgery have been used in more chronic cases. Obviously any measures taken to prevent injuries and to control fly populations will be beneficial.

Draschia

There is only one species of veterinary importance, *Draschia megastoma*, and this behaves in many ways like *Habronema*. However in the stomach the worms live in colonies around which develop large nodular lesions (Pl. IV). These occur in the fundus and seem to be well tolerated unless they protrude into the lumen sufficiently to interfere mechanically with stomach function. In all other respects *D. megastoma* can be considered similar to *Habronema*.

Parabronema

This genus in ruminants is the equivalent to *Habronema* in equines. It occurs in the abomasum of sheep and goats and has a wide distribution in Africa, Asia and some Mediterranean countries, notably Cyprus. The adult worms resemble *Haemonchus* spp. somewhat in gross form and size, but without the red spiral coloration, while the younger worms are closer to *Ostertagia* in appearance. Microscopically, the genus is readily distinguished from the other abomasal worms by the

IDENTIFICATION

Slender white worms 3.0–6.0 cm in length. Anteriorly, there are numerous papillae and circular ridges in the cuticle. In the female the vulva is situated anteriorly near the simple mouth opening.

Small embryonated eggs are laid on the skin surface where they hatch to release the microfilariae or L_1 which are about 200 μm in length.

LIFE CYCLE

Eggs or larvae present in exudates from bleeding points in the skin surface are ingested by muscid flies, for example *M. autumnalis* in Europe and *M. lusoria* and *M. xanthomelas* in Africa, in which they develop to L_3 within several weeks to months, depending on air temperature.

Transmission occurs when infected flies feed on lachrymal secretions or skin wounds in other cattle and the L_3 deposited then migrate and develop to the adult stage under the skin in 5–7 months (Fig. 64). Bleeding points develop 7–9 months after infection.

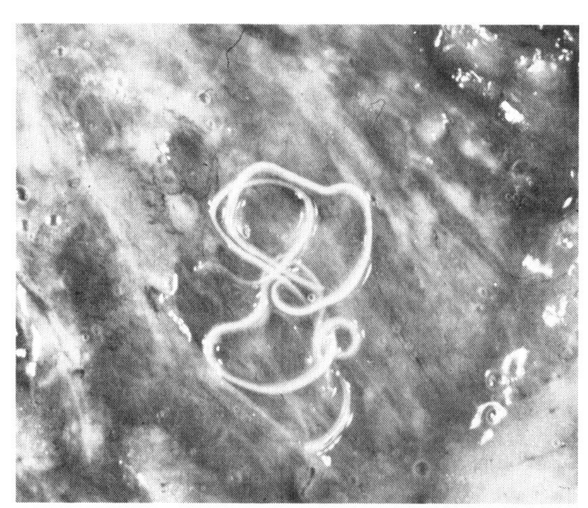

Fig 64 *Parafilaria bovicola* in the subcutaneous tissues.

PATHOGENESIS

When the gravid female punctures the skin to lay her eggs there is a haemorrhagic exudate or 'bleeding point' which streaks and mats the surrounding hairs and attracts flies. Individual lesions only bleed for a short time and healing is rapid.

At the sites of infection, which are predominantly on the shoulders, withers and thoracic areas, there is inflammation and oedema which, at meat inspection, resemble subcutaneous bruising in early lesions and have a gelatinous greenish-yellow appearance with a metallic odour in longer standing cases. Sometimes the lesions extend into the intermuscular fascia. The affected areas have to be trimmed at marketing and further economic loss is incurred by rejection or downgrading of the hides.

CLINICAL SIGNS

The signs of parafilariasis are pathognomonic. Active bleeding lesions are seen most commonly in warm weather, an apparent adaptation to coincide with the presence of the fly intermediate host.

EPIDEMIOLOGY

In Europe, bovine parafilariasis occurs in spring and summer, whereas in tropical areas it is seen mainly after the rainy season. A high prevalence of 36% in cattle has been reported from some endemic areas in South Africa and recently the disease has been diagnosed in Sweden, an area previously free from infection. *Parafilaria* infection may be introduced by the importation of cattle from endemic areas, but its spread will depend on the presence of specific fly vectors. It has been estimated in Sweden that one 'bleeding' cow will act as a source of infection for three other animals.

DIAGNOSIS

This is normally based on clinical signs, but if laboratory confirmation is required, the small embryonated eggs or microfilariae may be found on examination of exudate from bleeding points. The demonstration of eosinophils in smears taken from lesions is also considered a constant diagnostic feature. In Sweden, serodiagnosis using an ELISA technique has been developed.

TREATMENT

Patent infections in beef and non-lactating dairy cattle may be treated with ivermectin or nitroxynil. The former is given parenterally as a single dose whereas two doses of nitroxynil are required at an interval of three days. Neither drug is licensed for use in lactating cattle when the less effective levamisole may be tried.

These drugs produce a marked reduction in bleeding points and, due to resolution of the muscle lesions, a significant reduction in meat condemnation if slaughter is delayed for 70 days after treatment.

CONTROL

This is difficult, because of the long prepatent period during which drugs are thought not to be effective. In Sweden dairy cattle, and particularly heifers at pasture, are the main source of infection for *M. autumnalis*, which is an outdoor fly, active in spring and summer. However, infections in young beef cattle are the chief cause of economic loss through carcass damage.

Since neither ivermectin nor nitroxynil is effective against immature worms, treatment is only useful for patent infections recognisable by the clinical signs. However, because of restrictions on the use of ivermectin and nitroxynil in lactating cows, these are rarely treated and instead are kept indoors during the period of fly activity.

In endemic areas young beef cattle may be treated with an anthelmintic some time before slaughter as described above. In Sweden the use of insecticide-impregnated ear tags is recommended for vector control.

Parafilaria multipapillosa

This parasite, similar in appearance to *P. bovicola*, is transmitted by *Haematobia* spp. and occurs in the subcutaneous and intermuscular connective tissue of horses. It has been recorded from Asia, Africa, Europe and South America. In Britain it has only been recognised in imported horses.

The lesions are more nodular than those of *P. bovicola* in cattle and their distribution in the harness areas may make the animals unsuitable for work.

Clinically the condition is characterised by matting of the hair (Fig. 65) due to blood and tissue fluid exudates from ruptured nodules. The lesions are more prominent in the summer and particularly when the animals are hot, so that they appear to be 'sweating blood'.

Although the condition tends to disappear in cold weather it will periodically reappear during warmer months for up to four years in individual animals. Occasionally, lesions are mistaken for injuries caused by thorns and barbed wire. Treatment is difficult, but ivermectin may be tried.

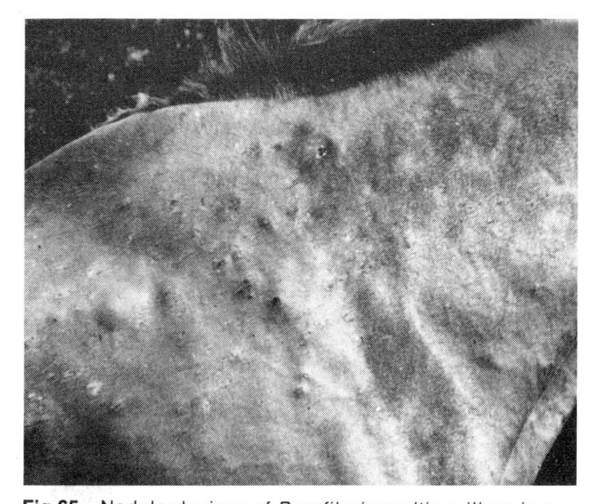

Fig 65 Nodular lesions of *Parafilaria multipapillosa* in a horse.

Stephanofilaria

The worms of this genus inhabit the dermis and are responsible for chroic dermatitis in cattle and buffalo in the tropics and subtropics.

Hosts:
Cattle and buffalo

Intermediate hosts:
Muscid flies

Site:
Skin; depending on the species, different regions of the body are preferred

SPECIES AND DISTRIBUTION

Stephanofilaria stilesi – lower abdomen – U.S.A., U.S.S.R.
S. assamensis – hump of zebu – Asia

In India and the Far East *S. zaheeri*, *S. kaeli* and *S. okinawaensis* occur mainly on the head, legs and teats

IDENTIFICATION

Very small worms, less than 1.0 cm in length. Microscopically the mouth opening is surrounded by a spiny collar.

LIFE CYCLE

The fly vectors are attracted to the open lesions in the skin caused by the adult parasites, and ingest the microfilariae in the exudate. Development to L_3 takes about three weeks, and the final host is infected when the flies deposit larvae on normal skin.

PATHOGENESIS AND CLINICAL SIGNS

Lesions begin to appear within two weeks of infection. In the case of *S. stilesi*, the flies congregate on the shady underside of the abdomen and it is in this area that the most severe damage occurs; in contrast, the lesion due to *S. assamensis* is commonly termed 'hump sore'. In all species, the lesions are usually localised to the preferred biting areas of the vectors. The skin is at first nodular, but later there is papular eruption with an exudate of blood and pus. In the centre of the lesion there may be sloughing of the skin, but at the margin there is often hyperkeratosis. The condition is essentially an exudative, often haemorrhagic, dermatitis which attracts the fly vectors.

EPIDEMIOLOGY

In endemic areas the incidence of infection may be as high as 90% and the occurrence is to a great extent influenced by the type of herbage. Succulent grazing produces soft, moist faeces which are more suitable breeding sites for the flies than the hard crumbly faeces deposited on sparse dry grazing. Hence irrigation of pasture may result in an increase of stephanofilariasis.

Though the lesions subside in cooler weather, the damage to the hide is permanent and may result in considerable economic loss. Milk yield may be severely diminished from the pain of the lesions and the irritation of cattle by the flies.

DIAGNOSIS

Though adult worms and microfilariae are present in the lesions they are often scarce and many scrapings prove negative. Diagnosis is therefore usually presumptive in endemic areas, and is based on the appearance and site of the lesions.

TREATMENT

Organophosphorus compounds applied topically as an ointment have proved effective. Ivermectin is likely to be effective.

CONTROL

This is rarely feasible because of the ubiquity of the vectors, but would have to be based on the use of insecticides or repellents.

Dirofilaria

Of the two species occurring in domestic carnivores, one, *Dirofilaria immitis*, is by far the more important. The adults which are found in the right side of the heart and adjacent blood vessels of dogs are responsible for a debilitating condition known as canine heartworm disease. Although primarily a problem of warm countries where the mosquito intermediate host abounds, the disease has become much more widespread in the past decade and the problem in North America is now so extensive that special heartworm clinics have been created.

Dirofilaria immitis

Hosts:
Dog, occasionally cat and rarely man

Intermediate hosts:
Mosquitoes

Site:
Cardiovascular system; adults in the right ventricle, pulmonary artery and posterior vena cava

DISTRIBUTION

Essentially in warm-temperate and tropical zones throughout the world including southern Europe and Canada. It is only found in imported dogs in Britain.

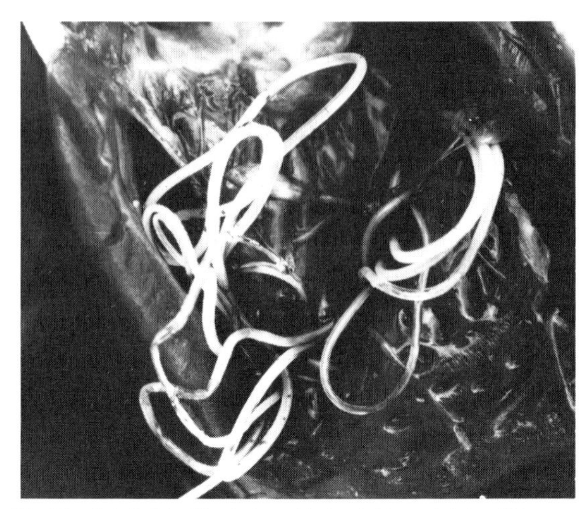

Fig 66 Adult *Dirofilaria immitis* in right ventricle of the heart.

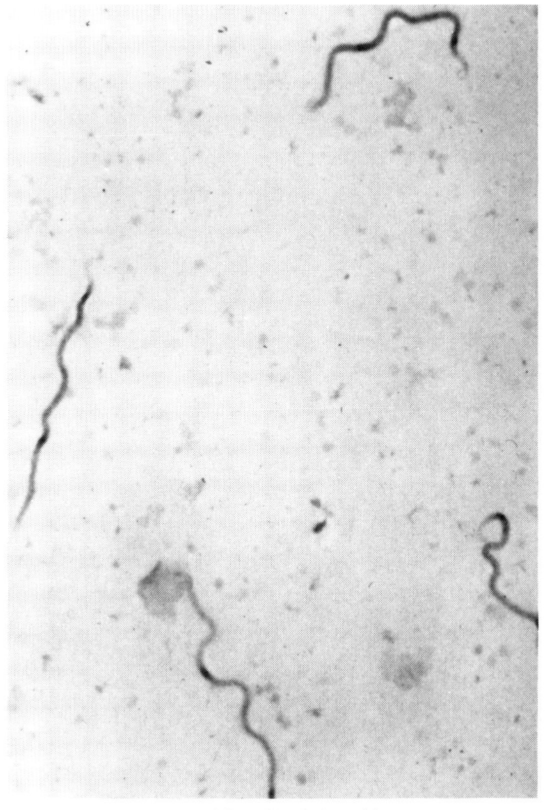

Fig 67 Microfilariae of *Dirofilaria immitis*.

IDENTIFICATION

Gross: Long slender worms 20–30 cm long (Fig. 66). The male tail has the typical loose spiral common to the filarioids. The size and site are diagnostic for *D. immitis*.
Microscopic: The microfilariae in the blood are not ensheathed and are 307–332 μm in length by 6.8 μm wide (Fig. 67). They have a tapered anterior end and blunt posterior end.

LIFE CYCLE

The adults live in the heart and adjacent blood vessels and the females release microfilariae directly into the bloodstream. These are ingested by female mosquitoes during feeding. Development to L_3 in the mosquito takes about two weeks, by which time the larvae are present in the mouthparts and the final host is infected when the mosquito takes a further blood meal. In the dog the L_3 migrate to the subcutaneous or subserosal tissues and undergo two moults over the next few months; only after the final moult do the young *D. immitis* pass to the heart via the venous circulation. The minimum prepatent period is six months. The adult worms survive for several years and patency has been recorded for over five years.

PATHOGENESIS

This is associated with the adult parasites. Many dogs infected with low numbers of *D. immitis* show no apparent ill-effects and it is only in heavy infections that circulatory distress occurs, primarily due to obstruction to normal blood flow leading to chronic congestive right-sided heart failure. The presence of a mass of active worms can cause an endocarditis in the heart valves and a proliferative pulmonary endarteritis, possibly due to a response to parasite excretory products. In addition, dead or dying worms may cause pulmonary embolism. After a period of about nine months the effect of the developing pulmonary hypertension is compensated for by right ventricular hypertrophy which may lead to congestive heart failure with the usual accompanying signs of oedema and ascites. At this stage the dog is listless and weak.

A mass of worms may lodge in the posterior vena cava and the resulting obstruction leads to an acute, sometimes fatal, syndrome known as the vena caval syndrome. This is characterised by haemolysis, haemoglobinura, bilirubinaemia, icterus, anorexia and collapse. Death may occur within 2–3 days. Very occasionally there is blockage of the renal capillaries by microfilariae leading to a glomerulonephritis, possibly related to the deposition of immune complexes.

CLINICAL SIGNS

Heavily infected dogs are listless and there is a gradual loss of condition and exercise intolerance. They have a chronic soft cough with haemoptysis and in the later stages of the disease become dyspnoeic and may develop oedema and ascites.

The acute vena caval syndrome described above is characterised by haemoglobinuria, icterus and collapse.

Lighter infections in working dogs may be responsible for poor performance during periods of sustained exercise.

EPIDEMIOLOGY

The important factors in the spread of heartworm disease can be divided into those affecting the host and those affecting the vector.

Host factors include a high density of dogs in areas where the vectors exist, the lengthy patent period of up to five years during which time circulating microfilariae are present, and the lack of an effective immune response against established parasites.

Vector factors include the ubiquity of the mosquito intermediate hosts, their capacity for rapid population increase and the short development period from microfilariae to L_3.

DIAGNOSIS

This is based on the clinical signs of cardiovascular dysfunction and the demonstration of the appropriate microfilariae in the blood. Affected dogs are seldom less than one year old and most are over two years. In suspected cases in which the microfilariae cannot be demonstrated thoracic radiography may show the thickening of the pulmonary artery, its tortuous course and right ventricular hypertrophy. Angiography may also be used to demonstrate more clearly the vascular changes.

In the near future, it is likely that simple serological kits, based on the ELISA test, will become available.

The identification of the microfilariae in the blood is aided by concentrating the parasites following lysis, filtration and then staining with methylene blue. Commercial kits are available for this technique. Alternatively one part of blood and nine parts of formalin are centrifuged and the sediment mixed with a blue stain and examined as a microscopic smear. The microfilariae have to be differentiated from those of *Dipetalonema reconditum*, a filarial parasite commonly found in the subcutis in dogs. Those of *D. immitis* are more than 300 μm in length and have a tapered head and a straight tail; those of *D. reconditum* are less than 300 μm length, and have a blunt head and a hooked posterior end. More precise differentiation may be achieved by using histochemical stains for acid phosphatase activity. *D. immitis* show distinct red acid-phosphate positive spots at the excretory pore and anus, while *D. reconditum* stains pink overall.

TREATMENT

Treatment should not be undertaken without a physical examination of the dog and an assessment of heart, lung, liver and kidney function. Where these functions are grossly abnormal it may be necessary to give prior treatment for cardiac insufficiency. The usual recommendation is that infected dogs are first treated intravenously with thiacetarsamide twice per day over a three day period to remove the adult worms; toxic reactions are not uncommon following this treatment due to the dying heartworms and resultant embolism; activity of the dog should be restricted for a period of 2–6 weeks. This drug should be used with extreme care.

A further treatment with a different drug is then given six weeks later to remove the microfilariae which are not susceptible to thiacetarsamide treatment. Several drugs are now available for this purpose; the traditional one was dithiazanine and either this or levamisole given orally over a 10–14 day period have proved effective. However, ivermectin achieves the same effect following a single treatment, but is unlicensed at present because of occasional toxic reactions in dogs at the usual therapeutic dose of 200 μg/kg.

With all of these drugs there is a risk of adverse reactions to dying microfilariae.

In some severe cases, heartworms have been removed surgically rather than risk adverse reactions following drug therapy.

Following treatment it is usual to place dogs on a prophylactic programme and this is considered under Control.

CONTROL

Mosquito control is difficult and therefore prophylaxis is based almost entirely on medication. The drug widely used for this has been diethylcarbamazine, which in endemic areas is given orally to pups daily from three months old. This kills developing larvae and so preempts the problems of treating patent infections and microfilaraemia. In tropical areas the drug is given all year round, but in more temperate zones, where the mosquito has a limited season, treatment commences one month prior to the mosquito season and ceases two months after it ends. Where prophylaxis is introduced in older dogs or after treatment of an infected dog, care must be exercised to ensure that the dog is free from microfilarial infection as anaphylactoid reactions may occur in infected dogs after diethylcarbamazine treatment. Once prophylaxis is introduced regular checks for microfilariae should be made every six months.

Ivermectin, at very low doses, has been shown to provide prophylaxis for at least one month after a single treatment and monthly therapy with this drug is likely to supplant daily therapy with diethylcarbamazine.

Dirofilaria repens

This parasite of the dog and cat is of little pathogenic significance. The adults are found in subcutaneous tissues and the microfilariae in the blood and lymph. It occurs in the Mediterranean basin, the Middle East, Africa and Asia.

Dipetalonema

Several species of *Dipetalonema* transmitted mainly by ticks and fleas occur in the subcutis of dogs in tropical and subtropical zones. They are not especially pathogenic except for causing occasional cutaneous ulceration and subcutaneous abscesses.

Dipetalonema reconditum often occurs in the same endemic area as *D. immitis* and the presence of its microfilariae may lead to misdiagnosis on blood examination. The morphological differences have been described in the description of *D. immitis*.

Onchocerca

Though onchocerciasis is an important filarial infection in human medicine, most species in domestic animals are relatively harmless.

Hosts:
Equines and cattle

Intermediate hosts:
Culicoides spp. and *Simulium* spp.

Sites:
Fibrous tissue, usually of ligaments and intermuscular connective tissue; one species occurs in the bovine aorta

SPECIES AND DISTRIBUTION

Several species occur in equines and cattle and they are conveniently discussed according to their host.

IDENTIFICATION

The slender worms range from 2.0–6.0 cm in length and lie tightly coiled in tissue nodules. In active lesions the presence of worms is readily established on section of these nodules.

LIFE CYCLE

The life cycle of *Onchocerca* spp. is typically filarioid, with the exception that the microfilariae occur in the tissue spaces of the skin, rather than in the peripheral bloodstream.

EQUINE ONCHOCERCIASIS

The single species *O. reticulata* (syn. *O. cervicalis*) has a worldwide distribution and occurs commonly in the ligamentum nuchae, and less frequently in the suspensory ligaments and flexor tendons of the lower limbs.

The ligamentum nuchae in the region of the withers is the preferential site. Following inoculation of L_3 by the vector, *Culicoides* (*C. nubeculosus* being most often incriminated), the arrival of the parasites in their final site results in host reaction in the form of a painless, diffuse swelling, which gradually increases in size to become a palpable soft lump, and then regresses to leave a calcified focus, the skin over the area remaining intact. Open purulent lesions, commonly called 'fistulous withers', may occur, but though *O. reticulata* has been found in these, there is no clear causal relationship between the worms and the condition, and it is thought that bacteria, possibly including *Brucella abortus*, are more likely to be involved.

In the lower limbs the reaction to the presence of the parasite is similar to that in the ligamentum nuchae, with a soft painless swelling succeeded by small fibrous nodules.

The general prevalence of equine onchocerciasis is high, most surveys in U.S.A. having shown rates of more than 50% though the highest so far recorded in Britain is 23%. Apart from the initial mild reaction no clinical signs attributable to the adult worms have been demonstrated. However in the U.S.A. a ventral midline dermatitis has been described apparently associated with the feeding of the hornfly *Haematobia irritans* on the skin over the linea alba, the predilection site of the microfilariae of *O. cervicalis*.

BOVINE ONCHOCERCIASIS

The main features of the species occurring in cattle are summarised below:

Species	Site	Distribution	Vector
O. gutturosa (syn. *O. lienalis*)	ligamentum nuchae and gastro-splenic ligament	Worldwide	*Simulium* spp.
O. gibsoni	subcutaneous and intermuscular nodules	Africa, Asia, Australasia	*Culicoides* spp.
O. armillata	wall of thoracic aorta	Middle East, Africa, India	Unknown

Lesser species include *O. dukei* in subcutaneous and muscle nodules and *O. ochengi* and *O. sweetae* in intradermal tissue. A further species, *O. stilesi*, found in the connective tissue of the stifle joint in cattle in U.S.A., is only recently described and has still to be generally accepted.

Depending on the site, the various species in cattle are associated with different changes. *O. gutturosa*, in the large ligaments, is of little clinical or economic importance; *O. gibsoni*, provoking a fibrous reaction in muscle tissue (Pl. IV), can be responsible for economic loss due to carcass trimming; *O. ochengi* and *O. sweetae* in the skin cause some economic loss from blemished hides. The nodules in muscle provoked by *O. dukei*, though of little importance in themselves, assume significance in some areas of Africa where they may be confused with *Cysticercus bovis* at meat inspection.

It is interesting that *O. armillata*, though occurring in a strategically important site in the bovine aorta, is never associated with clinical signs; it is usually only discovered at the abattoir, surveys in the Middle East having shown a prevalence as high as 90%. The worms are found in grossly visible nodules in the intima, media, and adventitia of the aorta, and atheromatous plaques are commonly seen on the intima (Fig. 68). Aortic aneurysms have been noted in about a quarter of infections.

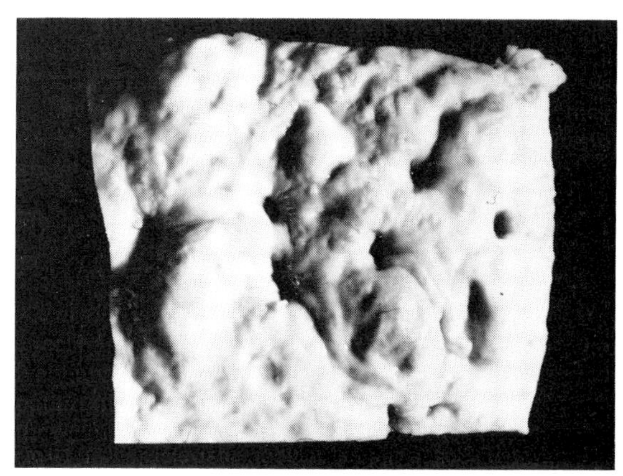

Fig 68 Confluent nodular aortic lesions due to *Onchocerca armillata* infection.

DIAGNOSIS

This is rarely called for and depends on the finding of microfilariae in skin biopsy samples. In most species the microfilariae are concentrated in the preferred feeding sites of the vectors, which for *Simulium* spp. and *Culicoides* spp. are usually the shaded lower parts of the trunk, and it is usually recommended that samples should be taken from the region of the linea alba. The piece of skin is placed in warm saline and teased to allow emergence of the microfilariae, and is then incubated for six hours or more. The microfilariae are readily recognised by their sinuous movements in a centrifuged sample of the saline.

TREATMENT

In the past this has consisted of daily administration of diethylcarbamazine over a period as a microfilaricide, but it now appears that a single dose of ivermectin is highly efficient in this respect, although the dying microfilariae may provoke local tissue reactions. In the case of equine ventral midline dermatitis local treatment with synthetic pyrethroids controls hornflies and aids resolution of the lesions.

CONTROL

With the ubiquity of the insect vectors there is little possibility of efficient control, though the use of microfilaricides will reduce the numbers of infected flies. In any case, with the relatively innocuous nature of the infection there is unlikely to be any demand for control.

Setaria

The members of this genus are usually harmless inhabitants of the peritoneal and pleural cavities.

Hosts:
Ruminants and equines

Intermediate hosts:
Many species of mosquito

Site:
Usually the peritoneal surface and free in the peritoneal cavity; less commonly the pleural cavity and, following erratic migration, the CNS

Species:

Setaria labiato-papillosa (syn. *S. digitata*)	cattle and wild ruminants
S. equina	horses and donkeys

Distribution:
Worldwide

IDENTIFICATION

Long slender worms, up to 12.0 cm in length. The site and gross appearance are sufficient for generic identification.

LIFE CYCLE

The microfilariae in the bloodstream are taken up by mosquitoes in which development to L_3 takes about 12 days. The prepatent period is 8–10 months.

PATHOGENESIS

The worms in their normal site are harmless and are only discovered at necropsy. *S. labiato-papillosa* may have an erratic migration in sheep and goats and enter the spinal canal causing cerebrospinal nematodiasis, 'lumbar paralysis', which is irreversible and often fatal; the condition has only been reported in the Middle and Far East.

CLINICAL SIGNS

There are no clinical signs when the worms are in their normal site, but when nervous tissue is involved there is locomotor disturbance, usually of the hind limbs, and if the parasites are high in the spinal canal there may be paraplegia.

EPIDEMIOLOGY

Since the worms are usually innocuous their epidemiology has received little study. The prevalence is higher in warmer countries, where there is longer seasonal activity of the mosquito vectors.

DIAGNOSIS

Infection with the adult worms is only accidentally discovered in the living animal by the finding of microfilariae in routine blood smears. In cases of cerebrospinal nematodiasis confirmatory diagnosis is only possible by microscopic examination of the spinal cord, since the parasites exist only as larval forms in their aberrant site.

TREATMENT

There is no treatment for setarial paralysis. Ivermectin has been reported to be effective against adult *S. equina*.

CONTROL

This would depend on control of the mosquito vectors, which is unlikely to be applied specifically for this parasite.

Elaeophora

These worms inhabit large blood vessels, but are only of local importance.

Hosts:
Ruminants and equines

Intermediate hosts:
Tabanid flies

Species	Hosts	Site	Distribution
Elaeophora schneideri	sheep, goat	carotid, mesenteric and iliac arteries	Southern U.S.A.
E. bohmi	equines	veins and arteries	Europe (Austria)
E. poeli	cattle, buffalo	aortic intima	Africa, Asia

IDENTIFICATION

Slender worms, up to 12.0 cm long.

LIFE CYCLE

Only the life cycle of *E. schneideri* has been studied, and this is typically filarioid. The microfilariae are ingested by the tabanid in feeding, and the L_3 when developed, are released into the wound made when the insect next feeds. Early development appears to be in the meningeal arteries, and the worms are mature and producing microfilariae about four and a half months from infection.

The adult worms are embedded in the arterial intima of blood vessels with only the anterior part of the female free in the lumen.

PATHOGENESIS

In cattle nodules, from which the worms protrude, form on the intima of the vessels but in other domestic animals the adults appear to provoke little reaction. In *E. schneideri* infection in sheep the circulating microfilariae are associated with a facial dermatitis which appears in the summer months; in severe cases there may be self-injury from rubbing, with abrasion, bleeding and scab formation. It is thought that the natural hosts of *E. schneideri* are deer, in which the infection is asymptomatic, and that sheep may be abnormal hosts.

CLINICAL SIGNS

Only the seasonal facial dermatitis in sheep is recognised as a clinical indication of elaeophoriasis.

EPIDEMIOLOGY

Because of the innocuous nature of the infection in cattle and equines, the distribution of the species in these hosts is not completely known. *Elaeophora schneideri* is distributed over the southern and western states of the U.S.A. The natural hosts appear to be deer of *Odocoileus* spp., the white-tail and the mule deer, and in these the infection is clinically inapparent. However, in American elk (*Cervus canadensis*) thrombosis due to the worms often results in necrosis of the muzzle, ears, and optic nerves, resulting in severe facial damage, blindness and death.

DIAGNOSIS

Only in sheep is diagnosis required, and though the obvious method is by examination of a skin biopsy, microfilariae are often scarce in samples, and diagnosis is usually presumptive, based on the locality, the facial lesions, and the seasonal appearance of the dermatitis.

TREATMENT

Repeated administration of diethylcarbamazine is effective, but the risk of fatalities from the presence of dead worms in the arteries should be recognised.

[**Filariasis in man:** Though they are probably the most important group of helminth infections in humans, these filarioidiases are of only marginal concern to the veterinarian, since domestic animals are of little significance in their epidemiology. The following are the most important species:
(a) *Onchocerca volvulus*. Human onchocerciasis due to *O. volvulus* occurs around the world in the equatorial zone, and is transmitted by *Simulium* spp. The adult worms live in subcutaneous nodules, and almost the entire pathogenic effect is caused by the microfilariae; dermatitis and elephantiasis are common, but the most important effect is 'river blindness', so-called because of its distribution along the habitats of *Simulium* spp. It has been estimated that in Africa there are about 20 million people affected by onchocerciasis. The only other animals to which it is transmissible are the higher primates, chimpanzee and gorilla.
(b) *Brugia* spp. are carried by many species of mosquito and occur in Asia, notably in Malaysia, causing elephantiasis. The most important species, *B. malayi*, is infective for monkeys and wild carnivores, and has been transmitted experimentally to the cat and dog. The lesser species occurring in man, *B. pahangi*, has a reservoir in many species of wild animals, and is also transmissible to the domestic cat.
(c) *Wuchereria bancrofti* is also mosquito-borne and causes elephantiasis in Africa, Asia and South America. It is exclusive to man.
(d) *Loa loa* is transmitted by *Chrysops* spp., and occurs in West and East Africa, where it causes the transient subcutaneous enlargements known as 'Calabar swellings'. It is confined to man, apes and monkeys.
(e) *Mansonella ozzardi*, carried by *Culicoides* spp. and

Simulium spp., occurs in the Caribbean, and in Central and South America. It lives in the fat and on the mesentery, and is usually considered to be nonpathogenic, though recently it has been associated with allergic signs. The prevalance is extremely high in endemic areas, where parasites closely resembling *M. ozzardi* are commonly found in monkeys and in horses and cattle. There is, however, reluctance to presume that these animals may be reservoir hosts until positive identification is made.]

Superfamily TRICHUROIDEA

The members of this superfamily are found in a wide variety of domestic animals. A common morphological feature is the 'stichosome' oesophagus which is composed of a capillary-like tube surrounded by a single column of cells.

There are three genera of interest:

The first, *Trichuris*, is found in the caecum and colon of mammals; the second, *Capillaria*, is most commonly present in the alimentary or respiratory tract of mammals or birds. Both lay eggs with plugs at both poles.

The adults of the third genus, *Trichinella*, are found in the small intestine of mammals and produce larvae which immediately invade the tissues of the same host.

Trichuris

The adults are usually found in the caecum but are only occasionally present in sufficient numbers to be clinically significant.

SPECIES AND HOSTS

The common species are:

Trichuris ovis	sheep and goat
T. globulosa	cattle
T. suis	pig
T. vulpis	dog

Less common are *T. discolor* in cattle, *T. skrjabini* in sheep, goat and camel and *T. serrata* in the cat

Site:
Large intestine, particularly the caecum

Distribution:
Worldwide

IDENTIFICATION

Gross: The adults are 4.0–6.0 cm long with a thick posterior end tapering rapidly to a long filamentous anterior end which is characteristically embedded in the mucosa. Because of their appearance the members of this genus are often called the 'whipworms'.

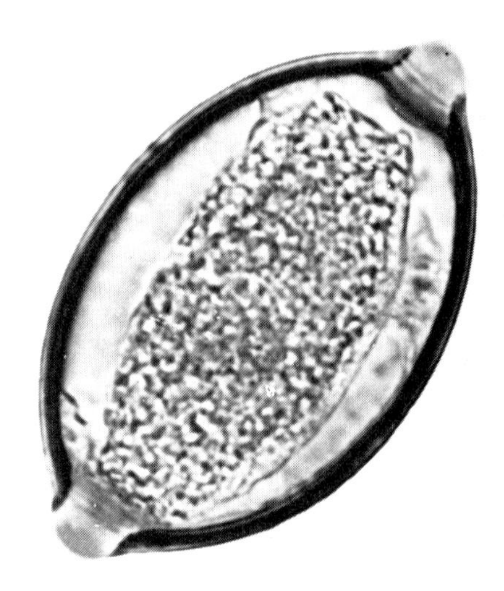

Fig 69 Typical appearance of egg of *Trichuris* species.

Microscopic: The male tail is coiled and possesses a single spicule in a sheath; the female tail is merely curved. The characteristic eggs are lemon shaped with a conspicuous plug at both ends; in the faeces these eggs appear yellow or brown in colour (Fig. 69).

LIFE CYCLE

The infective stage is the L_1 within the egg which develops in one or two months of being passed in the faeces depending on the temperature. Under optimal conditions these may subsequently survive for several years.

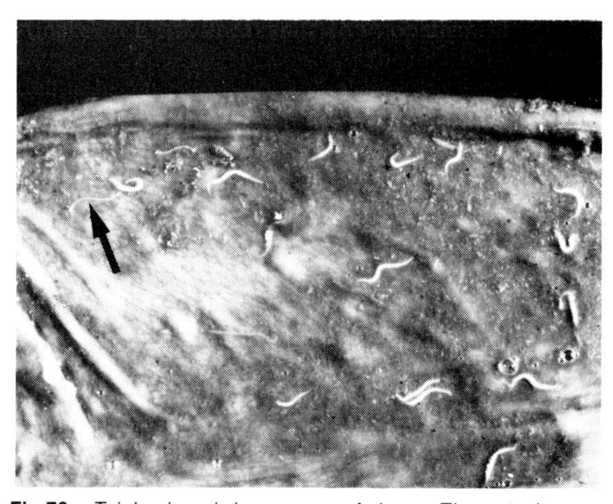

Fig 70 *Trichuris ovis* in caecum of sheep. The anterior end of the worm is usually embedded in the mucosa but is visible in the arrowed specimen.

After ingestion, the plugs are digested and the released L_1 penetrate the glands of the caecal mucosa. Subsequently all four moults occur within these glands, the adults emerging to lie on the mucosal surface with their anterior ends embedded in the mucosa (Fig. 70). The prepatent period ranges from 6–12 weeks depending on the species.

PATHOGENESIS

Most infections are light and asymptomatic. Occasionally when large numbers of worms are present they cause a diphtheritic inflammation of the caecal mucosa. This results from the subepithelial location and continuous movement of the anterior end of the whipworm as it searches for blood and fluid. In pigs, heavy infections are thought to facilitate the invasion of potentially pathogenic spirochaetes.

CLINICAL SIGNIFICANCE

Despite the fact that ruminants and to a lesser extent pigs have a high incidence of light infections, the clinical significance of this genus, especially in ruminants, is generally negligible although isolated outbreaks have been recorded. Sporadic disease due to heavy infections is more common in pigs and dogs and is associated with watery diarrhoea which usually contains blood.

EPIDEMIOLOGY

The most important feature is the longevity of the eggs which after three or four years may still survive as a reservoir of infection in piggeries or in kennels. On pasture this is less likely since the eggs tend to be washed into the soil.

DIAGNOSIS

Since the clinical signs are not pathognomonic, diagnosis may depend on finding numbers of *Trichuris* eggs in the faeces. However, since clinical signs may occur during the prepatent period, diagnosis in food animals may depend on necropsy and in dogs on a favourable response to anthelmintic treatment.

TREATMENT

In ruminants the pro-benzimidazoles, the modern benzimidazoles, ivermectin or levamisole by injection are very effective against adult *Trichuris*, but less so against larval stages. In pigs, these drugs or dichlorvos may be used while in the dog, mebendazole and dichlorvos are the drugs of choice.

CONTROL

Prophylaxis is rarely necessary, particularly in ruminants, but in the case of pigs or dogs attention should be given to areas where eggs might continue to survive for long periods. Such areas should be thoroughly cleaned and disinfected or sterilised by wet or dry heat.

[**Trichuriasis in man.** *Trichuris trichiura*, the whipworm of man, is morphologically indistinguishable from *T. suis*. However, it is generally considered that these two parasites are strictly host specific, although there have been reports of patent human infections with *T. suis*, after experimental ingestion of eggs.]

Capillaria

These very thin hair-like worms are not readily visible to the naked eye in unprepared gut contents. Although there are many species in mammals and birds only those in the latter are of general veterinary significance.

IDENTIFICATION

These are very fine filamentous worms between 1.0 to 5.0 cm long, the narrow stichosome oesophagus occupying half the body length. The males have a long thin single spicule and often possess a primitive bursa-like structure; the females contain eggs which resemble those of *Trichuris* in possessing bipolar plugs, but which are more barrel-shaped and colourless (Fig. 71).

Because of the large number of species, hosts and predilection sites and the fact that some of these parasites have a direct and others an indirect life cycle the important species in birds and mammals are summarised separately.

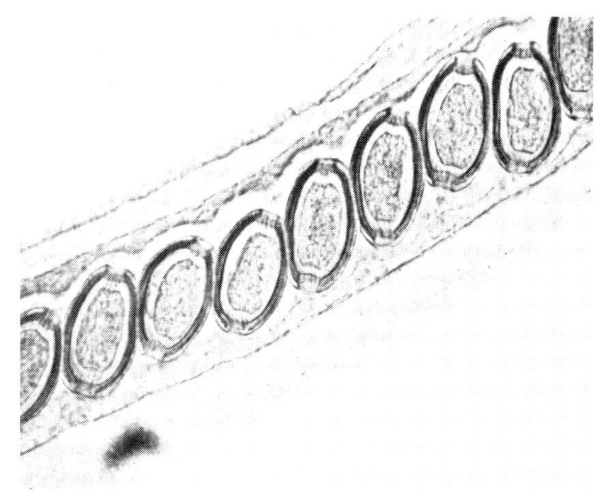

Fig 71 Typical barrel-shaped *Capillaria* eggs within the uterus of a female worm.

Capillaria in birds

Capillaria obsignata is present in the upper small intestine of chickens, turkeys and pigeons and has a direct life cycle. The infective L_1 develops within the egg in about a week.

C. caudinflata is also found in the small intestine of chickens and turkeys. The egg of this species requires to be ingested by an earthworm in which it hatches, the final host being infected by ingestion of the earthworm.

C. contorta (syn. *C. annulata*) occurs in the oesophagus and crop of the chicken, turkey, duck and wild birds and an earthworm intermediate host is also essential.

The prepatent period of these three avian species is 3–4 weeks.

PATHOGENESIS AND CLINICAL SIGNIFICANCE

Like *Trichuris* the anterior ends of the parsite are buried in the mucosa and in heavy infections cause diphtheritic inflammation leading to inappetence and emaciation, and in the intestinal infection, diarrhoea. Mortality in such cases may be high. There is also evidence that light infections of less than one hundred worms may cause poor weight gains and lowered egg production.

EPIDEMIOLOGY

Young birds are most susceptible to *Capillaria* infections while adults may serve as carriers. *C. obsignata* is perhaps most important since, having a direct life cycle, it occurs indoors in birds kept on deep litter. The epidemiology of the other two species is largely based on the ubiquity of the earthworm intermediate host.

DIAGNOSIS

Because of the non-specific nature of the clinical signs and the fact that, in heavy infections, these may appear before *Capillaria* eggs are present in the faeces, diagnosis depends on necropsy and careful examination of the oesophagus, crop or intestine for the presence of the worms. This may be carried out by microscopic examination of mucosal scrapings squeezed between two glass slides; alternatively the contents of the suspected organ should be gently washed through a fine sieve and the retained material resuspended in water and examined against a black background.

TREATMENT

Levamisole in the drinking water is highly effective.

CONTROL

Where the species of *Capillaria* involved have an earthworm intermediate host, complete control may be achieved by housing the birds after anthelmintic treatment. Otherwise, control depends on regular anthelmintic treatment accompanied if possible by moving the birds to fresh ground.

In the case of *C. obsignata* scrubbing and heat treatment of affected surfaces is essential as is the provision of fresh litter in chicken houses. Periodic treatment of the flock with an anthelmintic will also be of value.

Capillaria in mammals

C. aerophila is found embedded in the mucosa of the trachea, bronchi and nasal passages of foxes and occasionally dogs and cats. The parasite has a direct life cycle. The eggs can survive for months, and after ingestion, the larvae penetrate the intestine and migrate in the bloodstream to the lungs. The prepatent period is around 6 weeks.

The clinical signs are those of rhino-tracheitis and in this respect are similar to those caused by *Filaroides* or *Crenosoma* infection.

C. plica is common in the urinary bladder of foxes, dogs and more rarely cats. This parasite apparently requires an earthworm intermediate host, but it is rarely of pathogenic significance.

C. hepatica is primarily a parasite of wild rodents, but occurs occasionally in the dog, cat, and man. The predilection site is the liver and the eggs are laid in the parenchyma from which there is no natural access to the exterior. Infection is acquired by ingestion of either the liver, following predation, cannibalism or carrion feeding, or eggs on the ground which have been freed by decomposition of the host. The lesion in domestic animals and man is a granulomatous reaction to the egg masses in the liver, followed by cirrhosis. Infections are rarely fatal and most are discovered at routine necropsy.

Trichinella

The only member of this genus is *Trichinella spiralis*, a nematode with a very wide host range and the cause of an important worldwide zoonosis.

Hosts:
Most mammals. From the zoonotic aspect, the pig and man are the important hosts

Sites:
The adults occur in the small intestine and their larvae in the striated muscles; the diaphragmatic, intercostal and masseter muscles are considered to be predilection sites

Distribution:
Worldwide, with the apparent exceptions of Australia and Denmark

IDENTIFICATION

Because of their short lifespan, the adult worms are rarely found in natural infections.

The male is about 1.0 mm long, the oesophagus is at least one third of the total body length and the tail has two small cloacal flaps, but no spicule. The female is 3.0 mm long and the uterus contains developing larvae.

Trichinella infection is most easily identified by the presence of coiled larvae in striated muscle (Fig. 72).

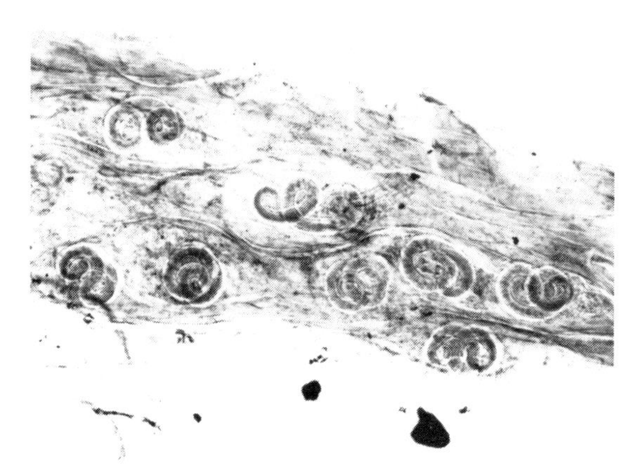

Fig 72 Muscle infected with *Trichinella spiralis* larvae.

LIFE CYCLE

The developing adults lie between the villi of the small intestine. After fertilisation, the males die while the females burrow deeper into the villi. Three days later, they produce L_1 which enter the lymphatic vessels and travel via the bloodstream to the skeletal muscles. There, still as L_1, they penetrate muscle cells where they are encapsulated by the host, grow and assume a characteristic coiled position; the parasitised muscle cell is sometimes known as a 'nurse cell'. This process is complete within seven weeks by which time the larvae are infective and may remain so for years.

Development is resumed when the larvae are ingested by another host, usually as a result of predation or carrion-feeding. The L_1 is liberated, and in the intestine undergoes four moults to become sexually mature within two days. Patent infections persist for only a few weeks at the most.

PATHOGENESIS AND CLINICAL SIGNS

Infection in domestic animals is invariably light, and clinical signs do not occur. However when hundreds of larvae are ingested, as occasionally happens in man and presumably also in predatory animals in the wild, the intestinal infection is often associated with enteritis, and 1–2 weeks later the massive larval invasion of the muscles causes acute myositis, fever, eosinophilia and myocarditis; periorbital oedema and ascites are also common in man. Unless treated with an anthelmintic and anti-inflammatory drugs, such infections may frequently be fatal, but in persons who survive this phase the clinical signs start to abate after 2–3 weeks.

It is important to realise that trichinosis is basically an infection of animals in the wild and that the involvement of man in these circumstances is accidental.

EPIDEMIOLOGY

The epidemiology of trichinosis depends on two factors. First, animals may become infected from a wide variety of sources, predation and cannibalism being perhaps the most common. Others include feeding on carrion, since the encapsulated larvae are capable of surviving for several months in decomposing flesh, and the ingestion of fresh faeces from animals with a patent infection. It is also thought that transport hosts such as crustaceans and fish, feeding on drowned terrestrial animals, may account for infection in some aquatic mammals such as seals.

The second factor is the wide host range of the parasite. In temperate areas rodents, brown bear, badger and wild pig are most commonly involved; in the arctic, polar bear, wolf and fox; in the tropics, lion, leopard, bushpig, hyaena and jackal.

In these wild cycles man and his animals are only occasionally involved. For example, the consumption of polar bear meat may cause infection in Eskimos and sledge-dogs, while in Europe the hunting and subsequent ingestion of wild pigs may also produce disease in man and his companion animals.

The domestic cycle in man and the pig is an 'artificial' zoonosis largely created by feeding pigs on food waste containing the flesh of infected pigs; more recently, tail-biting in pigs has been shown to be a mode of transmission. Rats in piggeries also maintain a secondary cycle which may on occasions pass to pigs or *vice versa* from the ingestion of infected flesh or faeces. Infection in man is acquired from the ingestion of raw or inadequately cooked pork or its by-products, such as sausages and salami. It is also important to realise that smoking, drying or curing pork does not necessarily kill larvae in pork products.

In areas such as Poland, Germany and the U.S.A. human trichinosis acquired from pork has, until recently, been an important zoonosis. Over the past few decades, prohibition of feeding uncooked food waste to pigs, improved meat inspection and public awareness have greatly diminished the significance of the problem. In Britain only eight outbreaks in man have been reported in the last 40 years, the last in 1953. In other countries in Europe and in the U.S.A. the numbers of outbreaks are similarly few and sporadic.

The decreasing prevalence is also reflected in the fact that inapparent infection in man, as shown by the presence of *T. spiralis* larvae in muscle samples at necropsy, has decreased from 10% to less than 1% in Britain and from 20% to under 5% in the U.S.A. over the past 50 years.

DIAGNOSIS

This is not relevant in live domestic animals. At meat inspection, heavy larval infections may occasionally be seen with the naked eye as tiny greyish white spots. For routine purposes small samples of pig muscle of about one gram are squeezed between glass plates, the apparatus being called a trichinoscope or a compressorium, and examined for the presence of larvae by direct microscopic examination or projection on a screen. Alternatively, small portions of muscle may be digested in pepsin/HCl and the sediment examined microscopically for the presence of larvae.

For mass screening purposes, designed to determine the incidence of trichinosis in pigs within regions, immunodiagnostic tests have been used. Of these, the ELISA appears to be the test of choice.

TREATMENT

Although rarely called for in animals, the adult worms and the larvae in muscles are susceptible to the benzimidazole anthelmintics.

CONTROL

Probably the most important factor in the control of trichinosis is a legal requirement that swill or waste human food intended for consumption by pigs must be boiled. In fact, this practice is mandatory in many countries to limit the potential spread of other diseases such as foot and mouth disease and swine fever.

Other essential steps include:

(i) **Meat inspection**, which plays an essential role in monitoring the detection of infected carcasses.
(ii) **Measures to eliminate rats** from piggeries and slaughterhouses.
(iii) **Regulations to ensure that larvae in pork are destroyed** by cooking or freezing. In the United States, for example, any pork or pork products, other than fresh pork, must be treated by heating or freezing before marketing and it is likely also that irradiation might soon be introduced as a further method of control.
(iv) **Consumer education**, and particularly the recognition that pork or pork products or the flesh of carnivorous game should be thoroughly cooked before consumption.

Superfamily DIOCTOPHYMATOIDEA

Dioctophyma

The only species of veterinary interest in this superfamily is the 'kidney worm', *Dioctophyma renale.*

Host:
Dog, fox and mink
Intermediate host:
The aquatic annelid, *Lumbriculus variegatus*
Site:
Kidney parenchyma
Species:
Dioctophyma renale

DISTRIBUTION

Temperate and subarctic areas. It occurs sporadically in Europe, but has not been recorded in Britain. Its main endemic area is the northern part of North America, chiefly Canada.

IDENTIFICATION

Dioctophyma is the largest parasitic nematode of domestic animals, the female measuring more than 60 cm in length, with a diameter of 1.0 cm. Its size and predilection site are sufficient for identification.

LIFE CYCLE

The worms are oviparous. The eggs, in the single-cell stage, are passed in the urine in clumps or chains, and are ingested by the annelid intermediate host in which the two preparasitic moults occur. The final host is infected by swallowing the annelid with the drinking water, or by the ingestion of a paratenic host, such as a frog or fish, which has itself eaten the infected annelid. The prepatent period is not known with certainty, but has been observed to be as long as two years.

PATHOGENESIS

The final effect of infection is destruction of the kidney. Usually only one kidney is affected, the right being more often involved than the left. The parenchyma is destroyed, leaving only the capsule as a sac containing the worms; though there may be three or four worms in a kidney, occasionally there is only one. Rarely, the worms may occur in the abdominal cavity and in the subcutaneous connective tissue.

CLINICAL SIGNS

The main signs are dysuria with some haematuria, especially at the end of micturition; in a few cases there is lumbar pain. Most cases, however, are completely asymptomatic, even when one kidney has been completely destroyed.

EPIDEMIOLOGY

As in many of the parasitic infections of domestic carnivores there is a large reservoir in wild animals from which the intermediate and paratenic hosts are infected. Ranch mink probably acquire infection from their fish diet, and domestic dogs by casual ingestion of infected annelids, frogs or fish.

DIAGNOSIS

The eggs are quite characteristic, being ovoid and brown, with pitted shells, and their occurrence in the urine, either singly or in clumps or chains, is diagnostic.

TREATMENT

This is rarely called for, although surgery may be attempted in confirmed cases.

CONTROL

Elimination of raw fish from the diet.

Phylum ACANTHOCEPHALA

This is a separate phylum, closely related to the Nematoda, which contains a few genera of veterinary importance. They are generally referred to as 'thorny-headed worms' due to the presence anteriorly of a hook covered proboscis and most are parasites of the alimentary tract of vertebrates. The hollow proboscis armed with recurved hooks, which aid in attachment, is retractable and lies in a sac. There is no alimentary canal, absorption taking place through the thick cuticle, which is often folded and invaginated to increase the absorptive surface. The sexes are separate males being much smaller than females. Posteriorly, the male has a muscular bursa and penis and after copulation, eggs, discharged by ovaries into the body cavity of the female, are fertilised and taken up by a complex structure called the uterine bell which only allows mature eggs to pass out. These are spindle-shaped, thick-shelled and contain a larva which has an anterior circlet of hooks and spines on its surface and is called an **acanthor**. The life cycle is indirect involving either an aquatic or terrestrial arthropod intermediate host. On ingestion by the intermediate host, the egg hatches and the acanthor migrates to the haemocoel of the arthropod where it develops to become a **cystacanth** after 1–3 months. The definitive host is infected by ingestion of the arthropod intermediate host, and the cystacanth, which is really a young adult, attaches and grows to maturity in the alimentary canal. The prepatent period varies from five to twelve weeks.

Although there are a number of genera parasitic in a wide variety of mammals and birds, the major genus of veterinary significance is *Macracanthorhynchus*. A few other genera are important as parasites of aquatic birds.

Macracanthorhynchus

Host:
Pig

Intermediate hosts:
Various dung beetles

Species:
Macracanthorhynchus hirudinaceus

Distribution:
Worldwide, but absent from certain areas, for example, parts of western Europe

IDENTIFICATION

Gross: Adults resemble *A. suum*, but taper posteriorly. The males are up to 10 cm and the females up to 65 cm in length. When placed in water the spiny proboscis is protruded, thus aiding differentiation from *Ascaris* (Fig. 73).

Microscopic: The egg is oval, 110 μm by 65 μm, with a thick brown shell and contains the acanthor larva when laid.

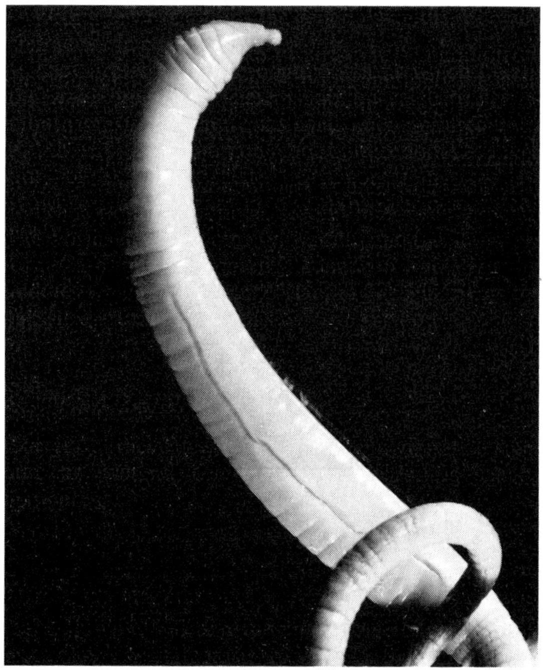

Fig 73 Anterior end of *Macracanthorhynchus hirudinaceus* showing the protruding spiny proboscis.

LIFE CYCLE

Adults, attached to the small intestinal mucosa, lay eggs which are passed in the faeces. These are produced in large numbers, are very resistant to extremes of climate and can survive for years in the environment. After ingestion by dung beetle larvae, the acanthor develops to the infective cystacanth stage in approximately three months. Infection of pigs occurs after ingestion of beetle grubs and adults and the prepatent period is 2–3 months.

PATHOGENIC SIGNIFICANCE

M. hirudinaceus produces inflammation and may provoke granuloma formation at the site of attachment in the small intestinal wall. Heavy infections may cause weight loss and, rarely, penetration of the intestinal wall results in a fatal peritonitis.

TREATMENT AND CONTROL

Pigs should be prevented from access to the intermediate hosts.

In modern management systems this may be easily achieved, but where pigs are kept in small sties the faeces should be regularly removed to reduce the prevalence of the dung beetle intermediate hosts. Although there is little information on treatment, levamisole and ivermectin are reported to be effective.

Acanthocephalans of aquatic birds

There are two genera which may cause enteritis in aquatic fowl, namely *Polymorphus* and *Filicollis*. The intermediate hosts in both cases are crustaceans and in all other respects they are typical of the group.

Phylum PLATYHELMINTHES

This phylum contains the two classes of parasitic flatworms, the TREMATODA and the CESTODA.

Class TREMATODA

The class Trematoda falls into two main subclasses, the Monogenea which have a direct life cycle and the **Digenea** which require an intermediate host. The former are found mainly as external parasites of fish, while the latter are found exclusively in vertebrates and are of considerable veterinary importance.

The adult digenetic trematodes, commonly called 'flukes', occur primarily in the bile ducts, alimentary tract and vascular system. Most flukes are flattened dorso-ventrally, have a blind alimentary tract, suckers for attachment and are hermaphrodite. Depending on the predilection site, the eggs pass out of the final host, usually in faeces or urine, and the larval stages develop in a molluscan intermediate host. For a few species, a second intermediate host is involved, but the mollusc is essential for all members of the group.

There are many families in the class Trematoda, and those which include parasites of major veterinary importance are the **Fasciolidae**, **Dicrocoeliidae**, **Paramphistomatidae** and **Schistosomatidae**. Of lesser importance are the **Troglotrematidae** and **Opisthorchiidae**. The most important group by far are the Fasciolidae and the discussion below of structure, function, and life cycle is largely orientated towards this group.

DIAGNOSIS

This is based on finding the typical eggs in the faeces.

Subclass DIGENEA

STRUCTURE AND FUNCTION OF DIGENETIC TREMATODES

The adult possesses two suckers for attachment. The oral sucker at the anterior end surrounds the mouth, and the ventral, as the name indicates, is on that surface. The body surface is a tegument which is absorptive and is often covered with spines. The muscles lie immediately below the tegument. There is no body cavity and the organs are packed in a parenchyma (Fig. 74).

The digestive system is simple, the oral opening leading into a pharynx, oesophagus and a pair of branched intestinal caeca which end blindly. Undigested material is presumably regurgitated. The excretory system consists of a large number of ciliated flame cells, which impel waste metabolic products along a system of tubules which ultimately join and open to the exterior. The nervous system is simple, consisting of a pair of longitudinal trunks connecting anteriorly with two ganglia.

The trematodes are usually hermaphrodite and both cross- and self-fertilisation may occur. The male reproductive system consists of a pair of testes each leading into a vas deferens; these join to enter the cirrus sac containing a seminal vesicle and the cirrus, a primitive penis which terminates at the common genital opening. The female system has a single ovary leading into an oviduct which is expanded distally to form the ootype. There the ovum acquires a yolk from the secretion of the vitelline glands and ultimately a shell. As the egg passes along the uterus, the shell becomes hardened and tough-

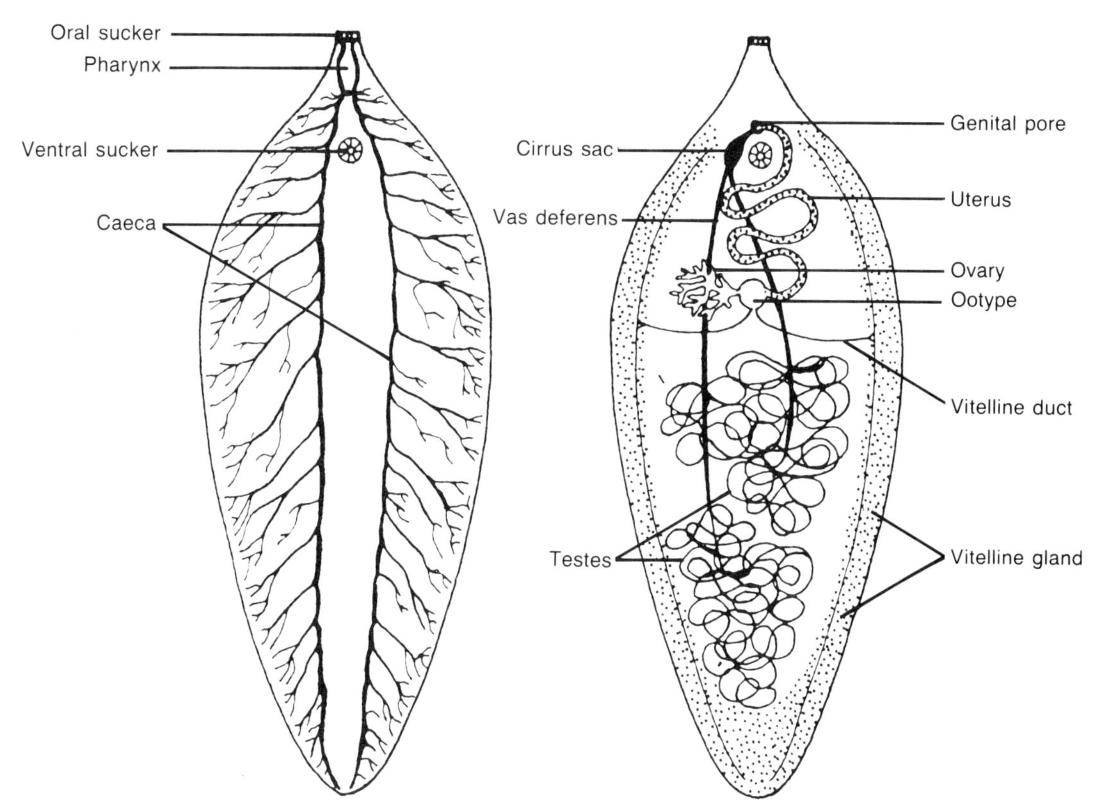

Fig 74 The structure of *Fasciola hepatica*.

ened and is finally extruded through the genital opening adjacent to the ventral sucker. The mature egg is usually yellow because of the tanned protein shell and most species have an operculum.

Food, generally blood or tissue debris is ingested and passed into the caeca where it is digested and absorbed. Metabolism appears to be primarily anaerobic.

THE LIFE CYCLE OF DIGENETIC TREMATODES

The essential point of the life cycle is that whereas one nematode egg can develop into only one adult, one trematode egg may eventually develop into hundreds of adults. This is due to the phenomenon of **paedogenesis** in the molluscan intermediate host, ie. the production of new individuals by single larval forms.

The adult flukes are always oviparous and lay eggs with an operculum or lid at one pole. In the egg the embryo develops into a pyriform (pear-shaped), ciliated larva called a **miracidium**. Under the stimulus of light, the miracidium releases an enzyme which attacks the proteinaceous cement holding the operculum in place. The latter springs open like a hinged lid and the miracidium emerges within a few minutes.

The miracidium, propelled through the water by its cilia, does not feed and must, for its further development, find a suitable snail within a few hours. It is believed to use chemotactic responses to 'home' on the snail and, on contact, it adheres by suction to the snail and penetrates its soft tissues aided by a cytolytic enzyme. The entire process of penetration takes about 30 minutes after which the cilia are lost and the miracidium develops into an elongated sac, the **sporocyst**, containing a number of germinal cells. These cells develop into **rediae** which migrate to the hepato-pancreas of the snail; rediae are also larval forms possessing an oral sucker, some flame cells and a simple gut. From the germinal cells of the rediae arise the final stages, the **cercariae**, although if environmental conditions for the snail are unsuitable, a second or daughter generation of rediae is often produced instead. The cercariae, in essence young flukes with long tails, emerge actively from the snail, usually in considerable numbers. The actual stimulus for emergence depends on the species, but is most commonly a change in temperature or light intensity. Once a snail is infected, cercariae continue to be produced indefinitely although the majority of naturally infected snails die prematurely from gross destruction of the hepato-pancreas.

Typically the cercariae swim for some time, utilising even a film of water, and within an hour or so attach themselves to vegetation, shed their tails and encyst. This stage is called a **metacercaria**.

Encysted metacercariae have great potential for survival extending to months. Once ingested, the outer cyst wall is removed mechanically during mastication. Rupture of the inner cyst occurs in the intestine and depends on a hatching mechanism, enzymatic in origin, triggered by a suitable oxidation-reduction potential and a CO_2 system provided by the intestinal environment. The emergent juvenile fluke then penetrates the intestine and migrates to the predilection site where it becomes adult after several weeks.

Fig 75 Adult *Fasciola hepàtica* showing characteristic leaf shape and conical anterior end.

Family FASCIOLIDAE

These are large leaf-shaped flukes. The anterior end is usually prolonged into the shape of a cone and the anterior sucker is located at the end of the cone. The ventral sucker is placed at the level of the 'shoulders' of the fluke. The internal organs are branched while the cuticle is covered in spines. There are three important genera: *Fasciola*, *Fascioloides* and *Fasciolopsis*.

Fasciola

The members of this genus are commonly known as liver flukes. They are responsible for widespread morbidity and mortality in sheep and cattle characterised by weight loss, anaemia and hypoproteinaemia. The two most important species are *F. hepatica* found in temperate areas and in cooler areas of high altitude in the tropics and subtropics, and *F. gigantica* which predominates in tropical areas.

Fasciola hepatica

Hosts:
Most mammals; sheep and cattle are the most important

Intermediate hosts:
Snails of the genus *Lymnaea*. The most common is *L. truncatula*, amphibious snail with a wide distribution throughout the world

Other important *Lymnaea* vectors of *F. hepatica* outside Europe are:

L. tomentosa	Australia, New Zealand
L. columella	North America, Australia, New Zealand
L. bulimoides	Southern U.S.A. and the Caribbean
L. humilis	North America
L. viator	South America
L. diaphena	South America

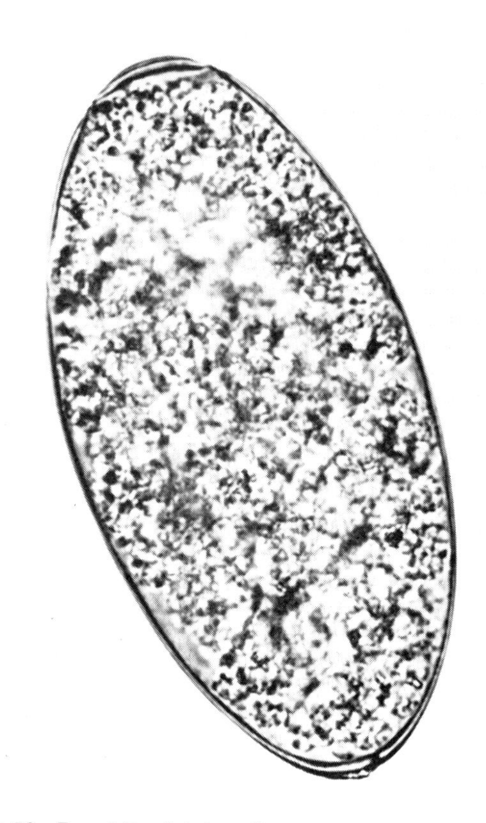

Fig 76 Egg of *Fasciola hepatica.*

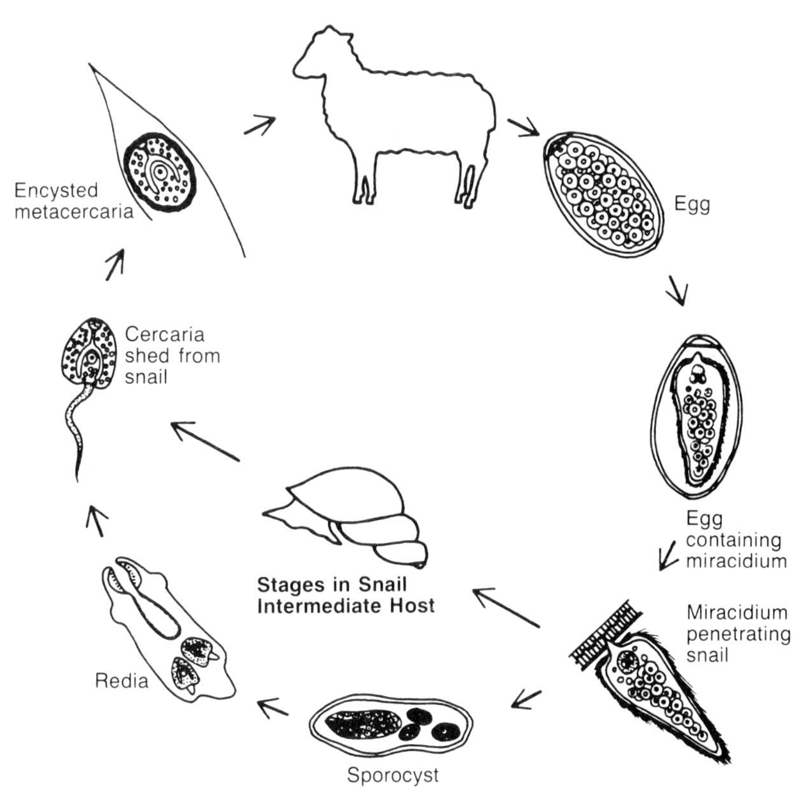

Fig 77 The life cycle of *Fasciola hepatica*.

Site:
The adults are found in the bile ducts and the immature flukes in the liver parenchyma. Occasionally aberrant flukes become encapsulated in other organs, such as the lung

Distribution:
Worldwide

IDENTIFICATION

Gross: The young fluke at the time of entry into the liver is 1.0–2.0 mm in length and lancet-like. When it has become fully mature in the bile ducts it is leaf-shaped, grey-brown in colour and is around 3.5 cm in length and 1.0 cm in width. The anterior end is conical and marked off by distinct shoulders from the body (Fig. 75).

Microscopic: The tegument is covered with backwardly projecting spines. An oral and ventral sucker may be readily seen.

The egg is oval, operculate, yellow and large ($150 \times 90\,\mu m$), and about twice the size of a trichostrongyle egg (Fig. 76).

LIFE CYCLE (Fig. 77)

Eggs passed in the faeces of the mammalian host develop and hatch releasing motile ciliated miracidia. This takes nine days at optimal temperatures of 22–26 °C and little development occurs below 10 °C.

The liberated miracidium (Fig. 78) has a short life span and must locate a suitable snail within three hours if successful penetration of the latter is to occur. In infected snails, development proceeds through the sporocyst and redial stages to the final stage in the intermediate host, the cercaria; these are shed from the snail as motile forms which attach themselves to firm surfaces, such as grass blades, and encyst there to form the infective metacercariae (Fig. 78). It takes a minimum of 6–7 weeks for completion of development from miracidium to metacercaria, although under unfavourable circumstances a period of several months is required. Infection of a snail with one miracidium can produce over 600 metacercariae.

Metacercariae ingested by the final host excyst in the small intestine, migrate through the gut wall, cross the peritoneum and penetrate the liver capsule. The young flukes tunnel through the parenchyma for 6–8 weeks, then enter the small bile ducts where they migrate to the larger ducts and occasionally the gall bladder (Fig. 78). The prepatent period is 10–12 weeks.

The minimal period for completion of one entire life cycle of *F. hepatica* is therefore 17–18 weeks.

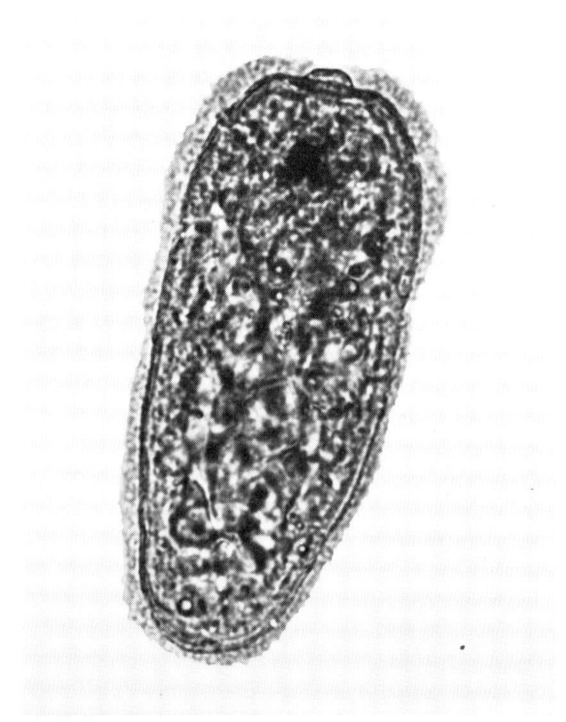

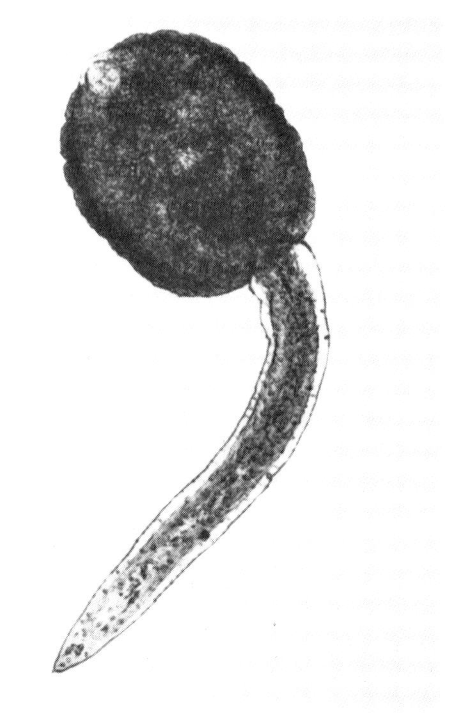

Fig 78 Stages in *Fasciola hepatica* life cycle.
(a) miracidium.

(c) cercaria.

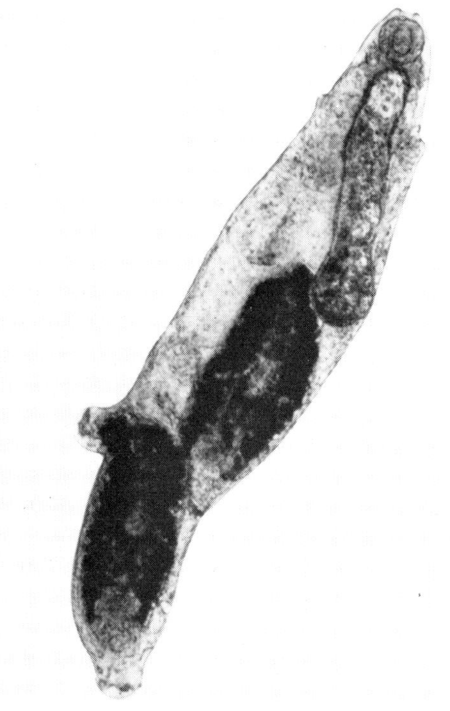

(b) mature redia containing cercariae.

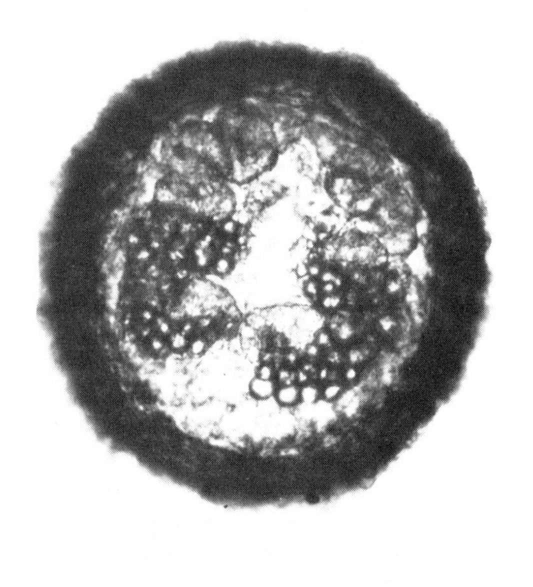

(d) encysted metacercaria.

(a)

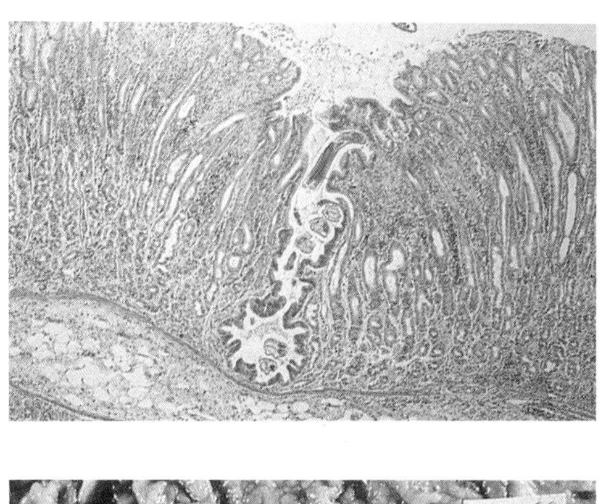

(d)

(b)

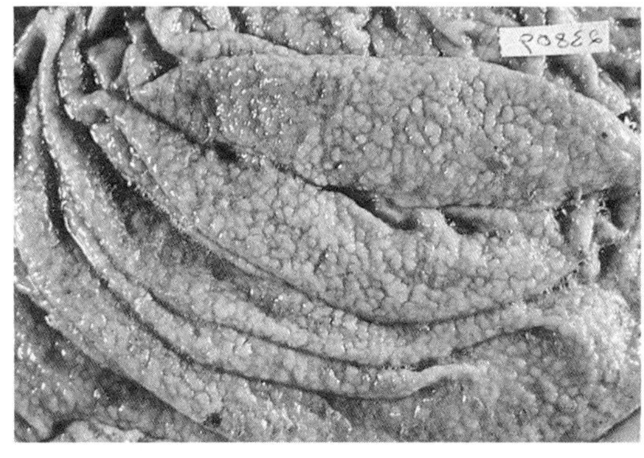

(e)

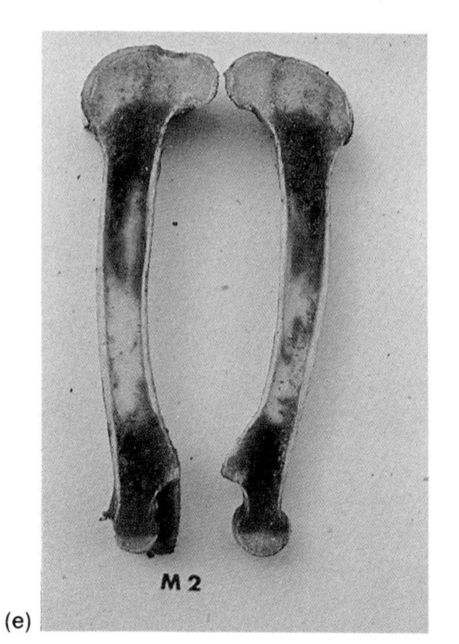

(c)

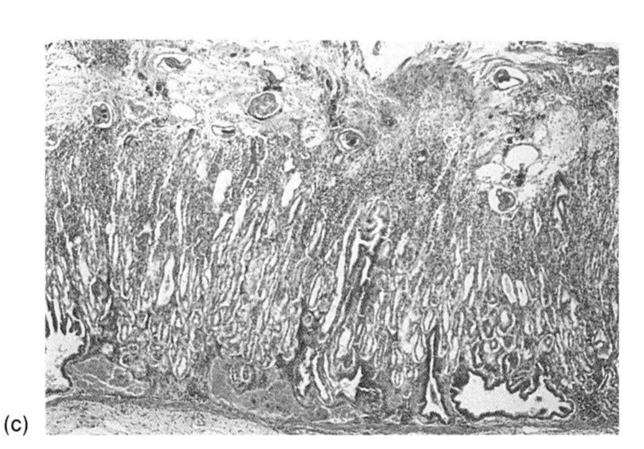

(f)

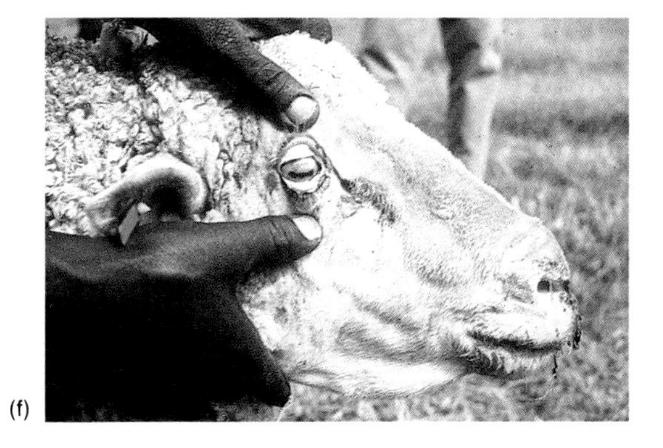

Plate I

(a) *Ostertagia ostertagi* emerging from a gastric gland.
(b) Characteristic gross lesions of ostertagiasis.
(c) Necrosis of mucosa in severe ostertagiasis.
(d) Abomasal haemorrhages in haemonchosis.
(e) Expansion of red marrow in tibia in acute haemonchosis.
(f) Anaemia and submandibular oedema characteristic of haemonchosis.

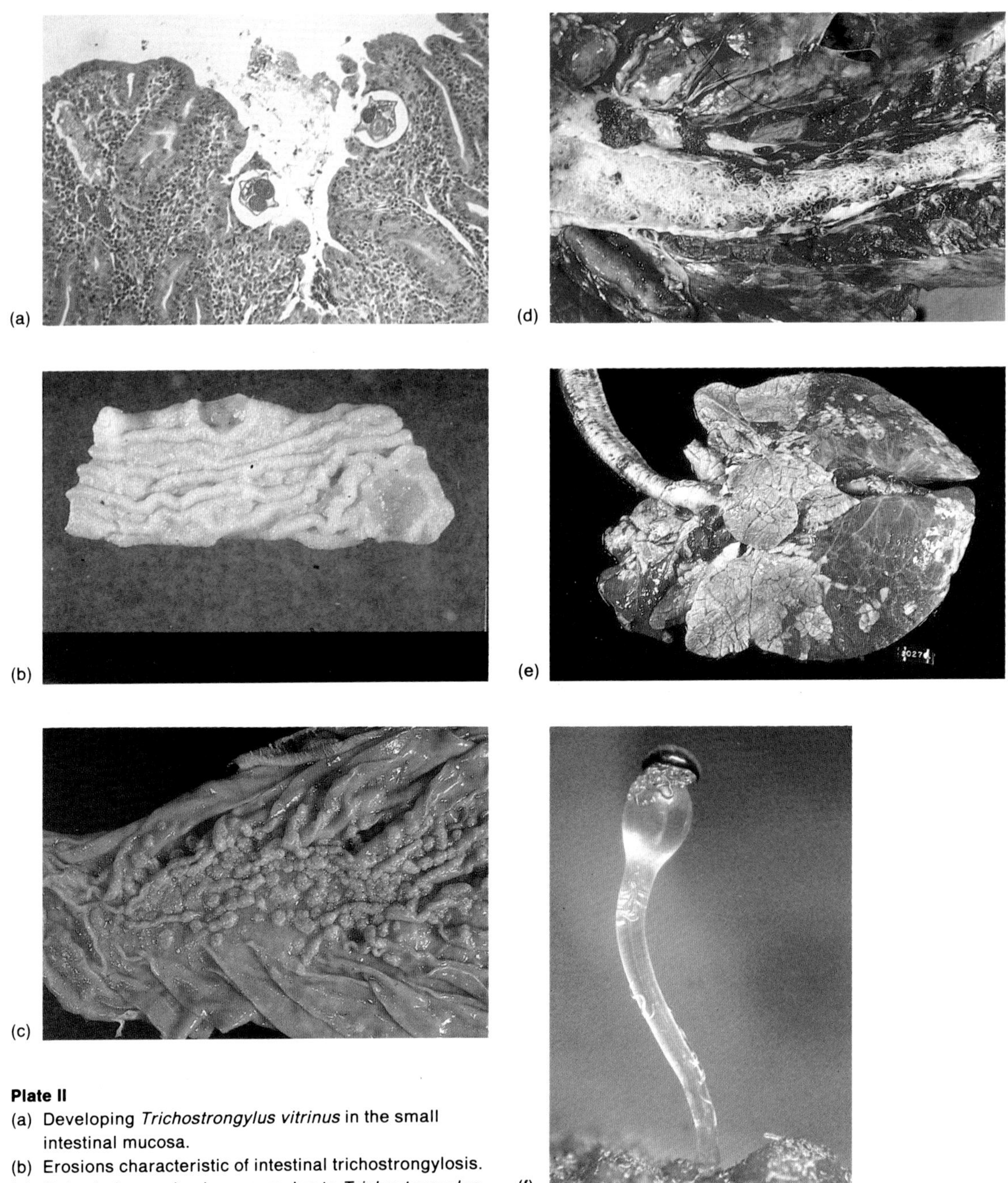

Plate II

(a) Developing *Trichostrongylus vitrinus* in the small intestinal mucosa.
(b) Erosions characteristic of intestinal trichostrongylosis.
(c) Raised plaques in abomasum due to *Trichostrongylus axei* infection.
(d) Adult *Dictyocaulus viviparus* in the bronchi.
(e) Typical distribution of pneumonic lesions of parasitic bronchitis.
(f) Larvae of *Dictyocaulus viviparus* on the fungus *Pilobolus*.

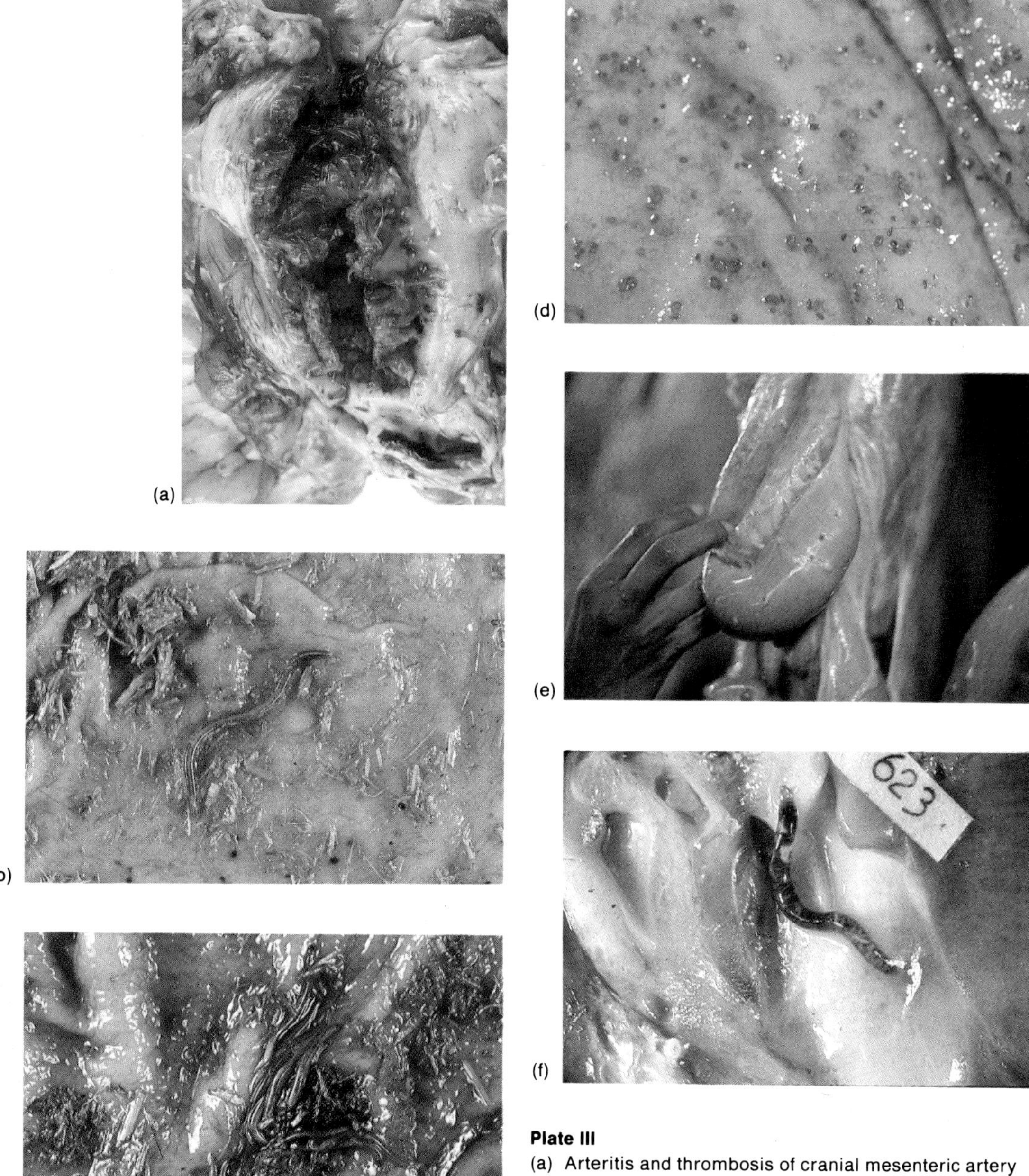

Plate III

(a) Arteritis and thrombosis of cranial mesenteric artery caused by *Strongylus vulgaris* larvae.
(b) *Strongylus edentatus* feeding on the mucosa of the large intestine.
(c) *Triodontophorus tenuicollis* adults feeding around the periphery of an ulcer in the ventral colon.
(d) Developing small strongyle larvae in the mucosa of the caecum.
(e) Intestinal nodules associated with developing *Oesophagostomum* spp.
(f) The 'kidney worm', *Stephanurus dentatus*.

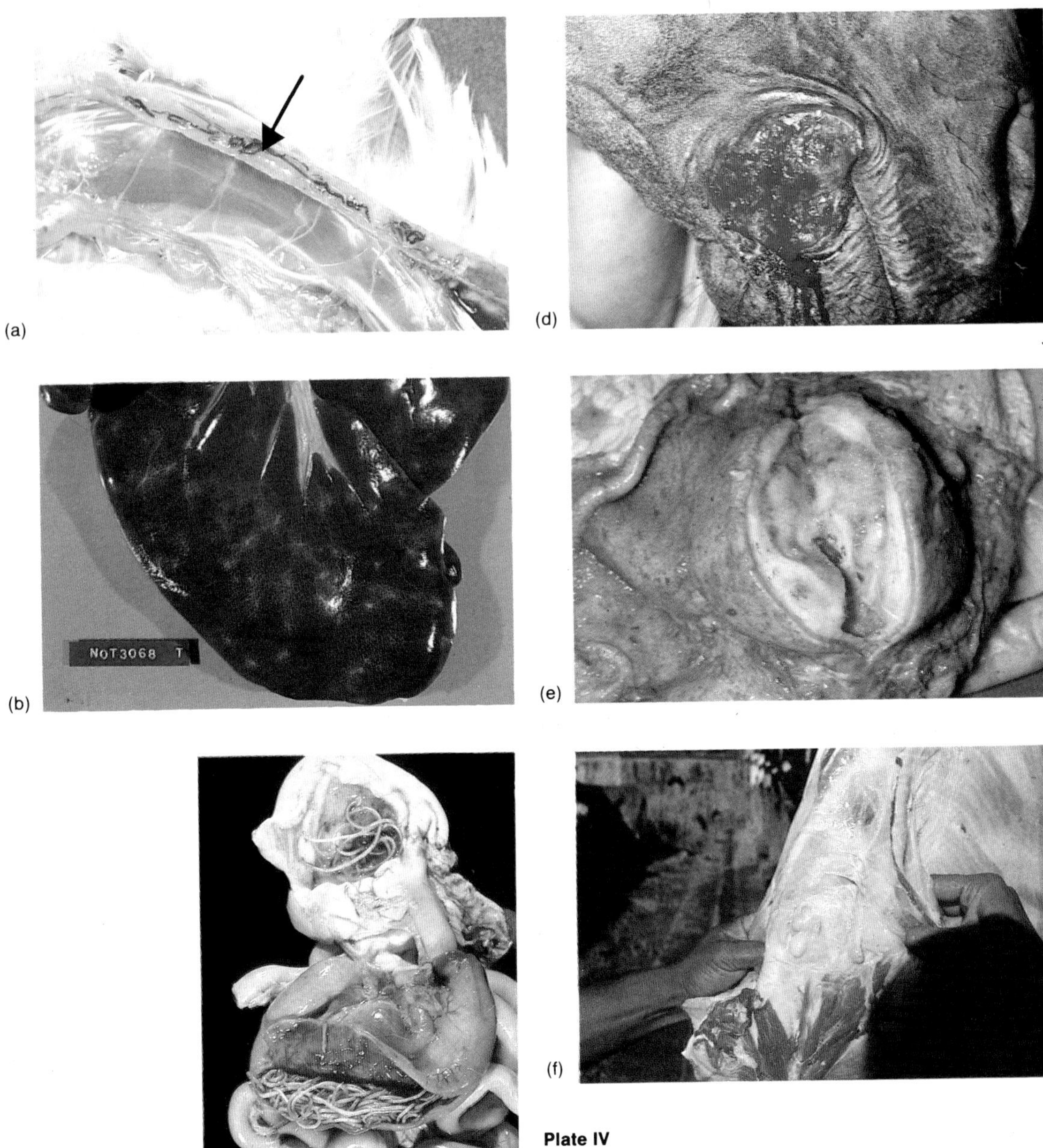

Plate IV

(a) Adult *Syngamus trachea in situ* (↑).

(b) 'Milk spot' lesions in the liver associated with *Ascaris suum* infection.

(c) Heavy *Toxocara canis* infection in the small intestine of a pup.

(d) Ulcerated granuloma on commissure of lips of horse due to cutaneous habronemiasis.

(e) Large nodule in equine stomach associated with *Draschia megastoma* infection.

(f) Nodules typical of bovine onchocerciasis.

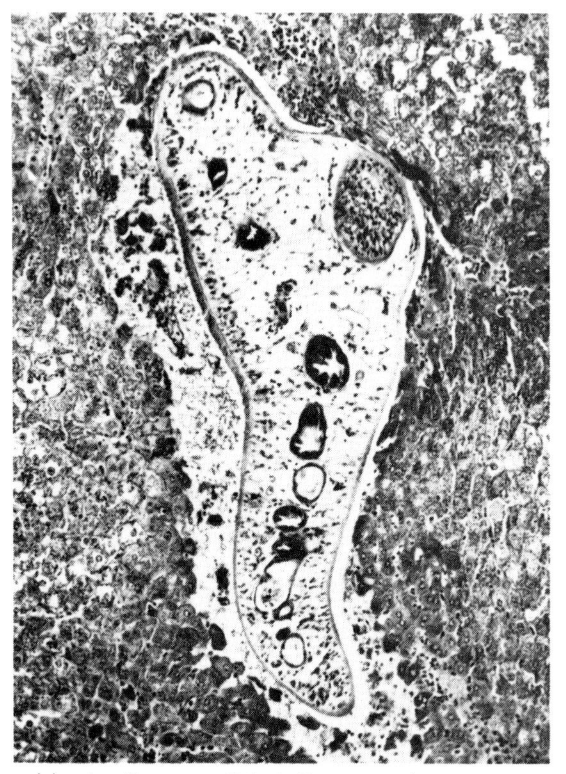

(e) migrating young fluke in liver parenchyma.

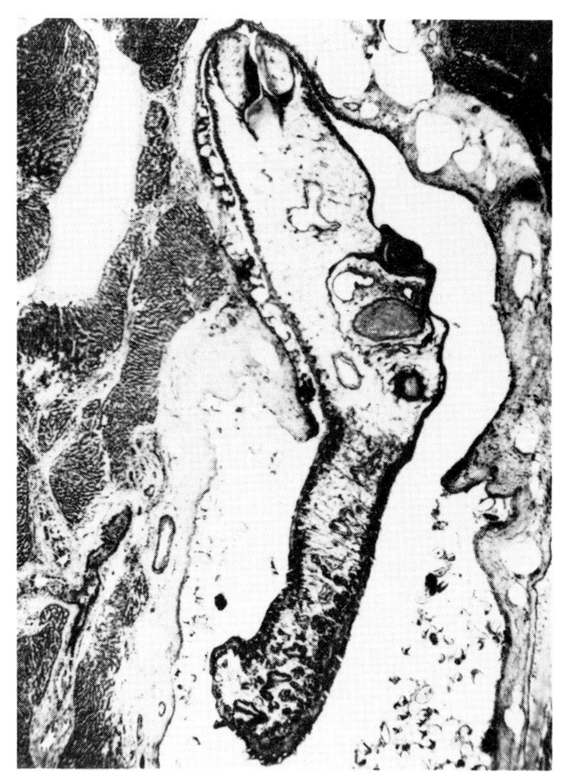

(f) fluke in bile duct.

The longevity of *F. hepatica* in untreated sheep may be years; in cattle it is usually less than one year.

ECOLOGY OF LYMNAEA SPECIES IN TEMPERATE CLIMATES

Since *L. truncatula* is the most widespread and important species involved in the transmission of *F. hepatica*, it is discussed in detail.

L. truncatula is a small snail, the adults being about 1.0 cm in length (Fig. 79). The shell is usually dark brown and has a turreted appearance, being coiled in a series of spiral whorls. When held with the turret upright and the aperture facing the observer, the latter is approximately half the length of the snail and is on the right hand side, and there are four and a half whorls. The snails are amphibious and although they spend hours in shallow water, they periodically emerge onto surrounding mud. They are capable of withstanding summer drought or winter freezing for several months by respectively aestivating or hibernating deep in the mud.

Optimal conditions include a slightly acid pH environment and a slowly moving water medium to carry away waste products. They feed mostly on algae and the optimum temperature range for development is 15–22 °C; below 5 °C development ceases. In Britain snails breed continuously from May to October, one snail being capable of producing up to 100,000 descendants over three months.

Fig 79 Adult *Lymnaea truncatula*; note encysted metacercariae on surface of shell.

EPIDEMIOLOGY

There are three main factors influencing the production of the large numbers of metacercariae necessary for outbreaks of fascioliasis.

1. **Availability of suitable snail habitats:** *L. truncatula* prefers wet mud to free water, and permanent habitats include the banks of ditches or streams and the edges of small ponds. Following heavy rainfall or flooding, temporary habitats may be provided by hoof marks, wheel ruts or rain ponds. Fields with clumps of rushes are often suspect sites. Though a slightly acid pH environment is optimal for *L. truncatula*, excessively acid pH levels are detrimental, such as occur in peat bogs, and ares of sphagnum moss.
2. **Temperature:** A mean day/night temperature of 10 °C or above is necessary both for snails to breed and for the development of *F. hepatica* within the snail, and all activity ceases at 5 °C. This is also the minimum range for the development and hatching of *F. hepatica* eggs. However, it is only when temperatures rise to 15 °C and are maintained above that level, that a significant multiplication of snails and fluke larval stages ensues.
3. **Moisture:** The ideal moisture conditions for snail breeding and the development of *F. hepatica* within snails are provided when rainfall exceeds transpiration, and field saturation is attained. Such conditions are also essential for the development of fluke eggs, for miracidia searching for snails and for the dispersal of cercariae being shed from the snails.

In temperate countries such as Britain, these factors usually only exist from May to October. A marked increase in numbers of metacercariae on pasture is therefore possible during two periods. First, from what is known as **the summer infection of snails**, in which metacercariae appear on pasture from August to October (Fig. 80). These snail infections arise from miracidia which either have hatched from eggs excreted in the spring/early summer by infected animals, or from eggs which have survived the winter in an undeveloped state. Development in the snail occurs during the summer and the cercariae are shed from August until October. Secondly, from **the winter infection of snails** in which metacercariae appear on the pasture in May to June (Fig. 81). These are derived from snails which were infected the previous autumn, and in which larval development had temporarily ceased during the period of winter hibernation of the snail host.

Both *F. hepatica* eggs and metacercariae can survive over the winter and play important parts in the epidemiology. The eggs, by hatching into miracidia in late spring, can infect snails. The metacercariae, by infecting stock in early spring, result in eggs being available by mid-summer at the optimal snail breeding season. However, survival of metacercariae is poor under conditions of high temperatures and drought and they rapidly lose their infectivity during processes such as silage making, although they may survive for several months on hay.

In most European countries, the summer infection of snails is the more important and an increase in the numbers of metacercariae occurs annually from August to October. The extent of this increase is highest in years when summer rainfall is heavy. The winter infection of snails is much less important, but occasionally gives rise to large numbers of metacercariae in late spring and early summer, particularly when the preceding months have been unduly wet.

Circulating antibodies to *F. hepatica* are readily detectable in sheep, but there is no evidence that, under

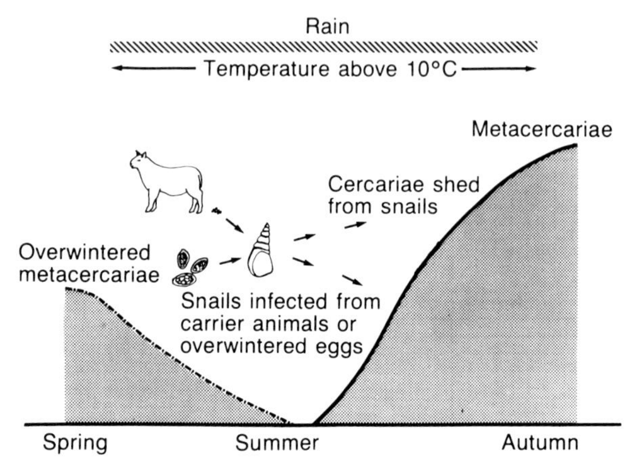

Fig 80 In the summer infection of snails metacercariae appear on pasture from August to October; overwintered metacercariae have died out by mid-summer.

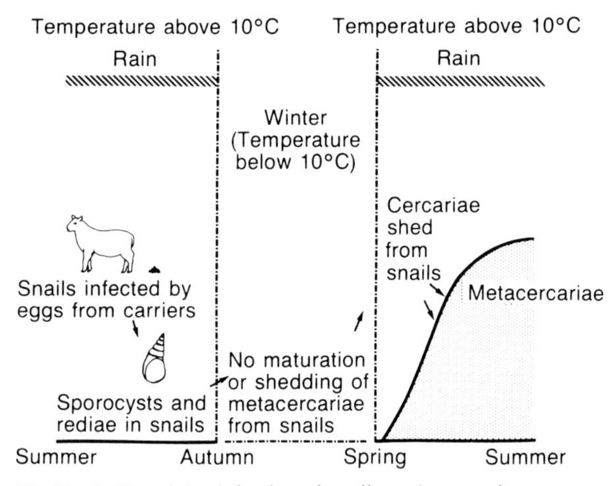

Fig 81 In the winter infection of snails metacercariae appear on pasture from May to June; these are derived from snails which have hibernated over the winter.

field conditions, sheep ever become immune to reinfection with *F. hepatica*, and in the absence of treatment, the flukes will live as long as the sheep. Severe outbreaks of ovine fascioliasis frequently involve adult sheep which have been previously exposed to infection. In contrast, although outbreaks do occur in young cattle, more usually an acquired immunity gradually develops; this limits the life span of the primary infection, slows the migration of secondary infection and eventually reduces the numbers of flukes established. Thus, in endemic areas, adult cattle often appear unaffected clinically whereas severe losses from fascioliasis may be occurring in adult sheep.

Finally it should be remembered that *F. hepatica* can infect a wide range of mammals including horses, donkeys, deer, pigs and rabbits and it is possible that on occasions these hosts may act as reservoirs of infection. Man may also become infected, especially from the consumption of watercress from unfenced beds.

Most of the above comments on the ecology of *L. truncatula* also applies to the other amphibian species of *Lymnaea* which transmit the parasite. Differentiation of *Lymnaea* species is a specialist task and is usually based on morphological characteristics, although biochemical and immunological methods are now also employed.

In warmer areas such as the southern U.S.A. or Australia, the sequence of events has a different seasonality, but the epidemiological principles are the same. For example, in both Texas and Louisiana snail activity is maximal during the cooler months of autumn with peak numbers of metacercariae appearing in the winter.

The situation differs with *L. tomentosa* which, although classed as an amphibian snail, is well adapted to aquatic life in swampy areas or irrigation channels and therefore temperature is the most important controlling biological factor. Thus, in most of Eastern Australia, *L. tomentosa* continues to produce egg masses throughout the year, although the rate of reproduction is controlled by temperature and is at its lowest during the winter. The lower winter temperatures also delay hatching of fluke eggs and larval development in the snail so that large numbers of metacercariae first appear in late spring. During the summer and autumn there is a second wave of metacercarial production derived from new generations of snails.

There is some evidence that the prevalence of fascioliasis in hot countries is higher after several months of drought, possibly because the animals congregate around areas of water conservation and so the chances of snails becoming infected are increased.

PATHOGENESIS AND CLINICAL SIGNS

These vary according to the phase of parasitic development in the liver and the species of host involved. Essentially the pathogenesis is two-fold; the first phase occurs during migration in the liver parenchyma and is associated with liver damage and haemorrhage. The second occurs when the parasite is in the bile ducts, and results from the haematophagic activity of the adult flukes and from damage to the mucosa by their cuticular spines.

Most studies have been in sheep and the disease in this host is discussed in detail. The seasonality of outbreaks is that which occurs in Western Europe.

OVINE FASCIOLIASIS

Fascioliasis may be acute, sub-acute and chronic.

The **acute disease** occurs 2–6 weeks after the ingestion of large numbers of metacercariae, usually over 2,000, and is due to the severe haemorrhage which results when the young flukes, migrating in the liver parenchyma, rupture blood vessels. Damage to the liver parenchyma is also severe.

At necropsy the liver is enlarged, haemorrhagic and honeycombed with the tracts of migrating flukes (Pl. V). The surface, particularly the ventral lobe, is frequently covered with a fibrinous exudate. Subcapsular haemorrhages (Pl. V) are common and these may rupture so that a quantity of blood-stained fluid is often present in the abdominal cavity.

Outbreaks of acute fascioliasis are generally presented as sudden deaths during autumn and early winter. On examination of the remainder of the flock, one may find some sheep which are weak, with pale mucous membranes, dyspnoeic and in some instances have palpable enlarged livers associated with abdominal pain and ascites.

Sometimes these outbreaks are complicated by concurrent infections with *Clostridium perfringens* resulting in 'black disease' although this is less common nowadays because of widespread vaccination against this pathogen.

In the **subacute disease**, metacercariae are ingested over a longer period and while some have reached the bile ducts, where they cause a cholangitis, others are still migrating causing lesions less severe, but similar to those of the acute disease; thus the liver is enlarged with numerous necrotic or haemorrhagic tracts visible on the surface and in the substance. Subcapsular haemorrhages are usually evident, but rupture of these is rare.

This form of the disease, occurring 6–10 weeks after ingestion of approximately 500–1,500 metacercariae, also appears in the late autumn and winter. It is presented as a rapid and severe haemorrhagic anaemia with hypoalbuminaemia, and if untreated, can result in a high mortality rate. However it is not so rapidly fatal as the acute condition and affected sheep may show clinical signs for 1–2 weeks prior to death; these include a rapid loss of condition, a marked pallor of the mucous membranes, and an enlarged and palpable liver. Submandibular or facial oedema and ascites may be present.

Chronic fascioliasis, which is seen mainly in late winter/early spring, is the most common form of the disease. It occurs 4–5 months after the ingestion of moderate numbers, 200–500, of metacercariae. The principal pathogenic effects are anaemia and hypoalbuminaemia and more than 0.5 ml blood per fluke can be lost into the bile ducts each day. Additional loss of plasma proteins occurs by leakage through the hyperplastic bilary mucosa and the pathogenic effect is exacerbated if the sheep is on a low plane of nutrition.

At necropsy the liver has an irregular outline and is pale and firm, the ventral lobe being most affected and reduced in size (Pl. V). The liver pathology is characterised by hepatic fibrosis and hyperplastic cholangitis.

Several different types of fibrosis are present. The first to occur is post-necrotic scarring, found mainly in the ventral lobe and associated with the healing of fluke tracts.

The second, often termed ischaemic fibrosis, is a sequel of infarction caused by damage and thrombosis of large vessels.

Thirdly, a peribiliary fibrosis develops when the flukes reach the small bile ducts.

Finally, a monolobular fibrosis occurs in which the portal canals become linked by strands of fibrous tissue which can be seen grossly as white streaks demarcating the hepatic lobule. This lesion is widespread throughout the liver and is of unknown etiology.

Sometimes fluke eggs provoke a granuloma-like reaction which can result in obliteration of the affected bile ducts.

The hyperplastic cholangitis in the larger bile ducts arises from the severe erosion and necrosis of the mucosa caused by the feeding flukes. It has also recently been demonstrated that the amino-acid proline, produced by flukes, may contribute to the biliary hyperplasia and anaemia.

Clinically, chronic fascioliasis is characterised by a progressive loss of condition and the development of anaemia and hypoalbuminaemia which can result in emaciation, pallor of the mucous membranes, submandibular oedema and ascites. The anaemia is hypochromic and macrocytic with an accompanying eosinophilia. *Fasciola* eggs can be demonstrated in the faeces.

In light infections, the clinical effect may not be readily discernible, but the parasites can have a significant effect on production due to an impairment of appetite and to their effect on post-absorptive metabolism of protein, carbohydrates and minerals.

BOVINE FASCIOLIASIS

Although acute and subacute disease may occasionally occur under conditions of heavy challenge, especially in young calves, the chronic form of the disease is by far the most important, and as in sheep, is seen in the late winter/early spring.

The pathogenesis is similar to that in sheep but has the added features of calcification of the bile ducts and enlargement of the gallbladder (Pl. V). Aberrant migration of the flukes is more common in cattle and encapsulated parasites are often seen in the lungs. On reinfection of adult cows, migration to the foetus has been recorded resulting in prenatal infection. There is some experimental evidence that fascioliasis increases the susceptibility of cattle to infection with *Salmonella dublin*.

In heavy infections, where anaemia and hypoalbuminaemia are severe, submandibular oedema frequently occurs. With smaller fluke burdens, the clinical effect is minimal and the loss of productivity is difficult to differentiate from inadequate nutrition. It must be emphasised that diarrhoea is not a feature of bovine fascioliasis unless it is complicated by the presence of *Ostertagia* spp. or other gastrointestinal pathogens.

Fasciola infections may cause a loss of production in milking cows during winter. Clinically, these are difficult to detect since the fluke burdens are usually low and anaemia is not apparent. The main effects are a reduction in milk yield and quality, particularly of the solids-not-fat component.

DIAGNOSIS

This is based primarily on clinical signs, seasonal occurrence, prevailing weather patterns, and a previous history of fascioliasis on the farm or the identification of snail habitats.

While diagnosis of ovine fascioliasis should present few problems, especially when a postmortem examination is possible, diagnosis of bovine fascioliasis can sometimes prove difficult. In this context, routine haematological tests and examination of faeces for fluke eggs are useful and may be supplemented by two other laboratory tests.

The first is the estimation of plasma levels of enzymes released by damaged liver cells. Two enzymes are usually measured. Glutamate dehydrogenase (GLDH) is released when parenchymal cells are damaged and levels become elevated within the first few weeks of infection. The other, gamma glutamyl transpeptidase (GGT) indicates damage to the epithelial cells lining the bile ducts; elevation of this enzyme takes place mainly after the flukes reach the bile ducts and raised levels are maintained for a longer period.

The second is the detection of antibodies against components of flukes, the ELISA and the passive haemaglutination test being the most reliable.

Meteorological forecasting of fascioliasis

The life cycle of the liver fluke and the prevalence of fascioliasis is dependent on climate. This has led to the development of forecasting systems, based on meteorological data, which estimate the likely timing and severity of the disease. In several western European countries, these forecasts are used as the basis for annual control programmes.

Two different formulae have been developed. One estimates 'ground surface wetness' which is the critical factor affecting the summer infection of snails. The formula is $M = n(R - P + 5)$ where M is the month, R the monthly rainfall in inches, P is the evapotranspiration in inches and n is the number of wet days per month. A value of 100 or more per month is optimal for parasite development and therefore values of more than 100 are registered as 100. The formula is applied over the months when temperatures are suitable for snail breeding and parasite development, ie. May–October in Europe, and the monthly values summated to give a seasonal index or Mt value. Since the temperatures are generally lower in May and October, the values for these months are halved prior to summation. Where the Mt exceeds 450, the prevalence of fascioliasis is likely to be high.

The forecast is used to issue an early warning of disease by calculating data from May to August so that control measure can be introduced prior to shedding of cercariae. The disadvantage of the forecast is that it may overestimate the prevalence where there is an autumn drought or underestimate the likely prevalence where the presence of drainage ditches allows the parasite life cycle to be maintained in dry summers. Although this technique is mainly applied to the summer infection of snails, it is also used for forecasting the winter infection of snails by summating the values for August, September and October; if these exceed 250 and the following May or June has a high rainfall then fascioliasis is forecast for the area.

The other technique used is a 'wet day' forecast. This compares the prevalence of fascioliasis over a number of years with the number of rain-days during the summers of these years. In essence, widespread fascioliasis is associated with 12 wet days (over 1.0 mm of rainfall) per month from June to September where temperatures do not fall below the seasonal normal.

TREATMENT

The older drugs such as carbon tetrachloride, hexachlorethane and hexachlorophene are still used in some countries, but these have been largely replaced by more efficient and less toxic compounds and only the latter will be discussed.

Acute ovine fascioliasis

Until recently, treatment was not highly successful due to the inefficiency of the older drugs against the early parenchymal stages. However, efficient drugs are now available and the ones of choice are diamphenethide and triclabendazole which remove all developing stages over one week old. Two other drugs are rafoxanide and nitroxynil, which, at increased dosage rates, will remove flukes over four weeks old.

A single dose of diamphenethide accompanied with a move to fluke-free pasture or well-drained, recently cultivated field should be adequate treatment. However with rafoxanide or nitroxynil a second treatment may be necessary 2–3 weeks after moving to fluke-free ground. Where sheep cannot be moved to clean ground, treatment should be repeated at three weekly intervals until six weeks after deaths have ceased.

Subacute ovine fascioliasis

The drugs recommended for acute fascioliasis can be used, rafoxanide and nitroxynil being given at normal dosage rates; at these levels they are effective against the older flukes responsible for subacute fascioliasis. Movement to fluke-free pasture is again advisable following treatment, and where this is not possible treatment should be repeated at four and eight weeks to eliminate maturing flukes.

In addition to the above drugs, brotianide and acedist are also effective.

Chronic ovine fascioliasis

Outbreaks of chronic fascioliasis can be successfully treated with a single dose of any of a range of drugs (rafoxanide, nitroxynil, brotianide, acedist, oxyclozanide and triclabendazole) and following treatment, the anaemia usually regresses within 2–3 weeks. Diamphenethide is not recommended for the treatment of chronic fascioliasis since this drug is less effective against adult flukes. The roundworm anthelmintic, albendazole, is also effective against adult flukes albeit at increased dosage rates.

Bovine fascioliasis

At present there is only one drug, namely triclabendazole, which will remove the early parenchymal stages, diamphenethide not being available for cattle. Apart from triclabendazole, the two drugs most commonly used for subacute or chronic fascioliasis are rafoxanide and nitroxynil and several others, such as acedist, clorsulon and niclofolan, are also marketed in some countries; albendazole is also effective at an increased dosage rate. In lactating cows, where the milk is used for human consumption, the above drugs are either banned or have extended withdrawal periods in most countries. An exception is oxyclozanide which is licensed for use in lactating animals in many countries and has a zero milk-withholding time.

CONTROL

Control of fascioliasis may be approached in two ways; by reducing populations of the intermediate snail host or by using anthelmintics.

Reduction of snail populations

Before any scheme of snail control is undertaken a survey of the area for snail habitats should be made to determine whether these are localised or widespread.

The best long-term method of reducing mud-snail populations such as *L. truncatula* is drainage, since it ensures permanent destruction of snail habitats. However, farmers are often hesitant to undertake expensive drainage schemes, although in many countries special drainage grants are available.

When the snail habitat is limited a simple method of control is to fence off this area or treat annually with a molluscicide. Currently copper sulphate is most widely used and although more efficient molluscicides, such as N-trityl morpholine, have been developed, none are generally available.

In Europe experimental evidence indicates that a molluscicide should be applied either in the spring (May), to kill snail populations prior to the commencement of breeding, or in summer (July/August) to kill infected snails. The spring application should ensure better contact with the snails, because pasture growth is limited, but in practice is often impractical because the saturated nature of the habitat makes vehicular access difficult. In the summer this is less of a problem although molluscicide/snail contact may be reduced because of the increase in herbage growth. The application of a molluscicide should be combined with anthelmintic treatment to remove existing fluke populations and thus the contamination of habitats with eggs.

When the intermediate snail host is aquatic, such as *L. tomentosa*, good control is possible by adding a molluscicide, such as N-trityl morpholine, to the water habitat of the snail, but there are many environmental objections to the use of molluscicides in water or irrigation channels.

Use of anthelmintics

The prophylactic use of fluke anthelmintics is aimed at:

(i) **Reducing pasture contamination** by fluke eggs at a time most suitable for their development, ie. April to August.

(ii) **Removing fluke populations** at a time of heavy burdens or at a period of nutritional and pregnancy stress to the animal. To achieve these objectives, the following control programme for sheep in the British Isles is recommended for years with normal or below average rainfall. Since the timing of treatments is based on the fact that most metacercariae appear in autumn and early winter, it may require modification for use in other areas.

In late April/early May treat all adult sheep. Fasciolicides such as rafoxanide or nitroxynil will be effective against adult and late immature stages.

In October, treat the entire flock using a drug effective against early parenchymal stages, such as diamphenethide or triclabendazole.

In January, treat the flock with any drug effective against adult stages.

In wet years further doses may be necessary as follows:

In June, 4–6 weeks after the April/May dose, all adult sheep should be treated with rafoxanide or nitroxynil.

In October/November, 4 weeks after the early October dose, treat all sheep with a drug effective against parenchymal stages.

The precise timing of the spring and autumn treatments will depend on lambing and service dates. Although very good results have been obtained on sheep farms by anthelmintic prophylaxis, the same has not been achieved for the control of disease in cattle. This is because most drugs available for cattle are ineffective against the early parenchymal stages.

Prophylactic treatment in cattle is therefore directed at reducing the fluke burdens in winter at a time when parasites are susceptible to available drugs and when the nutritional status of the animals is at its lowest. It is unlikely that their faeces will remain free from eggs for any length of time as the parenchymal stages not removed by the anthelmintic will soon develop to the adult stage. In the British Isles it is usual to treat cattle in fluke areas in mid-winter.

This situation may change with the increased use of drugs such as triclabendazole in cattle.

Fasciola gigantica

Hosts:
Ruminants

Intermediate hosts:
Snails of the genus *Lymnaea*; in southern Europe it is *L. auricularia* which is also the important species in the southern U.S.A., the Middle East and the Pacific Islands

Other important *Lymnaea* vectors of *F. gigantica* are:

L. natalensis	Africa
L. rufescens and *L. acuminata*	India and Pakistan
L. rubiginosa	Malaysia

All these snails are primarily aquatic snails and are found in streams, irrigation channels and marshy swamps

Site:
The adults are found in the bile ducts and the immature flukes in the liver parenchyma

Distribution:
Most continents. Does not occur in western Europe

IDENTIFICATION

Gross: It is larger than *F. hepatica* and can reach 7.5 cm in length. The shape is more leaf-like, the conical anterior end is very short and the shoulders characteristic of *F. hepatica* are barely perceptible.
Microscopic: The eggs are larger than those of *F. hepatica* measuring 190 × 100 μm.

LIFE CYCLE

This is similar to *F. hepatica*, the main differences being in the time scale of the cycle. Most parasitic phases are longer and the prepatent period is 13–16 weeks.

PATHOGENESIS AND CLINICAL SIGNS

Similar to those of *F. hepatica*.

EPIDEMIOLOGY

The snails which carry the larval stages of *F. gigantica* are primarily aquatic and as a result the disease is associated with animals grazing on naturally or artificially flooded areas or around permanent water channels or dams.

In subtropical or tropical countries with distinct wet and dry seasons, it appears that optimal development of eggs to miracidia occurs at the start of the wet season and development within the snail is complete by the end of the rains. Shedding of cercariae then commences at the start of the dry season when the water level is still high and continues as the water level drops. Under laboratory conditions, a large number of metacercariae simply encyst on the surface of the water rather than on herbage, and under natural conditions this could have a very significant effect on the dissemination of infection. Metacercariae are acquired by animals utilising such areas during the dry season and clinical problems, depending on the rate of infection, occur at the end of that season or at the beginning of the next wet season.

Like *F. hepatica*, *F. gigantica* is capable of infecting man.

Diagnosis and treatment: As for *F. hepatica*

CONTROL

The principles are the same as for the control of *F. hepatica* and are based on the routine use of anthelmintics together with measures to reduce populations of the snail intermediate host. There is, however, the important difference that the latter are water snails whose control depends on a different approach from that for the mud snail *L. truncatula*.

Routine anthelmintic treatment of animals at seasons when heavy infections of adult flukes accumulate in the host is recommended using, if possible, nitroxynil or rafoxanide. This should prevent serious losses in production, but for optical benefit should be accompanied by snail control.

When watering of stock is from a reservoir or stream, complete control can be achieved by fencing the water source and leading a pipe to troughs. To do this effectively from streams, the water may require to be pumped and in remote areas simple water-driven pumps whose power source depends on the water flow have been found useful. It is important that the water troughs be cleaned out regularly since they can become colonised by snails.

When grazing depends on the dry season use of marshy areas around receding lake beds, snail control is difficult. Molluscicides are usually impractical because of the large body of water involved and their possible effect on fish which may form an important part of the local food supply. Apart from repeated anthelmintic treatment to prevent patency of acquired infections of *F. gigantica*, there is often little one can do. Ideally, such areas are often best suited to irrigation and the growing of cash crops, the profit from which can be used to improve the dry season food and water supply to cattle.

Other genera of Fasciolidae

Fascioloides magna

This is primarily a parasite of deer which can infect other ruminants. It is found in North America and central and S.W. Europe. Transmitted by *Lymnaea* species this very large fluke measures up to 10 cm and has no anterior cone.

In deer and cattle, the parasite can cause hepatic damage on reaching the liver, but it rapidly becomes encapsulated by the host reaction and clinical effects are minimal. However in sheep this response is negligible and the damage to the liver can be severe or even fatal; because of this sheep rearing is difficult in areas where the parasite is prevalent.

Fasciolopsis buski

This fluke, found in India, Pakistan, S.E. Asia and China, is primarily a parasite of man, but can occur in the pig which may act as a reservoir host. Unlike *Fasciola* and *Fascioloides* it is located in the small intestine where it causes severe ulceration.

Family DICROCOELIIDAE

These trematodes are small lancet-like flukes occurring in the biliary and pancreatic ducts of vertebrates. Miracidia are present in the eggs when they are passed in the faeces; there is no redial stage during development in the snail.

Dicrocoelium

Hosts:
Sheep, cattle, deer and rabbits

Intermediate hosts:
Two are required.
(i) **Land snails** of many genera
(ii) **Brown ants** of the genus *Formica*

Site:
Bile ducts and gall bladder

Distribution:
Worldwide. The prevalence in the British Isles is very low being confined to small foci in western Scotland and Ireland. In contrast the prevalence in many other European countries is high

Species:
Dicrocoelium dendriticum

IDENTIFICATION

Gross: There is no possibility of confusion with other flukes in the bile ducts of ruminants as *Dicrocoelium* is less than 1.0 cm long, distinctly lanceolate and semi-transparent (Fig. 82).

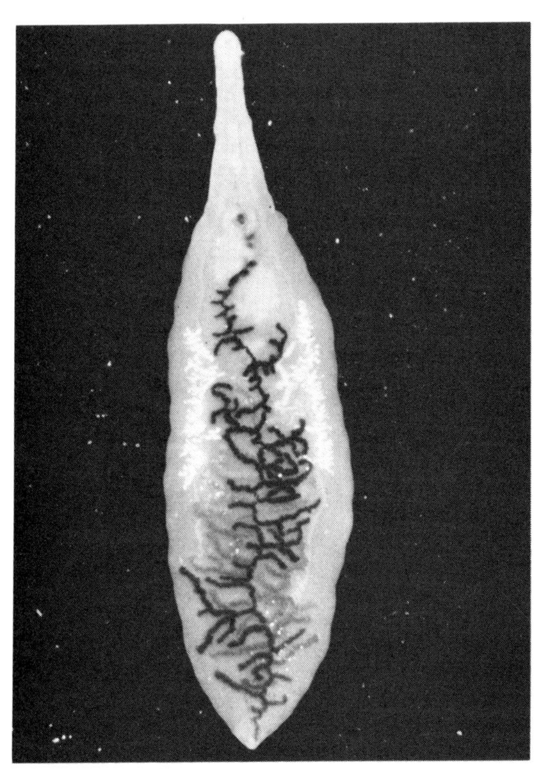

Fig 82 *Dicrocoelium dendriticum*, a small lanceolate fluke of ruminants. The dark uterus and white vitellaria can be clearly seen.

Microscopic: The gut is simple, consisting of two branches and resembles a tuning fork. Behind the ventral sucker the testes lie in tandem with the ovary immediately posterior. There are no spines on the cuticle (cf. *Fasciola*). The egg is small, 45 × 30 µm, dark brown and operculate, usually with a flattened side (Fig. 83). It contains a miracidium when passed in the faeces.

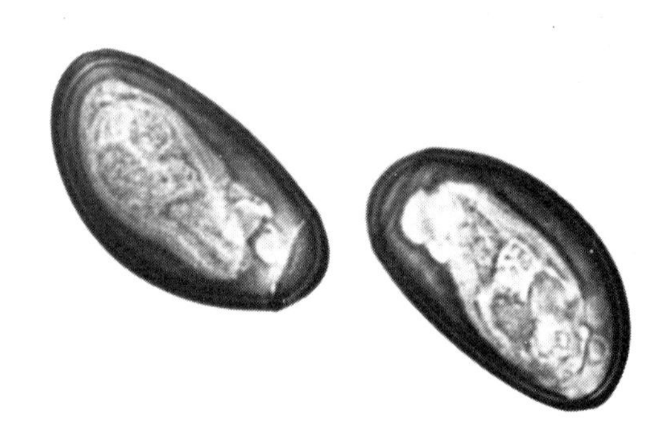

Fig 83 The egg of *Dicrocoelium dendriticum* contains a miracidium when passed in the faeces.

LIFE CYCLE

The egg does not hatch until ingested by the first intermediate host, the land snail, in which two generations of sporocysts develop which then produce cercariae. The latter are extruded in masses cemented together by slime. This phase of development takes at least three months.

The slime balls of cercariae are ingested by ants in which they develop to metacercariae mainly in the body cavity and occasionally the brain. The presence of a brain lesion in the ant, induced by metacercariae, impels the ant to climb up and remain on the tips of the herbage thus increasing the chance of ingestion by the final host. This phase in the ant is completed in just over one month in summer temperatures.

In the final host, the metacercariae hatch in the small intestine and the young flukes migrate up the main bile duct and thence to the smaller ducts in the liver. There is no parenchymal migration and the prepatent period is 10–12 weeks. The flukes are long-lived and can survive in the final host for several years.

PATHOGENESIS

Although several thousand *D. dendriticum* are commonly found in the bile ducts, the livers are relatively normal; this is presumably due to the absence of a migratory phase. However, in heavier infections there is fibrosis of the smaller bile ducts and extensive cirrhosis can occur (Pl. V); sometimes the bile ducts become markedly distended.

CLINICAL SIGNS

In many instances these are absent. Anaemia, oedema and emaciation have been reported in severe cases.

EPIDEMIOLOGY

There are two important features which differentiate the epidemiology of *Dicrocoelium* from that of *Fasciola*.

(i) **The intermediate hosts are independent of water** and are evenly distributed on the terrain.

(ii) **The egg can survive for months on dry pasture,** presenting a reservoir additional to that in the intermediate and final hosts.

DIAGNOSIS

This is entirely based on faecal examination for eggs and necropsy findings.

TREATMENT

High doses of anthelmintics are required for effective removal of *Dicrocoelium*. The benzimidazole, albendazole, given at three times the roundworm dosage rate, is very effective, as is praziquantel at twice the rate used for tapeworms. Other drugs such as thiabendazole and fenbendazole are also effective, but at very high dose rates. Recently netobimin has been shown to be highly effective.

CONTROL

This is difficult because of the longevity of *D. dendriticum* eggs, the wide distribution of the intermediate hosts and the number of reservoir hosts. Control depends almost entirely on regular anthelmintic treatment.

Note: In sub-Saharan Africa a closely related species *D. hospes* occurs, ants of the genus *Campanotus* being the second intermediate host.

Eurytrema

The fluke *Eurytrema pancreaticum* is found in the pancreatic ducts of ruminants in parts of Asia, Brazil and Venezuela. Like *D. dendriticum* it has two consecutive intermediate hosts, a land snail followed by a grasshopper or tree cricket. Infection of the final host is by ingestion of the grasshopper and migration of the fluke from the small intestine to the final site in the pancreatic duct. Heavy infections are reported as causing fibrosis and atrophy of the pancreas. At present there is no known effective treatment.

Platynosomum

The parasite *Platynosomum fastosum* is found in the bile duct of cats in parts of South America, the Caribbean, southern U.S.A., West Africa, Malaysia and the Pacific Islands. Three consecutive intermediate hosts, a land snail, a crustacean and lizard, are required.

The cat is infected by ingesting the lizard. Most infections are well tolerated by the cat causing only a mild inappetence but in heavy infestations, so-called 'lizard poisoning', cirrhosis and jaundice have been reported with diarrhoea and vomiting in terminal cases. Praziquantel and nitroscanate are reported to be effective treatments.

Family PARAMPHISTOMATIDAE

Adult paramphistomes are mainly parasitic in the forestomachs of ruminants, although a few species occur in the intestine of ruminants, pigs and horses. Their shape is not typical of the trematodes, being conical rather than flat. All require a water snail as an intermediate host. There are several genera of which *Paramphistomum* is the most common and widespread.

Paramphistomum

Hosts:
Ruminants

Intermediate hosts:
Water snails; principally *Planorbis* and *Bulinus*

Site:
Adults in the rumen and reticulum and immature stages in the duodenum

Species:
There are at least 14 species of which *Paramphistomum cervi* and *P. microbothrium* are the most common

Distribution:
Worldwide. However, they are of little veterinary significance in Europe and America, but are occasionally the cause of disease in the tropics and subtropics

IDENTIFICATION

Gross: The adults are small, conical, maggot-like flukes about 1.0 cm long. One sucker is visible at the tip of the cone and the other at the base. The larval stages are less than 5.0 mm, fresh specimens having a pink colour.

Microscopic: The egg resembles that of *F. hepatica* being large and operculate, but is clear rather than yellow.

LIFE CYCLE

Development in the snail intermediate host is similar to that of *Fasciola* and under favourable conditions

(26–30 °C) can be completed in four weeks. After ingestion of encysted metacercariae with herbage, development in the final host occurs entirely in the alimentary tract. Following excystment in the duodenum the young flukes attach and feed there for about six weeks before migrating forward to the fore-stomachs where they mature. The prepatent period is between 7–10 weeks.

PATHOGENESIS

Any pathogenic effect is associated with the intestinal phase of the infection. The young flukes are plug feeders and this results in severe erosions of the duodenal mucosa. In heavy infections these cause an enteritis characterised by oedema, haemorrhage and ulceration. At necropsy the young flukes can be seen as clusters of brownish pink parasites attached to the duodenal mucosa and occasionally also in the jejunum and abomasum. The adult parasites (Pl. V) in the fore-stomachs are well tolerated, even when many thousands are present and feeding on the wall of the rumen or reticulum (Fig. 84).

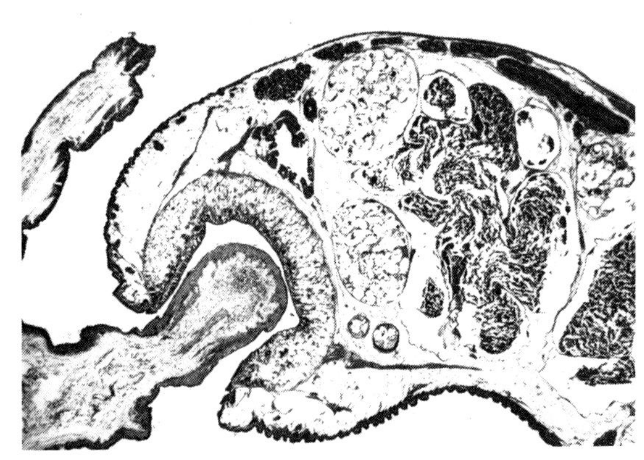

Fig 84 Section of adult *Paramphistomum* feeding on the ruminal epithelium.

CLINICAL SIGNS

In heavy duodenal infections, the most obvious sign is diarrhoea accompanied by anorexia and intense thirst. Sometimes in cattle, there is rectal haemorrhage following a period of prolonged straining. Mortality in acute outbreaks can be as high as 90%.

EPIDEMIOLOGY

Paramphistomiasis often depends for its continuous endemicity on permanent water masses, such as lakes and ponds, from which snails are dispersed into previously dry areas by flooding during heavy rains. Paramphistome eggs deposited by animals grazing these areas hatch and infect snails. Subsequent production of cercariae often coincides with receding water levels making them accessible to grazing ruminants. In other areas, the situation is complicated by the ability of the snails to aestivate on dry pastures and become reactivated on the return of rainfall.

A good immunity develops in cattle, and outbreaks are usually confined to young stock. However, adults continue to harbour low burdens of adult parasites and are important reservoirs of infection for snails. In contrast, sheep and goats are relatively susceptible throughout their lives.

DIAGNOSIS

This is based on the clinical signs usually involving young animals in the herd and a history of grazing around snail habitats during a period of dry weather. Faecal examination is of little value since the disease occurs during the prepatent period. Confirmation can be obtained by a postmortem examination and recovery of the small flukes from the duodenum.

TREATMENT

For the treatment of outbreaks due to the immature stages, rafoxanide and niclofolan are recommended at dosage rates 2–3 times the level recommended for fascioliasis. The flukicide, oxyclozanide, is effective against adult paramphistomes.

CONTROL

As in *F. gigantica* the best control is achieved by providing a piped water supply to troughs and preventing access of the animals to natural water. Even then snails may gain access to watering troughs and regular application of a molluscicide at source or manual removal of snails may be necessary.

Other paramphistomes of veterinary importance

Apart from *Paramphistomum* there are many genera found in domestic stock. Among the best known of those occurring in the fore-stomachs are *Cotylophoron* and *Ceylonocotyle* which are responsible for outbreaks of paramphistomiasis in buffalo and cattle in Asia and cattle in Australasia and the southern U.S.A. Of those occurring in the lower alimentary tract, *Gastrodiscus* has a high prevalence in the small and large intestine of equines in the tropics and *Homologaster* in the large intestine of ruminants in some parts of Asia. Another genus, *Gigantocotyle*, occurs in the bile ducts of buffaloes in the Middle and Far East.

Family TROGLOTREMATIDAE

Two genera are of local veterinary interest. The first, *Paragonimus*, commonly referred to as the 'lung fluke', is found in cats, dogs and other carnivores and in man in North American and Asia. The cycle involves a water snail and a crayfish or fresh water crab. The water snail, previously infected by a miracidium in the same way as in other flukes, is ingested by a crustacean. Infection of the final host occurs by ingestion of the metacercariae in the liver or muscles of the crustacean. The young flukes migrate to the lungs where they are encapsulated by fibrous cysts connected by fistulae to the bronchioles to facilitate egg excretion. Pulmonary signs are comparatively rare in cats or dogs and the veterinary interest is in the potential reservoir of infection for man.

The second genus, *Nanophyetus*, is a fluke found mainly in the small intestine of dogs, mink and other fish-eating mammals. It occurs in the N.W. United States and parts of Siberia and is of importance because the flukes are vectors of the rickettsia *Neorickettsia helminthoeca* which causes a severe haemorrhagic enteritis of dogs, the so-called 'salmon poisoning'. This name is derived from the cycle of the fluke which involves a water snail and a fish which is often one of the salmonid type.

The rickettsial infection is presumed to pass via the fluke egg to the water snail and thence to the fish where the cercariae locate the kidney. The dog is infected by ingesting fish containing the metacercariae and the associated rickettsia.

Family OPISTHORCHIIDAE

The members of this family require two intermediate hosts, the first being water snails and the second a wide variety of fish, in which the metacercariae are encysted. The final hosts are fish-eating mammals in which they inhabit the bile ducts.

Opisthorchis, by far the most important genus, is a small lanceolate fluke, grossly resembling *Dicrocoelium*. Two species have man as their primary host. *O. sinensis* (formerly *Clonorchis sinensis*) is the 'Chinese liver fluke' which, in spite of its name, occurs in most of Asia east of, and including, the Indian subcontinent; the dog and cat are occasional hosts. *O. viverrini*, though discovered in the civet cat, which is therefore the type host, is essentially a parasite of humans; it is found sufficiently often in the dog and cat for these animals to be important reservoir hosts. It occurs in the Far East. *O. tenuicollis* (syn. *O. felineus*) has the cat, and less often the dog, as primary hosts; man is an occasional host. This species occurs in eastern Europe, the Middle East, Asia Minor, and Asiatic Russia; it has been reported in Canada.

Humans and animals acquire infection by ingestion of raw, or insufficiently-cooked fish, and the young flukes travel to the liver by the bile ducts. Most infections are inapparent, though *Opisthorchis* is very common in its endemic areas. In very heavy infections the main damage is caused by the young flukes whose cuticular spines abrade the bile ducts, causing thickening with papilloma formation and the development of cysts, containing flukes, adjacent to the ducts. In these infections the symptoms are emaciation, jaundice and ascites; it is sometimes possible to palpate the enlarged nodular liver.

Control of opisthorchiasis is based on thorough cooking of fish.

A second genus, *Metorchis*, occurs mainly in sledge dogs and hunting dogs, and has a mostly Arctic and subarctic distribution in Europe, Asia and America; it has also been recorded in Yugoslavia. The pathogenesis is similar to that of opisthorchiasis and there is a reservoir of infection in wild fish-eating carnivores.

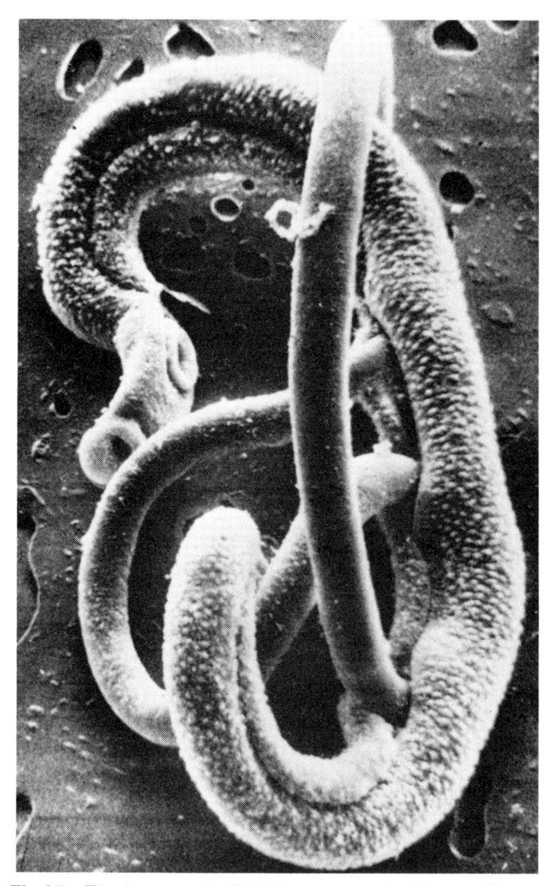

Fig 85 The large male *Schistosoma* carries the smaller female in the gynaecophoric canal.

Family SCHISTOSOMATIDAE

This family is primarily parasitic in the blood vessels of the alimentary tract and bladder. In man, schistosomes are often responsible for severe and debilitating disease and veterinary interest lies in the fact that they can cause a similar disease in animals some of which may act as reservoirs of infection for man. The schistosomes differ from other flukes in that the sexes are separate, the small adult female lying permanently in a groove, the gynaecophoric canal, in the body of the male (Fig. 85). The most important genus is *Schistosoma* and this will be discussed in some detail.

Schistosoma

Hosts:
All domestic mammals. Mainly important in sheep and cattle

Intermediate hosts:
Water snails. *Bulinus* and *Physopsis* spp. are particularly important in the transmission of bovine and ovine schistosomiasis

Site:
Usually mesenteric veins; one species occurs in nasal veins

Species and distribution:
Major:

Schistosoma bovis	Ruminants in Africa, Middle East, Asia, southern Europe
S. mattheei	Ruminants and occasionally man in Africa
S. japonicum	Man and most domestic animals in the Far East

Minor:

S. leiperi	Cattle in Africa
S. spindale	Ruminants, horses and pigs in Asia and the Far East
S. incognitum	Pigs and dogs in India and Pakistan
S. mansoni	Man and wild animals in Africa, South America and the Middle East
S. (syn. *Orientobilharzia*) *turkestanica*	Ruminants in Asia
S. nasalis	Ruminants and horses in India and Pakistan

Distribution:
Tropics and subtropics. Occurs sparsely in southern Europe

IDENTIFICATION

Gross: The sexes are separate, the male, which is broad and flat and about 2.0 cm long, carrying the female in the hollow of its inwardly curved body. This characteristic and the vascular predilection site are sufficient for generic identification.
Microscopic: The egg is 100 μm long, spindle-shaped and has a lateral or terminal spine (Fig. 86). There is no operculum.

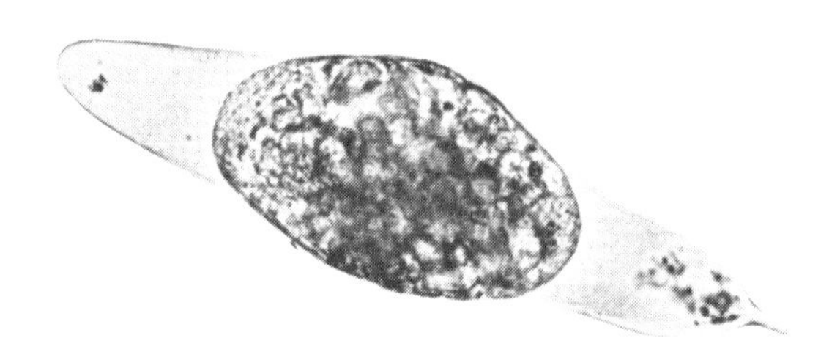

Fig 86 Thin shelled *Schistosoma* egg with terminal spine.

LIFE CYCLE

The female in the mesenteric vein inserts her tail into a small venule and since the genital pore is terminal, the eggs are deposited, or even pushed, into the venule. There, aided by their spines and by proteolytic enzymes secreted by the unhatched miracidia, they penetrate the endothelium to enter the intestinal submucosa and ultimately the gut lumen; they are then passed out in the

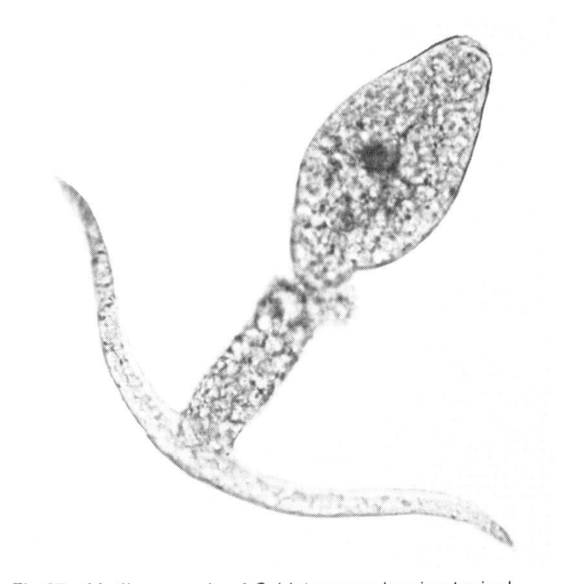

Fig 87 Motile cercaria of *Schistosoma* showing typical forked (furcocercous) tail.

faeces. The eggs hatch in minutes in water and the miracidia penetrate appropriate snails. Development to the cercarial stage occurs without a redial form and there is no metacercarial phase, penetration of the final host by the motile cercariae (Fig. 87) occurring via the skin or by ingestion in drinking water. The developmental period in the snail can be as short as five weeks. After penetration or ingestion the cercariae lose their forked tails, transform to schistosomula, or young flukes, and travel via the blood stream through the heart and lungs to the systemic circulation. In the liver they locate in the portal veins and become sexually mature before migrating to their final site, the mesenteric veins. The prepatent period is 6–7 weeks.

PATHOGENESIS

Schistosomiasis is generally considered to be a much more serious and important infection in sheep than in larger ruminants and even where a high prevalence of the parasite is detected in slaughtered cattle clinical signs of the disease are seen only rarely.

Acute disease characterised by diarrhoea and anorexia occurs 7–8 weeks after heavy infection and is entirely due to the inflammatory and granulomatous response to the deposition of eggs in the mesenteric veins and their subsequent infiltration in the intestinal mucosa. Following massive infection death can occur rapidly, but more usually the clinical signs abate slowly as the infection progresses. As this occurs, there appears to be a partial shift of worms away from the intestinal mucosa and reactions to these migrating parasites and their eggs can occur in the liver (Fig. 88).

At necropsy during the acute phase of the disease there are marked haemorrhagic lesions in the mucosa of the intestine, but as the disease progresses the wall of the intestine appears greyish, thickened and oedematous due to confluence of the egg granulomata and the associated inflammatory changes; on sections of the liver there is also evidence of egg granulomata and of portal fibrosis provoked by eggs which have, inadvertently, been swept into small portal vessels.

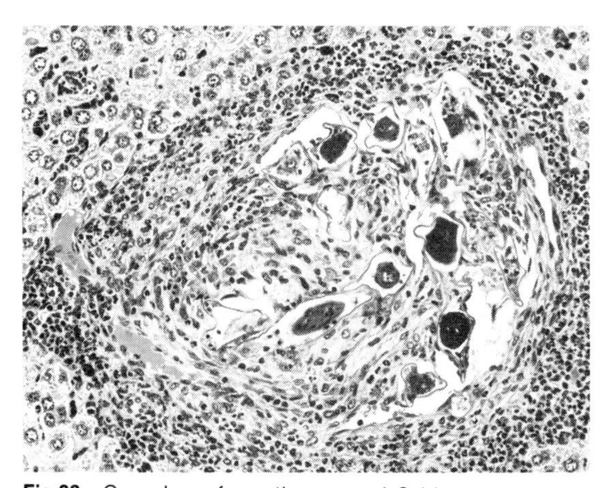

Fig 88 Granuloma formation around *Schistosoma* eggs in the liver.

In sheep, anaemia and hypoalbuminaemia have been shown to be prominent during the clinical phase apparently as a result of mucosal haemorrhage, dyshaemopoeisis and an expansion in plasma volume.

The significance of low level infection is not known, but it has been suggested that this may have a considerable effect on productivity.

There is evidence, experimentally, of acquired resistance to reinfection by homologous species and, from natural infections, that resistance may develop as a result of prior exposure to a heterologous species.

CLINICAL SIGNS

These are diarrhoea, sometimes blood-stained and containing mucus, anorexia, thirst, anaemia and emaciation.

EPIDEMIOLOGY

The epidemiology is very similar to that of *F. gigantica* and *Paramphistomum* spp., *Schistosoma* spp. being totally dependent upon water as a medium for infection of both the intermediate and final host. The fact that percutaneous infection may occur encourages infection where livestock are obliged to wade in water.

DIAGNOSIS

This is based mainly on the clinico-pathological picture of diarrhoea, wasting and anaemia, coupled with a history of access to natural water sources. The relatively persistent diarrhoea, often blood stained and containing mucus, may help to differentiate this syndrome from fascioliasis.

The demonstration of the characteristic eggs in the faeces or in squash preparations of blood and mucus from the faeces is useful in the period following patency but less useful as egg production drops in the later stages of infection.

In general, when schistosomiasis is suspected, diagnosis is best confirmed by a detailed post-mortem examination which will reveal the lesions and, if the mesentery is stretched, the presence of numerous schistosomes in the veins. In epidemological surveys, serological tests may be of value.

TREATMENT

Care has to be exercised in treating clinical cases of schistosomiasis since the dislodgement of the damaged flukes may result in emboli being formed and subsequent occlusion of major mesenteric and portal blood vessels with fatal consequences. The drugs still widely used are the antimonial preparations, tartar emetic, antimosan and stibophen although these are being superseded by niridazole and trichlorfon, all of which have to be given over a period of days at high dosage rates. Fatalities associated with the use of these drugs are not uncommon. The recent development of praziquantel for the treatment of human schistosomiasis is promising, but has still to be evaluated in animals.

CONTROL

This is similar to that outlined for *F. gigantica* and *Paramphistomum* infections. Since the prevalence of snail populations varies according to temperature, local efforts should be made to identify the months of maximum snail population, and cattle movements planned to avoid their exposure to dangerous stretches of water at these times.

Nasal schistosomiasis

In parts of the India sub-continent *Schistosoma nasalis* occurs in the nasal mucosal veins of cattle, buffalo and horses. In heavy infections there is a copious mucopurulent discharge, snoring and dyspnoea. The main pathogenic effects are associated with the eggs which cause abscess formation in the mucosa. Fibrous granulomatous growths occur which may occlude the nasal passages. Infection is confirmed by the presence of the spindle-shaped eggs in the nasal discharge.

[**Schistosomiasis in man:** Schistosomiasis, still often called by the old name of bilharziasis, is of major importance and together with ancylostomiasis and filariasis are the most important helminth diseases of man. The two most common intestinal species are *S. mansoni* and *S. japonicum*, the latter occasionally occuring in domestic livestock. The third, *S. haematobium*, is found in the veins of the bladder.

A form of cutaneous larva migrans, often called 'swimmers itch', occurs in man and is thought to be caused by cercariae of avian and animal schistosomes which have a limited migration in human skin.]

Family DIPLOSTOMATIDAE

The family Diplostomatidae includes the genus *Alaria* of which several species occur in the small intestine of dogs, cats, foxes and mink in most continents although they have not been recorded in Britain. The life cycle involves two intermediate hosts, namely freshwater snails and frogs. The definitive host becomes infected by eating frogs containing encysted metacercariae (mesocercariae) and the flukes migrate extensively, including passage through the lungs, before returning to the small intestine. Heavy infections may cause a severe duodenitis in dogs and cats. A fatal case has been recorded in man through eating inadequately cooked frogs' legs; the principal lesions were in the lungs.

Treatment with praziquantel is recommended.

Class CESTODA

This class differs from the Trematoda in having a tapelike body with no alimentary canal. The body is segmented, each segment containing one and sometimes two sets of male and female reproductive organs. Almost all the tapeworms of veterinary importance are in the Order Cyclophyllidea, the two exceptions being in the Order Pseudophyllidea.

Order CYCLOPHYLLIDEA

STRUCTURE AND FUNCTION

The adult cestode (Fig. 89) consists of a head or **scolex** bearing attachment organs, a short unsegmented neck and a chain of segments. The chain is known as a **strobila** and each segment as a **proglottid**.

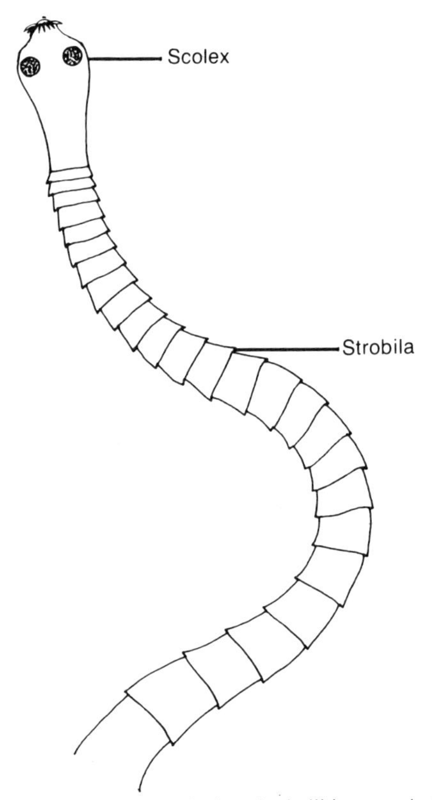

Fig 89 Structure of a typical cyclophyllidean cestode.
(a) head, neck and strobila.

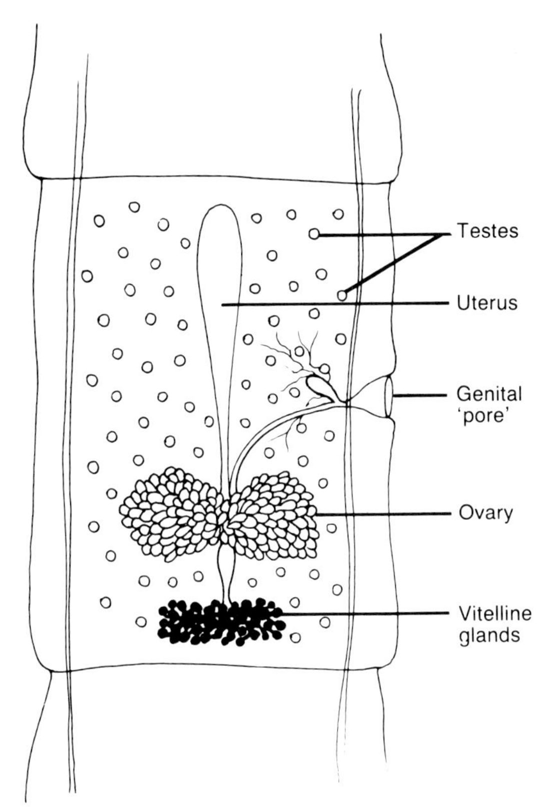

(b) mature segment showing reproductive organs.

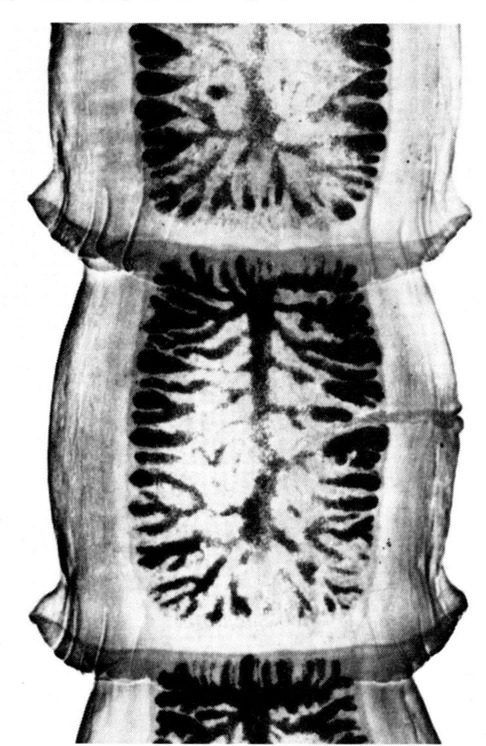
(c) gravid segment.

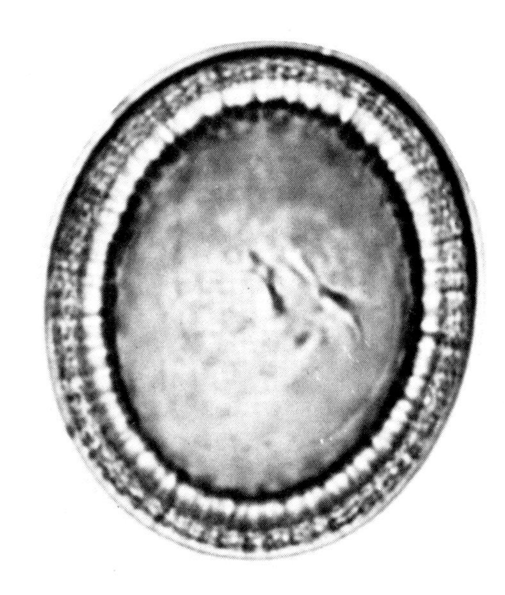
(d) typical taeniid egg.

The organs of attachment are four suckers on the sides of the scolex and these may bear hooks.

The scolex usually bears anteriorly a mobile protrusible cone or rostellum and in some species this may be also armed with one or more concentric rows of hooks which aid in attachment.

The proglottids are continuously budded from the neck region and become sexually mature as they pass down the strobila. Each proglottid is hermaphrodite with one or two sets of reproductive organs, the genital pores usually opening on the lateral margin or margins of the segment (Fig. 89); both self-fertilisation and cross-fertilisation between proglottids may occur. The structure of the genital system is generally similar to that of the trematodes. As the segment matures, its internal structure largely disappears and the fully ripe or gravid proglottid eventually contains only remnants of the branched uterus packed with eggs. The gravid segments are usually shed intact from the strobila and pass out with the faeces. Outside the body the eggs are liberated by disintegration of the segment or are shed through the genital pore.

The egg (Fig. 89) consists of:

(i) The hexacanth (6-hooked) embryo or **onchosphere**.

(ii) A thick, dark, radially striated 'shell' called the **embryophore**.

(iii) A true shell which is a delicate membrane and is often lost while still in the uterus.

The tegument of the adult tapeworm is highly absorptive, the worm deriving all its nourishment through this structure. Below the tegument are muscle cells and the parenchyma, the latter a syncytium of cells which fills the space between the organs. The nervous system consists of ganglia in the scolex from which nerves enter the strobila. The excretory system, as in the Trematoda, is composed of flame cells leading to efferent canals which run through the strobila to discharge at the terminal segment.

LIFE CYCLE

The typical life cycle of these cestodes is indirect with one intermediate host. With few exceptions, the adult tapeworm is found in the small intestine of the final host, the segments and eggs reaching the exterior in the faeces.

When the egg is ingested by the intermediate host, the gastric and intestinal secretions digest the embryophore and activate the onchosphere. Using its hooks, it tears through the mucosa to reach the blood or lymph stream or, in the case of invertebrates, the body cavity. Once in its predilection site the onchosphere loses its hooks and develops, depending on the species, into one of the following larval stages, often known as **metacestodes** (Fig. 90).

Cysticercus: fluid-filled cyst containing an attached single invaginated scolex, sometimes called a protoscolex.

Coenurus: similar to a cysticercus, but with numerous invaginated scolices.

Strobilocercus: the scolex is evaginated and is connected to the cyst by a chain of asexual proglottids. The latter are digested away after ingestion by the final host, leaving only the scolex.

Hydatid: this is a large fluid-filled cyst lined with germinal epithelium from which are produced invaginated scolices which lie free or in bunches, surrounded by germinal epithelium (brood capsules). The contents of the cysts other than the fluid, ie. scolices and brood capsules, are frequently described as 'hydatid sand'.

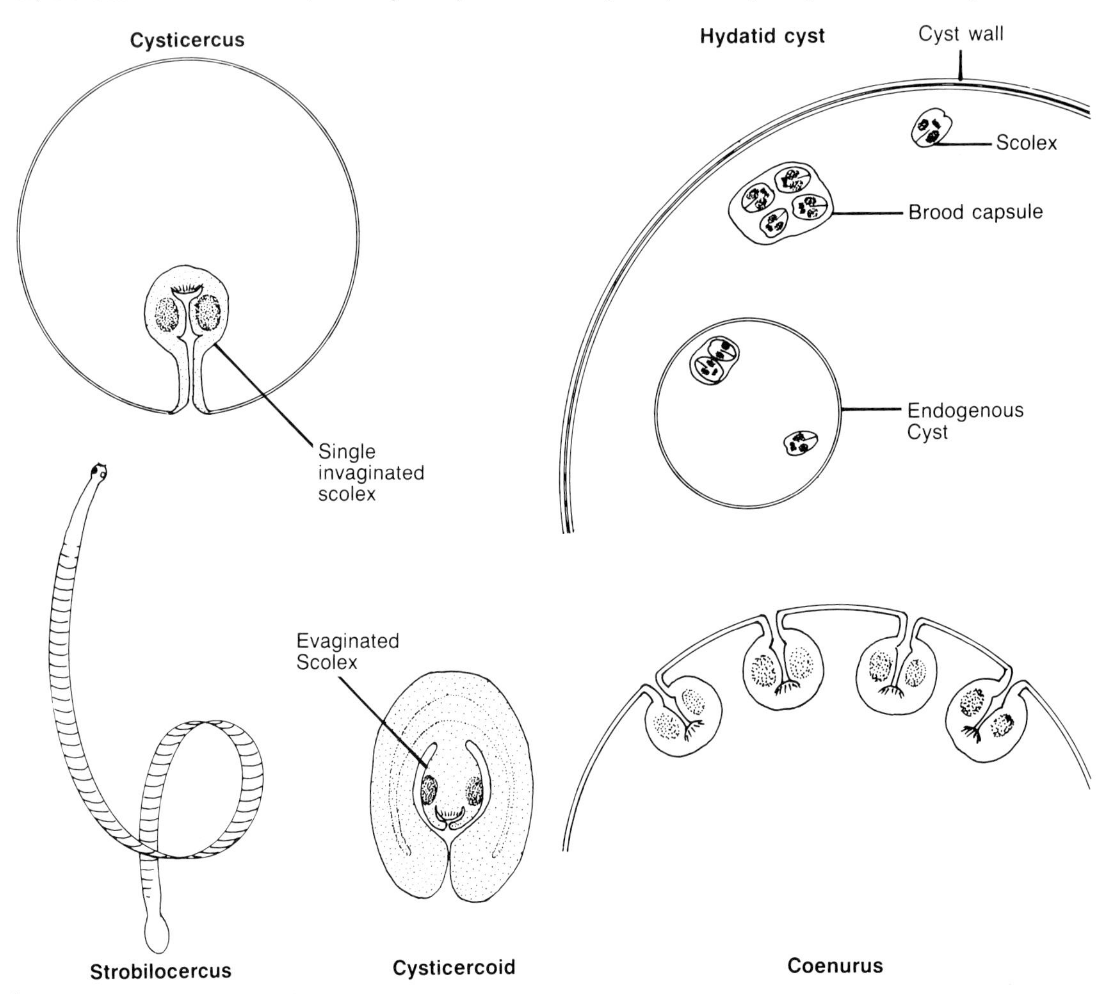

Fig 90 Larval stages of cyclophyllidean cestodes.

Occasionally also, daughter cysts complete with cuticle and germinal layer are formed endogenously or, if the cyst wall ruptures, exogenously.
Cysticercoid: a single evaginated scolex embedded in a small solid cyst. Typically found in very small intermediate hosts such as arthropods.
Tetrathyridium: worm-like larva with an invaginated scolex; found only in Mesocestoididae.

When the metacestode is ingested by the final host the evaginated scolex attaches to the mucosa, the remainder of the structure is digested off, and a chain of proglottids begins to grow from the base of the scolex.

The seven main families of veterinary interest in the order Cyclophyllidea are: the *Taeniidae*, *Anoplocephalidae*, *Dilepididae*, *Davaineidae*, *Hymenolepididae*, *Mesocestoididae* and *Thysanosomidae*.

Family TAENIIDAE

The adults are found in domestic carnivores and man. The scolex has an armed rostellum with a concentric double row of hooks (the important exception is *Taenia saginata* whose scolex is unarmed). The gravid segments are longer than they are wide.

The intermediate stage is a cysticercus, strobilocercus, coenurus or hydatid cyst and these occur only in mammals.

Taenia

This is a most important genus, both the adult and larval stages being of importance in human health and veterinary medicine.

The final host, larval stages, intermediate host and larval predilection site of the major species are given in Table 2.

Taenia saginata

The intermediate stage of this tapeworm, found in the muscles of cattle, frequently presents economic problems to the beef industry and is a public health hazard.

DISTRIBUTION

Worldwide

IDENTIFICATION

The adult tapeworm, found only in man, ranges from 5.0–15.0 m in length. The scolex, exceptional among the species of *Taenia*, has neither rostellum nor hooks; the uterus of the gravid segment has 15–30 lateral branches on each side of the central stem in contrast to that of *T. solium* with only 7–12 lateral branches.

Table 2 Final and intermediate hosts of *Taenia* spp.

Adult tapeworm	Final host	Larva	Intermediate host	Larval site
T. saginata	Man	*Cysticercus bovis*	Cattle	Muscle
T. solium	Man	*Cysticercus cellulosae*	Pig, man	Muscle
T. multiceps	Dog	*Coenurus cerebralis*	Sheep, cattle	Central nervous system
T. hydatigena	Dog	*Cysticercus tenuicollis*	Sheep, cattle, pig	Peritoneum
T. ovis	Dog	*Cysticercus ovis*	Sheep	Muscle
T. pisiformis	Dog	*Cysticercus pisiformis*	Rabbit	Peritoneum
T. serialis	Dog	*Coenurus serialis*	Rabbit	Connective tissue
T. taeniaeformis	Cat	*Cysticercus fasciolaris* (strobilocercus)	Mouse, rat	Liver
T. krabbei	Dog	*Cysticercus tarandi*	Reindeer	Muscle

In the bovine animal the mature cysticercus, *C. bovis*, is greyish white, about 1.0 cm in diameter and filled with fluid in which the scolex is usually clearly visible. As in the adult tapeworm it has neither rostellum nor hooks. Although it may occur anywhere in the striated muscles the predilection sites, at least from the viewpoint of routine meat inspection, are the heart, the tongue and the masseter and intercostal muscles.

LIFE CYCLE

An infected human may pass millions of eggs daily, either free in the faeces or as intact segments each containing about 250,000 eggs, and these can survive on pasture for several months. After ingestion by a susceptible bovine the onchosphere travels via the blood to striated muscle. It is first grossly visible about two weeks later as a pale, semi-transparent spot about 1.0 mm in diameter, but is not infective to man until about 12 weeks later when it has reached its full size of 1.0 cm (Pl. VI). By then it is enclosed by the host in a thin fibrous capsule but despite this the scolex can usually still be seen. The longevity of the cysts ranges from weeks to years. When they die they are usually replaced by a caseous, crumbly mass which may become calcified. Both living and dead cysts are frequently present in the same carcass.

Man becomes infected by ingesting raw or inadequately cooked meat. Development to patency takes 2–3 months.

PATHOGENESIS AND CLINICAL SIGNS

Under natural conditions the presence of cysticerci in the muscles of cattle is not associated with clinical signs although, experimentally, calves given massive infections of *T. saginata* eggs have developed severe myocarditis and heart failure associated with developing cysticerci in the heart.

In man, the adult tapeworm may produce diarrhoea and hunger pains, but the infection is usually asymptomatic and is mainly objectionable on aesthetic grounds.

EPIDEMIOLOGY

There are two quite distinct epidemiological patterns found in developing countries and developed countries respectively.

Developing countries

In many parts of Africa, Asia and Latin America cattle are reared on an extensive scale, human sanitation is poorly developed and cooking fuel is expensive. In these circumstances the incidence of human infection with *T. saginata* is high, in certain areas being well over 20%. Because of this, calves are usually infected in early life, often within the first few days after birth, from infected stockmen whose hands are contaminated with *Taenia* eggs. Prenatal infection of calves may also occur, but is rare. Of the cysts which develop, a proportion persist for years even although the host has developed an acquired immunity and is completely resistant to further infection.

Based on routine carcass inspection, the infection rate is often around 30–60%, although the real prevalence is considerably higher.

Developed countries

In areas such as Europe, North America, Australia and New Zealand, the standards of sanitation are high and meat is carefully inspected and generally thoroughly cooked before consumption. In such countries, the prevalence of cysticercosis is low, being less than 1% of carcasses inspected. Occasionally however, a cysticercosis 'storm', where a high proportion of cattle are infected, has been reported on particular farms. In Britain and in Australia, this has been associated with the use of human sewage on pasture as a fertiliser in the form of sludge, ie. sedimented or bacterial-digested faeces. Since *T. saginata* eggs may survive for more than 200 days in sludge, the occurrence of these 'storms' is perhaps not surprising. Other causes of a sudden high incidence of infection on particular farms are due to a tapeworm infection in a stockman occurring either as a random event or, as has been reported from feedlots in some of the southern states of the U.S.A., as a result of the use of migrant labour from a country with a high prevalence of infection.

As distinct from these 'storms', the cause of the low but persistent prevalence of infection in cattle is obscure, but is thought to be due to the access of cattle to water contaminated with sewage effluents, to the carriage and dispersal of *T. saginata* eggs by birds which frequent sewage works or feed on effluent discharged into rivers or the sea, and to occasional fouling of pasture by itinerant infected individuals.

In contrast to the epidemiology in developing countries cattle of any age are susceptible to infection since they generally possess no acquired immunity. There is also evidence that when cattle are first infected as adults the longevity of the cysticerci is limited, most being dead within nine months.

DIAGNOSIS

Individual countries have different regulations regarding the inspection of carcasses, but invariably the masseter muscle, tongue and heart are incised and examined and the intercostal muscles and diaphragm inspected; the triceps muscle is also incised in many countries.

The inspection is inevitably a compromise between detection of cysticerci and the preservation of the economic value of the carcass.

TREATMENT

As yet there is no drug available which will effectively destroy all of the cysticerci in the muscle.

CONTROL

In developed countries the control of bovine cysticercosis depends on a high standard of human sanitation, on the general practice of cooking meat thoroughly since the thermal death point of cysticerci is 57 °C, and on compulsory meat inspection.

Regulations usually require that infected carcasses are frozen at −10 °C for at least 10 days which is sufficient to kill the cysticerci although the process reduces the economic value of the meat. Where relatively heavy infections of more than 25 cysticerci are detected, it is usual to destroy the carcass.

In agricultural practice the use of human sludge as a fertiliser should be confined to cultivated fields or to those on which cattle will not be grazed for at least two years.

In developing countries the same measures are necessary, but are not always economically feasible, and at present the most useful step would appear to be the education of communities in both sanitary and culinary hygiene.

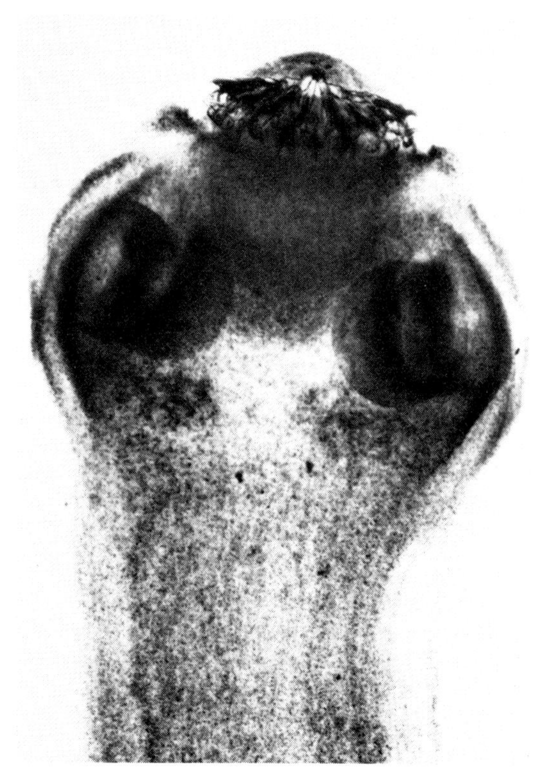

Fig 91 Typical taeniid head showing four suckers and rostellum bearing hooks.

Taenia solium

This is the other *Taenia* species of man, the larval stage, *Cysticercus cellulosae*, occurring in the muscles of the pig. On occasions the cysticerci may also develop in man and the disease, human cysticercosis, is the most serious aspect of this zoonosis.

DISTRIBUTION

This cestode is most prevalent in Latin America, India, Africa and parts of the Far East, apart from areas where there are religious sanctions on the eating of pork. It is now uncommon in developed countries.

IDENTIFICATION

The adult tapeworm is similar to *T. saginata* except that the scolex is typically taeniid, having a rostellum armed with two concentric rows of hooks (Fig. 91), while the uterus of the gravid segment, as noted previously, has fewer lateral branches.

LIFE CYCLE

This is also similar to that of *T. saginata* with the important differences that man, the final host, may also become infected with cysticerci. This may occur either from the accidental ingestion of *T. solium* eggs or, apparently, in a person with an adult tapeworm, from the liberation of oncospheres after the digestion of a gravid segment which has entered the stomach from the duodenum by reverse peristalsis; this is known as autoinfection.

PATHOGENESIS AND CLINICAL SIGNS

As in *T. saginata* infection, clinical signs are inapparent in pigs naturally infected with cysticerci and generally insignificant in humans with adult tapeworms.

However, when man is infected with cysticerci, various clinical signs may occur depending on the location of the cysts in the organs, muscles or subcutaneous tissue. Most seriously, cysticerci may develop in the central nervous system producing mental disturbances or clinical signs of epilepsy or intracranial hypertension; they may also develop in the eye with consequent loss of vision. In Latin America alone it is estimated that almost 0.5 million people are affected, either by the nervous or ocular forms of cysticercosis.

EPIDEMIOLOGY

This is basically similar to that of *T. saginata* in developing countries except, of course, that it depends primarily

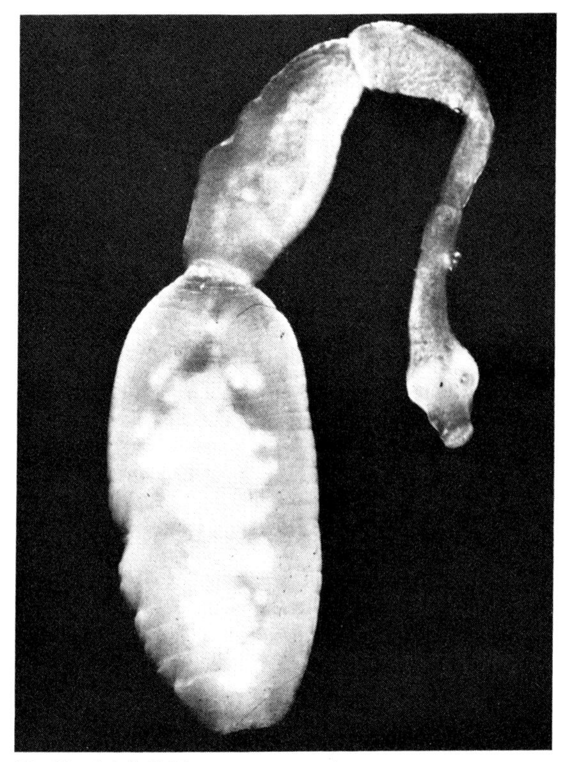

Fig 94 Adult *Echinococcus granulosus.*

(Fig. 94), and is therefore difficult to find in the freshly opened intestine, It consists of a scolex and three or four segments, the terminal gravid one occupying about half the length of the complete tapeworm.

Microscopic: The scolex is typically taeniid, and each segment has a single genital opening. The embryophore is similar to that of *Taenia* spp., radially striated and containing a six-hooked onchosphere.

LIFE CYCLE

The prepatent period in the final host is around 40–50 days, after which only one gravid segment is shed by the tapeworm per day. The onchospheres are capable of prolonged survival outside the host, being viable on the ground for about two years. After ingestion by the intermediate host, the onchosphere penetrates the gut wall and travels in the blood to the liver, or in the lymph to the lungs. These are the two commonest sites for larval development, but occasionally onchospheres escape into the general systemic circulation and develop in other organs and tissues.

Growth of the hydatid is slow, maturity being reached in 6–12 months. In the liver (Pl. VI) and lungs the cyst may have a diameter of up to 20 cm, but in the rarer sites such as the abdominal cavity, where unrestricted growth

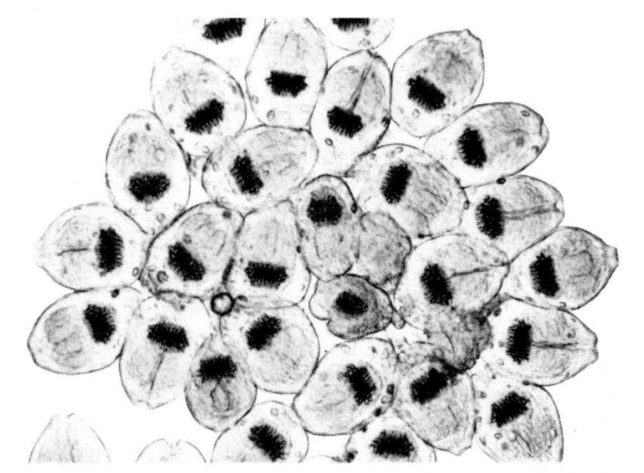

Fig 95 Scolices of *Echinococcus granulosus* from hydatid cyst.

is possible, it may be very large, and contain several litres of fluid. The cyst capsule consists of an outer membrane and an inner germinal epithelium from which, when cyst growth is almost complete, brood capsules each containing a number of scolices are budded off (Fig. 95). Many of these brood capsules become detached and exist free in the hydatid fluid; collectively these and the scolices are often referred to as 'hydatid sand'.

Sometimes, complete daughter cysts are formed either inside the mother cyst or externally; in the latter case they may be carried to other parts of the body to form new hydatids.

In sheep about 70% of hydatids occur in the lungs, about 25% in the liver, and the remainder in other organs. In horses and cattle more than 90% of cysts are usually found in the liver. Little local reaction is shown by most animals to the growing hydatid, which appears as a thin-walled cyst, partially embedded in the organ, but in horses a thick fibrous capsule develops around the cyst. The cysts may be few, and as large as tennis balls, or numerous and small and showing on the liver as small white patches.

PATHOGENESIS AND CLINICAL SIGNS

The adult tapeworm is not pathogenic, and thousands may be present in a dog without clinical signs.

In domestic animals the hydatid in the liver or lungs is usually tolerated without any clinical signs, and the majority of infections are only revealed at the abattoir. Where onchospheres have been carried in the circulation to other sites, such as the kidney, pancreas, CNS or marrow cavity of long bones, pressure by the growing cyst may cause a variety of clinical signs.

In contrast, when man is involved as an intermediate host the hydatid in its pulmonary or hepatic site is often of pathogenic significance. One or both lungs may be affected causing respiratory symptoms and if several hydatids are present in the liver there may be gross abdominal distension. If a cyst should rupture there is a risk of death from anaphylaxis or if the person survives, released daughter cysts may resume development in other regions of the body.

EPIDEMIOLOGY

E.g. granulosus

Only a few countries, notably Iceland and Eire, are free from this strain of *E. granulosus*. It is customary to consider the epidemiology as being based on two cycles, pastoral and sylvatic.

In the pastoral cycle the dog is always involved, being infected by the feeding of ruminant offal containing hydatid cysts. The domestic intermediate host will vary according to the local husbandry but the most important is the sheep, which appears to be the natural intermediate host, scolices from these animals being the most highly infective for dogs. In parts of the Middle East the camel is the main reservoir of hydatids, while in northern Europe and northern U.S.S.R. it is the reindeer.

The pastoral cycle is the primary source of hydatidosis in man, infection being by accidental ingestion of onchospheres from the coats of dogs, or from vegetables and other foodstuffs contaminated by dog faeces.

The sylvatic cycle occurs in wild canids and ruminants and is based on predation or carrion feeding. It is less important as a source of human infection, except in hunting communities where the infection may be introduced to domestic dogs by the feeding of viscera of wild ruminants.

E.g. equinus

Equine hydatidosis is commonest in Europe, and in other parts of the world most cases have been recorded in imported European horses. The strain is highly specific for the horse and the eggs do not develop in the sheep. The domestic dog and the red fox are the final hosts, and the cycle in countries of high prevalence depends on access by dogs to infected equine viscera. On mainland Europe, the most likely source is offal from horse abattoirs and in Britain the viscera of hunting horses which are fed to foxhounds.

DIAGNOSIS

The presence of hydatids as a clinical entity is rarely suspected in domestic animals, and specific diagnosis is never called for. In man, the methods most commonly used are serological tests such as complement fixation or immunoelectrophoresis. Scanning techniques may be used to locate the cysts.

Diagnosis of infection in dogs with adult tapeworms is difficult, because the segments are small and are only shed sparsely. When found, identification is based on their size of 2.0–3.0 mm, ovoid shape, and single genital pore.

In some countries control regimes involve the administration of purgative anthelmintics, such as arecoline hydrochloride, so that the whole tapeworm is expelled in mucus and can be searched for in the faeces. If a necropsy is available the small intestine should be opened and immersed in shallow water, when the attached tapeworms will be seen as small slender papillae.

TREATMENT

Echinococcus tapeworms are more difficult to remove than *Taenia*, but several drugs, notably praziquantel, are now available which are highly effective. After treatment it is advisable to confine dogs for 48 hours to facilitate the collection and disposal of infected faeces.

In man, hydatid cysts may be excised surgically although, more recently, albendazole therapy has been reported to be effective.

CONTROL

This is based on the regular treatment of dogs to eliminate the adult tapeworms and on the prevention of infection in dogs by exclusion from their diet of animal material containing hydatids. This is achieved by denying dogs access to abattoirs, and where possible, by proper disposal of sheep carcasses on farms. In some countries these measures have been supported by legislation, with penalties when they are disregarded.

In countries where no specific measures for hydatid control exist, it has been found that an incidental benefit from the destruction of stray dogs for rabies control has been a great reduction in the incidence of hydatid infection in humans.

Echinococcus multilocularis

Hosts:
Wild canids, domestic dog and cat

Intermediate hosts:
Mainly microtine rodents, such as voles and lemmings, and insectivores; some of the larger mammals, including man, are also susceptible

Site:
The adults occur in the intestine and the hydatids mainly in the liver

Distribution:
Northern Hemisphere, including North America, Greenland, Scandinavia, Central Europe, Middle East, India and Japan

IDENTIFICATION

Generally similar to *E. granulosus*, but usually with four or five segments.

LIFE CYCLE AND PATHOGENESIS

The intermediate host is infected by ingestion of the onchosphere, and the larval stage develops primarily in the liver as the so-called multilocular or alveolar cyst, a diffuse growth with many compartments containing a gelatinous matrix into which the scolices are budded off (Fig. 96). Growth of the intermediate stage is invasive, extending locally and capable of systemic metastases.

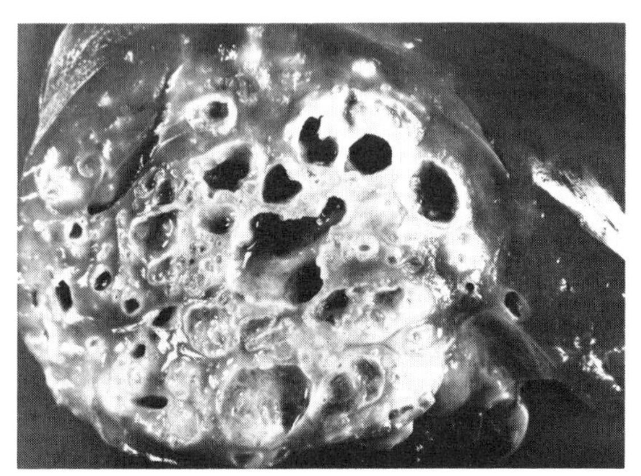

Fig 96 Human liver infected with larval *Echinococcus multilocularis*.

EPIDEMIOLOGY AND SIGNIFICANCE

Though *E. multilocularis* has a wide distribution in the northern hemisphere, it is essentially a parasite of tundra regions with its greatest prevalence in the subarctic regions of Canada, Alaska and the U.S.S.R. Its basic epidemiological cycle in these regions is in the arctic fox and wolf, and their prey, small rodents and insectivores. In North America, its range is extending south from Canada into the United States where the red fox and coyote act as final hosts. The cycle is therefore sylvatic, and most cases in humans occur in trappers and their families following contact with the contaminated fur of foxes and wolves. However, occasionally suburban man may become infected by eating vegetables or fruit contaminated by infected foxes seeking garden voles. The invasive growth in man simulates malignant neoplasia, and because of its infiltrative spread in tissues and its readiness to develop metastatically the surgical success rate is poor.

The adult tapeworm occurs in the sledge-dogs of trappers, and the fact that these and other domestic dogs, as well as the domestic cat, are potential carriers of infection for man brings it into the veterinary sphere of interest.

Because of the large sylvatic reservoir control of *E. multilocularis* is unlikely ever to be achieved.

Family ANOPLOCEPHALIDAE

Anoplocephala

Essentially tapeworms of herbivores. The scolex has neither rostellum nor hooks and the gravid segments are wider than they are long. The intermediate stage is a cysticercoid.

Two species of this genus are parasitic in the intestine of the horse and although little is known of their precise pathogenic significance they are associated with erosive changes in the intestinal mucosa.

Hosts:
Horses and donkeys

Intermediate hosts:
Various forage mites in the soil and pasture, of the family Oribatidae

Site:
The adults are found in the small and large intestine, and the cysticercoids in forage mites

Species:
Anoplocephala perfoliata and *A. magna*

Distribution:
Worldwide, *A. perfoliata* being the more common

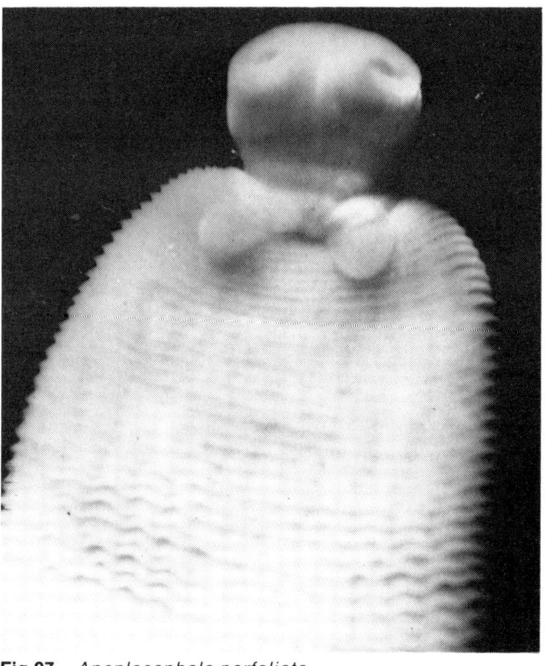

Fig 97 *Anoplocephala perfoliata*.
(a) head showing lappets.

(b) a group of tapeworms attached to caecal mucosa.

IDENTIFICATION

These together with the related but less common and smaller tapeworm, *Paranoplocephala mamillana*, are the only adult cestodes found in the horse. *A. perfoliata*, whose overall appearance (Fig. 97) is that of a large white fluke, is up to 20.0 cm in length and has a rounded scolex with a lappet behind each of the four suckers: it has a very short neck and the strobila widens rapidly, individual proglottids being much wider than they are long. *A. magna* is similar morphologically but much longer, up to 80.0 cm, and there are no lappets on the scolex.

Eggs are irregularly spherical or triangular and vary from 50–80 μm in diameter; the onchosphere is supported by a pair of projections called the pyriform apparatus (Fig. 98).

LIFE CYCLE

Mature segments are passed in the faeces and disintegrate, releasing the eggs. These are ingested by forage mites in which they develop to the cysticercoid stage in 2–4 months. One or two months after the ingestion of infected mites in the herbage the adult tapeworms are found in the intestine of horses.

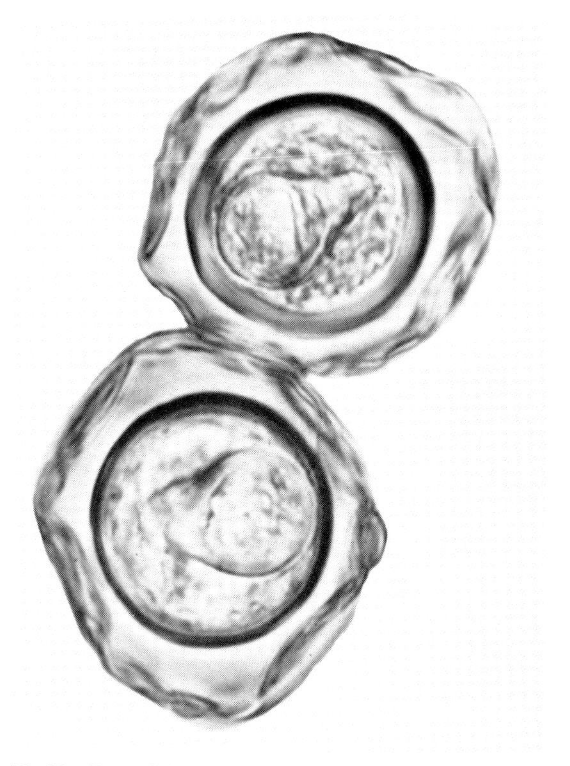

Fig 98 Eggs of *Anoplocephala perfoliata*.

PATHOGENESIS

Anoplocephala is usually considered to be relatively non-pathogenic but there is some evidence that heavy infections may cause severe clinical signs and may even prove fatal.

A. perfoliata is usually found around the ileo-caecal junction and causes ulceration of the mucosa at its site of attachment; these lesions have been incriminated as a cause of intussusception. *A. magna* is more commonly found in the jejunum and when present in large numbers may result in catarrhal or haemorrhagic enteritis.

Cases of intestinal obstruction and perforation of the

intestinal wall have been recorded associated with massive infections with both species.

CLINICAL SIGNS

In most infections there are no clinical signs. However, when there are significant pathological changes in the intestine there may be unthriftiness, enteritis and colic. Perforation of the intestine will prove rapidly fatal.

EPIDEMIOLOGY

Horses of all ages may be affected, but clinical cases have been reported mainly in animals up to 3–4 years of age. There appears to be a slight seasonal fluctuation in *A. perfoliata*, worm numbers being lowest in the spring and then accumulating until winter.

DIAGNOSIS

Where clinical signs occur they may be difficult to differentiate from more common causes of unthriftiness and digestive upsets. However, it may be possible to confirm the presence of *Anoplocephala* by the demonstration of the typical eggs on faecal examination.

TREATMENT AND CONTROL

Specific treatment for *Anoplocephala* infection is rarely called for but a number of compounds have been reported as efficient including niclosamide, and pyrantel or mebendazole both at increased dosage rates.

Control is difficult, since forage mites are widespread on pasture. Treatment with an effective anthelmintic before the animals enter new grazing may help to control *Anoplocephala* infections in areas where problems have arisen.

Paranoplocephala

Paranoplocephala mamillana is a non-pathogenic small tapeworm, up to 5.0 cm long and 0.5 cm wide, which occurs in the duodenum of horses in most parts of the world. There are no lappets on the scolex and the suckers are slit-like. The life cycle is similar to that of *Anoplocephala*.

Moniezia

This genus of cestodes is common in ruminants and resembles, in most respects, *Anoplocephala* of the horse.

Hosts:
Ruminants

Intermediate hosts:
Forage mites, mainly of the family Oribatidae

Site:
Adults in small intestine; cysticercoids in mites

Species:

Moniezia expansa	sheep, goats, occasionally cattle
M. benedeni	chiefly cattle

Distribution:
Worldwide. The only tapeworms of ruminants in western Europe

IDENTIFICATION

These are long tapeworms, 2m or more, which are unarmed possessing only suckers (Fig. 99). Segments are broader than they are long and contain two sets of genital organs grossly visible along the lateral margin of each segment. *M. expansa* is up to 1.5 cm wide whereas *M. benedeni* may be up to 2.5 cm wide. Microscopically there is a row of interproglottidal glands at the posterior border of each segment which may be used in species differentiation (Fig. 100); in *M. expansa* they extend along the full breadth of the segment, whereas in *M. benedeni* they are confined to a short row close to the middle of the segment.

The irregularly triangular eggs have a well-defined pyriform apparatus and vary from 55–75 μm in diameter.

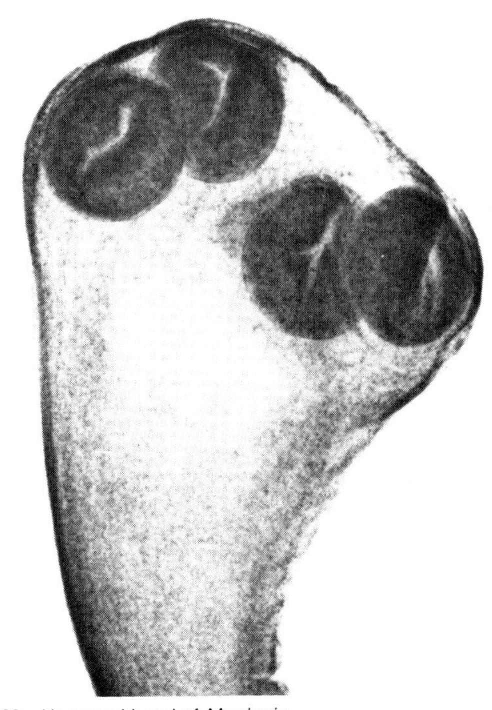

Fig 99 Unarmed head of *Moniezia*.

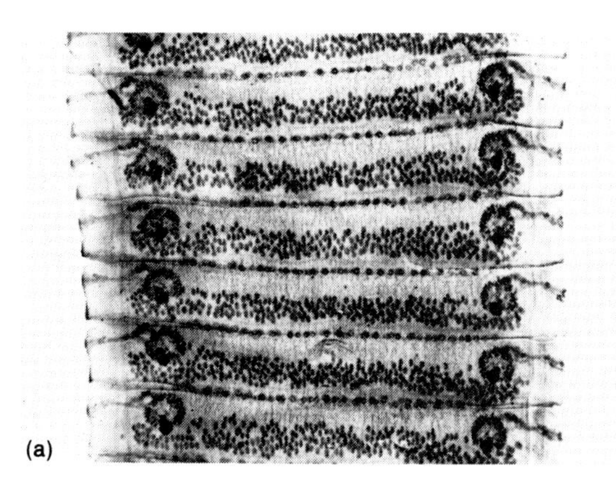

(a)

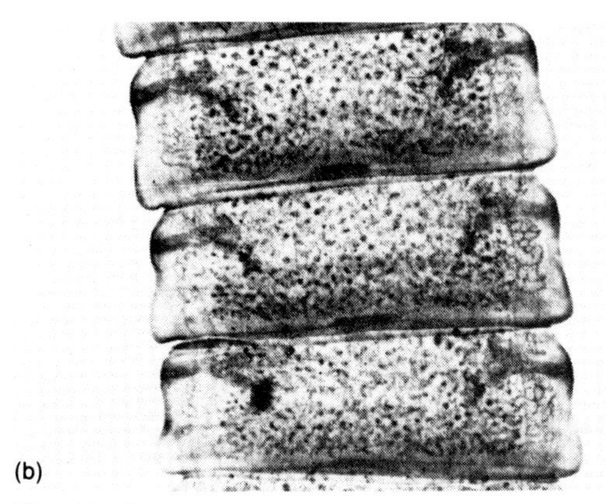

(b)

Fig 100 Segments of (a) *Moniezia expansa*. (b) Moniegia *benedeni*. Note different distribution of interproglottidal glands.

LIFE CYCLE

This is similar to *Anoplocephala*. Mature proglottids or eggs are passed in the faeces and on pasture the onchospheres are ingested by forage mites. The embryos migrate into the body cavity of the mite where they develop to cysticercoids in 1–4 months and infection of the final host is by ingestion of infected mites during grazing. The prepatent period is approximately six weeks, but the adult worms appear to be short-lived, patent infections persisting for only three months.

PATHOGENESIS

Although generally regarded as of little pathogenic significance there are a number of reports, especially from eastern Europe and New Zealand, of heavy infections causing unthriftiness, diarrhoea and even intestinal obstruction. However, *Moniezia* infections are so obvious, both in life, because of the presence of proglottids in the faeces, and at necropsy, that other causes of ill-health may be overlooked. It is interesting that experimental studies have failed to demonstrate substantial clinical effects even with fairly heavy worm burdens.

CLINICAL SIGNS

While a great variety of clinical signs including unthriftiness, diarrhoea, respiratory signs and even convulsions have been attributed to *Moniezia*, infection is generally symptomless. Subclinical effects remain to be established.

EPIDEMIOLOGY

Infection is common in lambs, kids and calves during their first year of life and less common in older animals. A seasonal fluctuation in the incidence of *Moniezia* infection can apparently be related to active periods of the forage mite vectors during the summer in temperate areas. The cysticercoids can overwinter in the mites.

DIAGNOSIS

This is based largely on the presence of mature proglottids in the faeces.

TREATMENT AND CONTROL

In many countries a variety of drugs including niclosamide, praziquantel, bunamidine and a number of broad spectrum benzimidazole compounds, which have the advantage of also being active against gastrointestinal nematodes, are available for the treatment of *Moniezia* infection. If this is carried out in calves and lambs in late spring, in temperate areas, the numbers of newly infected mites on pasture will be reduced. Ploughing and reseeding, or avoiding the use of the same pastures for young animals in consecutive years, may also prove beneficial.

Family DILEPIDIDAE

Tapeworms of the dog, the cat and the fowl. The scolex usually has an armed rostellum with several rows of hooks. The intermediate stage is a cysticercoid.
mediate stage is a cysticercoid.

Dipylidium

This is the commonest tapeworm genus of the domestic dog and cat.

Hosts:
Dog and cat; rarely man

Intermediate hosts:
Fleas (*Ctenocephalides canis*, *C. felis* and *Pulex irritans*), and lice (*Trichodectes canis*)

Site:
Small intestine; cysticercoid in fleas and lice

Species:
Dipylidium caninum

Distribution:
Worldwide

IDENTIFICATION

Dipylidium is a much shorter tapeworm than *Taenia*, the maximum length being about 50 cm. The scolex has a protrusible rostellum which is armed with four or five rows of small hooks (Fig 101). The proglottid is easily recognised, being elongate, like a large rice grain, and has two sets of genital organs, with a pore opening on each margin.

LIFE CYCLE

The newly passed segments are active, and can crawl about on the tail region of the animal. The onchospheres are contained in egg packets or capsules (Fig. 102), each with about twenty eggs, and these are either expelled by the active segment or released by its disintegration.

After ingestion by the intermediate host, the onchospheres travel to the abdominal cavity where they develop into cysticercoids. All stages of the biting louse can ingest onchospheres, but the adult flea, with its mouthparts adapted for piercing, cannot do so, and infection is only acquired during the larval stage which has chewing mouthparts.

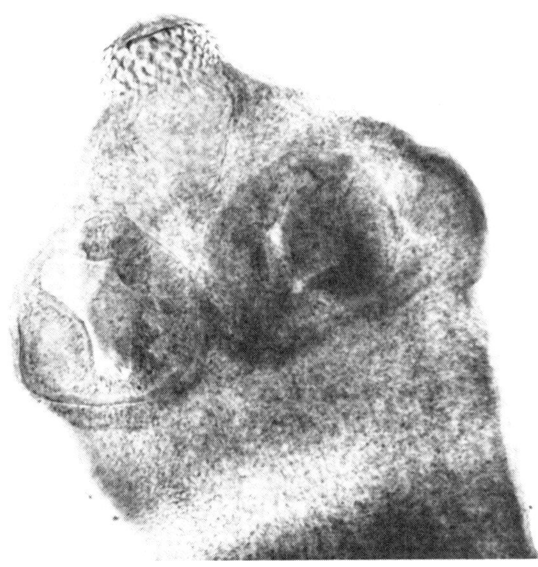

Fig 101 *Dipylidium caninum.*
(a) head showing rostellum armed with several rows of hooks.

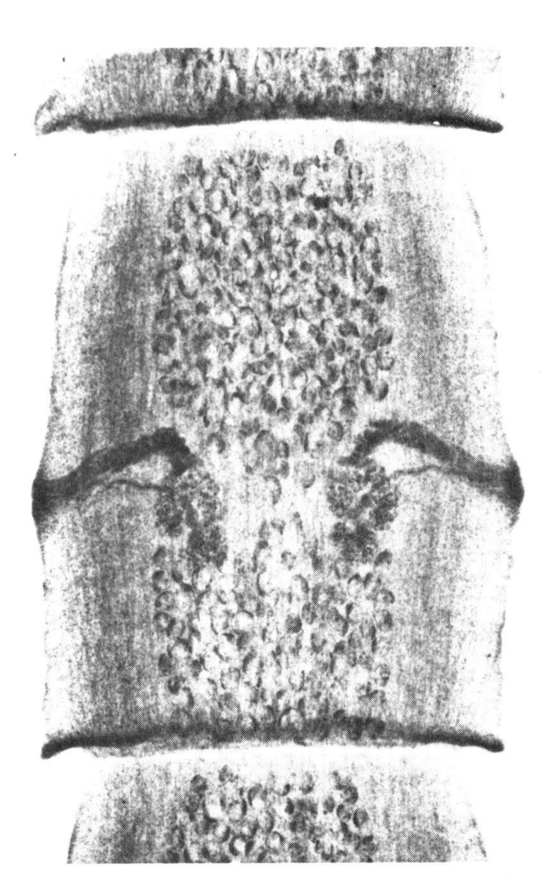

(b) mature segment with genital pore on each margin.

Development in the louse, which is permanently parasitic and therefore enjoys a warm habitat, takes about 30 days, but in the flea larva and the developing adult in the cocoon, both of which are on the ground, development may extend over several months.

The final host is infected by ingestion of the flea or louse containing the cysticercoids and development to patency, when the first gravid segments are shed, takes about three weeks.

PATHOGENESIS AND CLINICAL SIGNS

The adult is non-pathogenic and several hundreds can be tolerated without clinical effect. They shed segments, which, as they crawl actively from the anus, may cause some discomfort, and a useful sign of infection is excessive grooming of the perineum. It has been suggested that infected dogs form the habit of rubbing the anus along the floor, but impacted anal glands are a more common cause of this behaviour.

EPIDEMIOLOGY

Dipylidium infection is very common and, being dependent on the continuous presence of ectoparasites for

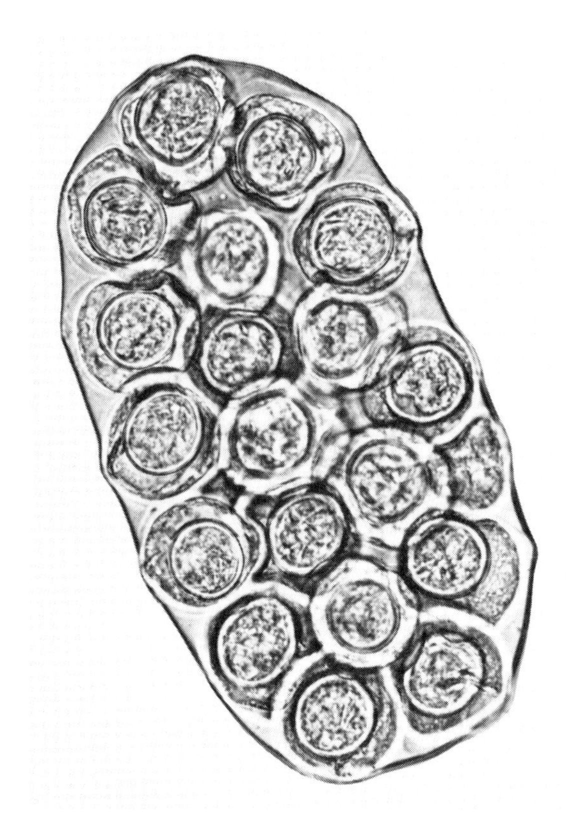

Fig 102 Egg packet of *Dipylidium caninum*.

its local endemicity, it is more prevalent in neglected animals, though infestations are also seen in well kept dogs and cats.

DIAGNOSIS

Often the first indication of infection is the presence of a segment on the coat around the perineum. If the segment is freshly passed, preliminary identification may be made on the elongate shape, and the double genital organs which may be seen with a hand lens. If it is dried and distorted it will be necessary to break it up with mounted needles in water, where the egg packets are easily seen under the microscope, thus differentiating the segment from that of *Taenia* spp. which contains only numerous single onchospheres.

TREATMENT AND CONTROL

In *Dipylidium* infection, treatment and control must be instituted together, for it is clearly of no value to eliminate the adult tapeworm while leaving a reservoir in the animal's ectoparasites. Hence, administration of anthelmintics such as nitroscanate, niclosamide, bunamidine and praziquantel should be accompanied by the use of insecticides. It is also imperative that the animal's bedding and customary resting places should receive attention with insecticides to eliminate the immature stages of the flea, which are many times more numerous than the adult parasites feeding on the dog or cat.

Amoebotaenia

Amoebotaenia sphenoides, a small tapeworm of the small intestine of domestic fowls, is up to 4.0 mm long, has up to 20 proglottids and is roughly triangular in shape. The intermediate stage, a cysticercoid, is found in earthworms.

It is not normally pathogenic unless present in very large numbers.

Choanotaenia

Choanotaenia infundibulum, a relatively large tapeworm of the small intestine of the fowl, is up to 20.0 cm in length. Each segment is wider posteriorly, giving the margin of the tapeworm a 'saw-edge' appearance. The cysticercoid is found in the house-fly, *Musca domestica*, and in beetles. Like *Amoebotaenia*, it is not normally pathogenic.

Family DAVAINEIDAE

Mainly parasites of birds. These tapeworms usually have rows of hooks on both rostellum and suckers. The intermediate stage is a cysticercoid.

Davainea

This genus contains *Davainea proglottina*, the most pathogenic cestodes of poultry.

Hosts:
Domestic fowl and pigeon

Intermediate hosts:
Slugs and land snails

Site:
Adults in small intestine; cysticercoids in slugs and snails

Species:
Davainea proglottina

Distribution:
Worldwide

IDENTIFICATION

D. proglottina is a small cestode up to 4.0 mm long, and unlike *Amoebotaenia*, possesses only 6–9 segments (Fig. 103). Both the rostellum and suckers bear hooks.

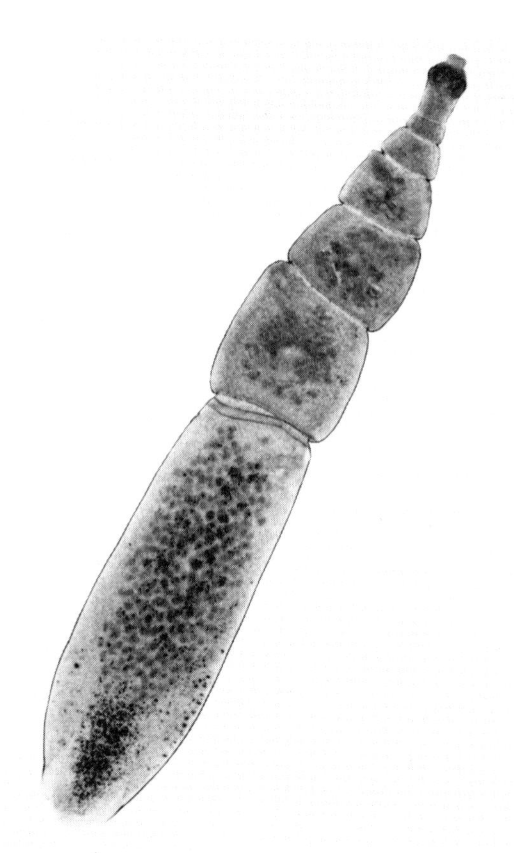

Fig 103 *Davainea proglottina* is less than 4 mm long and possesses only 6–9 segments.

PATHOGENESIS AND CONTROL

This is the most pathogenic of the poultry cestodes, the doubly armed scolex penetrating deeply between the duodenal villi. Heavy infections may cause haemorrhagic enteritis, and light infections retarded growth and weakness.

Control depends on the treatment of infected birds with a suitable anthelmintic such as niclosamide or butynorate and the destruction of slugs and snails when possible.

Raillietina

The numerous species of this genus are found in the small intestine of the fowl and turkey and the cysticercoid intermediate stage, depending on the species, in ants or beetles.

Perhaps the most important species is *Raillietina echinobothrida* which is up to 25 cm long, has hooks on both rostellum and suckers and whose eggs are contained in egg capsules in the gravid segments. In heavy infections, the embedded scolices of this parasite produce large caseous nodules in the wall of the small intestine.

Control depends on anthelmintic treatment as for *Davainea* and the destruction of the intermediate hosts with insecticides.

Family HYMENOLEPIDIDAE

Of minor veterinary importance. Members of this family, which has a characteristically slender strobila, infect birds, man and rodents. The intermediate host is a cysticercoid.

Hymenolepis nana is of peripheral veterinary importance in that it is a common tapeworm of man and of laboratory and wild rodents. It is of interest in that the life cycle can be direct, the cysticercoids developing in the villi of the small intestine of the final host and then emerging to develop to the adult tapeworm, about 30.0 mm long, in the intestinal lumen. Otherwise flour beetles or fleas can serve as intermediate hosts.

In laboratory colonies of rodents, eradication depends on strict hygiene, preferably with caesarean-derived stock. Treatment depends on niclosamide or mebendazole.

Other species of *Hymenolepis* are recorded in domestic poultry.

Family MESOCESTOIDIDAE

Also of minor veterinary importance, these cestodes of carnivorous animals and birds have two metacestode stages. The first is a cysticercoid in an insect or mite, and the second a solid larval form, a tetrathyridium, in a vertebrate.

Mesocestoides

The adult cestodes are found in the small intestine of dogs, cats and wild carnivores in parts of Europe, Asia, Africa and North America. They are up to 40 cm in length, each segment having a central genital pore, and the unarmed scolex has four suckers.

The life cycle of the parasite apparently requires two intermediate stages and hosts, the first a cysticercoid in a mite and the second a tetrathyridium in the peritoneal or pleural cavity of a wide variety of vertebrates.

Their veterinary interest lies in the fact that the dog or cat as well as being definitive hosts may also harbour tetrathyridia in their peritoneal cavity. These tetrathyridia, each 1.0 cm or longer, have the capacity to multiply asexually and the resulting massive infections may produce severe ascites.

Family THYSANOSOMIDAE

Closely related to the Anoplocephalidae, this family contains three tapeworms of veterinary importance.

Stilesia

Stilesia hepatica is extremely common in sheep and other ruminants in Africa and Asia. Large numbers of these tapeworms are often found in the bile ducts of sheep at slaughter and although they cause neither clinical signs nor significant hepatic pathology, the liver condemnations are a source of considerable economic loss, on aesthetic grounds.

The eggs possess a pyriform apparatus and the intermediate host is probably an oribatid mite.

S. globipunctata, another species, occurs in the small intestine of ruminants in southern Europe as well as Africa and Asia.

Praziquantel is recommended for treatment.

Thysanosoma

Thysanosoma actinioides is known as the 'fringed tapeworm' since each segment has a row of large, grossly visible, papillae on the posterior border of each segment. Like *S. hepatica* it is found in the bile ducts of sheep and other ruminants and its significance is largely concerned with liver condemnation at meat inspection. However its geographic distribution is different, being confined to North and South America. The intermediate hosts are thought to be oribatid mites or psocid lice (bark-lice).

Niclosamide has been shown to be effective.

Avitellina

Avitellina centripunctata is found in the small intestine of sheep and other ruminants in southern Europe, Africa and Asia. This tapeworm resembles *Moniezia* on gross inspection except that the segmentation is so poorly marked that it appears somewhat ribbon-like.

Of negligible pathogenicity, the intermediate hosts are thought to be oribatid mites or psocid lice.

Treatment is as for *Thysanosoma*.

Order PSEUDOPHYLLIDEA

The morphology of the Pseudophyllidea is generally similar to that of the Cyclophyllidea, but there are two distinct features. First, the scolex has no suckers and instead has two longitudinal grooves or **bothria** (Fig. 104) which become flattened to form organs of attachment. Secondly, the egg shell is thick, brown and operculate, and the **coracidium** which emerges after hatching is an onchosphere with an embryophore which is ciliated for mobility in water.

The pseudophyllidean life cycle utilises two intermediate hosts. The coracidium must first be ingested by a crustacean in whose body cavity a larval **procercoid** develops. Subsequently, if the crustacean is eaten by a freshwater fish the procercoid is liberated, and in the muscles of the new host develops into a second larval stage, a **plerocercoid**, which possesses the characteristic scolex; it is only this stage which is infective to the final host.

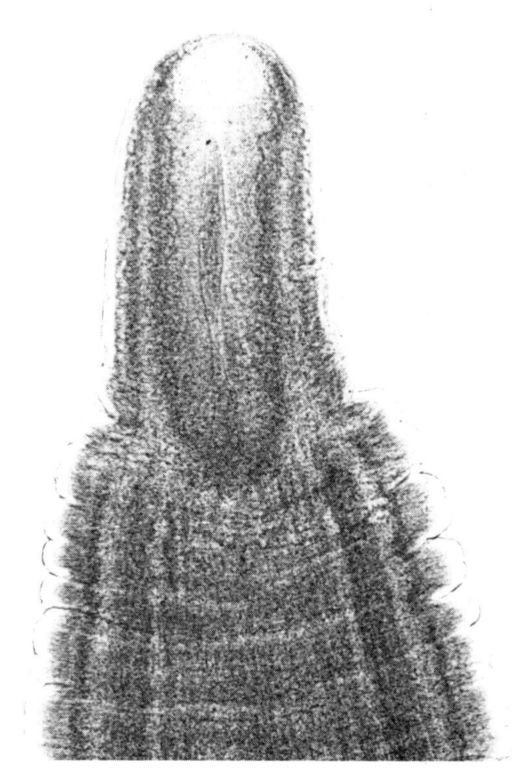

Fig 104 Head of *Diphyllobothrium latum*.

This order contains only two genera of veterinary importance, *Diphyllobothrium* and *Spirometra*.

Diphyllobothrium

Diphyllobothrium latum is an important cestode parasite of the small intestine of man in northern climates such as parts of Scandinavia, the U.S.S.R. and North America; it may also infect other fish-eating mammals.

Hosts:

Man and fish-eating mammals such as the dog, cat, pig and polar bear

Intermediate hosts:

Two are required, a copepod crustacean such as *Cyclops*, followed by a freshwater fish

IDENTIFICATION

A very long tapeworm up to 20 m in length. The scolex is unarmed with two muscular longitudinal grooves or

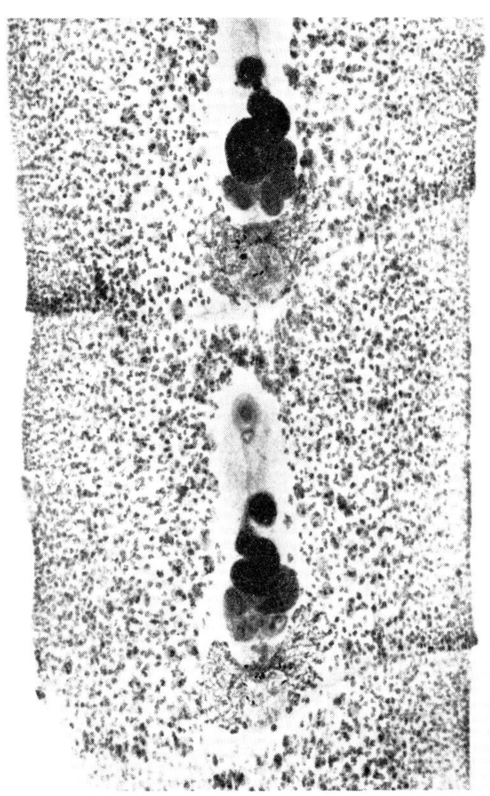

Fig 105 Segment of *Diphyllobothrium latum* showing square shape and central genital pore.

bothria as organs of attachment. The mature and gravid segments are square-shaped with a central genital pore (Fig. 105).

LIFE CYCLE

Eggs are continuously discharged from the genital pores of the attached gravid segments of the strobila and pass to the exterior in the faeces. They resemble *F. hepatica* eggs being yellow and operculate, but are smaller.

The eggs must develop in water and within a few weeks each hatches to liberate a motile ciliated coracidium which, if ingested by a copepod, develops into the first parasitic larval stage, a procercoid. When the copepod is ingested by a freshwater fish such as pike, trout or perch, the procercoid migrates to the muscles or viscera to form the second larval stage, the plerocercoid; this solid larval metacestode is about 5.0 mm long and possesses the characteristic scolex. The life cycle is completed when the infected fish is eaten raw, or insufficiently cooked, by the final host. Development to patency is rapid, occurring within four weeks of ingestion of the plerocercoid. However, if the infected fish is eaten by a larger fish, the plerocercoid has the ability to establish itself in its new host.

PATHOGENESIS AND CLINICAL SIGNS

In man the tapeworm sometimes causes a macrocytic anaemia, resembling pernicious anaemia, due to its uptake of vitamin B_{12} from the intestine.

EPIDEMIOLOGY

D. latum is essentially a parasite of man since in other hosts the cestode produces few fertile eggs. The epidemiology is therefore largely centred around two factors, the access of human sewage to freshwater lakes and the ingestion of uncooked fish. Domestic animals such as dogs or pigs become infected by eating raw fish or fish offal.

DIAGNOSIS

This depends on the detection of the characteristic eggs in the faeces.

TREATMENT

Praziquantel and niclosamide are effective against the adult tapeworm.

CONTROL

In areas where infection is common, domestic animals should not be fed fish products unless these have been thoroughly cooked or deep-frozen.

Spirometra

Adult *Spirometra* are found in dogs, cats and wild carnivores in North and South America, Australia and the Far East. The morphology and life cycle of these tapeworms is similar to that of *D. latum*, the procercoids being found in crustaceans and the plerocercoids in a wide variety of hosts including amphibia, birds and mammals.

Occasionally, man may become infected with plerocercoids either through drinking water containing procercoid-infected crustacea or from eating a plerocercoid-infected host such as a pig. This zoonosis, known as sparganosis (*Sparganum* was the old name for these plerocercoids), is characterised by the presence of larvae up to 35 mm long in the muscles and subcutaneous tissues, particularly the periorbital area, causing oedema and inflammation.

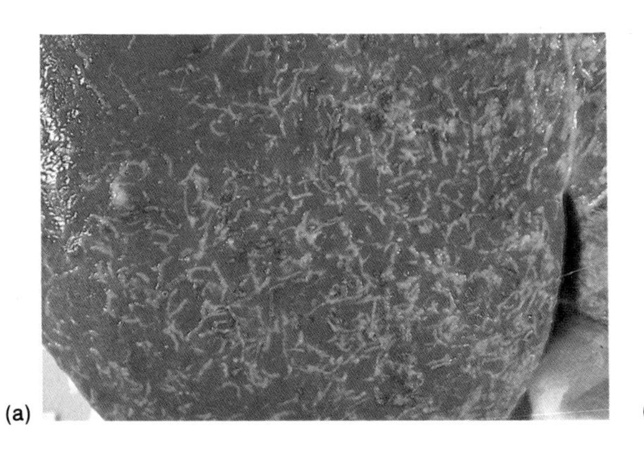
(a)

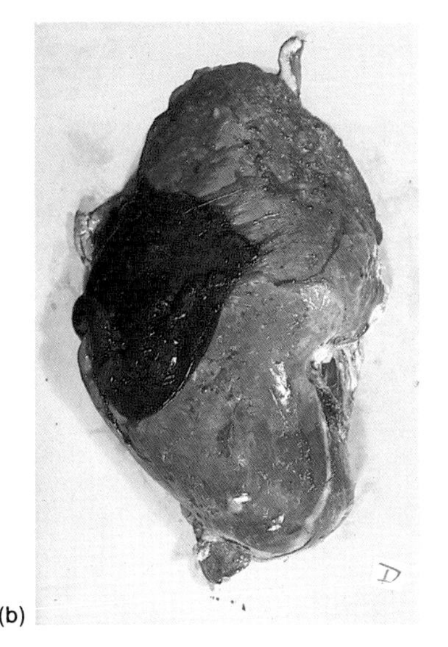
(b)

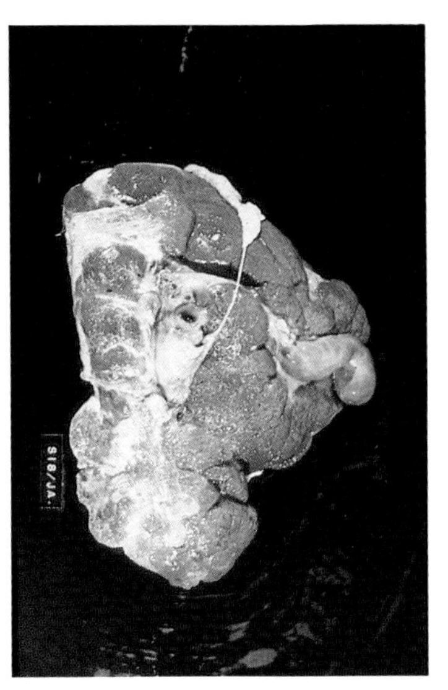
(c)

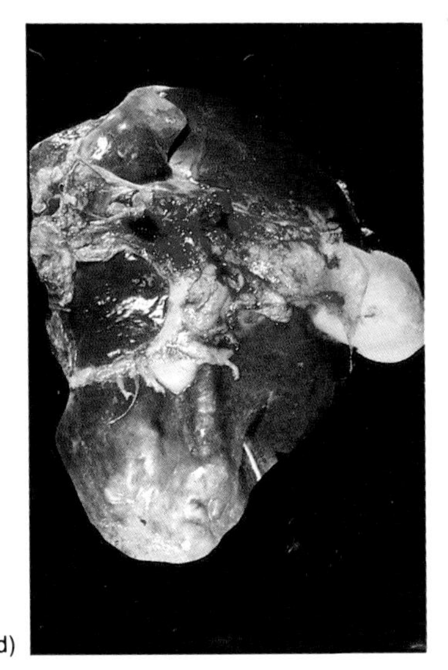
(d)

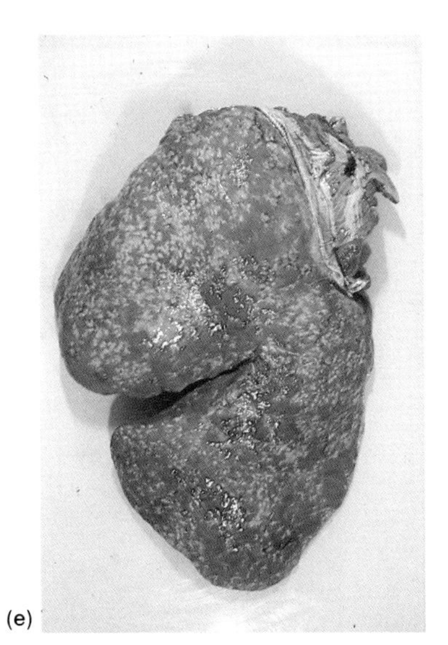
(e)

PlateV

(a) Liver lesions associated with acute ovine fascioliasis.
(b) Massive subcapsular haemorrhage frequently seen in acute ovine fascioliasis.
(c) Extensive cirrhosis associated with chronic ovine fascioliasis.
(d) Gross appearance of the liver in bovine fascioliasis.
(e) Liver lesions caused by severe *Dicrocoelium dendriticum* infection.
(f) Adult paramphistomes attached to the wall of the rumen.

(f)

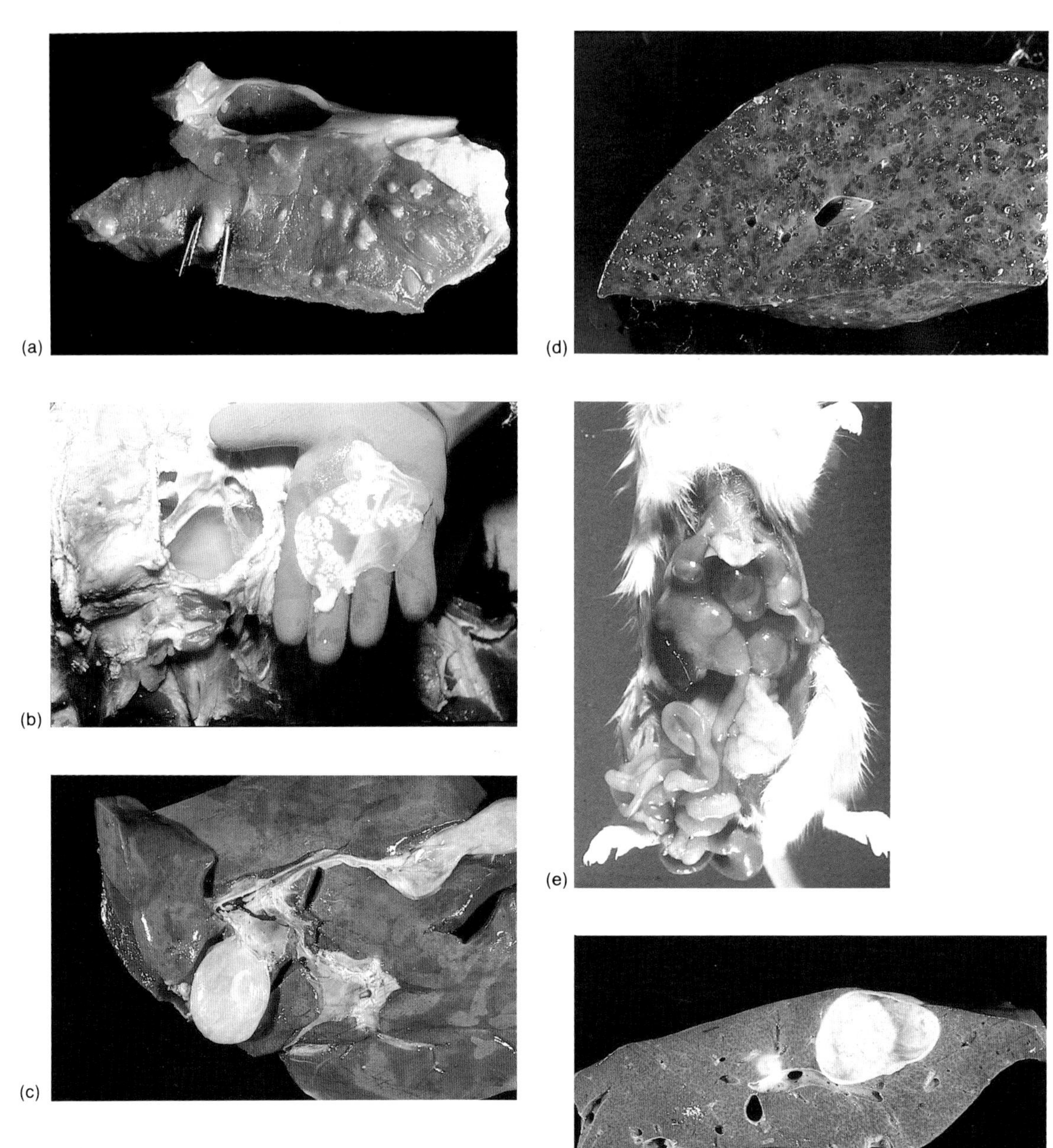

Plate VI

(a) *Cysticercus bovis* in skeletal muscle.
(b) *Coenurus cerebralis* cyst from brain of sheep.
(c) Large fluid-filled *Cysticercus tenuicollis* attached to the liver.
(d) 'Hepatitis cysticercosa' caused by massive infection with *Cysticercus tenuicollis*.
(e) Mouse liver infected with *Cysticercus fasciolaris.*
(f) Section of sheep liver showing hydatid cyst.

VETERINARY ENTOMOLOGY

Veterinary entomology, in its literal sense, means a study of insects of veterinary importance. This term, however, is commonly used to describe the wider study of all arthropods parasitic on animals, including arachnids such as ticks and mites.

Phylum ARTHROPODA

The phylum Arthropoda contains over 80% of all known animal species and consists of invertebrates whose major characteristics are a hard chitinous exoskeleton, a segmented body and jointed limbs.

STRUCTURE AND FUNCTION

The hard exoskeleton of arthropods is secreted by an underlying epidermis and consists of numerous segments which are often clearly separated into three regions, the head, thorax and abdomen (Fig. 106). These segments are made up of thick chitinous plates called sclerites, which may be variously adapted or fused, thus obscuring segmentation, but each usually consists of a dorsal tergum, a ventral sternum and two lateral pleurons connected by small flexible pieces of chitin which function as joints.

There is great diversity in the morphology of the alimentary canal, but generally it can be divided into three regions, namely, the fore-, mid- and hind-gut.

The fore-gut begins with the frequently complex and varied mouthparts which lead into the buccal cavity, pharynx, oesophagus and proventriculus. In many cases the oesophagus is dilated posteriorly and referred to as the crop, while the muscular proventriculus or gizzard acts as a valve, to prevent regurgitation, and may have 'teeth' to aid the disintegration of food particles.

The mid-gut stores food and secretes enzymes necessary for digestion. Opening into the alimentary canal at the junction of the mid- and hind-gut are a variable number of excretory tubules termed **malpighian tubules**. These act as filters, extracting waste products from the blood which are subsequently discharged into the gut.

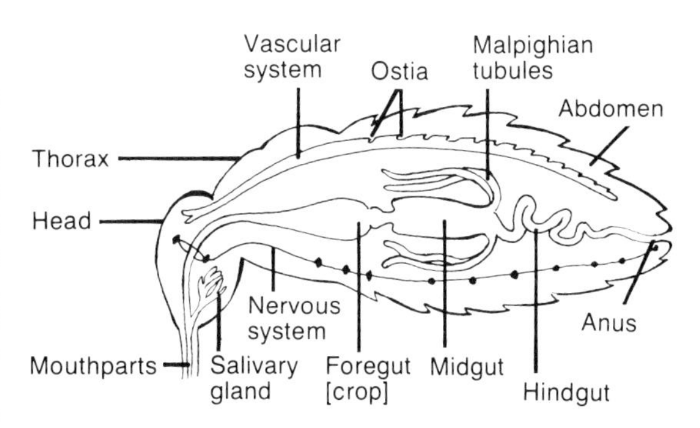

Fig 106 Some general features of an arthropod.

The hind-gut consists of an anterior ileum and a dilated rectum posteriorly, the latter often having papillae or glands involved in the reabsorption of water from the faeces.

The body cavity or **coelom** (the space between the gut and the body wall) is often called the **haemocoele** since it contains blood or haemolymph whose primary function is the transport of metabolites. The internal organs are bathed in this blood which is continuously circulated by a dorsally situated, primitive, tubular heart. Blood enters through openings in the heart wall, called **ostia**, and is expelled through short vessels into the haemocoele.

Respiration in arthropods is simple, oxygen reaching the tissues by direct gaseous diffusion. Small circular openings in the exoskeleton called **spiracles** allow air into the body. This then enters a system of branching tracheae and tracheoles which ramify through most parts of the body. Oxygen diffuses from the tracheoles into the cells, and conversely, carbon dioxide from the cells passes to the exterior via the tracheoles, tracheae and spiracles. The tracheal system in some arthropods is also involved in regulation of water loss. Direct cuticular respiration also occurs in some aquatic arthropods.

The nervous system consists of a ventrally situated ganglionated nerve cord which connects in the head region with a large supraoesophageal ganglion often called the 'brain'. Associated with the nervous system are the sensory organs, including the eyes and various tactile and auditory organs. In some arthropods eyes are absent or reduced, as in the ticks and lice, whereas in others such as some blood-sucking flies, whose sight is important in locating their hosts, the eyes are well developed. Often two types of eyes co-exist in the same animal. The most complex is the compound eye which is particularly adapted for the perception of movement; in the female of some species the eyes are distinctly separated (**dichoptic**) while in the males they may be very close together (**holoptic**). Simple eyes or **ocelli** are also frequently present on the top of the head, and although their function is not well understood, they can apparently differentiate between light and dark. Other sensory organs include antennae, palps and various receptors in the body which respond to temperature, humidity, food stimuli and host odours. Tactile sense in some arthropods also depends on the displacement of hairs situated on the body surface.

In the arthropods the sexes are separate. The external genital structures, such as the penis and paired claspers of the male, can be of taxonomic value in identification at the species level. Internally the male reproductive organs consist of a pair of testes, each with a vas deferens which is expanded distally to form a common or paired seminal vesicle which stores the spermatozoa. The external genital apparatus of the female is usually inconspicuous although a few species may have well developed ovipositors. Internally the female reproductive system comprises paired ovaries and oviducts leading into a 'uterus', sometimes referred to as the common oviduct, and, posteriorly, the vagina. An important accessory organ opening into the common oviduct is the **spermatheca** which consists of several receptacles which are filled with spermatozoa from the male during copulation. Eggs passing into the vagina are fertilised by the spermatozoa which remain viable in the spermatheca often throughout the female's life. Most arthropods are oviparous.

CLASSIFICATION

There are two major classes of arthropods of veterinary importance namely the Insecta and Arachnida and the important orders in these classes are shown in Table 3.

The two major classes can be differentiated by the following general characteristics:

INSECTA: These have 3 pairs of legs, the head, thorax

Table 3 Classification of the Arthropoda

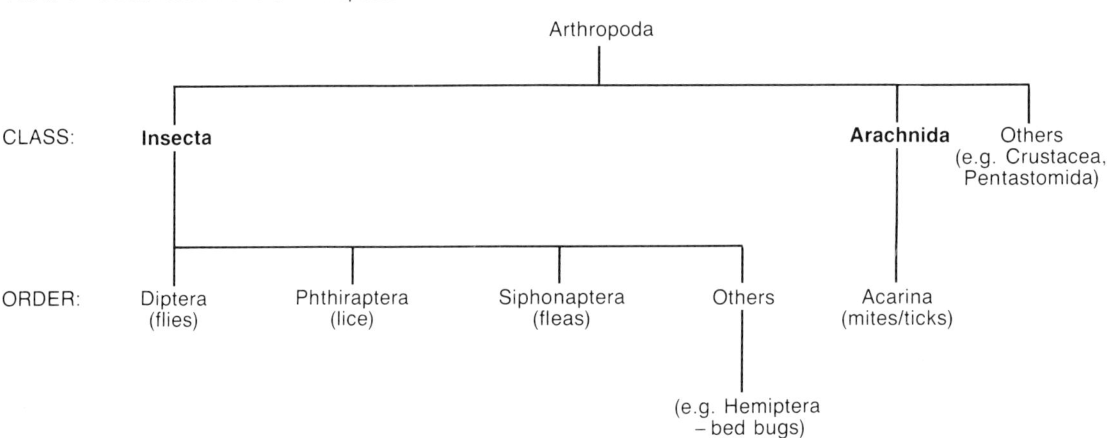

and abdomen are distinct, and they have a single pair of antennae.

ARACHNIDA: The adults have 4 pairs of legs, the body is divided into a cephalo-thorax and abdomen, and there are no antennae.

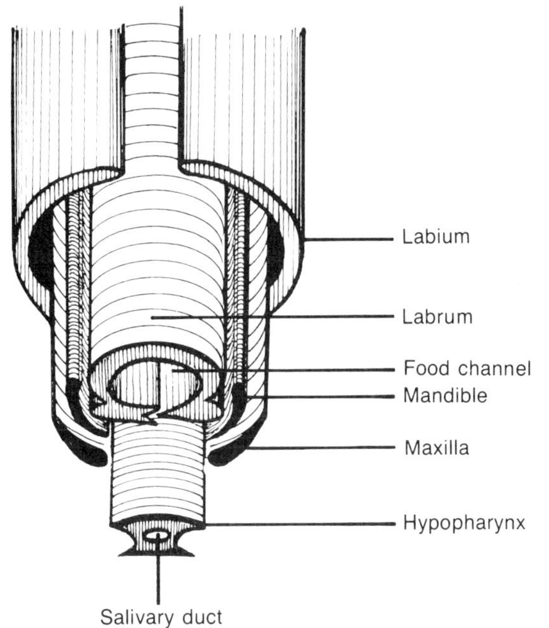

Fig 107 Basic elements of insect mouthparts.

Class INSECTA

GENERAL MORPHOLOGY AND LIFE CYCLE

The head of an insect generally comprises six fused segments with a single pair of antennae. There is great variation in the structure of the mouthparts, depending on feeding habits, with adaptations for chewing-biting, sponging or piercing-sucking. Basic elements of insect mouthparts are illustrated in Fig. 107 and consist of the following:

(i) The **labrum** or upper lip is a hinged plate attached to the face or clypeus.
(ii) The paired **mandibles** and **maxillae** or jaws have areas of their surfaces adapted for cutting, slashing or grinding. The maxillae may also carry maxillary palps which are sensory in function and used in the monitoring of food.
(iii) A **hypopharynx** which arises from the floor of the mouth, bears the external opening of the salivary glands and is similar to a tongue.
(iv) A **labium** or lower lip, which may be extensively modified, especially in the flies, and sometimes bears two sensory labial palps.

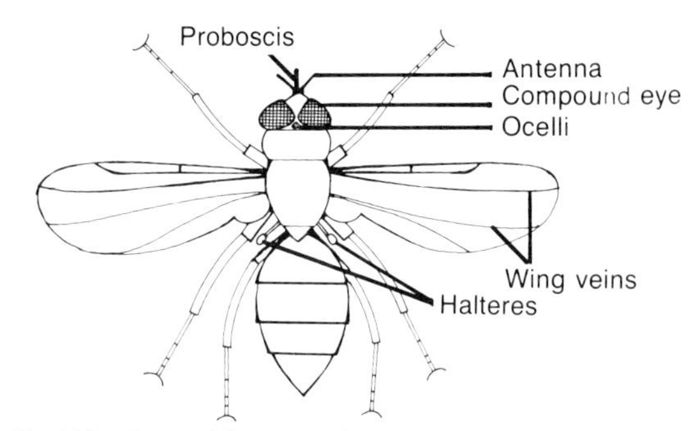

Fig 108 General features of a dipteran fly.

The three segments in the thorax (pro-, meso- and meta-thorax) each bear a pair of jointed legs. The thorax of many insects also bears two pairs of wings, but in the winged insects of veterinary significance, ie. the Diptera (Fig. 108), only one pair is functional, the second being reduced to small knob-like sensory structures, called **halteres**, which apparently have a balancing function. Wings are outgrowths of the thoracic tegument supported by hollow tubes called **veins** which run longitudinally and crosswise, the intervening areas of tegument being known as **cells**. The arrangement of the veins and the shape of the cells are important in identification.

The abdomen of insects consists of up to 11 segments with terminal modifications to form the genitalia.

In insects the sexes are separate and after fertilisation either eggs or larvae are produced. Development often involves three or more larval stages followed by the formation of a pupa and a marked transformation or metamorphosis to the adult stage as in all the Diptera, ie. a **holometabolous** life cycle (Fig. 109). In other insects

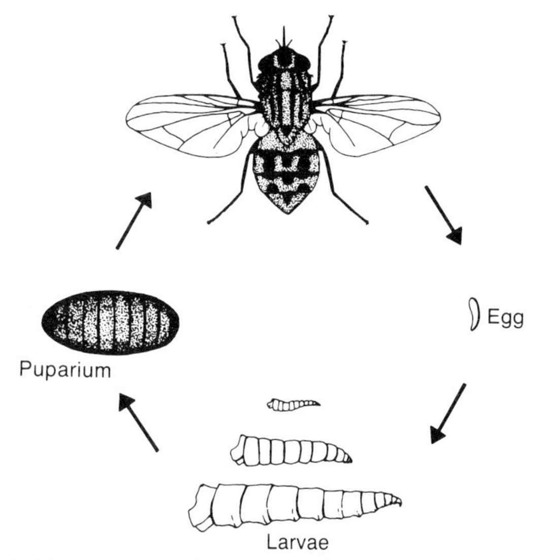

Fig 109 A holometabolous life cycle characterised by metamorphosis is typical of many flies.

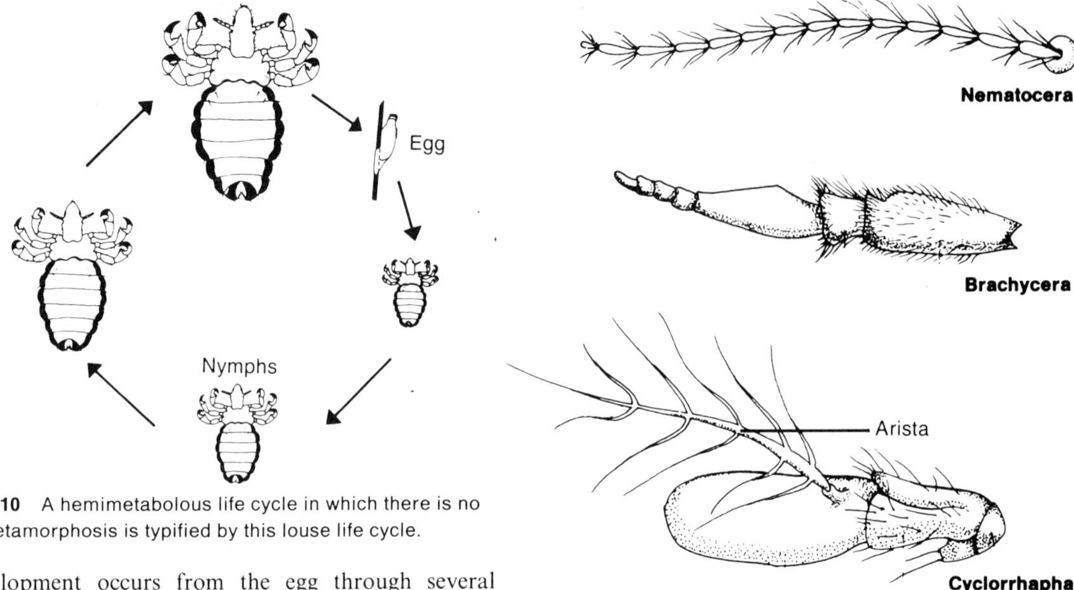

Fig 110 A hemimetabolous life cycle in which there is no metamorphosis is typified by this louse life cycle.

Fig 111 The three suborders of the Diptera each possess characteristic antennae.

development occurs from the egg through several nymphal stages which resemble the adult, as in lice, ie. a **hemimetabolous** life cycle (Fig. 110). The different stages in the life cycle are known as instars.

Order DIPTERA

This Order of insects contains all of the flies of veterinary importance. These are generally characterised by having a single pair of membranous wings and a pair of halteres. Some are important as external parasites, while in others the larvae parasitise the tissues of the host. Many members of this group are also important as vectors of disease.

The Diptera can be conveniently divided into three suborders, namely, the Nematocera, Brachycera and Cyclorrhapha and a simplified classification showing the various families is given in Table 4.

Suborder NEMATOCERA

These are small flies and the adults are characterised by having a pair of long, jointed antennae (Fig. 111) and segmented maxillary palps. The wings generally have no cross-veins (Fig. 112). Only the females are parasitic and have piercing-sucking mouthparts.

Eggs are laid in or near water and develop into aquatic larvae and pupae: both of these stages have well-developed heads and are mobile.

Suborder BRACHYCERA

These are large flies with stout antennae often consisting

Table 4 Classification of the Diptera

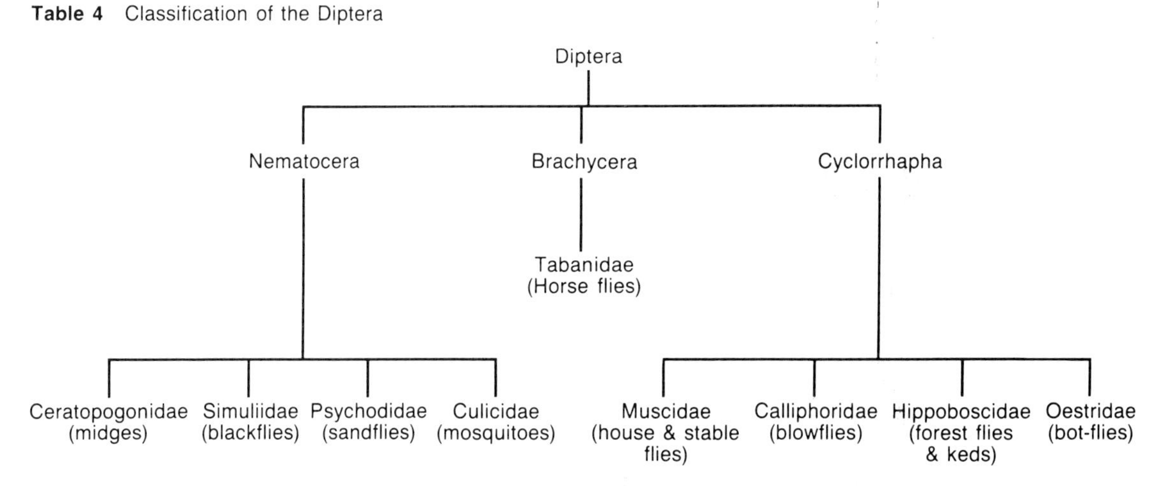

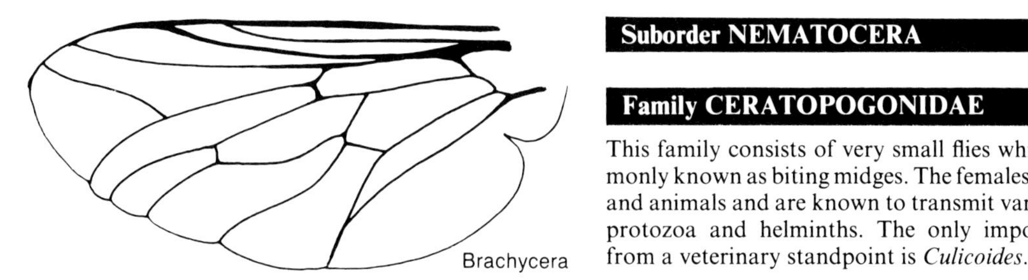

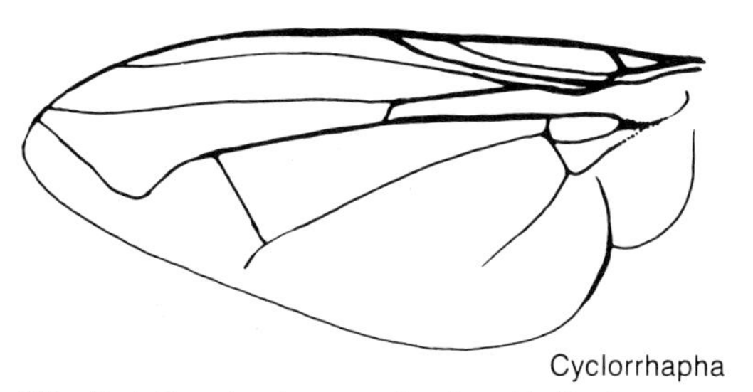

Fig 112 Variations in wing venation found in the three suborders of the Diptera.

of only three segments (Fig. 111), the last segment frequently bearing annulations. The maxillary palps are usually held forwards and cross-veins are present on the wings (Fig. 112).

Using their slashing-sponging mouthparts, the females feed on blood and the eggs, which are laid on vegetation overhanging mud or shallow water, develop into large carnivorous larvae with ill-defined but usually retractile heads. Like the Nematocera, both larvae and pupae are mobile and aquatic, and are usually found in mud.

Suborder CYCLORRHAPHA

These are small to medium sized flies with short, three-segmented antennae, the last of which often bears an attachment, the arista (Fig. 111). The maxillary palps are small and the wings show limited cross-venation (Fig. 112).

Both males and females may feed on animals, but many members of this group are not parasitic as adults and have either vestigial or sponging mouthparts. The larvae have a poorly defined head, and are mobile and worm-like, often being referred to as 'maggots'. The mature larva pupates on the ground within a hard pupal case or puparium, which is completely immobile. When the young fly is ready to emerge, it does so by inflating a membranous bladder called the ptilinal sac situated on the front of the head which then pushes off a circular piece of the anterior end of the puparium.

Suborder NEMATOCERA

Family CERATOPOGONIDAE

This family consists of very small flies which are commonly known as biting midges. The females feed on man and animals and are known to transmit various viruses, protozoa and helminths. The only important genus from a veterinary standpoint is *Culicoides*.

Culicoides

Hosts:
All domestic animals and man

Species:
There are over 800 species of *Culicoides*, commonly known as midges

Distribution:
Worldwide

MORPHOLOGY

These flies are 1.5–5.0 mm long with the thorax humped over a small head, and wings, generally mottled, which are held at rest like a closed pair of scissors over the grey or brownish-black abdomen. The antennae are prominent, the legs relatively short, and the small mouthparts hang vertically.

The short piercing proboscis consists of a sharp labrum, two maxillae, two mandibles, a hypopharynx and a fleshy labium which does not enter the skin during feeding by the adult female. In the male, the long antennae are feathery or plumose whereas those of the female possess only short hairs, and are known as pilose antennae. Microscopic hairs cover the wings.

LIFE CYCLE

The eggs, which are brown or black, are cylindrical or banana-shaped and 0.5 mm in length; these are laid in damp marshy ground or in decaying vegetable matter near water. Hatching occurs in 2–9 days depending on the species and temperature, but temperate species may overwinter as eggs. There are four larval stages and these are characterised by having small dark heads, segmented bodies and terminal anal gills. They have a serpentine swimming action in water and feed on decaying vegetation. Larval development is complete in warm countries in 14–25 days, but in temperate areas this may be delayed for periods of up to seven months. The less active brown pupae, 2.0–4.0 mm long, are found at the surface or edges of water and are characterised by a pair of respiratory trumpets on the cephalothorax and a pair of terminal horns which enable the pupa to move. Adult flies emerge from the pupae in 3–10 days and the females suck blood.

PATHOGENIC SIGNIFICANCE

Since these flies may be present in vast numbers, they can be a serious source of annoyance. In addition they may transmit virus diseases such as bluetongue and African horse sickness as well as filarioid nematodes such as *Dipetalonema* spp. and *Onchocerca reticulata* and *O. gibsoni*. Several species have also been incriminated as the causative agents of a seasonally occurring, intensely pruritic, skin disease of horses called 'sweet itch' (Fig. 113), and, in Australia, 'Queensland itch'. This affects mainly the withers and base of the tail and has been shown to be due to an immediate-type hypersensitivity reaction to the bites of the flies.

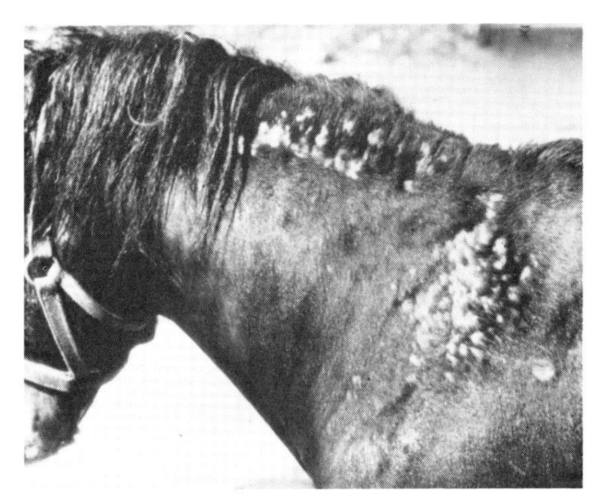

Fig 113 A severe case of 'sweet itch' in a pony.

CONTROL

This is difficult because of the usually extensive breeding habitat and depends on the destruction of breeding sites by drainage or spraying with insecticides since the adults normally fly only a few hundred metres. However, wind dispersion of these small flies may be particularly important in the spread of some virus diseases. Repellents or screens may be used, but the latter have to be so fine that they may reduce air flow and, instead, impregnation with insecticides of screens designed to exclude larger flies has been recommended. For 'sweet itch', antihistamine treatment may give immediate relief and the application of synthetic pyrethroid dressings every six days may help prevent recurrence of the condition. It is also recommended that animals are housed when fly activity is maximal, usually in late afternoon and early morning.

Family SIMULIIDAE

Of the 12 genera belonging to this family of small flies, *Simulium* is the most important. Commonly referred to as 'Blackflies' or 'Buffalo gnats' they have a wide host range, feeding on a great variety of mammals and birds and causing annoyance due to their painful bites. In man however, they are most important as vectors of *Onchocerca volvulus*, the filarioid nematode which causes 'river blindness' in Africa and Central and South America.

Simulium

Hosts:
All domestic animals and man

Species:
Numerous and often divided into sub-species

Distribution:
Worldwide except New Zealand, Hawaii and some minor island groups

MORPHOLOGY

As their common names indicate these flies are usually black with a humped thorax (Fig. 114). The adults are

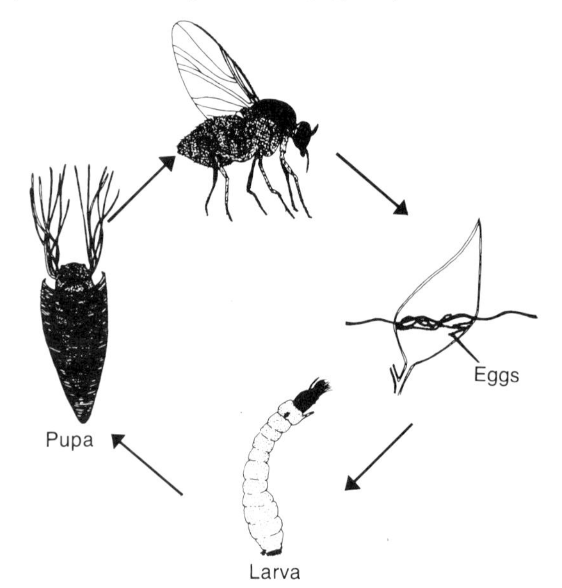

Fig 114 The life cycle of *Simulium*.

1.5–5.0 mm long, relatively stout-bodied, with colourless wings which show indistinct venation and are held at rest like the closed blades of a pair of scissors. Morphologically, adult male and female flies are similar, but can be differentiated by the fact that in the female the eyes are distinctly separated (dichoptic) whereas in males the eyes are very close together (holoptic).

Compared with other nematoceran flies, the antennae, although segmented, are relatively short and stout and do not bear hairs. Basically the mouthparts resemble those of the biting midges except for the presence of conspicuous segmented maxillary palps.

LIFE CYCLE

Eggs, 0.1–0.4 mm long, are laid in sticky masses of several hundred on partially submerged stones or vegetation in flowing water (Fig. 114). Hatching takes only a few days in warm conditions, but may take weeks in temperate areas and in some species the eggs can overwinter. There may be up to eight larval instars, the mature larvae, which are 5.0–13.0 mm long, light-coloured and poorly segmented, being distinguished by a blackish head which bears a prominent pair of feeding brushes. The body is swollen posteriorly and just below the head is an appendage called the proleg which bears hooks. Larvae normally remain attached to submerged vegetation or rocks by a circlet of posterior hooks, but may change their position in a looping manner by alternate use of the proleg and the posterior hooks. Larval maturation takes several weeks to several months and in some species larvae can overwinter. Mature larvae pupate in a slipper-shaped brownish cocoon fixed to submerged objects and the pupa has prominent respiratory gills projecting from the cocoon. The pupal period is normally 2–6 days and a characteristic feature of many species is that there is simultaneous mass emergence of the adult flies which gain the surface of the water and take flight.

PATHOGENIC SIGNIFICANCE

Only the adult females suck blood and different species have different preferred feeding sites and times. Generally they feed on the legs, abdomen, head and ears, and most species are particularly active during the morning and evening in cloudy warm weather. Although flies may be active throughout the year there may be a large increase in their numbers in the tropics during the rainy season. In temperate and arctic regions the biting nuisance may be seasonal, since adults die in the autumn with new generations in spring and summer. In domestic animals, especially cattle, mass attack by these flies may be associated with an acute syndrome characterised by generalised petechial haemorrhages, particularly in areas of fine skin, together with oedema of the larynx and abdominal wall. The painful bites of swarms of *Simulium* may interfere with grazing and cause production loss and in certain areas of Central Europe it is often impossible to graze cattle during the spring due to the activity of these flies. Horses are often affected by the flies feeding inside the ears and poultry may become anaemic from blood loss when attacked.

Simulium spp. may transmit the viruses causing Eastern equine encephalitis and vesicular stomatitis, the avian protozoan *Leucocytozoon* and filarioid helminths such as *Onchocerca gutturosa* of cattle.

CONTROL

The most practical control method is the application of insecticides to breeding sites to kill larvae. This technique has been developed for the control of *Simulium* species which are vectors of 'river blindness' in man in Africa, and entails the repeated application of organochlorine or organophosphorus insecticides to selected water courses at intervals throughout the year. The insecticide is then carried downstream and kills larvae over long stretches of water.

Alternatively, bush clearing will remove adult resting sites and aerial application of insecticides may help in areas where breeding occurs in networks of small streams and watercourses. In horses, insecticides or repellents may be applied topically and poultry can be provided with insecticidal dust baths.

Family PSYCHODIDAE

The flies of this family are called the 'sandflies' and *Phlebotomus* is the only genus of any veterinary importance. Since, in some areas of the world, the term 'sandflies' includes some biting midges and blackflies a better term is 'phlebotomine sandflies'. These flies are important as vectors of *Leishmania*.

Phlebotomus

Hosts:
Many mammals, reptiles, birds and man

Species:
There are over 600 species of phlebotomine sandflies

Distribution:
Widely distributed in the tropics, subtropics and the Mediterranean area. Most species prefer semi-arid and savannah regions to forests

MORPHOLOGY

These small flies, up to 5.0 mm long, are characterised by their hairy appearance, their large black eyes and long stilt-like legs. The wings, which, unlike those of other biting flies, are lanceolate in outline, are also covered in hairs and are held erect over the body at rest.

As in many other nematoceran flies the mouthparts are of short to medium length, hang downwards, and are adapted for piercing and sucking. The maxillary palps are relatively conspicuous and consist of five segments. In both sexes the very long antennae of up to 16 segments bear many short hairs.

LIFE CYCLE

Up to 100 ovoid, 0.3–0.4 mm long, brown or black eggs may be laid at each oviposition in small cracks or holes in the ground, the floors of animal houses or in leaf litter. Although not laid in water the eggs need moist conditions for survival, as do the larvae and pupae. Under optimal conditions the eggs can hatch in 1–2 weeks, but this may be prolonged in cold weather. The larvae, which resemble small caterpillars, scavenge on organic matter and can survive flooding. There are four larval instars, maturation taking three weeks to several months depending on species, temperature and food availability and in temperate regions these flies overwinter as mature larvae. The major characteristics of the mature larvae, which are 4.0–6.0 mm long, are a black head and a segmented greyish body covered in bristles. The adults emerge from pupation after 1–2 weeks. The whole life cycle takes 30–100 days, or even longer in cool weather.

PATHOGENIC SIGNIFICANCE

As in many other small biting flies, only the females suck blood. They prefer to feed at night, resting in shaded areas during the day. Since they are capable of only limited flight, nuisance due to biting may be confined to certain areas near the breeding sites.

There is some seasonality in activity, the numbers of flies increasing during the rainy season in the tropics whereas they are only present during the summer months in temperate zones. Apart from their biting nuisance in localised areas, phlebotomine sandflies are important as the sole known vectors of *Leishmania tropica* and *L. donovani*, which cause cutaneous and visceral leishmaniasis in man, dogs being important reservoir hosts in some regions.

CONTROL

There have been few serious attempts to control phlebotomine sandflies, probably due to the fact that leishmaniasis has merited insufficient attention as a disease and also because little is known in detail of the biology and ecology of the developing stages of these flies. The adults are, however, susceptible to most insecticides and where there have been spraying campaigns to control the mosquito vectors of malaria these have effectively controlled *Phlebotomus*. Man has protected himself from the bites of these flies by using residual house-sprays, repellents and very fine mesh fly screens.

Family CULICIDAE

The Culicidae (not to be confused with midges of the genus *Culicoides*) are the mosquitoes, small slender flies with long legs. Although their bites are a severe nuisance to man and animals they are principally important as vectors of malaria (*Plasmodium* spp.), filarial nematodes and viruses. Primarily because of their importance as vectors of human malaria there is a vast literature on their classification, behaviour and control, but the family is of limited veterinary significance and only general aspects of morphology, significance and control need be discussed.

Hosts:
A wide variety of mammals, including man; reptiles and birds

Species:
This family contains over 3,000 species belonging to 34 genera, the most important of which are *Anopheles*, *Culex* and *Aedes*

Distribution:
Worldwide

GENERAL MORPHOLOGY

Mosquitoes vary from 2.0–10.0 mm in length and have slender bodies, prominent eyes and long legs (Fig. 115). The long narrow wings bear scales, which project as a fringe on the posterior margin, and are held crossed flat over the abdomen at rest.

The mouthparts consist of a conspicuous, forward-projecting, elongated proboscis adapted for piercing and sucking. Individual elements comprise a long U-shaped fleshy labium containing paired maxillae, mandibles and a hypopharynx which carries a salivary duct which delivers anticoagulant into the host's tissues (Fig. 116). The labrum forms the roof of the proboscis and all the elements, with the exception of the labium, enter the skin during feeding by the females, forming a tube through which blood is sucked. In the non-parasitic males the maxillae and mandibles are reduced or absent. The maxillary palps of different species are variable in length and morphology. Both sexes have long filamentous segmented antennae, pilose in females and plumose in males.

GENERAL LIFE CYCLE

After a blood meal the gravid female lays up to 300 eggs on the surface of water either singly or, in the case of *Culex*, in groups forming egg-rafts (Fig. 115). The eggs

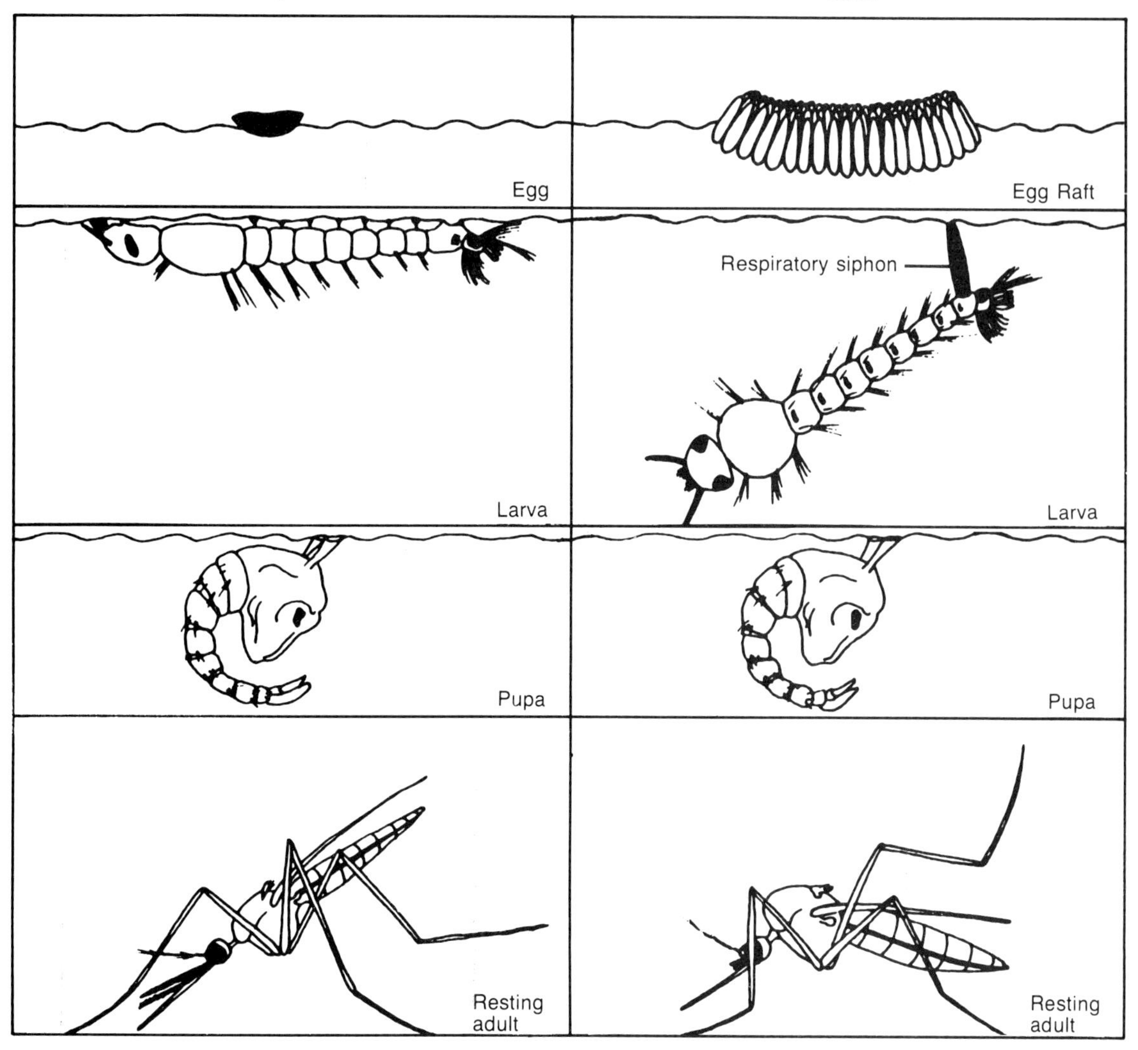

Fig 115 The comparative life cycles of anopheline and culicine mosquitoes.

are dark-coloured, elongate or ovoid, and in the genus *Anopheles*, boat-shaped, and cannot survive desiccation. Hatching is temperature-dependent and occurs after several days to weeks, but in some temperate species eggs may overwinter. All four larval instars are aquatic. There is a distinct head with one pair of antennae, compound eyes and prominent mouth brushes, used in feeding on organic material. Most larvae take in air through a pair of spiracles on the penultimate abdominal segment, but in *Culex* spp. these are situated at the end of a small tube called the respiratory siphon (Fig. 115). Maturation of larvae can extend from one week to several months and several species overwinter as larvae in temperate areas. Larval habitats vary tremendously and range from small temporary collections of water to extensive areas such as marshes, but they are usually absent from large tracts of uninterrupted water, such as lakes, and from fast-flowing streams or rivers.

All mosquito pupae are aquatic, motile and comma-shaped with a distinct cephalothorax which bears a pair of respiratory trumpets (Fig. 115). The tegument of the cephalothorax is transparent and the eyes, legs and other structures of the developing adult are readily visible. The tapering abdominal segments have short hairs and terminally there is a pair of oval paddle-like extensions which enable the pupa to move up and down in the water. Generally the pupal stage is short, being only a few days in the tropics and several weeks or

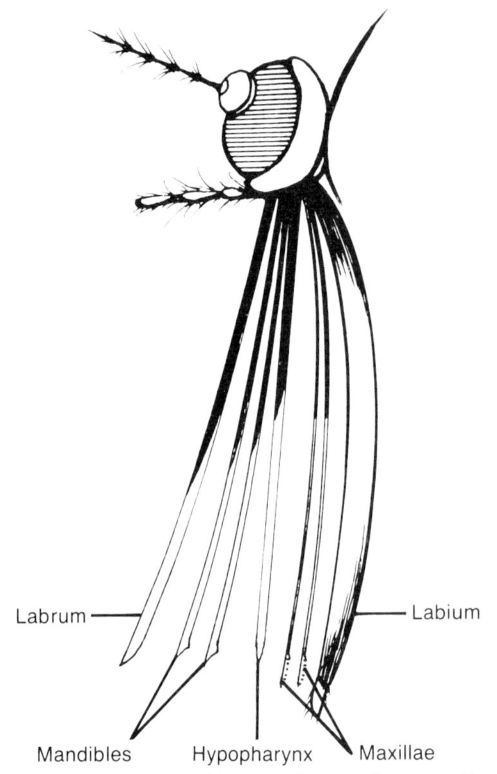

Fig 116 Piercing and sucking mouthparts of a mosquito.

longer in temperate regions, the adult emerging through a dorsal split in the pupal tegument. Adults usually fly only up to a few hundred metres from their breeding sites, but may be dispersed long distances by winds. Although the life-span of adult flies is generally short some species can overwinter by hibernating.

PATHOGENIC SIGNIFICANCE

Most species of mosquitoes are nocturnal feeders and may cause considerable annoyance by biting, their long mouthparts allowing them to bite man even through clothing. More importantly, species of *Anopheles*, *Culex* and *Aedes* transmit both the dog heartworm, *Dirofilaria immitis*, and one form of avian malaria caused by *Plasmodium*. Mosquitoes are also important in the transmission of the arboviruses (*AR*thropod-*BO*rne) causing Eastern, Western and Venezuelan encephalitis in horses and other arbovirus diseases of man and animals.

The only known vectors of human malaria belong to the genus *Anopheles* while yellow fever is transmitted by *Aedes* spp. All three genera transmit the human filarial nematodes *Wuchereria* and *Brugia*.

CONTROL

Measures, largely developed for the control of human malaria, are directed either against the developing larvae or adults, or against both simultaneously.

The various measures used against larvae include the removal or reduction of available breeding sites by drainage or other means which make these sites unsuitable for larval development. This is not always practicable, economical or acceptable and the feasibility of these methods must always be assessed locally. Biological control has been attempted by, for example, introducing predatory fish into marshy areas and rice fields, but these methods are unsuitable for those mosquito species breeding in small temporary collections of water. The isolation and development of mosquito pathogens including micro-organisms, protozoa and nematodes is mainly experimental at present as are genetic methods of control.

Probably the most widely used measures against mosquito larvae are those which involve the repeated application to breeding sites of toxic chemicals, mineral oils or insecticides, but these have to be continuously applied. Since such measures may lead to environmental pollution and may also accelerate the development of insecticide resistance, the only permanent solution is the destruction of breeding sites.

Insecticides with a residual action are effective against the adult stages, particularly if applied indoors, and these have been widely used to control the *Anopheles* vectors of malaria in man. Organophosphorous compounds and carbamates are recommended for this purpose as well as the use of the residual organochlorines where resistance is not present.

Although synthetic pyrethroids have been available for some time as short acting space sprays, some are now being developed as residual insecticides. Fly-screens, nets and repellents are available for the protection of man.

Suborder BRACHYCERA

Family TABANIDAE

These large robust flies are commonly known throughout the world as horseflies, although they will attack and feed on a wide variety of large animals and man. The pain caused by their bites leads to interrupted feeding, and as a consequence, flies may feed on a succession of hosts and are therefore important in the mechanical transmission of pathogens such as trypanosomes.

There are many genera of tabanids, but only three are of veterinary significance, namely *Tabanus*, *Haematopota* and *Chrysops*. Since these are closely related in behaviour and pathogenic significance they will be considered as a group.

Hosts:
Generally large domestic or wild animals and man, but small mammals and birds may also be attacked

Species:
There are over 3,000 species of tabanids

Distribution:
Worldwide, although certain genera are absent from large areas; for example, there are no *Haematopota* species in Australia or North and South America

MORPHOLOGY

These are medium to large biting flies, up to 2.5 cm in length, with wing spans of up to 6.5 cm. They are generally dark coloured, but may have various stripes or patches of colour on the abdomen or thorax and even the large eyes, which are dichoptic in the female and holoptic in the male, may be coloured. The coloration of the wings is useful in differentiating the three major genera. Thus *Tabanus* has clear or brownish wings, while there are often dark bands across the wings in *Chrysops;* in contrast *Haematopota* has mottled or speckled wings (Fig. 117). Also useful in generic differentiation are the characteristics of short, stout, three-segmented antennae which, unlike large cyclorrhaphan flies, have no arista (Fig. 111).

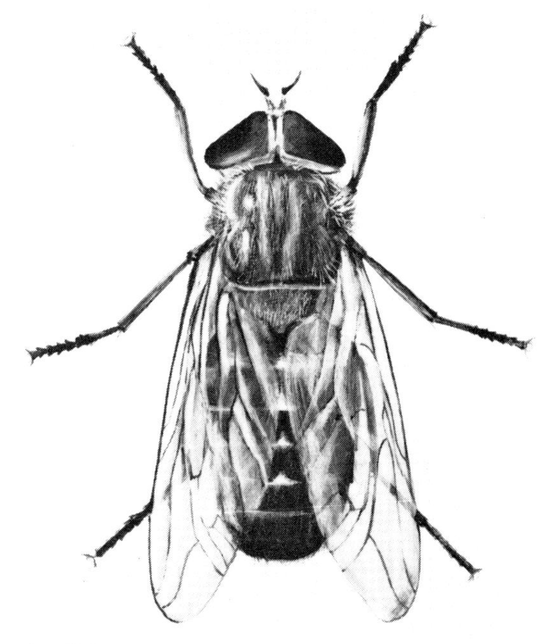

(b) *Tabanus*

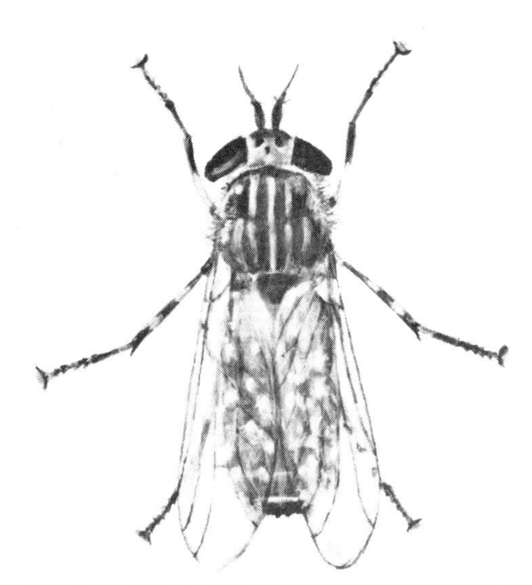

Fig 117 Differentiation of three common genera of tabanid flies.
(a) *Haematopota*
(c) *Chrysops*

The mouthparts are short and strong and always point downwards. Most prominent is the stout labium which is grooved dorsally to take the other mouthparts, collectively termed the biting fascicle; the labium is also expanded terminally as paired large labella which carry tubes called pseudotracheae through which blood or fluid from wounds is aspirated (Fig. 118). The biting fascicle, which creates the wound, consists of six elements, the upper sharp labrum, the hypopharynx with its salivary duct, paired rasp-like maxillae and paired broad pointed mandibles. Male flies have no mandibles and therefore cannot feed on blood.

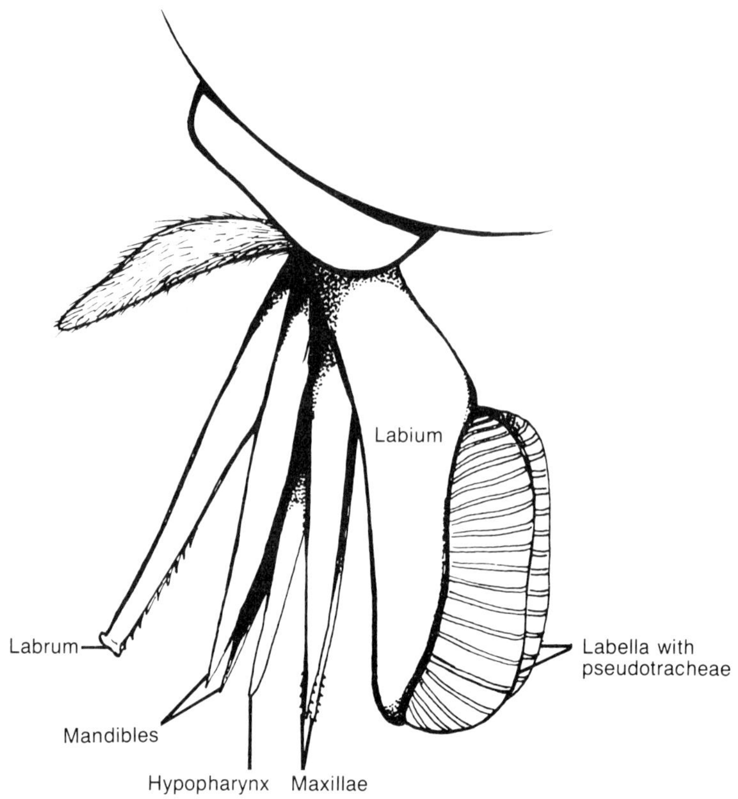

Fig 118 Slashing and sponging mouthparts of a tabanid fly.

LIFE CYCLE

After a blood meal the female lays batches of several hundred creamy-white or greyish cigar-shaped eggs, 1.0–2.5 mm long, on the underside of vegetation or on stones, generally in muddy or marshy areas. The eggs hatch in 1–2 weeks and the cylindrical, poorly differentiated larvae drop into the mud or water. The larvae, 1.0–6.0 cm long, are recognised as tabanids by their small black retractable heads, the prominent raised rings around the segments, most of which bear pseudopods and a structure in the last segment, unique to tabanid larvae, known as Graber's organ, the function of which may be sensory. They are sluggish and feed either by scavenging on decaying organic matter or by predation on small arthropods including other tabanid larvae. Optimally, larval development takes three months, but if hibernation occurs, may extend for up to three years. Mature larvae pupate partially buried in mud or soil and the adult fly emerges after 1–3 weeks. The whole life cycle takes a minimum of 4–5 months or longer if larval development is prolonged.

Populations of adult flies show seasonal fluctuations in both temperate and tropical areas. In temperate climates, adults die in the autumn and are replaced by new populations the following spring and summer, whereas in tropical areas their numbers are merely reduced during the dry season with an increase at the start of the rainy season.

PATHOGENIC SIGNIFICANCE

These powerful flies may disperse many kilometres from their breeding areas and are most active during hot, sunny days. The adult females locate their prey mainly by sight and their bites are deep and painful. They feed every 3–4 days causing a great deal of annoyance, and because their feeding is often disturbed, are efficient mechanical vectors of the organisms responsible for such diseases as anthrax, pasteurellosis, trypanosomiasis, anaplasmosis and the human filarial disease, loiasis.

It is difficult to assess the real significance of this group of flies, but there are estimates of annual losses in the U.S.A. of approximately $40 million due to both nuisance effect and disease transmission.

CONTROL

This poses a special problem since breeding places are both diffuse and difficult to detect. Insecticidal sprays with a residual effect are used in animal houses and on the animals themselves. There have also been recent trials in the U.K. using dark panels with sticky adhesive as traps and there are a number of electrocution grids which may prove useful in animal houses.

Suborder CYCLORRHAPHA

This Suborder consists of a number of families of flies which are important as parasites or as vectors of disease in animals. Since the classification of a number of genera is variable they will be discussed, for simplicity, under four major families, the Muscidae (syn. Anthomyidae), Calliphoridae (syn. Tachinidae), Oestridae and Hippoboscidae.

Family MUSCIDAE

This family comprises many biting and non-biting genera, the latter commonly referred to as nuisance flies. As a group they may be responsible for 'fly-worry' in livestock and a number of species are vectors of important bacterial, helminth and protozoal diseases of animals.

The major genera of veterinary importance include *Musca* (house flies and related flies), *Hydrotaea* (head fly), *Stomoxys* (stable fly) and *Haematobia* (horn fly). An atypical genus included here is *Glossina* (tsetse fly) which is given family status in some classifications.

Musca

Hosts:
Members of this non-biting genus are not obligatory parasites, but they can feed on a wide variety of animal secretions and are especially attracted to wounds

Species:

Musca domestica The house fly

M. autumnalis The face fly

A number of other species and sub-species occur in the tropics and subtropics, eg. the *Musca sorbens* group

Distribution:
Worldwide

MORPHOLOGY

Adults are about 5.5–7.5 mm in length and vary in colour from light to dark grey. There are four distinct dark longitudinal stripes on the thorax and the greyish abdomen has various light and dark markings (Pl. VII). The complex mouthparts, which are adapted for sponging, are obvious only when extended during feeding (Fig. 119). Wing venation is of taxonomic importance in the differentiation of *Musca* from similar flies belonging to other genera such as *Fannia*, *Morellia* and *Muscina* and in the identification of different *Musca* species, but is beyond the scope of this text.

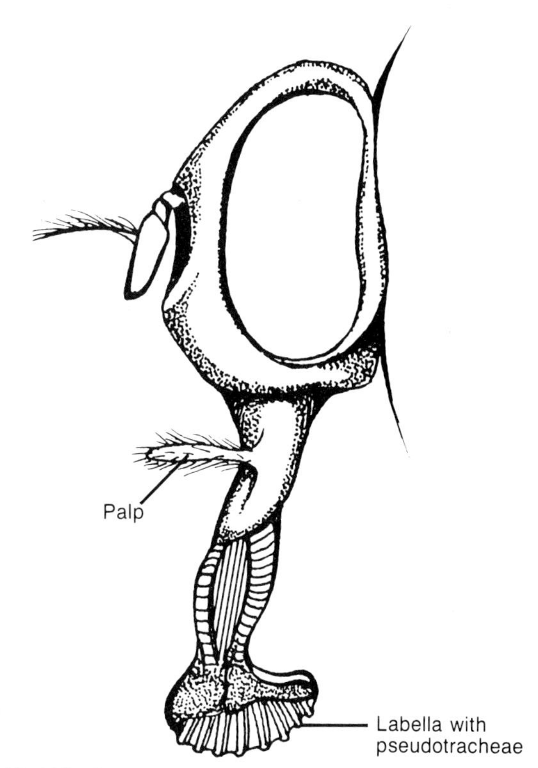

Fig 119 Sponging mouthparts of the house fly *Musca domestica* extended for feeding.

An important morphological feature is the presence of sticky hairs on pad-like structures at the end of the clawed legs. These enable the fly to adhere to smooth surfaces, but more importantly, they are responsible for the transmission of pathogenic bacteria when, for example, flies feed on septic wounds and decaying organic matter.

LIFE CYCLE

Female flies lay batches of up to 100 creamy-white, 1.0 mm long, banana-shaped eggs in faeces or rotting organic material. Eggs hatch, under optimal temperatures, in 12–24 hours to produce whitish, segmented, cylindrical larvae (maggots), which, anteriorly, are pointed and have a pair of small hooks. At the blunt posterior end of the larvae there are paired respiratory spiracles, the shape and structure of which allow generic and specific differentiation.

The three larval instars feed on decomposing organic material and mature to 1.0–1.5 cm long maggots in 3–7 days under suitable conditions. These then move to drier areas around the larval habitat and pupate in the final larval skin which contracts and becomes rigid and dark brown, to form the 6.0 mm long barrel-shaped puparium or pupal case. The adult fly emerges after 3–26 days depending on temperature.

Total development time from egg to adult fly may therefore be as little as eight days at 35 °C, but is extended at lower temperatures, eg. to 49 days at 16 °C. In temperate areas a small proportion of pupae or larvae may survive the winter, but apparently more frequently the flies overwinter as hibernating adults.

PATHOGENIC SIGNIFICANCE

Houseflies, as their name suggests, are closely associated with buildings inhabited by animals and man. They are not only a source of annoyance, but may also mechanically transmit viruses, bacteria, helminths and protozoa due to their habit of visiting faecal and decaying organic material; pathogens are either carried on the hairs of the feet and body or regurgitated as salivary vomit during subsequent feeding. A number of *Musca* spp. have been incriminated in the spread of diseases including mastitis, conjunctivitis and anthrax. In man they are probably most important in the dissemination of *Shigella* and other enteric bacteria. Eggs of various helminths may be carried by flies which feed on faeces and they also may act as intermediate hosts of a number of helminths such as *Habronema* spp. and *Raillietina* spp. Deposition of *Habronema* larvae in wounds may give rise to skin lesions commonly termed 'summer sores' in horses.

The 'face fly' *M. autumnalis* tends to feed on secretions from the eyes, nose and mouth as well as on wounds left by biting flies and this is often the most numerous of the flies which worry cattle at pasture. The eggs of *M. autumnalis* are usually laid in bovine faeces and if conditions are suitable the resultant large fly populations can cause serious annoyance and so interfere with grazing. These flies are considered to be important in the transmission of infectious bovine keratoconjunctivitis ('pink eye' or New Forest disease) due to *Moraxella bovis* and they are also intermediate hosts of *Parafilaria bovicola* and the eyeworm *Thelazia*. In North America in recent years, increases in the number of these flies has led to a high incidence of eye disorders such as conjunctivitis.

There are a number of closely related genera of non-biting muscid flies, namely *Fannia, Morellia* and *Muscina*, which may, in some areas, make a substantial contribution to 'fly-worry' in livestock, but the life cycles and control of these are similar to that described for *Musca* spp.

CONTROL

Various types of screens and electrocution grids for buildings are available to reduce fly nuisance to humans, but the best methods of control are those aimed at improving sanitation and reducing breeding places (source reduction). For example, in stables and farms, manure should be removed or stacked in large heaps when the heat of fermentation will kill the developing stages of flies, as well as eggs and larvae of helminths. In addition, insecticides applied to the surface of manure heaps may prove beneficial.

A range of insecticides and procedures are available for the control of adult houseflies. Aerosol space sprays, residual insecticides applied to walls and ceilings and insecticide-impregnated cards and strips may reduce fly numbers indoors. Insecticides may also be incorporated in solid or liquid fly baits using attractants such as various sugary syrups or hydrolysed yeast and animal proteins.

Outdoors, insecticide impregnated ear tags, tail bands and halters, mainly containing synthetic pyrethroids, together with pour-on and spray preparations, are widely used to reduce fly annoyance in cattle and horses.

Previously, insecticide dust bags ('backrubbers') were used to reduce the numbers of muscid flies associated with fly-worry. These consist of sacking impregnated with or containing insecticide, which is suspended between two posts at a height which allows cattle to rub and thus apply the insecticide to the skin.

Hydrotaea

This genus of non-biting flies, which closely resembles *Musca*, contains one species of special veterinary importance, the headfly *Hydrotaea irritans*, which is responsible for a serious condition in sheep. In many areas it is also the most numerous muscid species found on cattle and horses and has been incriminated in the transmission of summer mastitis and infectious bovine keratoconjunctivitis.

Hosts:
Sheep, cattle and horses

Species:
Hydrotaea irritans

A number of other *Hydrotaea* spp. may also be associated with fly-worry in livestock, for example *H. albipuncta* which is found round the eyes of cattle

Distribution:
Mainly northern Europe

MORPHOLOGY

H. irritans, not unlike *Musca* in size, is characterised by an olive-green abdomen and an orange yellow coloration at the base of the wings (Pl. VIII). Specific identification of non-biting muscid flies requires specialist advice.

LIFE CYCLE

Adult flies prefer still conditions, and are found near woodlands and plantations with peak numbers occurring in mid-summer. Eggs are laid in decaying vegetation or faeces and these hatch and develop into mature larvae by the autumn: these larvae then go into diapause (a temporary cessation of development) until the following spring when pupation and development is completed with emergence of a new generation of adults in early summer. Thus there is only one generation of headflies each year.

PATHOGENIC SIGNIFICANCE

Although not obligatory parasites, headflies are attracted to animals and feed on lachrymal secretions and wounds, such as those incurred by fighting rams. As in *Musca*, the labella are adapted for sponging, but in addition they possess small teeth and the rasping effect of these during feeding leads to skin damage. Horned breeds of sheep, such as the Swaledale and Scottish Blackface, are most susceptible to attack and swarms of these flies around the head lead to intense irritation and annoyance and result in self-inflicted wounds which then attract more flies. Clusters of flies feeding at the base of the horns (Pl. VIII) lead to extension of these wounds and the condition may be confused with blowfly myiasis. Secondary bacterial infection of wounds is common, and this may encourage blowfly strike.

In cattle, large numbers of *H. irritans* have been found on the ventral abdomen and udder and since the bacteria involved in 'summer mastitis' (*Corynebacterium pyogenes*, *Streptococcus dysgalactiae* and *Peptococcus indolicus*) have been isolated from these flies there is strong presumptive evidence that they may transmit the disease.

The economic losses due to headfly infection are difficult to assess, but are thought to be substantial.

CONTROL

This has proved difficult. It has been traditional in some areas to provide rams with protective canvas head caps. Reduction in the use of fields bordering woodlands has also been advised as has the introduction of polled breeds of sheep. Affected animals require to be housed and treated to prevent further damage.

A number of protective or repellent creams are available for application around the base of the horns, but many of these only prevent skin contact, not annoyance. Recent trials using synthetic pyrethroids either as sprays applied every 14 days during the fly season, or as impregnated ear tags have given promising results.

In cattle the use of one or two insecticide-impregnated ear tags, or tail-bands, usually containing synthetic pyrethroids, is beneficial in easing fly-worry. Where summer mastitis is a problem tail-bands are preferable.

Stomoxys

The commonest species in this genus is *Stomoxys calcitrans*, commonly known as the stable fly or biting housefly. The bites of this fly are painful and it is a vector of several protozoal and helminth diseases of animals.

Hosts:
Most animals and man

Species:
Stomoxys calcitrans

Distribution:
Worldwide

MORPHOLOGY

Superficially, *S. calcitrans* resembles the housefly *M. domestica*, being similar in size and grey with four longitudinal dark stripes on the thorax. Its abdomen, however, is shorter and broader than *Musca* with three dark spots on the second and third abdominal segments (Pl. VII). Probably the simplest method of distinguishing stable flies from *Musca* and other genera of non-biting muscid flies is by examination of the proboscis, which in *Stomoxys* is conspicuous and forward projecting. Stable flies can be distinguished from biting muscid flies of the genus *Haematobia* by the larger size and the shorter palps of the former.

Larvae of *Musca* and *Stomoxys* can be differentiated by examination of the posterior spiracles.

LIFE CYCLE

Both male and female flies feed on blood and the female lays batches of 25–50 eggs, resembling those of houseflies, in moist, decaying vegetable matter such as hay and straw contaminated with urine. Eggs hatch in 1–4 days, or longer in cold weather, and the larvae, which again resemble housefly larvae are mature in 6–30 days. After emergence the adult females require several blood meals before the ovaries mature and egg laying can start.

The complete life cycle from egg to adult fly may take 12–60 days depending mainly on temperature.

In temperate areas flies may overwinter as larvae or pupae whereas in tropical climates breeding is continuous throughout the year.

PATHOGENIC SIGNIFICANCE

When feeding, the proboscis swings downwards and skin penetration is achieved by the rasping action of fine

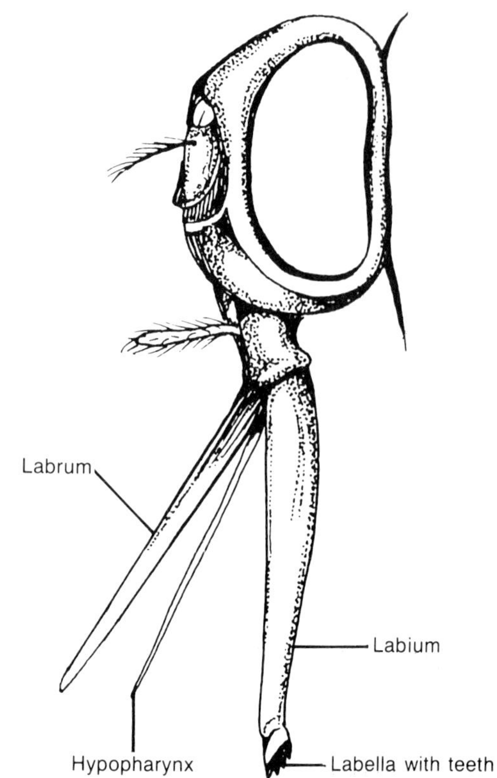

Fig 120 Piercing-sucking mouthparts of the stable fly *Stomoxys calcitrans* in a downward feeding position.

teeth on the end of the labium (Fig. 120). This is painful and stable flies may be a serious pest of animals and man. Approximately three minutes is required for a blood meal and feeding is often interrupted thus allowing mechanical transmission of pathogenic microorganisms and protozoa, such as trypanosomes. *S. calcitrans* also acts as an intermediate host of *Habronema*.

Adult flies live for about one month and are abundant around farm buildings and stables in late summer and autumn in temperate areas. They prefer strong sunlight and they bite mainly out of doors although they will follow animals inside to feed. In large numbers these flies are a great source of annoyance to grazing cattle and in some areas there are estimates of milk and meat production losses of up to 20%.

CONTROL

Many of the control measures outlined for *Musca* are applicable to *Stomoxys*. Source reduction is important and potential breeding sites should be avoided by the regular removal and stacking of moist bedding, hay and food wastes from stables and cattle accommodation. Aerosol insecticide sprays in and around stables and farm buildings may give good local control of *S. calcitrans*.

Haematobia

Several species of this genus of blood-sucking muscid flies occur in various countries and may be a serious nuisance on cattle. One species, the horn fly, is known as *Haematobia irritans* in the U.S.A. and sometimes as *Lyperosia irritans* in Britain and parts of Europe.

There is additional taxonomic confusion in that certain other species assigned to the genus *Haematobia* are sometimes referred to under the genera *Haematobosca* and *Siphona*. For simplicity all of the biting muscid flies other than *Stomoxys* and *Glossina* will be considered under *Haematobia*.

Hosts:
Cattle and buffalo

Species:

Haematobia (syn. Lyperosia) irritans	the horn fly
H. exigua	the buffalo fly
H. (syn. *Haematobosca) stimulans*	

Distribution:
Horn flies occur in many parts of the world including Europe, U.S.A. and Australia, whereas the buffalo fly is restricted to the Far East and northern Australia, while *H. stimulans* appears to be distributed mainly in Europe

MORPHOLOGY

The adults are up to 4.0 mm long and are the smallest of the blood-sucking muscids (Plate VII). They are usually grey, often with several dark stripes on the thorax. Unlike *Musca* the proboscis is held forwards and unlike *Stomoxys* the palps are stout and as long as the proboscis. In contrast to other muscids *Haematobia* spp. generally remain on their hosts leaving only to fly to another host or, in the case of females, to lay eggs in freshly passed faeces.

LIFE CYCLE

Eggs, 1.0–1.5 mm long, are laid in fresh faeces. These hatch quickly and larvae may be mature in as little as four days given adequate moisture and temperatures of around 27 °C. Low temperatures and dry conditions delay larval development and kill the eggs. The pupal period is around 6–8 days and on emergence the adult flies seek and remain on their cattle or buffalo hosts.

PATHOGENIC SIGNIFICANCE

Horn flies may be found in thousands feeding along the back, sides and ventral abdomen of cattle. Their common name is derived from the fact that they tend to cluster around the horns or poll region when not feeding. Large numbers cause intense irritation and the skin wounds made in feeding may attract other muscids and myiasis-producing flies. It is difficult to assess the precise economic effect of these flies, but their control on grazing cattle can result in significant increases in production. The buffalo fly and other *Haematobia* spp. have a similar pathogenic effect.

Although less important than many other muscid flies in disease transmission, species of *Haematobia* transmit *Stephanofilaria*, the skin filarioid of cattle.

CONTROL

Since *Haematobia* spp. spend much time on their hosts, control is easy compared to the control of other stock-visiting muscids. Recently developed insecticide-impregnated ear tags or tail-bands are particularly beneficial in the control of horn flies; alternatively, animals may be repeatedly sprayed with insecticides.

Glossina

This genus is considered here as a member of the family Muscidae, although in some classifications it is discussed as the sole member of the family Glossinidae.

Members of this group of biting flies are commonly termed tsetse flies. They are distributed over 10 million square kilometres of Africa and are extremely important as vectors of African trypanosomiasis which is a serious disease of domestic animals and man.

Hosts:
Various mammals, reptiles and birds

Species:
There are around 30 species and sub-species of the genus *Glossina*. Identification of individual species and sub-species is a matter for the specialist

DISTRIBUTION

These flies are confined to a belt of tropical Africa extending from the southern Sahara (Lat. 15 °N) in the north to Zimbabwe and Mozambique in the south (Lat. 20–30 °S). The species are restricted to various geographical areas according to habitat, the three main groups, named after the commonest species in each group, being *fusca*, *palpalis* and *morsitans*, found respectively in forest, riverine and savannah areas. The last two groups, because of their presence in the major livestock-rearing areas, are the most important from a veterinary standpoint.

MORPHOLOGY

In general the adults are narrow, yellow to dark brown flies, 6–15 mm in length, and have a long, rigid and forward projecting proboscis (Pl. VII). When at rest, the wings are held over the abdomen like a closed pair of scissors. If in doubt they are easily distinguished from all other flies by the characteristic cleaver (hatchet) cell in the wings.

There are no maxillae or mandibles in the mouthparts although the proboscis is adapted for piercing and sucking, and like *Stomoxys*, consists of a lower U-shaped labium with rasp-like labella terminally, an upper narrower sharp labrum and between these a food channel containing the slender hypopharynx which carries saliva and anticoagulant into the wound formed during feeding. Tsetse flies become infected with trypanosomes during feeding and these then undergo multiplication within the fly before they are infective for other hosts during subsequent feeding.

LIFE CYCLE

Both male and female flies suck blood and although they may have some host preferences, they will feed on a wide variety of animals. The females, in contrast to other muscids, are viviparous and produce only one larva at a time, up to a total of 8–12 larvae. Maturation in the uterus from fertilised egg to the mobile, 8.0–10 mm long, third stage larva deposited by the adult takes approximately 10 days. At this stage the larva is creamy white, segmented and posteriorly has a pair of prominent dark ear-shaped protuberances known as polypneustic lobes (Fig. 121): these have a respiratory function similar to the posterior spiracles of other muscid larvae. After deposition the larva wriggles into loose soil to a depth of a few centimetres and forms a rigid dark brown, barrel-shaped puparium. The pupal period is relatively long, taking 4–5 weeks, or more in cool weather. On emergence the female fly requires several blood meals over a period of 16–20 days before producing her first larva.

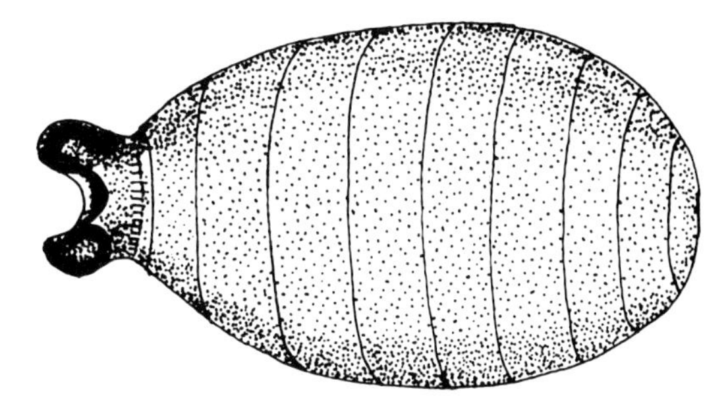

Fig 121 Mature larva of *Glossina* showing segmented appearance and polypneustic lobes.

Breeding generally continues throughout the year with peak fly numbers occurring at the end of the rainy season.

PATHOGENIC SIGNIFICANCE

Although the bites of tsetse flies are very painful and cause marked irritation, their main significance is in the transmission of animal and human trypanosomiasis.

CONTROL

In the past, campaigns against tsetse flies to control trypanosomiasis both in humans and in animals have depended mainly on large scale killing of game animals which act as reservoirs of trypanosome infection and as a source of blood for the flies. It was also common to clear large areas of bush in order to destroy the habitats of the adult flies. These methods were fairly successful, but are now largely unacceptable on ecological and economic grounds.

Currently, most anti-tsetse measures rely on the use of insecticides applied from the ground or by aircraft. When the objective is complete eradication of *Glossina*, residual formulations of insecticides are used. When the objective is merely control of fly populations to reduce the risk of trypanosomiasis to man and animals, or periodically, to ensure that a fly-free area remains so, non-residual insecticides may be applied aerially.

Eradication is preferable but, because of the inevitable reinvasion of tsetse from surrounding untreated areas, is uneconomic unless the selected area is on the edge of a tsetse belt where the fly population is already under stress because of relatively unfavourable climatic conditions. It is also essential that the area to be sprayed has economic potential and that agricultural development of the cleared area should proceed contemporaneously.

Control with non-residual insecticides is expensive because of the necessity to repeat the operation regularly and is justified only where the area has great economic potential.

Advocates of insecticidal spraying argue that, since *Glossina* is highly susceptible to the insecticides used, the sophisticated and selective use of modern chemicals, usually on one occasion only, has no major and permanent effects on the environment; in fact, the changes in land use which should ensue from successful control are much more significant in this respect.

Recently much attention has been focused on the possibility of reducing or eradicating tsetse flies in localised areas by the use of traps. These have the advantages of being cheap, can be used by local labour and are harmless to the environment. Essentially they depend on the presentation of material, such as dark cloth, which attracts the flies and leads into a trap which often incorporates an insecticide. Currently, experiments carried out using odoriferous extracts from cattle placed in or near traps to attract flies to the area have given promising results.

Family CALLIPHORIDAE

This family together with the Sarcophagidae and the Oestridae contain the species responsible for the most important myiases of domestic animals and man.

Myiasis is defined as the infestation of living animals with the larvae of dipteran flies. It may be **facultative** (optional), as in the calliphorids, or **obligatory**, as in the oestrids. It also may be cutaneous (eg. *Lucilia*), nasal (eg. *Oestrus*), or somatic (eg. *Hypoderma*).

A common term for myiasis caused by members of the Calliphoridae is 'Blow-Fly Strike', the laying of eggs by the fly being termed the 'blow' and the development of the larvae (maggots) and the damage they cause the 'strike'.

BLOWFLY MYIASIS

Hosts:
Mainly sheep, but any other animal may be affected. It is important to note that only the larvae are responsible for myiasis

Distribution:
Worldwide

MAJOR SPECIES IN EUROPE

Lucilia sericata	(greenbottles)
Phormia terrae-novae	(blackbottles)
Calliphora erythrocephala *C. vomitoria*	(bluebottles)

OTHER IMPORTANT SPECIES IN TROPICS AND SUBTROPICS

Lucilia cuprina
L. caesar
L. illustris
Phormia regina
Calliphora stygia
C. australis
C. fallax
Chrysomyia albiceps
C. chlorophyga
C. micropogon
C. rufifacies

MORPHOLOGY

Adults: Blowflies measure up to 1.0 cm in length and on microscopic examination all show distinct dorsal bristles on the thorax. Some genera such as the greenbottles are relatively slender, while others like the bluebottles are stout.

All are characterised by having a metallic blue or green sheen on the body; thus *Lucilia* is greenish to bronze, *Phormia* is black with an overlying blue-green sheen, while *Calliphora* is blue (Pl. IX) and *Chrysomyia* bluish-green. Identification of individual species can be made according to local colour differences mainly on the thorax and abdomen.

Larvae: Apart from some *Chrysomyia* spp. all of these are smooth, segmented and measure 10–14 mm in length. They possess a pair of oral hooks at the anterior extremity, spiracles on the anterior segment, and posteriorly, stigmatic plates also bearing spiracles. The arrangement of the spiracles on these plates serves to differentiate the species.

LIFE CYCLE

The gravid female blowfly lays clusters of yellowish-cream eggs on wounds, soiled fleece or dead animals, being attracted by the odour of the decomposing matter.

In temperate areas under summer conditions, the eggs hatch into larvae in about 12 hours; the larvae then feed, grow rapidly and moult twice to become fully mature maggots in 3–10 days. These then drop to the ground and pupate. The pupal stage is completed in 3–7 days in summer and the emergent female fly, after a protein meal, reaches sexual maturity. The fertilised female can lay up to 3,000 eggs, usually in batches of 100–200. Adult flies can live for about 30 days, and up to four generations can develop between May–September. The final generation overwinters in the soil, usually as pupae, to emerge in the following spring.

In warmer climates the number of generations per annum is greater and up to nine or ten have been recorded in southern Africa and Australia.

EPIDEMIOLOGY

The blowflies which attack sheep fall into two main categories:

(i) **Primary flies**, which are capable of initiating a strike on living sheep. These include *Lucilia* and *Phormia* spp. and some *Calliphora* spp.
(ii) **Secondary flies** which cannot initiate a strike, but attack an area already struck or otherwise damaged. They frequently extend the injury rendering the strike one of great severity. Examples include many *Calliphora* spp. and in warmer climates *Chrysomyia* spp.

There is some evidence that the development of larvae of the secondary flies results in competition for food with larvae of primary flies. This battle is usually won by the secondary larvae particularly in countries where *Chrysomyia* spp. are prevalent as larvae of this genus are carnivorous and feed on the larvae of the primary flies.

In Australia a further group of 'tertiary' flies composed of the *Musca* group are found mainly when the lesions of the 'struck' carcass are becoming dry.

The epidemiology of cutaneous myiasis in sheep depends on factors which affect the prevalence of blowflies and those which affect host susceptibility. The three principal factors are:

(i) **Temperature.** In temperate areas, the warmer temperatures of late spring and summer stimulate hibernating pupae to complete their development and the first wave of adult blowflies then appears. High ambient temperatures, provided the relative humidity is also high, will favour the creation of suitable areas of microclimate in the fleece which attract the adult flies to lay their eggs.
(ii) **Rainfall.** Persistent rain can create 'wool rot' which makes the fleece attractive to the adult flies. The latter are not active during rainfall, but become so immediately the rain ceases, taking advantage of the fleece conditions produced by the rain. Breeds of sheep with long fine wool are particularly susceptible to wool or fleece rot.
(iii) **Host susceptibility.** This is increased where putrefactive odours develop on the fleece and usually originate due to bacterial decomposition of organic matter. The commonest causes of this are soiling of the hindquarters due to urine or diarrhoea and injuries due to shearing, fighting or barbed wire.

Certain breeds of sheep, such as the Merino, possess a narrow breech area with excessive wrinkling of skin, which favours constant soiling by urine or faeces and so attracts blowflies. In rams and wethers with a narrow opening of the penile sheath the accumulation of urine favours strike in this area.

From a consideration of the above factors it is clear that in Europe the blowfly season occurs from the beginning of June until the end of September. In hill sheep, there occurs a first phase in June which affects the unshorn adults, the lamb fleece being insufficiently grown. Immediately following shearing, there is little or no strike, but a second phase occurs in August/September and affects the lambs as their fleeces grow. In lowland flocks where shearing is earlier the first phase is less conspicuous and the second, affecting mainly the lambs, occurs earlier, in July.

Since the number of fly generations is greater in warmer countries, the period of risk is more prolonged and is especially high in warm, moist weather.

PATHOGENESIS

After the eggs are deposited on the wool by the primary adult fly, the larvae emerge and crawl down the wool on to the skin, which they lacerate with their oral hooks, and secrete proteolytic enzymes which digest and liquefy the tissues. Secondary blowflies are then attracted by the odour of the decomposing tissues and their larvae extend and deepen the lesion. The situation is often complicated by secondary bacterial infection.

The irritation and distress caused by the lesion is extremely debilitating and sheep can rapidly lose condition. The latter is often the first obvious sign of strike as the lesion occurs at the skin surface and is sometimes observed only on close examination. Where death occurs, it is often due to septicaemia.

Strike may be classified according to the area of the body affected, ie. breech, tail, body, pollor penile sheath ('pizzle rot'). In Europe body strike, emanating from wool-rot created by heavy rainfall, is the most common.

In Australasia, South Africa and South America where the Merino breed is prevalent, breech and tail strike are the most common due to the conformation of this breed and the wrinkled skin in the breech area which favours the accumulation of urine and faeces.

CLINICAL SIGNS

Affected sheep are anorexic, appear dull and usually stand away from the main flock. In body strike the fleece in the affected area is darker, has a damp appearance and a foul odour. However, except for advanced cases, nothing else can be seen until the fleece is parted, revealing the damaged skin (Pl. VIII) and the maggot larvae.

DIAGNOSIS

This is based on the clinical signs and recognition of maggots in the lesion.

TREATMENT

Once the problem is diagnosed in a flock, all affected sheep should be separated and the area surrounding the lesion clipped, where possible larvae removed, and the lesion dressed with a suitable insecticide such as diazinon or coumaphos.

CONTROL

This is based primarily on the prophylactic treatment of sheep with insecticides. The problems associated with this are the relatively short period spent by the larvae on the sheep, the repeated infestations which occur throughout the season and the rapidity with which severe damage occurs. Any insecticide used must therefore not only kill the larvae, but persist in the fleece. In this respect the chlorinated hydrocarbon, dieldrin, proved particularly effective and gave protection for at least 20 weeks. However this product has been largely withdrawn on safety grounds and replaced mainly by organophosphorous compounds which have a persistence of 10–16 weeks, unless resistance supervenes when this period becomes much shorter. The carbamate, butocarb, is also successfully used for the control of strike.

Application of these insecticides is made by plunge dipping or, more rarely in Europe, in a spray race or by jetting. In Europe the high prevalence of body strike makes whole body protection necessary and therefore the use of dips is more effective. In practice an annual dip, usually in June, should give protection for the remainder of the fly season, but a second dipping in August may be necessary in order to ensure complete protection.

Other measures which should be taken to aid control are the prevention of diarrhoea by effective worm control and the removal of excess wool from the groin and perineal area to prevent soiling, a technique known as crutching. Burial or burning of carcasses, which otherwise offer an excellent alternative breeding place for blowflies, is also recommended.

In Australia where the Merino breed predominates, selective breeding of sheep with plain rather than wrinkled breech areas has been tried but progress has been slow. An alternative has been the use of Mule's operation in which prophylactic surgery is used to remove a part of the breech skin so that when the excised areas heal the skin is flat rather than wrinkled.

Screw-worm myiasis

The name screw-worm is given to the larvae of certain species of *Cochliomyia* (syn. *Callitroga*) including *C. hominivorax* and *C. macellaria*, and to that of a single species of *Chrysomyia*, *C. bezziani*, which cause myiasis in animals and occasionally man. *Cochliomyia* is found in the New World while *C. bezziana* is confined to Africa and Asia.

These bluish-green flies have longitudinal stripes on the thorax and orange-brown eyes (Pl. IX). They occur primarily in tropical areas and lay their eggs on wounds, the larval stages characteristically feeding as a colony and penetrating the tissues creating a large and foul-smelling lesion. *C. hominivorax* was such a problem in the southern U.S.A. that a mass eradication campaign using biological control was undertaken. This involved the release of up to 1,000 male flies, sterilised by irradiation, per square mile. Since the female fly mates only once, control proved very successful except where the flies, which are capable of flying up to 200 miles, migrated from across the Mexican border.

Tumbu fly myiasis

Larvae of *Cordylobia anthropophaga*, the 'tumbu fly', are responsible for myiasis in man and animals in sub-Saharan Africa. The larvae develop under the skin and produce a painful swelling with a small central opening.

Family SARCOPHAGIDAE

The Sarcophagidae or 'flesh flies' such as *Sarcophaga* and *Wohlfahrtia* are widely distributed, and lay their larvae in sores, wounds or decomposing flesh; they primarily affect man and only occasionally animals. The larvae can cause great disfiguration.

Family OESTRIDAE

This is an important family consisting of several genera of large, usually hairy, flies whose larvae are obligatory parasites of animals. The adults have primitive, non-functional mouthparts and are short-lived whereas the highly host-specific larvae spend a considerable time feeding and developing in their animal hosts.

Three genera which behave similarly are considered here, namely *Hypoderma*, *Oestrus* and *Gasterophilus* (the latter is currently often classified in a separate family, the Gasterophilidae).

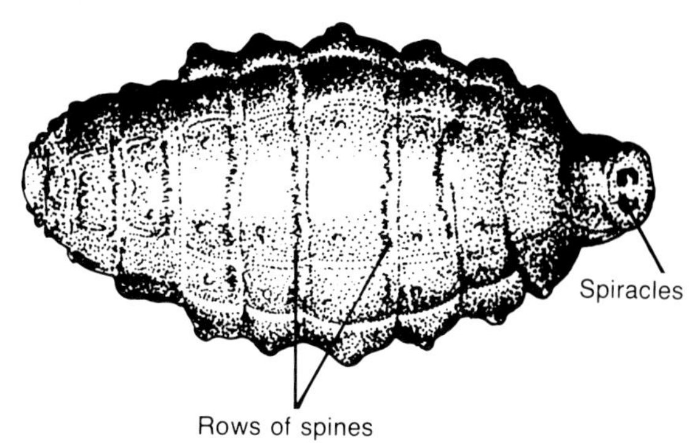

Fig 122 *Hypoderma* larva showing segmented appearance, short spines and posterior spiracles.

Hypoderma

The members of this genus are the 'warble flies' and their economic importance is reflected in the national eradication schemes which have been undertaken in several countries.

Hosts:
Cattle; the larvae occur erratically in other animals including equines, sheep and, very rarely, man

Species:
Hypoderma bovis
H. lineatum

Distribution:
Northern hemisphere. However, *Hypoderma* is absent from extreme northern latitudes, including Scandinavia, and it has occasionally been found sparsely south of the equator in Argentina, Chile, Peru, and southern Africa, following accidental introduction in imported cattle

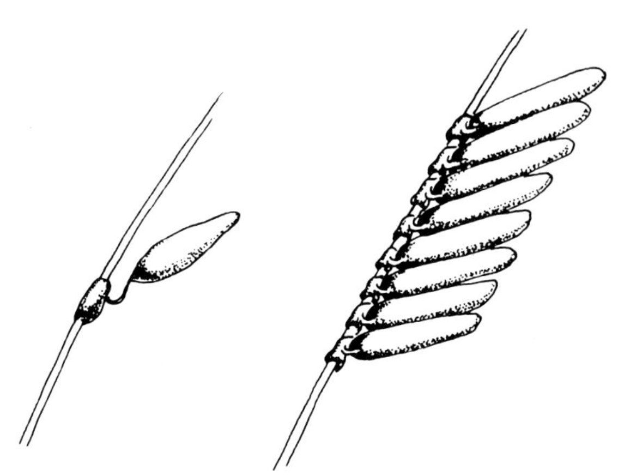

Fig 123 Eggs of *Hypoderma*.
(a) *H. bovis*
(b) *H. lineatum*

MORPHOLOGY

Adults: *H. bovis* and *H. lineatum* resemble bees, but, being Diptera, have only one pair of wings; the abdomen is covered with yellow-orange hairs with a broad band of black hairs around the middle (Pl. X).
Larvae: The mature larvae are thick and somewhat barrel-shaped, tapering anteriorly; when mature they are 2.5–3.0 cm long, and most segments bear short spines (Fig. 122). The colour is dirty white when newly emerged from the host, but rapidly turns to dark brown; the pupa is almost black.

LIFE CYCLE

The adult flies are active only in warm weather, and in Europe the peak period is in June and July. The females attach their eggs to hairs, *H. bovis* singly on the lower parts of the body and on the legs above the hocks, and *H. lineatum* in rows of six or more on individual hairs below the hocks (Fig. 123). Below 18 °C there is no fly activity.

The first stage larvae, which are less than 1.0 mm long, hatch in a few days and crawl down the hairs, penetrate the hair follicles and migrate towards the region of the diaphragm. Migration is aided by the use of paired mouth hooks and the secretion of proteolytic enzymes, and the larvae feed as they travel to the resting sites where they will spend the winter, *H. lineatum* to the submucosa of the oesophagus and *H. bovis* to the epidural fat in the spinal canal. These sites are reached in late autumn, usually after the end of November in Europe, and the moult to the second stage occurs there. In February and March migration is resumed, and the L_2 arrive under the skin of the back where they moult to the L_3, which can be palpated as distinct swellings ('warbles'). A cutaneous perforation is made by the L_3 and the larvae breathe by applying their spiracles to the aperture. After about 4–6 weeks in this site they emerge, *H. lineatum* in March–May, and *H. bovis* in May–June, and fall to the ground where they pupate under leaves and loose vegetation for about five weeks. The adults then emerge, copulate and the females lay their eggs and die, all within 1–2 weeks. Oviposition can take place as soon as 24 hours after emergence from the puparium.

PATHOGENIC SIGNIFICANCE

By far the most important feature of this genus is the economic loss caused by down-grading and condemnation of hides perforated by larvae.

In addition, however, the adult flies themselves are responsible for some loss. When they approach animals to lay their eggs their characteristic buzzing noise, which appears to be instantly recognisable, causes the animals to panic, or 'gad', sometimes injuring themselves on posts, barbed wire, and other obstacles. Dairy cows show reduced milk yield, and beef animals have reduced weight gains as a result of interrupted feeding. *H. bovis* is the most important in this respect, since it lays its eggs singly on the upper body, pursuing animals for some distance and making repeated buzzing attacks. *H. lineatum* reaches the animals by a series of hops along the ground and remains on the lower limb for a time while it lays its row of eggs, so that the animal may be unaware of its presence. In parts of the U.S.A. this species is appropriately termed the 'heel fly'.

The L_3 under the skin damage the adjacent flesh and this necessitates trimming from the carcass the greenish, gelatinous tissue called 'butcher's jelly', also seen in the infested oesophageal submucosal tissues.

Finally, if larvae of *H. bovis* should die in the spinal canal, the release of a highly toxic proteolysin which they contain may cause paraplegia, while the death of *H. lineatum* larvae in the oesophageal wall may cause bloat through oesophageal stricture and faulty regurgitation. Larval death in other regions may, in very rare cases, lead to anaphylaxis in sensitised animals.

CONTROL

Hypoderma is susceptible to systemically active organophosphorus insecticides and to ivermectin. The organophosphorus preparations are applied as 'pour-ons' to the backs of cattle and are absorbed systemically from there; ivermectin is given by subcutaneous injection. In control schemes in Europe, a single annual treatment is usually recommended, or is mandatory, preferably in September, October or November, before the larvae of *H. bovis* have reached the spinal canal, so that there is no risk of spinal damage from disintegration of dead larvae. Treatment in the spring, when the larvae have left their resting sites and arrived under the skin of the back, though effective in control, is less desirable since the hide has then been perforated by the breathing L_3. However, in some countries, such as the United Kingdom, such treatment is mandatory if warbles are present on the backs of cattle.

Hypoderma has a great capacity for population regeneration, and hence any control measures must have total eradication as their object, with safeguards against reintroduction. For this reason the most successful schemes, supported by legislation, such as restriction of cattle movement on infected farms and compulsory treatment in the autumn, have been undertaken on islands, the United Kingdom and Eire being notable examples. For example, in the former the prevalence of infected cattle has dropped during the past decade from around 40% to a fraction of 1%. Other areas which have had successful eradication, such as Denmark and the Netherlands, are clearly at greater risk of reintroduction.

Hypoderma infection in other animals

DEER

In these animals *H. diana* is equivalent to *H. bovis* in cattle, using the spinal canal as its larval resting place. The fly is most active in May and June, but it is not recognised as a cause of 'gadding' in deer. The mature larvae occur subcutaneously along the back, and the hide damage is similar to that in cattle, with linear perforations. With the success of control measures against warbles in cattle, it is important to realise that *H. diana*, though capable of infecting many species of deer, will not infect cattle so that even in areas where, as is commonly the case, almost all the deer carry the parasitic larvae, cattle are not at risk. Like other species, *H. diana* is susceptible to the organophosphorus insecticides and to ivermectin.

EQUINES

These animals may become infected with *Hypoderma* species of cattle and deer as erratic parasites, and a proportion of larvae will develop to maturity in the back, causing problems if they occur in the saddle region. In such cases treatment is by minor surgery or the topical application of insecticide. In some areas of the U.S.A. where horses are regularly infected, trials have shown that the annual use of pour-on systemic insecticides is an effective prophylactic measure.

SHEEP

There have been a few reports of warble infection in sheep in the United Kingdom, with *H. diana* as the likely species involved, but in these infections the larvae have never developed fully, simply forming so-called 'blind warbles', without skin perforation.

Oestrus

Larvae of this genus spend the parasitic period in the air passages of the hosts and are commonly referred to as 'nasal bots'.

Hosts:
Sheep and goats

Species:
Oestrus ovis

Distribution:
Worldwide

MORPHOLOGY

Adults: Grey flies about 1.0 cm long, with small black spots on the abdomen and a covering of short brown hairs (Pl. X).
Larvae: Mature larvae in the nasal passages are about 3.0 cm long, yellowish-white, tapering anteriorly with a prominent 'step' posteriorly. Each segment has a dark transverse band dorsally (Fig. 124).

LIFE CYCLE

The females are viviparous and infect the sheep by squirting a jet of liquid containing larvae at the nostrils during flight, up to 25 larvae being delivered at a time. The newly deposited L_1 are about 1.0 mm long, and migrate through the nasal passages to the frontal sinuses feeding on mucus whose secretion is stimulated by their movements. The first moult occurs in the nasal passages, and the L_2 crawl into the frontal sinuses where the final moult to L_3 takes place. In the sinuses, the larvae complete their growth and then migrate back to the nostrils. Where flies are active throughout the year, two or three generations are possible, but in cool or cold weather the small L_1 and L_2 become dormant and remain in recesses of the nasal passages over winter; they move to the frontal sinuses only in the warmer spring weather, and then complete their development, the L_3 emerging from the nostrils and pupating on the ground to give a further generation of adults. The females survive only two weeks, but during this time each can deposit 500 larvae in the nasal passages of sheep.

PATHOGENIC SIGNIFICANCE

Most infections are light, sheep showing nasal discharge, sneezing, and rubbing their noses on fixed objects. In the rare heavier infections, there is unthriftiness and sheep may circle and show incoordination, these signs being often termed 'false gid'. If a larva dies in the sinuses there may be secondary bacterial invasion and cerebral involvement.

The most important effects, however, are due to the activity of the adult flies. When they approach sheep to deposit larvae the animals panic, stamp their feet, bunch together and press their nostrils into each others' fleeces and against the ground. There may be several attacks each day, so that feeding is interrupted and animals may fail to gain weight.

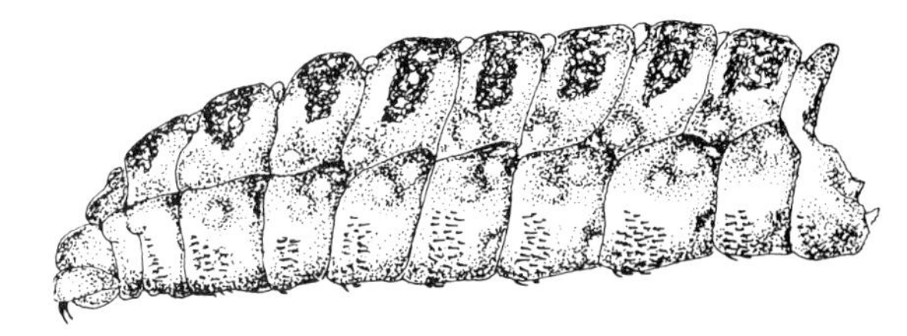

Fig 124 *Oestrus ovis* larva showing posterior 'step' and dorsal transverse bands on segments.

Oestrus can occasionally also infect man. Larvae are usually deposited near the eyes, where a catarrhal conjunctivitis may result, or around the lips, leading to a stomatitis. Such larvae never fully develop.

TREATMENT AND CONTROL

Where the numbers of larvae are small, it may not be economically worthwhile to treat. However, in heavy infections, nitroxynil, rafoxanide and ivermectin are highly effective as are the organophosphates, trichlorfon and dichlorvos.

Should a control scheme be necessary it has been suggested by South African workers that flock treatment should be given twice in the year, the first at the beginning of summer to kill newly-acquired larvae, and the second in midwinter to kill any overwintering larvae. Russian control schemes, in which 23 million sheep have been treated annually, have been based on trichlorfon as an in-feed medication, and on dichlorvos as an intranasal aerosol.

Gasterophilus

Members of this genus are commonly referred to as 'bot flies'. Their larvae, termed 'bots', spend most of their time developing in the stomach of equines, but they are generally considered of little pathogenic significance.

Hosts:
Horses and donkeys.

SPECIES AND DISTRIBUTION

Major Species:

Gasterophilus intestinalis	Worldwide
G. nasalis	Worldwide
G. haemorrhoidalis	Worldwide
G. pecorum	Europe, Africa, Asia

Minor Species:

G. inermis	N. Europe, N. Asia, N. Africa and south to Zululand
G. nigricornis	Spain, Middle East, southern U.S.S.R., China

MORPHOLOGY

Adults: Bot flies are robust dark flies 1–2 cm long (Pl. X). The most common species, *G. intestinalis*, has irregular, dark, transverse bands on the wings, but species differentiation of adult flies is rarely necessary.
Larvae: When mature and present in the stomach or passed in faeces these are cylindrical, 16–20 mm long and reddish-orange with posterior spiracles, the morphology of which is different from those of *Oestrus* and *Hypoderma* larvae. Differentiation of mature larvae of the various species can be made on the numbers and distribution of the spines present on various segments.

LIFE CYCLE

The life cycles of the various species differ only slightly. In temperate areas adult flies are most active in late summer and in the case of the common bot fly, *G. intestinalis*, eggs are laid on the hairs of the fore legs and shoulders, whereas the throat bot fly, *G. nasalis*, and the nose bot fly, *G. haemorrhoidalis*, lay their eggs in the intermandibular area and around the lips respectively. The eggs are easily seen being 1.0–2.0 mm long and usually creamy white in colour; they either hatch spontaneously in about five days or are stimulated to do so by warmth which may be generated during licking and self-grooming. Larvae either crawl into the mouth or are transferred to the tongue during licking. These then penetrate the tongue or buccal mucosa and wander in these tissues for several weeks before passing via the pharynx and oesophagus to the stomach where they attach to the gastric epithelium. An apparent exception to this is *G. pecorum* which lays eggs on pasture and these are ingested by horses during grazing.

In the stomach the red larvae of *G. intestinalis* favour the cardiac region (Fig. 125) whereas the yellow *G. nasalis* larvae attach around the pylorus and sometimes the duodenum. Larvae remain and develop in this site for periods of 10–12 months and when mature in the following spring or early summer, they detach and are passed in the faeces: in a few species, notably *G. haemorrhoidalis*, the larvae re-attach in the rectum for a few days before being passed out. Pupation takes place on the ground and after 1–2 months the adult flies emerge. These do not feed and live for only a few days or weeks during which time they mate and lay eggs. There is therefore only one generation of flies per year in temperate areas.

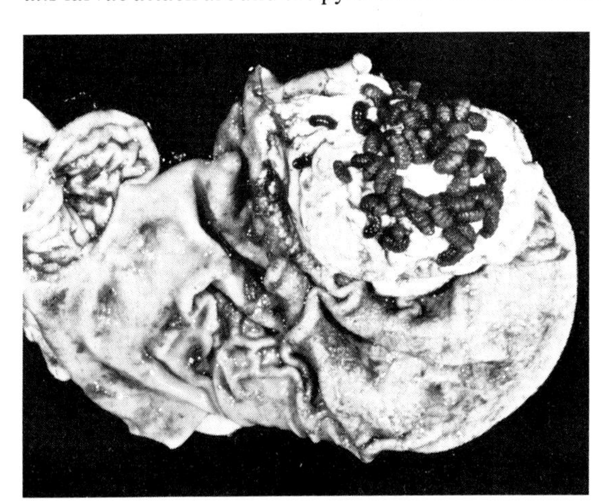

Fig 125 *Gasterophilus intestinalis* larvae attached to cardiac area of the stomach.

PATHOGENIC SIGNIFICANCE

Adult flies are often a source of great annoyance when they approach horses to lay their eggs, especially those species which lay their eggs around the head.

The presence of larvae in the buccal cavity may lead to stomatitis with ulceration of the tongue, but this is very rare. On attachment by their oral hooks to the stomach lining, larvae provoke an inflammatory reaction with the formation of funnel-shaped ulcers surrounded by a rim of hyperplastic epithelium. These are commonly seen at post-mortem examination of horses in areas of high fly prevalence and although dramatic in appearance their true pathogenic significance remains obscure. Larvae of *G. haemorrhoidalis* can cause irritation when they re-attach in the rectum. Despite the lack of detail on the pathogenic effect of bots, treatment is usually recommended as owners are concerned when larvae appear in the faeces. Treatment, however, does reduce fly populations and thus the fly-worry associated with egg laying. Rarely, abnormal hosts such as man or other animals may be infected with a few bots, but migration of larvae is usually limited to the skin causing a 'creeping eruption'.

CONTROL

From the life cycle it is obvious that in temperate areas, almost the entire *Gasterophilus* population will be present as larvae in the stomach during the winter since adult fly activity ceases with the advent of the first frosts in autumn. A single treatment during the winter, therefore, should effectively break the cycle. In certain areas, where adult fly activity is prolonged by mild conditions, additional treatments may be required. The most widely used specific drugs include carbon disulphide and trichorfon; the broad spectrum insecticide/anthelmintics, dichlorvos and ivermectin are also very effective against bots.

If, during the summer and autumn, eggs are found on the coat, subsequent infection can be prevented by vigorously sponging with warm water containing an insecticide. The warmth stimulates hatching and the insecticide kills the newly-hatched larvae.

Dermatobia

This fly, often placed in a separate family from the other oestrids, is a serious problem in human and animal medicine in Central and South America.

Hosts:
Man, most domestic and wild mammals, and many types of bird

Species:
Dermatobia hominis

Distribution:
Latin America from Mexico to northern Argentina, and the island of Trinidad

MORPHOLOGY

Adult: The fly resembles *Calliphora*, the abdomen having a bluish metallic sheen, but there are only vestigial mouthparts covered by a flap.
Larvae: Mature larvae measured up to 2.5 cm long and are somewhat oval.

LIFE CYCLE

Dermatobia is most common in forest and bush regions, the latter known in many parts of south America as the 'monte'. The female has a sedentary habit, resting on leaves, and when oviposition is imminent she catches an insect, usually a mosquito, and attaches to its abdomen a batch of up to 25 eggs. While attached to this transport host the L_1 develop within the eggs in about a week, but do not hatch until the insect lands on a warm-blooded animal. The larvae then penetrate the skin, migrate to the subcutis, and develop to the L_3, breathing through a skin performation in the fashion of *Hypoderma*. The mature larvae (Fig. 126) emerge after about three months and pupate on the ground for a further month before the adult flies emerge. There are up to three generations each year.

Fig 126
(a) Mature *Dermatobia* larvae being expressed from skin.

(b) General view of skin affected by *Dermatobia*.

PATHOGENIC SIGNIFICANCE

The larvae occur in swellings in various parts of the body (Fig. 126), and these may suppurate and cause severe pain; in Latin America the condition is often known as 'Ura'.

In man the most common larval sites are the extremities of the limbs and the scalp. Fatal cerebral damage has occurred in children when larvae have migrated through the fontanelle into the cranial cavity.

Dermatobia is a major problem in cattle in South America. Lesions are most numerous on the upper body, neck, back, flanks and tail, and are often grouped together, forming large, confluent, and often purulent swellings (Fig. 126). As well as hide damage, the pain and distress of the lesions result in retarded growth and lowered meat and milk production, a loss compounded by the fact that cattle spend much time standing on bare, dry ground, in preference to grazing in the forest and scrubland habitats of the fly.

CONTROL

Though the larvae are susceptible to systemic insecticides, the rapid turnover of generations and the lack of seasonality demand that any control regime must involve repeated treatments, and will therefore be expensive. However, if the cost is acceptable, control can be achieved by frequent sprays, dips or pour-on preparations of organophosphorus compounds such as trichlorfon, and by parenteral administration of ivermectin or closantel.

Lesser oestrid genera

The following oestrids, though of limited geographical distribution, are locally of veterinary importance.

Przevalskiana (syn. *Crivellia*)

This genus parasitises domestic goats and less commonly sheep, with gazelles as wild reservoirs over much of its range. The major species is *P. silenus*, occurring in Asia, the Middle East, North Africa and southern Europe; *P. aegagri* has a more limited distribution in Turkey, Israel, Crete and Cyprus.

The fly has much in common with *Hypoderma*, the third stage larva occurring under the skin of the back. The eggs are laid in short rows on the hairs of the legs and chest, and, after penetrating the skin, the L_1 migrate in the subcutis directly to the back, so that there is no risk of complications from resting larvae in strategic sites, as can happen with *Hypoderma*.

Though heavy infestations can result in loss of weight, the chief importance of *Przevalskiana* is in hide damage. In parts of the Punjab more than 90% of goats are infected, and since the Indian sub-continent produces about one third of the world supply of goat skins the parasite has some influence on the region's economy.

Cephenemyia

The third stage larvae of this fly inhabit the nasopharynx of deer. One species, *C. trompe*, has always been recognised as a problem in reindeer, but with increased domestication of red deer, *C. auribarbis*, the species occurring in these animals, is now of some economic importance. Several other species occur in wild deer in Europe and North America.

The adult flies are active from June to September, and, like *Oestrus*, the females are viviparous. The fly hovers close to the animal, then darts in and ejects larvae in fluid. Further development occurs in the nasopharynx, and the L_3, which may be 4.0 cm long, are sneezed out. The pupation period is about four weeks.

The adult flies cause 'gadding' in deer, with loss of condition. Although the larvae occasionally cause death from suffocation their general effect is loss of condition. In summer, keratitis and blindness may occur in reindeer when larvae are deposited in the eye.

Oedemagena

In the reindeer and caribou of northern Europe, Asiatic Russia and America *O. tarandi* replaces *Hypoderma diana* of more southern deer.

The adult flies resemble *Hypoderma*, with reddish-yellow hairy abdomens, and are active in July and August, each female laying betwen 500 and 700 eggs, which are attached to the downy undercoat, rather than the outer hair. The flanks, legs, and brisket are preferential laying sites. Unlike *Hypoderma*, however, the L_1 migrate directly to the back in the subcutaneous connective tissue.

The adult flies cause gadding, and the newly hatched larvae may cause a dermatitis with local oedema when they penetrate the skin. The main importance of this genus, however, is economic, from damage to hides by the L_3, and in Sweden this loss can amount to a fifth of the total income from reindeer herds. Two hundred holes are commonly produced in typically infested reindeer skins in the U.S.S.R.

Insecticides effective against the bovine warble appear able to limit this infection.

Gedoelstia

In southern Africa this oestrid fly is responsible for an oculo-vascular myiasis causing extrusion of the eyeball in sheep and, rarely, cattle. The larvae are deposited by the adult flies in the orbit of the natural hosts, which are antelopes, and travel by a vascular route to the nasopharynx where they mature, thus showing some affinity with *Cephenemyia*. Some larvae appear to include the lungs in this migration. In these hosts, the infection is tolerated without clinical signs, and the infection becomes of veterinary importance when domestic ruminants are grazed closed to, or among, the wild hosts.

In sheep the larvae begin their migration, and many arrive in the brain, ocular tissues and heart. It is in the eye that the signs are most prominent, with glaucoma, extrusion, and even rupture of the eyeball, but myocardial, pulmonary and renal infarction may occur, as well as encephalomalacia, from vascular thrombosis.

Flocks may have a 30% morbidity, of which a third will die, and in some areas sheep farming has had to be abandoned to cattle farming because of this parasite.

Domestic stock can safely graze with antelope during winter, when the flies are inactive (June–August). They should be removed from such areas in early spring since flies then begin to emerge from puparia with the rising temperature.

Organophosphates such as trichlorfon are effective against the larvae and flock treatment will reduce the blindness and mortality.

Cephalopina

This is the nasal bot fly of camels, equivalent to *Cephenemyia* in deer and *Oestrus* in sheep, and occurs over the entire range of both species of camel. A synonym which is still encountered is *Cephalopsis*.

The fly deposits its larvae in the nostrils, from which they migrate to the nasopharynx and nasal sinuses. The larval phase usually occupies about eleven months, and is associated with inflammation, sometimes purulent, of the nasopharyngeal mucosa. Camels snort and sneeze and are restless, and may even stop feeding, especially during the emergence of mature larvae from the nostrils. When large numbers of larvae are present the animals' breathing and working capacity may be severely impaired. Unlike many oestids adult *Cephalopina* do not panic the animals, and large numbers are often seen resting on the camels' heads and around the nostrils. They are easily recognised by the irregular blotches of black and white hairs on the abdomen.

Family HIPPOBOSCIDAE

Members of this family were formerly classified with *Glossina* in a separate suborder, the Puparia, because the adult females produce mature larvae which are ready to pupate. However they and the closely related *Glossina* are now regarded as groups within the Cyclorrhapha.

The Hippoboscidae are unusual in being flattened dorsoventrally and having an indistinctly segmented abdomen which is generally soft and leathery. They have piercing bloodsucking mouthparts, are parasitic on mammals and birds and have strong claws on the feet which allow them to cling to hair or feathers. They tend to be either permanent ectoparasites or to remain on their hosts for long periods. The two major genera of veterinary importance are *Hippobosca* and *Melophagus*.

Hippobosca

Members of this genus have wings and are commonly known as 'forest flies'.

Hosts:
Mainly horses and cattle, but other domestic animals and birds may be attacked

Species and distribution

Hippobosca equina	Worldwide
H. rufipes	Africa
H. maculata	Tropics and subtropics

MORPHOLOGY

Adult flies are approximately 1.0 cm long and are generally pale reddish brown with yellow spots on the indistinctly segmented abdomen.

They have wings and the major part of the piercing proboscis is usually retracted into the head except during feeding. Forest flies remain on their hosts for long periods and their preferred feeding sites are the perineum and between the hind legs.

LIFE CYCLE

Female flies leave their hosts and deposit mature larvae singly in dry soil or humus. Each female can produce only 5–6 larvae in its lifetime. These larvae pupate almost immediately and change from a yellow to black colour; maturation to the adult stage is temperature-dependent and in temperate areas flies are most abundant in the summer months.

PATHOGENIC SIGNIFICANCE

Some animals apparently become accustomed to attack by these flies and several hundred may be seen clustered around the perineum of horses without the animal showing a great deal of annoyance. However, they may be a source of great irritation to animals not accustomed to attack.

Since they pierce the skin to suck blood they may be mechanical vectors of blood parasites such as the non-pathogenic *Trypanosoma theileri* in cattle.

CONTROL

This is best achieved by topical application of insecticides preferably those with some repellent and residual effect such as the synthetic pyrethroids, permethrin and deltamethrin.

Melophagus

Hosts:
Sheep

Species:
Melophagus ovinus

Distribution:
Worldwide but most common in temperate areas

MORPHOLOGY

Commonly called the sheep 'ked', *M. ovinus* is a hairy, wingless insect approximately 5.0 mm long with a short head and broad, flattened, brownish thorax and abdomen. It has strong legs provided with claws and is a permanent ectoparasite (Pl. VIII).

LIFE CYCLE

Adults live for several months and the larvae produced by the females adhere to the wool. These are immobile and pupate immediately, the 3.0–4.0 mm long brown pupae being easily visible on the fleece. Adult keds emerge in approximately three weeks in summer, but this period may be extended considerably during winter. Ked populations build up slowly since only one larva is produced by each female every 10–12 days, up to a total of 15. Adults and pupae can only live for short periods off their hosts. *M. ovinus* is the vector of the non-pathogenic *Trypanosoma melophagium*. Heavy infestations of keds are most commonly seen in autumn and winter.

PATHOGENIC SIGNIFICANCE

Since keds suck blood, heavy infections may lead to loss of condition and anaemia. They are spread by contact and long-wooled breeds appear to be particularly susceptible. The irritation caused by these parasites also results in animals biting and rubbing with resultant damage to the fleece.

CONTROL

Specific measures are rarely undertaken, since the routine use of insecticides for the control of blowflies and ticks usually also results in the efficient control of keds.

Other hippoboscids

There are a number of other genera of hippoboscids of limited veterinary importance. One genus *Pseudolynchia* parasitises birds and closely resembles the sheep ked, but has wings. *P. canariensis* may be a problem in young pigeons due to their painful bites and blood sucking. Some members of another genus *Lipoptena* are parasites of deer and are similar to *Hippobosca* except that they lose their wings after locating their hosts. The behaviour and control of all hippoboscids is similar.

Order PHTHIRAPTERA

These are the lice.

Insects of this order are highly host-specific, and are permanently ectoparasitic, most being unable to survive away from the host for more than a day or two.

They are of variable size and colour, but all are flattened dorsoventrally. Most are blind, but a few species have primitive eyes which are merely photosensitive spots. The legs terminate in claws, the lice of mammals having one claw on each leg, while those of birds have two (Fig. 127).

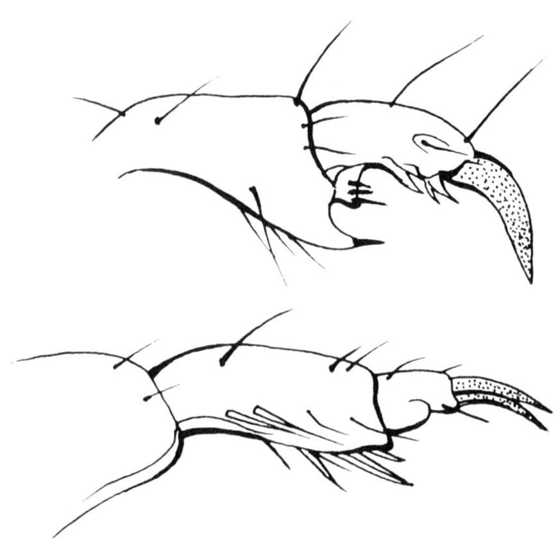

Fig 127 The legs of mammalian lice each possess one claw while those of birds have two.

There are two suborders:

Anoplura: the sucking lice; these occur only on mammals.

Mallophaga: the biting lice; these occur on both mammals and birds.

Heavy infestations of sucking lice can cause severe anaemia, while both sucking and biting lice are a source of irritation and skin damage which may lead to a loss of production and damage to hides.

The following are the more common genera of the suborders found on domestic animals. Individual species are dealt with under the appropriate hosts.

Suborder ANOPLURA

The sucking lice are usually large, up to 5 mm, with small, pointed heads and terminal mouthparts. They are generally slow-moving, and have powerful legs, each with a single large claw. They occur exclusively on mammals.

Haematopinus: The short-nosed louse. This is the largest louse of domestic mammals, up to 0.5 cm in length. It is yellow or greyish-brown with a dark stripe on each side (Fig. 128)

Hosts: cattle, pigs, equines

Linognathus: This, the 'long-nosed' louse (Fig. 128), is bluish black and the eggs are exceptional in being dark blue, and are less easy to see on hair

Hosts: cattle, sheep, goats, dogs

Solenopotes: Small bluish lice, which tend to occur in clusters

Hosts: cattle

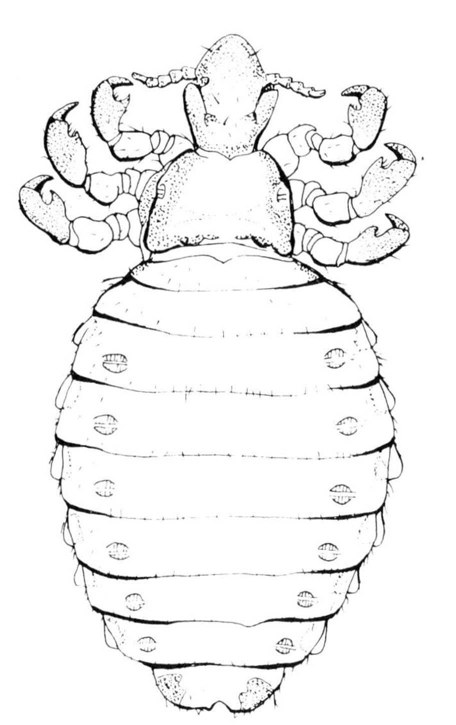

Fig 128 Two common genera of sucking lice of mammals are
(a) *Haematopinus*

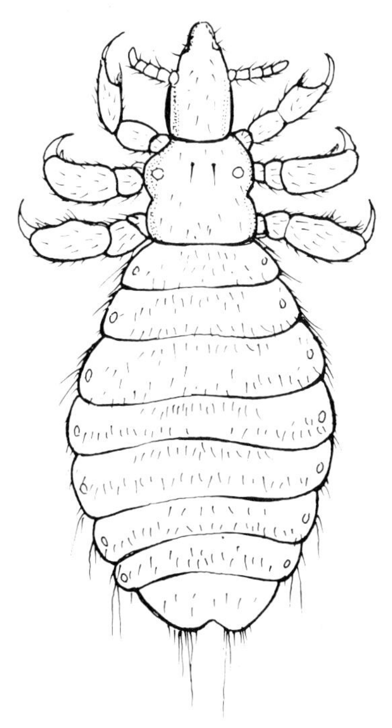

(b) *Linognathus*

Suborder MALLOPHAGA

Though the biting lice are generally smaller, up to 3 mm, than the anoplurans the head is relatively much larger, occupying the width of the body, and is rounded anteriorly, with the mouthparts ventral. The claws are small, the genera on mammals having one on each leg, and those on birds, two.

BITING LICE OF MAMMALS

Damalinia: These lice are a reddish-brown colour (Fig. 129)
Hosts: cattle, sheep, goats, equines

Felicola: Distinctive among the mallophagans in having a pointed head, somewhat resembling the anoplurans. It is, nevertheless, a true biting louse, with ventral mouthparts (Fig. 129)
Host: cat

Trichodectes: This louse is short, broad and yellowish, and is important as a vector of the tapeworm, *Dipylidium caninum*
Host: dog

Heterodoxus: A slender, yellowish louse, confined to tropical and subtropical regions
Host: dog

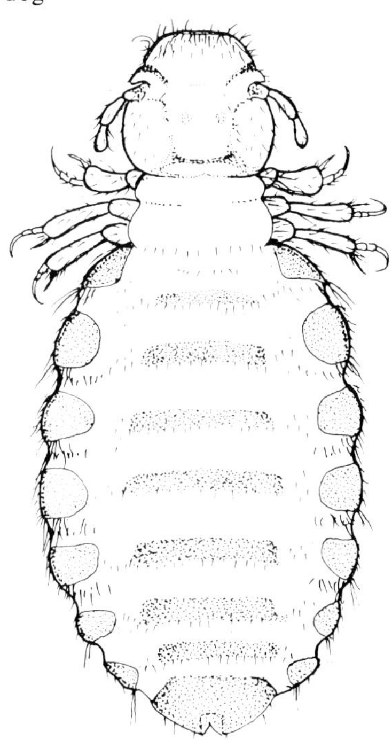

Fig 129 Two common genera of biting lice of mammals are
(a) *Damalinia*

Fig 129 (continued)

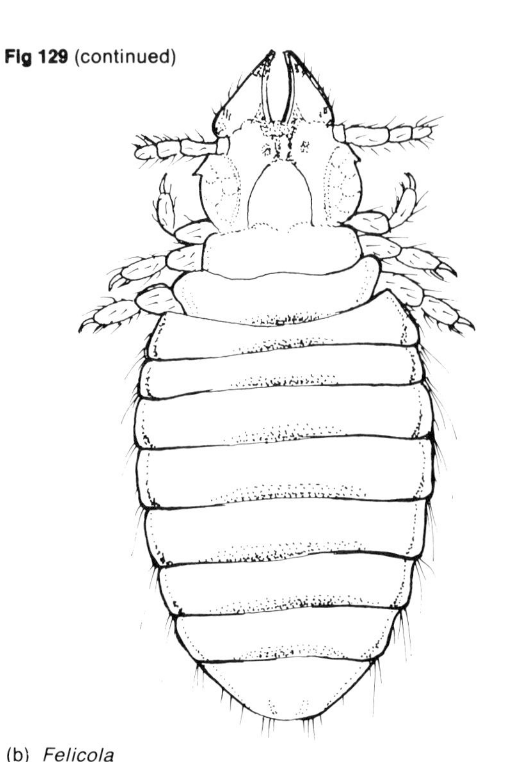

(b) *Felicola*

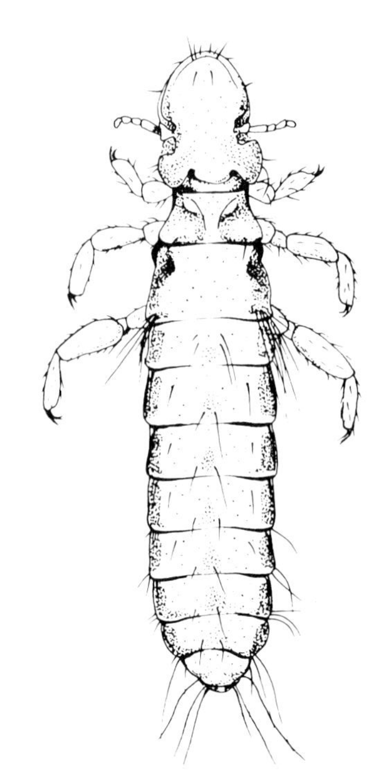

Fig 130 Two common genera of biting lice of birds are (a) *Lipeurus*

BITING LICE OF BIRDS

Birds harbour a great many genera, and the following list comprises a few of the more common. Many of these have acquired names, relating to their preferred sites on the body.

Major Genera: (Fig. 130)

Lipeurus	'wing louse'	domestic fowl
Cuclotogaster	'head louse'	
Menacanthus	'body louse'	

Minor Genera:

Goniocotes	'fluff louse'	domestic fowl
Menopon	'shaft louse'	
Goniodes		
Columbicola		pigeons and doves
Holomenopon		ducks

GENERAL LIFE CYCLE OF LICE

Lice of the two suborders have very similar life cycles.

During a life span of about a month the female lays 200–300 operculate eggs ('nits'). These are usually whitish, and are glued to the hair or feathers where they may be seen with the naked eye.

There is no true metamorphosis and from the egg hatches a nymph, similar to, though much smaller than,

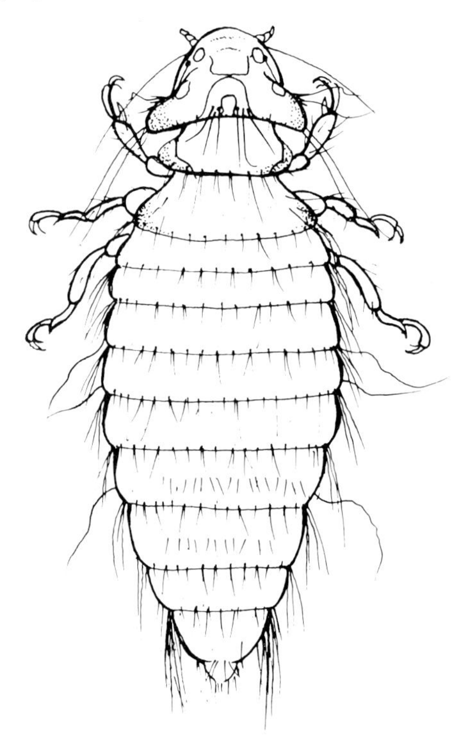

(b) *Menacanthus*

(a)

(c)

(b)

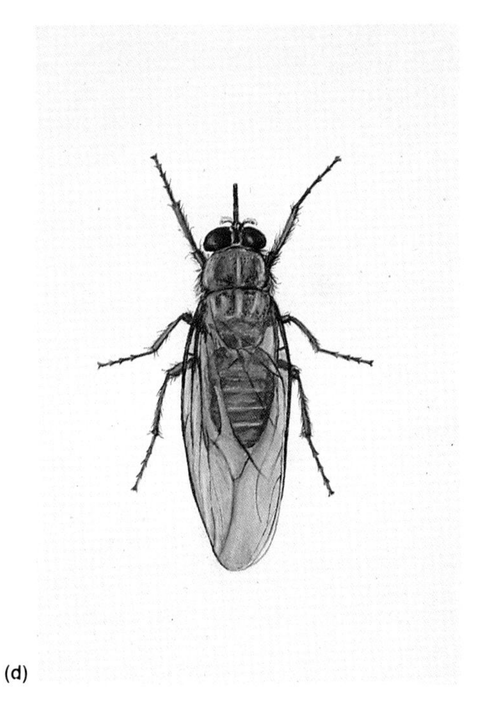
(d)

Plate VII
(a) *Musca domestica*
(b) *Stomoxys calcitrans*
(c) *Haematobia (Lyperosia)* spp.
(d) *Glossina* spp.

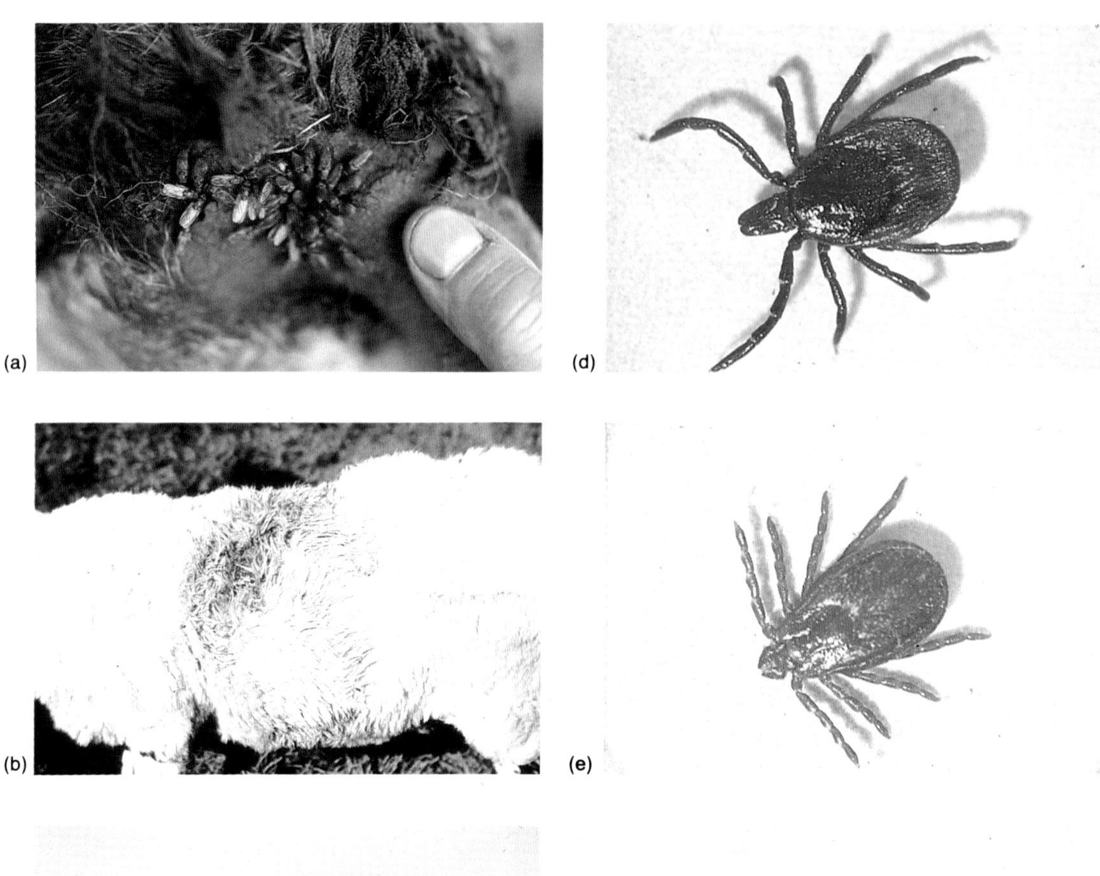

(a) (d) (b) (e)

(c)

(f)

Plate VIII

(a) *Hydrotaea irritans* clustered around the base of the horns in a sheep.

(b) Characteristic lesions of body strike due to blowfly myiasis.

(c) The sheep ked *Melophagus ovinus.*

(d) Female *Ixodes ricinus*: note elongated mouthparts.

(e) Female *Haemaphysalis punctata*: note short mouthparts and festoons.

(f) Male and female *Dermacentor reticulatus*: this is an ornate tick and possesses festoons.

(a)

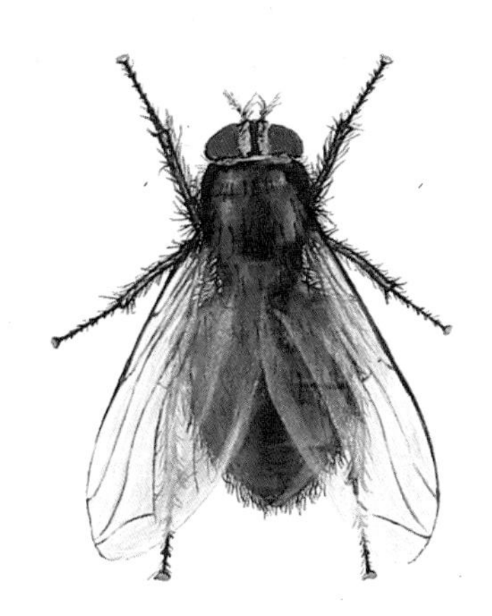

(c)

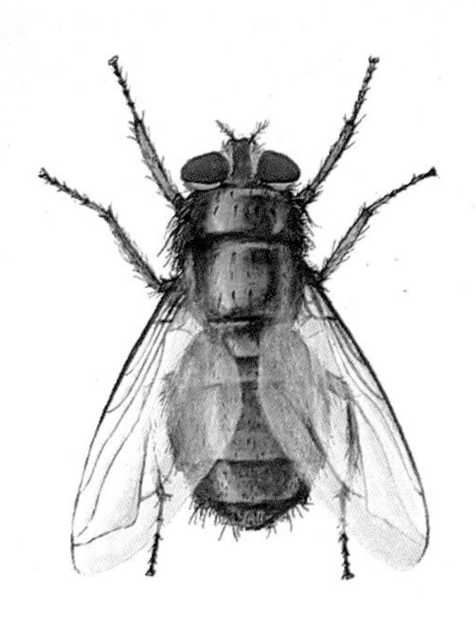

(b)

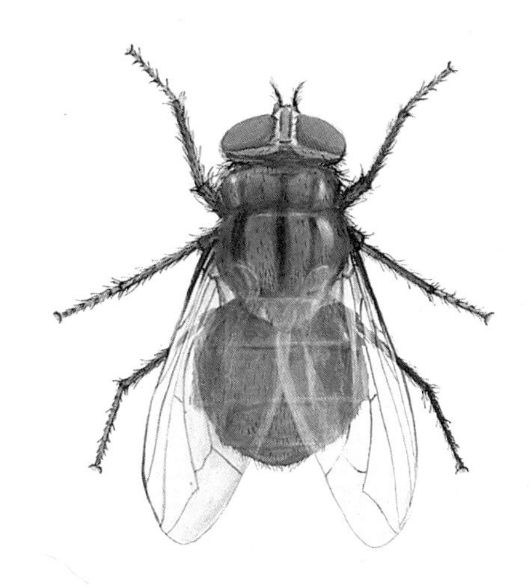

(d)

Plate IX

(a) *Calliphora* spp.
(b) *Lucilia* spp.
(c) *Phormia* spp.
(d) *Cochliomyia* spp.

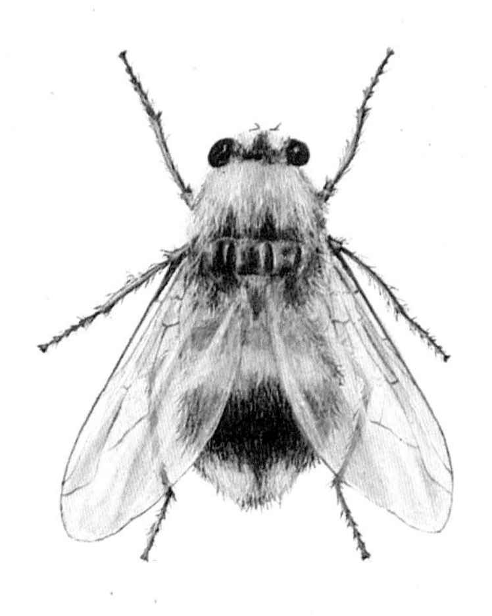

(a)

(c)

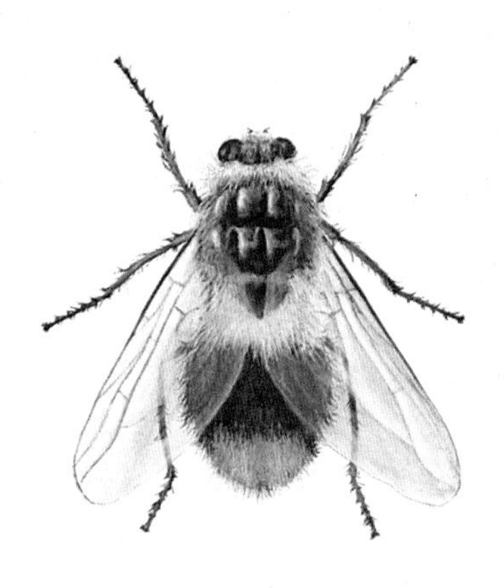

(b)

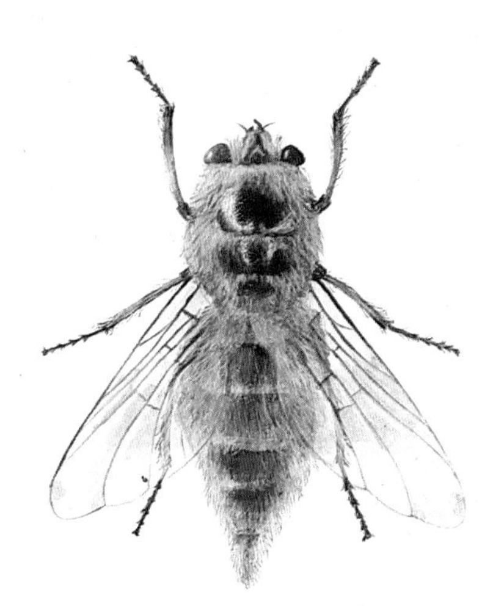

(d)

Plate X

(a) *Hypoderma bovis*
(b) *Hypoderma lineatum*
(c) *Oestrus ovis*
(d) *Gasterophilus* spp.

the adult. After three moults the fully grown adult is present. The whole cycle from egg to adult takes 2–3 weeks.

The anoplurans, with their piercing mouthparts, feed on blood, but the mallophagans, equipped for biting and chewing, have a wider range of diet. Those on mammals ingest the outer layers of the hair shafts, dermal scales, and blood scabs; the bird lice also feed on skin scales and scabs, but unlike the mammalian species, they can digest keratin, so that they also eat feathers and down.

Some genera are capable of rapid population expansion by changing to asexual reproduction by parthenogenesis, the most notable example in domestic stock being *Damalinia*.

LOUSE INFESTATION (PEDICULOSIS) IN CATTLE

Pediculosis in cattle occurs throughout the world, and is seen in these animals more commonly than in any other domestic mammals.

The various genera have preferential sites on the animal, but in heavy infections they spread from these reservoir areas to involve the whole body.

Damalinia, the solitary biting genus, favours the top of the head, especially the curly hair of the poll and forehead, the neck, shoulders, back, and rump, and occasionally the tail switch. Of the sucking lice, *Linognathus* and *Solenopotes* prefer the head, neck and dewlap, while each species of *Haematopinus* has its own preference; *H. eurysternus* occurs on the poll and at the base of the horns, in the ears, and around the eyes and nostrils (Pl. XI), and even in mild infestations is found in the tail switch, while *H. quadripertusus* is limited to the tail region.

Some species, notably *H. eurysternus* and *L. vituli*, are gregarious in habit, forming dense, isolated clusters.

With the exception of *H. quadripertusus*, which is confined to warmer areas such as Africa and Australia, all the cattle lice mentioned above are found throughout the world, from arctic to tropical zones.

EPIDEMIOLOGY

In warm countries there is no marked seasonality of bovine pediculosis, but in cold and temperate regions the heaviest infestations are in late winter and early spring, when the coat is at its thickest, giving a sheltered, bulky and humid habitat for optimal multiplication. The most rapid annual increase in louse populations is seen when cattle are winter-housed, and *Damalinia* especially, with its faculty for parthenogenesis, can build up in numbers very quickly.

In late spring there is an abrupt fall in the numbers of lice, most of the parasites and eggs being shed with the winter coat. Numbers generally remain low throughout the summer, partly because the thinness of the coat provides a restricted habitat, but partly also because high skin surface temperatures and direct sunlight limit multiplication and may even be lethal, especially to *Damalinia*.

PATHOGENESIS

Moderate infestations are associated only with a mild chronic dermatitis, and are well tolerated.

Biting lice, in large numbers, cause intense irritation leading to rubbing against posts, wire and other objects, with loss of hair, and more important economically, extensive hide damage, but they have less effect on the health of the animal. On the other hand the sucking lice, and especially *H. eurysternus*, can cause serious anaemia and loss of weight.

CLINICAL SIGNS

Light infestations are usually only discovered accidentally and should not be considered of any pathogenic importance, lice being almost normal inhabitants of the dermis and coat of many cattle, especially in winter.

In heavier infestations there is pruritus, more marked in *Damalinia* infestation, with rubbing and licking, while if sucking lice are present in large numbers there may be anaemia and weakness. In these infections the lice and eggs are easily found by parting the hair, especially along the back, the lice being next the skin and the eggs scattered like coarse powder throughout the hair.

It is important to remember that a heavy louse infestation may itself be merely a symptom of some other underlying condition such as malnutrition or chronic disease, since debilitated animals do not groom themselves and leave the lice undisturbed: in such animals the shedding of the winter coat may be delayed for many weeks, retaining large numbers of lice.

TREATMENT AND CONTROL

The organophosphorus insecticides applied as a 'pour-on' or as a dusting powder, especially if combined with gamma HCH, are effective in killing all lice. A second treatment is recommended two weeks later to kill newly emergent lice. Alternatively, the more recent 'pour-on' synthetic pyrethroids, such as cypermethrin, or parenteral ivermectin may be used; the latter is particularly effective against sucking lice.

In Europe, louse control is usually undertaken when cattle are housed for the winter. Treatment with pour-on organophosphates or ivermectin at this time also controls *Hypoderma* and mange mites.

LOUSE INFESTATION IN SHEEP

The two species of sucking lice in sheep are essentially parasites of the haired regions of the body, invading the woolled areas only when the population is expanding rapidly. They are not very active, and have a gregarious habit, feeding in swarms.

EPIDEMIOLOGY

Generally, for the transfer of louse infestation, close bodily contact is necessary, and while this is unusual in the grazing animal in Europe, it does happen at gatherings and in sale-yards, and especially when sheep are housed for the winter since the heavy fleece provides a habitat which is readily colonised by lice.

Linognathus pedalis, the 'foot louse', inhabits mainly the lower region of the hind limbs, from the feet to below the hocks, and spreads from there to the crutch, scrotum and belly. In Merinos and other heavily woolled breeds, it is usually first detected at crutching. In its normal habitat on the legs it is exposed to great fluctuations in temperature, and having adapted to survive in these conditions, it is one of the few lice which can live away from the host's body for more than a day or two and is viable on pasture for about a week.

L. ovillus, the 'face louse', occurs on the face and ears, spreading from there to the cheeks, neck and body, and is usually detected in the heavily woolled breeds when excess wool is removed from the poll ('wigging'). In contrast to *L. pedalis*, this species, more used to even temperatures, can only survive off the sheep for 1–2 days. In very hot weather, when exposed to sun, the temperature of the sheep's back may reach 48 °C, and in these conditions *L. ovillus* is killed in about an hour, its persistence depending on the part of the population inhabiting the cooler skin of the face and ears.

Damalinia ovis, the biting louse of sheep, sometimes called the 'body louse', is much more active than *Linognathus*, roaming in the wool over the whole body. Like the others, *Damalinia* is susceptible to high temperatures, but it is also intolerant of moisture. In a damp fleece, with a relative humidity of more than 90%, it will die in six hours, and when covered by water it will drown in an hour.

PATHOGENESIS

Though *Linognathus* spp. can cause anaemia, it is *Damalinia* which is usually considered to be the more pathogenic. Being highly active it can cause great irritation, so that sheep are restless and have their grazing interrupted, with consequent loss of condition. In response to the irritation the sheep rub against posts and wire, with damage to the fleece and some loss of wool. When these lice bite there is an exudate of serum from the damaged skin on which the lice also feed; in heavy infestations the amount of exudate is great enough to cause matting of the wool. Reduction in the value of the wool clip is economically the most important consequence of ovine pediculosis, but an additional hazard in warm countries is that the fleece and skin, damaged by rubbing and soiled by louse faeces, is an attractant for blowflies, and places the animal at risk from strike.

CLINICAL SIGNS

In common with other animals, sheep with light to moderate infestations show no signs, and lice are usually only detected when wool is being removed.

In heavy infestations, and especially when in full fleece, the intense pruritus causes restlessness and scratching, the fleece showing bare patches and being stained. On parting the wool the reddish *Damalinia* and the bluish *Linognathus* will be found, the latter being present also on haired areas. The louse eggs, appearing as a powder, will be found attached to the wool fibres close to the skin.

TREATMENT AND CONTROL

Lice on sheep are generally treated by dipping or spraying with insecticides containing organochlorines, organophosphates or a combination of both. Since organochlorines have a greater persistence a single dip with these will usually suffice; with the organophosphates it may be necessary to treat two weeks later.

A recent innovation is the introduction of the synthetic pyrethroids, 'pour-on' cypermethrin and the 'spot-on' deltamethrin, which act by diffusion over the body surface in the sebum and give protection for 8 to 14 weeks.

The treatment of lice infestations in goats, due to species of *Damalinia* and *Linognathus*, is similar to that of sheep.

LOUSE INFESTATION IN PIGS

Only one species occurs on pigs, the sucking louse, *Haematopinus suis*. It is very highly host-specific, and will not even establish in wild pigs, the ancestral stock of our domestic species. *Haematopinus* is a large, greyish-brown louse, which is most often present on skin folds of the neck and jowl, the flanks and the insides of the legs on thin-coated animals.

EPIDEMIOLOGY

Infection is transferred between pigs mainly by contact, in closely confined fattening animals and in suckling sows penned with their piglets, but lice may also be acquired when animals are put into recently vacated dirty accommodation.

PATHOGENESIS AND CLINICAL SIGNS

This louse is very common, and is usually tolerated without any signs, apart from occasional mild irritation. In heavy infestations pigs are restless and fail to thrive, but though *Haematopinus* is a blood-sucker, anaemia is hardly ever seen. Economically, the most important feature of pediculosis in pigs is probably skin damage from scratching, with reduction in hide value.

This louse is said to be a vector of African swine fever, *Eperythrozoon suis* and the virus of swine pox.

Until recently control has been based on the application of insecticides either as a powder or as a wash. These included the organochlorine, gamma HCH, and the organophosphates such as diazinon and malathion. Lately, ivermectin given parenterally or the organophosphate, phosmet, administered as a 'pour-on' have both proved highly effective as a single treatment.

For herd prophylaxis, gilts and sows should be treated before farrowing to prevent spread of infection to their piglets, and boars treated twice annually.

LOUSE INFESTATION IN EQUINES

Two species are common on equines, the sucking louse, *Haematopinus asini*, and the biting louse, *Damalinia equi*.

EPIDEMIOLOGY

In normal light infestations, both species occupy the same reservoir sites in the dense hair of the mane, the base of the tail, and submaxillary space, and also on the fetlocks of rough-legged breeds. From these sites, spread occurs over the whole body, and the numbers are greatest in winter and early spring when the winter coat is at its most dense. As in cattle, the shedding of the winter coat is important in ridding animals of the greater part of their louse burden in spring.

In hot countries the skin temperature of the animal's back may be high enough, as has been noted in sheep, to kill lice in the exposed fine-coated areas.

Equine pediculosis spreads by contact and via contaminated grooming equipment, blankets, rugs and saddlery.

PATHOGENESIS

As in other animals, equine lice, and in particular the more active *Damalinia*, may cause intense irritation, resulting in rubbing and scratching, with matting and loss of hair and sometimes excoriation, almost the whole body being involved in extreme cases. Animals are restless and lose condition, and in heavy *Haematopinus* infestations, there may also be anaemia.

CLINICAL SIGNS

It is possible that, as in cattle, heavy louse infestations in equines are themselves symptomatic of some other disorder which may be disease or, more likely, simple neglect. It is true that animals in a debilitated condition will fail to shed their winter coats and harbour very large numbers of lice, but it is also the case, with horses especially, that if neglected and left ungroomed the undisturbed louse population will rapidly multiply.

Restlessness, rubbing, and damage to the coat would suggest that lice are present, and when the hair is parted the parasites will be found. *Haematopinus*, large and yellow-brown, is very easily seen, and in temperate countries on warm sunny days this louse will often move on to the surface of the coat. *Damalinia* appears as small yellowish specks in the hair and the small pale eggs are readily found, scattered throughout the coat.

TREATMENT AND CONTROL

Horses may be treated with an organochlorine, such as gamma HCH or bromocyclen, or an organophosphate such as ronnel. Two dressings with a 14-day interval between may be sufficient, but four may be required in heavily infected animals. All of the horses in the establishment should be treated.

Grooming equipment should be scalded, blankets and rugs thoroughly washed, and saddlery thoroughly cleaned. Ideally, animals should have individual grooming equipment, and saddlery should not be interchanged, but this may not be economically feasible on some establishments. Regular and thorough grooming is, of course, the essence of control.

LOUSE INFESTATION IN THE DOG AND CAT

In dogs, the biting louse, *Trichodectes canis*, and the sucking louse, *Linognathus setosus*, are by far the commonest and most widespread.

The sole louse of any importance in cats is the biting species, *Felicola subrostratus*, which has a worldwide distribution.

EPIDEMIOLOGY

Though pediculosis in these animals is essentially a disease of neglect, some types are especially prone to infestations; in dogs, the long ears of such breeds as the spaniel, basset, and Afghan hound provide an extensive and sheltered habitat in which the lice can multiply, and in cats the long-haired breeds, which cannot groom so thoroughly as the shorthairs, can harbour reservoir populations deep in the fur.

PATHOGENESIS

Heavy louse infestations are most often found in animals which are neglected and underfed, though as in other animals, they may mask underlying disease. They are in some cases associated with senility, but the majority of severely affected animals are young.

In dogs *Trichodectes* is more harmful, though *Linognathus* is a cause of anaemia. The former is a very active louse, moving rapidly through the coat and causing intense pruritus, and it provokes self-inflicted injury by scratching, with loss of hair and excoriation of the skin. In heavy combined infestations with the two genera, pups may die from anaemia and debility.

In long-haired cats pathogenic populations of *Felicola* may develop under thickly matted neglected fur.

CLINICAL SIGNS

With most infestations animals are restless and scratch almost continuously, but in heavy infestations may become severely debilitated. The louse eggs are easily seen in the coat, and the two common dog lice are easily differentiated, *Trichodectes* being small and yellow, while *Linognathus* is bluish and larger.

TREATMENT AND CONTROL

Louse infestations are normally treated with powder, washes or shampoos of organochlorine, organophosphates or carbamate insecticides. The older preparations such as pyrethrum, rotenone and benzyl benzoate are also effective. Some of these drugs are also available as aerosols or 'trigger' sprays which are often convenient for the owner to apply. Treatment is best repeated at an interval of 14 days to kill newly hatched lice. For prophylaxis, dog and cat collars, impregnated with dichlorvos or diazinon, are often used, although dog collars should not be used on cats since the concentration of insecticides may be toxic to cats.

LOUSE INFESTATION IN BIRDS

More than 40 louse species, all mallophagan, occur on domestic birds.

The genera *Lipeurus* and *Menacanthus* contain the most pathogenic species of poultry lice. *Lipeurus* spp. are grey, slow-moving lice, which are found close to the skin. *L. caponis*, the 'wing louse', prefers the bases of the wing and tail feathers, while *Cuclotogaster (Lipeurus) heterographus*, the 'head louse', occurs on the head and neck; in this species the eggs are laid singly on the feathers, and not in clusters as is customary in poultry lice. These lice can infect all domestic fowls including turkeys, game birds, and ducks.

Menacanthus has one important species, *M. stramineus*, the 'yellow body louse'. Like *Lipeurus* it infects domestic fowls and favours the skin surface as a habitat, being found in greatest numbers on the thinly covered skin of the breast and thighs and around the anus. It is a very active louse, and lays its eggs in clusters mainly in the anal region. Though a biting louse it can cause severe anaemia by puncturing small feathers and feeding on the blood which oozes out. Being active, and a voracious feeder, it causes severe irritation, and the skin is inflamed and eventually covered by scabs, especially in the region of the vent, and in young birds, on the head and throat. It is the most pathogenic louse of adult birds, but has also been responsible for fatalities in chicks. Cage birds, and in particular canaries, are occasionally infested, and suffer irritation, restlessness and debility.

MINOR GENERA

Common, but less pathogenic, genera of bird lice include the following:

Goniocotes gallinae, the 'fluff louse', occurs in the fluff at the bases of feathers, its preferred sites being the back and rump. It is one of the smallest lice of poultry.

Goniodes has several species, including *G. gigas* and *G. dissimilis* in the domestic hen, *G. meleagridis* in turkeys and guinea fowl, and *G. pavonis* in peacocks. These are all very large lice, inhabiting the skin surface and body feathers and are commonest in adult birds. They are never present in large numbers.

Menopon gallinae, the 'shaft louse', is a pale yellow, rapidly moving louse which does not develop to appreciable populations in young birds until they are well feathered. It feeds only on feathers, and though common, is never a serious pathogen. Its main host is the domestic hen, but it will spread to other fowl, such as turkeys and ducks, which are in contact.

Holomenopon, occuring in ducks, is also sometimes called the 'shaft louse' of these birds. It is a small, rapidly moving louse, which favours especially the preen gland, inhibiting production of the oily secretion. Partly due to irritation, birds preen continuously, but without the secretion the feathers cannot be waterproofed. Unable to repel water and injured by constant preening, the plumage becomes tattered and dirty, with the feathers broken. Water can penetrate to the skin, and when much of the body is affected the birds are soaked, and may die of pneumonia following chilling. Though the damaged plumage may be replaced at the annual moult it soon degenerates, with the excessive preening, into its former sodden condition.

Columbicola columbae, parasitic on pigeons and

doves, is very common. Its preferred site is the anterior part of the body, where it may cause a mild pruritus, and in common with most pediculoses, heavy infestations are usually seen only in diseased and debilitated birds.

PATHOGENICITY AND EPIDEMIOLOGY

The bird lice can digest keratin, biting off pieces of feather, breaking these up with comb-like structures in their crops, and digesting them with secretions aided by bacterial action. They will ingest not only the sheaths of growing feathers, but also down and skin scabs.

Though there are differences in pathogenicity between genera the effects of avian pediculosis are broadly similar, varying only in degree. Birds are unable to rest, cease feeding and may injure themselves by scratching and feather-plucking, with results often more serious than any immediate damage by the lice.

In general young birds suffer more severely, with loss of body weight, debility, and perhaps death. In adult laying birds the effect on body weight is slight, and the main loss is in depression of egg production. As would be expected, the practice of de-beaking allows an increase in infestations by preventing birds from preening and grooming.

As in the other pediculoses, the condition in domestic birds is often itself a symptom of ill-health from other causes, such as other infection, malnutrition or inadequate, overcrowded and unhygienic housing.

TREATMENT AND CONTROL

Although methods such as dusting the litter or providing insecticide-treated laying boxes are used to avoid undue handling of birds the results obtained from dusting individual birds are undoubtedly better. Organophosphorus insecticides such as malathion and a carbamate, carbaryl, are recommended. Two treatments should be given at 14 days interval.

A more recent measure is the use of dichlorvos-impregnated leg bands, which give control for up to four months, and, for caged birds, the same material in strips, threaded through the bottom of the cage, gives control for about a month.

Order SIPHONAPTERA

These are the fleas. They are of veterinary significance not only because of their effects on their hosts, but as carriers of disease. Though most important in dogs, cats and poultry, their readiness to parasitise humans as alternative hosts gives the fleas of these domestic animals a relevance in public health. Fleas do not occur on ruminants, horses or pigs.

MORPHOLOGY

Fleas (Fig. 131) are dark brown, wingless insects, with laterally compressed bodies which have a glossy surface, allowing easy movement through hairs and feathers.

Fig 131
(a) Lateral view of a flea.

Fig 131 (continued)

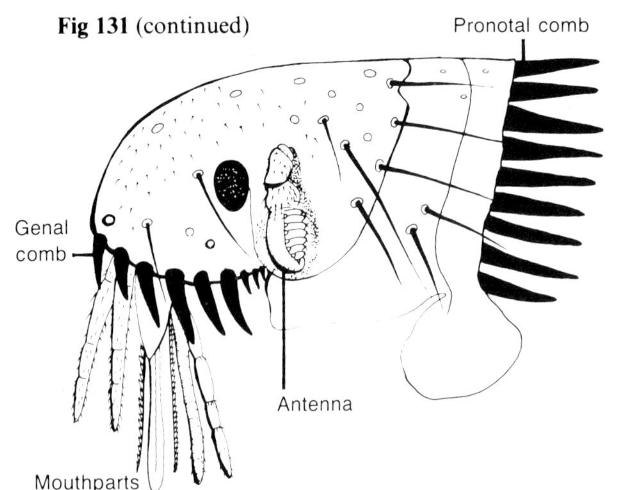

(b) Head of flea showing piercing, sucking mouthparts and showing genal and pronotal 'combs'.

Eyes, when present, are simply dark, photosensitive spots, and the antennae, which are short and club-like, are recessed into the head. The third pair of legs are much longer than the others, an adaptation for leaping on and off their hosts.

The head may bear at its posterior (pronotal) or ventral (genal) borders rows of dark spines called ctenidia or 'combs', and these are the most important features used in identification (Fig. 131).

LIFE CYCLE

Both sexes are bloodsuckers, and only the adults are parasitic. The ovoid eggs have smooth surfaces, and may be laid on the ground or on the host from which they soon drop off. Hatching occurs in two days to two weeks, depending on the temperature of the surroundings. The larvae are maggot-like and have a coat of bristles. They have chewing mouthparts and feed on debris and on the faeces of the adult fleas, which contain blood and give the larvae a reddish colour. The larva moults twice, the final stage being about 5.0 mm long, and then spins a cocoon, a form of woolly puparium, from which the adult emerges. Moulting and pupation are dependent on temperature, and though in warm conditions the whole cycle may be completed in about three weeks, in low temperatures it may extend to two years.

It is important to recognise that most of the flea's life cycle is spent away from the host. This includes not only the eggs, larvae and cocoon, but also, if necessary, the adult flea which can survive for as long as six months between feeds. The usual life span is 1–2 years.

Most fleas feed for only a few minutes before moving to another part of the host, or leaping to the ground or to a fresh host.

A few genera remain permanently attached throughout adult life. These are the burrowing, or 'stickfast', fleas, whose females are embedded in the skin, within nodules. Only the posterior part of these fleas communicates with the surface, allowing the eggs to drop to the ground and develop in the usual manner.

Though each species of flea has its own host preferences, casual feeding is common, and most will feed on a wide range of mammals and birds.

The following are the more important genera occurring on domestic mammals and birds, and their distribution may be taken as worldwide unless otherwise stated. Their differential morphology is given in Fig. 132.

FLEAS OF MAMMALS

Ctenocephalides

This is the only important genus in the dog and cat. *Ctenocephalides canis* and *C. felis* occur on the dog and cat, but *C. felis* is much the more widespread, and in many areas it is the dominant species on dogs and on man, as well as cats. Both species can act as intermediate hosts for the common tapeworm of dogs and cats, *Dipylidium caninum*, and for the filarioid of dogs, *Dipetalonema reconditum*. Though the adult flea can acquire the filarioid infection by intake of microfilariae in a blood meal, the specialised mouthparts do not allow the ingestion of the relatively large egg packets of *Dipylidium*, and this infection can only be acquired by the flea larva, which has chewing mouthparts. Development of the cestode occurs concurrently with that of the flea, so that the adult contains the cysticercoid. *Ctenocephalides* is the genus largely responsible for provoking allergic flea-bite dermatitis in dogs and cats.

Pulex

Pulex irritans is primarily parasitic on man, but in some areas it is common on dogs and cats. It can act as intermediate host of *Dipylidium caninum*, and is sometimes involved in flea-bite dermatitis.

Spilopsyllus

Spilopsyllus cuniculi occurs on the ears of rabbits and is the main vector of myxomatosis. It has a more sedentary habit than most fleas, and will remain on the ear even when it is handled. It is quite commonly found near the edges of the ear pinna of dogs and cats which frequent rabbit habitats.

Xenopsylla

Though this genus has little immediate importance for the veterinarian it requires mention because one species, *Xenopsylla cheopis*, is the main vector of *Yersinia pestis*, the cause of bubonic plague in man. *X. cheopis* is a rat flea, and acquires *Y. pestis* when feeding on its usual hosts. When the bacilli multiply in its gut the proventriculus becomes blocked so that blood cannot be taken in; the hungry flea moves from host to host in attempts to feed, and in its wanderings the infection may be transferred from its endemic base in rodents to the human population. Though now rare in humans, plague still exists in wild rodents ('sylvatic plague') in parts of Africa, Asia, South America and the western states of the U.S.A.

WITHOUT CTENIDIA

Frons ('forehead') angled anteriorly .... *Echidnophaga*

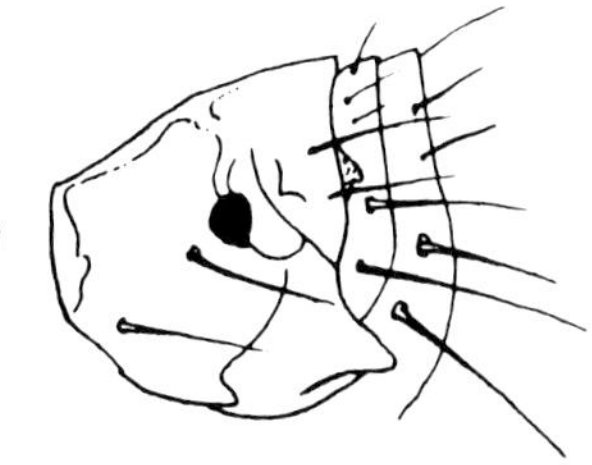

Frons rounded anteriorly .... *Pulex*

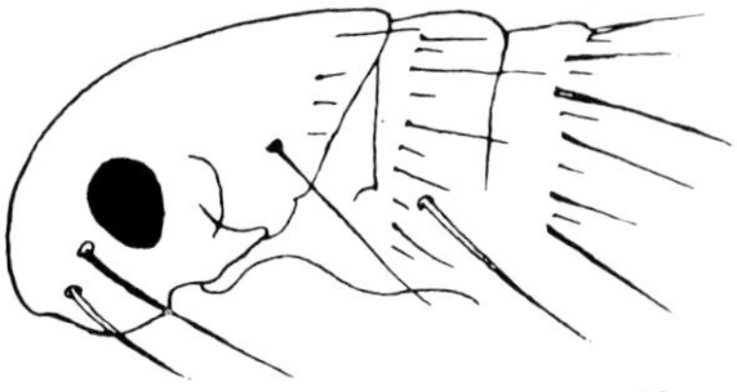

WITH CTENIDIA

Pronotal ctenidium only .... *Ceratophyllus*

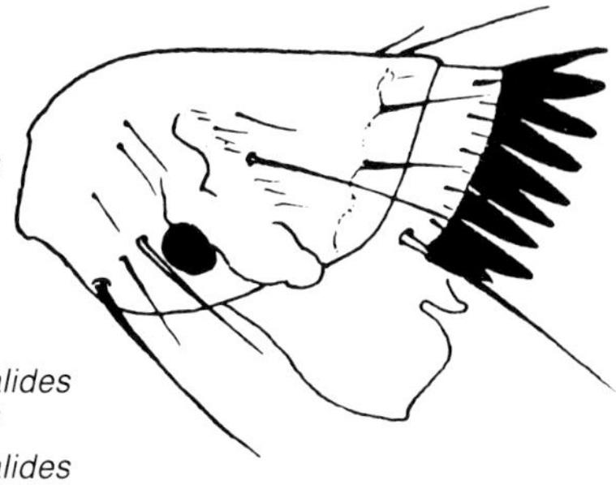

Both genal and pronotal ctenidia .... *Ctenocephalides*
Spilopsyllus

Genal ctenidium horizontal .... *Ctenocephalides*

Head length *less than twice* height. Spine 1 of genal ctenidium *shorter* than spine 2 — *C. canis*

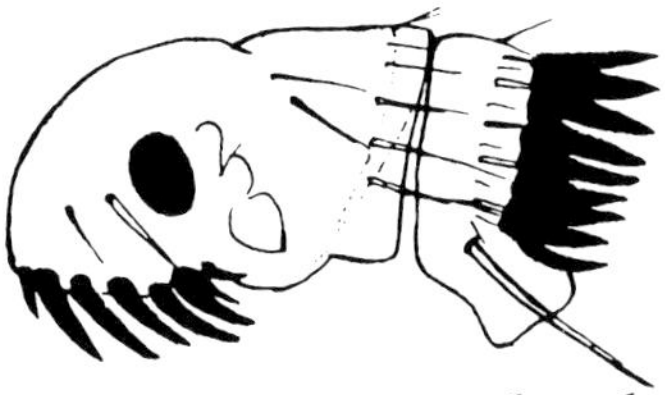

Head length *twice* height. Spine 1 of genal ctenidium *equal to* spine 2 — *C. felis*

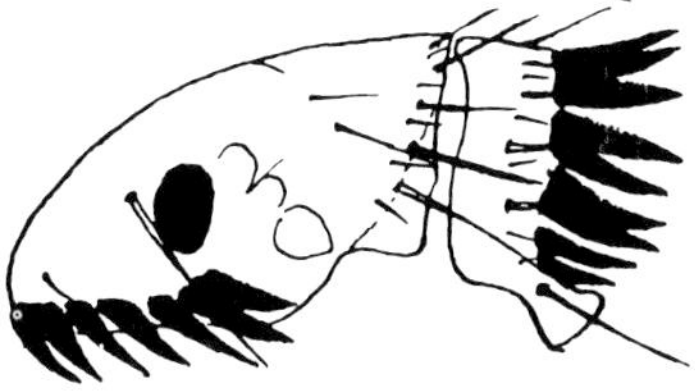

Genal ctenidium oblique, with 4–6 elements .... *Spilopsyllus*

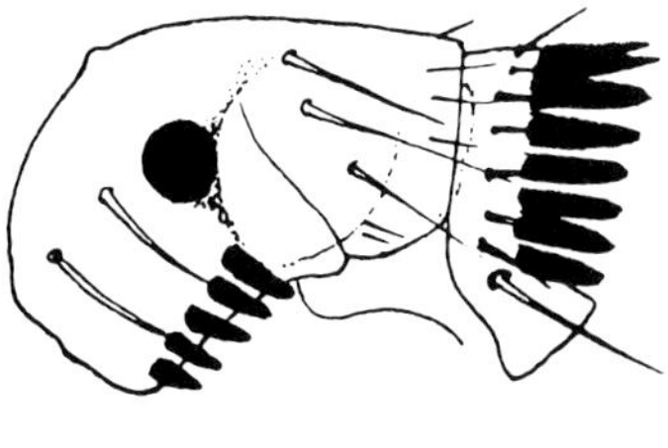

Fig 132 Key to the differentiation of fleas of veterinary importance.

Tunga

Tunga penetrans is the representative in mammals of the burrowing fleas and occurs in man and rarely pigs. The popular name for this flea in humans is the 'jigger'. Its distribution includes parts of Africa, Asia, and North and South America, but it does not occur in Europe. The female burrows into the skin, where its abdomen becomes enormously distended and filled with eggs, forming a distinct nodule. This flea occurs mainly on the feet of humans, causing severe irritation. In pigs the reported sites are the feet and scrotum, but these animals tolerate the infection with no signs of distress.

PATHOGENIC SIGNIFICANCE

The response to a flea bite is a raised, slightly inflamed weal on the skin, associated with mild pruritus, but though the animal will scratch intermittently there is little distress. However, after repeated flea bites over a period of several months a proportion of dogs and cats develop flea-bite allergy which is often associated with profound clinical signs.

Flea-bite allergy is a hypersensitive reaction to the flea saliva released into the skin during feeding. In the saliva is a hapten (an incomplete antigen) which combines with the host's skin collagen to form a complete allergen. The resulting allergy is most commonly a combination of immediate and delayed-type responses.

The allergy shows a seasonality in temperate areas, appearing in summer when flea activity is highest, though in centrally heated homes exposure may be continuous. In warmer regions, such as the western states of the U.S.A., the problem occurs throughout the year.

As would be expected, the most commonly affected areas in both dogs and cats are the preferential biting sites of the fleas which are the back, the ventral abdomen, and the inner thighs. In the dog the primary lesions are discrete crusted papules which cause intense pruritus. The most important damage, however, is subsequently inflicted by the animals themselves, in scratching and biting the affected areas, to produce areas of alopecia or of moist dermatitis ('wet eczema'). In older dogs, which have been exposed for many years, the skin may become thickened, folded and hairless, and in these animals the pruritus is much less intense.

In the cat, flea-bite allergy produces the condition commonly known as miliary dermatitis or eczema, readily detectable on palpation, in which the skin is covered with innumerable small, brown, crusty papules which cause marked pruritus (Pl. XI).

DIAGNOSIS

When the signs are indicative of flea infestation, but no parasites can be found, the host should be sprayed with an insecticide, placed on a large sheet of plastic or paper, and vigorously combed. The combings and debris should be examined for fleas or flea faeces which show as dark brown to black crescentic particles. Consisting almost entirely of blood, these will produce a spreading reddish stain when placed on moist tissue.

Another technique is the use of a vacuum cleaner with fine gauze inserted behind the nozzle; the latter is applied to the host or its habitat and the fleas are retained on the gauze.

TREATMENT AND CONTROL

In flea-bite allergy, where there is much distress, corticosteroids may be used topically or systemically as palliative treatment.

For specific treatment, insecticides, in the form of a dust, spray or shampoo, should be applied to the animal at weekly intervals. These may be organophosphorus compounds, pyrethrum and its derivatives, or organochlorines. Of the organophosphorus compounds a proprietary combination of dichlorvos and fenitrothion, or alternatively, iodofenphos is commonly used, and of the synthetic pyrethroids, permethrin. Of the organochlorines, gamma HCH is suitable for dogs, but is toxic to cats, while bromocyclen is suitable for either host. Since in-contact animals may also harbour fleas without developing allergy these should also be treated.

Flea collars containing dichlorvos or diazinon are also used to control fleas in both dogs and cats; the latter require a special collar with a low concentration of insecticide. In both species occasional cases of contact dermatitis have occurred in the neck region when collars have been used.

Since the greater part of the flea population is not on the animal itself, but in its environment, it is important that insecticides are also applied to its living quarters and general indoor habitat, and that bedding should be destroyed where possible. Fitted carpets should be thoroughly vacuum-cleaned. A useful technique, giving continuous exposure to insecticide, is to place dichlorvos impregnated strips near or under the animal's bed; in this case, insecticidal collars should not be used.

Recently, an insect growth regulator, methoprene, has been marketed as an aerosol for direct application to bedding, carpets and other habitats of flea larvae. When ingested by the larvae the chemical prevents the emergence of adult fleas from pupae. Protection against reinfestation may persist for up to four months.

FLEAS OF BIRDS

Ceratophyllus

Ceratophyllus gallinae is the commonest flea of domestic poultry and may be responsible for irritation, restlessness and even anaemia. It feeds readily on humans and domestic pets, and is often acquired in the handling of poultry and from injured wild birds brought into houses. It has also been known to migrate into rooms from nests under adjacent eaves. When such nests are removed they should be incinerated; otherwise the hungry fleas may parasitise domestic pets and humans.

Echidnophaga

Echidnophaga gallinacea, the 'stick-tight' flea, is one of the burrowing fleas. After fertilisation the female burrows into the skin of the fowl, usually on the comb and wattles, resulting in the formation of nodules in which the eggs are laid. Hatching occurs within the nodules, and the larvae drop to the ground to complete development. The skin over the nodules often becomes ulcerated, and young birds may be killed by heavy infections. *Echidnophaga* also attacks mammals, principally dogs, the nodules being formed around the eyes and between the toes.

TREATMENT AND CONTROL

Several organophosphorus insecticides such as malathion or carbaryl are effective. These are used as dusts for *Ceratophyllus* and as a solution, in which the head is dipped, in the case of *Echidnophaga*.

Should fleas become established in a poultry house, then drastic measures may have to be adopted to get rid of them. All litter should be removed and burnt and the poultry house sprayed with an insecticide. A 1% solution of ronnel used twice at an interval of 14 days has given good results.

Class ARACHNIDA

This class includes the ricks and mites which are of considerable veterinary importance and also the spiders and scorpions. They differ from the Insecta in that the adult has four pairs of legs and the body is composed of a cephalo-thorax and abdomen. The mouthparts are extensively modified and carry two pairs of appendages, the first called chelicerae and the second the palps. Antennae are absent.

Order ACARINA

The acarines are small, sometimes microscopic, arthropods. They are characterised by their arachnid structure, the mouthparts (Fig. 133), which are borne on the basis capituli, consisting of a pair of **chelicerae** with mobile digits adapted for cutting and a pair of sensory **palps**. Ventromedially there is a hypostome with recurved teeth for maintaining position; it bears a dorsal groove to permit the flow of saliva and host blood. The acarines of veterinary importance are the ticks and mites

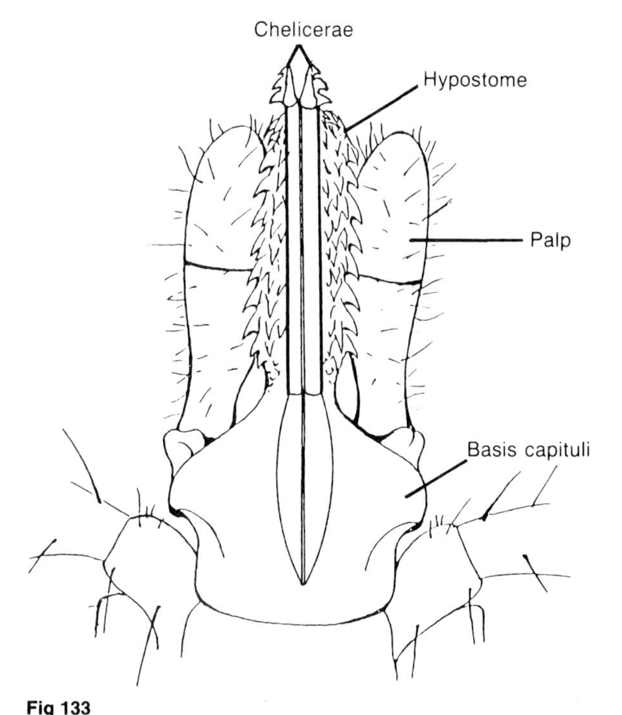

Fig 133
(a) Typical mouthparts of ixodid tick.

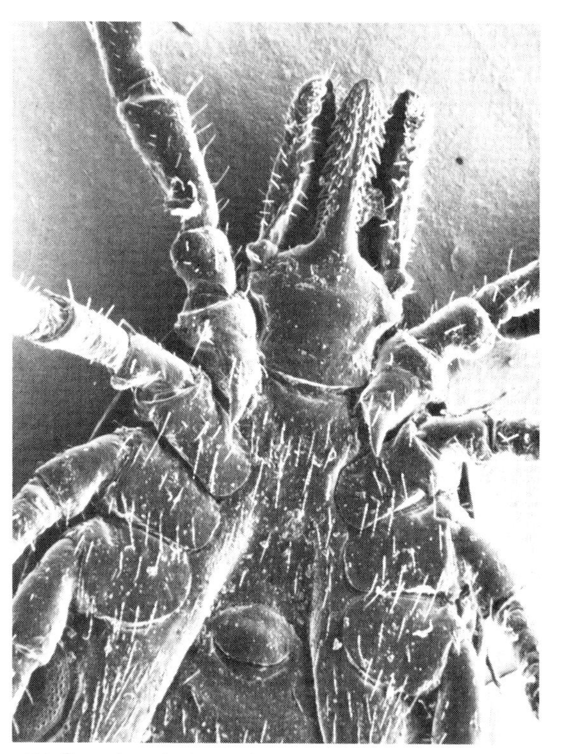

(b) Scanning electron micrograph of *Ixodes ricinus*.

and their life cycle consists of development from egg through the larva, which rather resembles the adult, to nymph to adult. Usually there is only one larval instar.

THE TICKS

Two families, the **Ixodidae** and **Argasidae** are commonly known as ticks. The most important is the Ixodidae, often called the **hard ticks**, because of the presence of a rigid chitinous scutum which covers the entire dorsal surface of the adult male; in the adult female and in the larva and nymph it extends for only a small area which permits the abdomen to swell after feeding.

The other family is the Argasidae or **soft ticks**, so-called because they lack a scutum; included in this family are the bird ticks and the tampans.

Family IXODIDAE

The ixodids are important vectors of protozoal, viral and rickettsial diseases. Although there are many genera of Ixodidae only three occur in W. Europe; these are *Ixodes*, *Haemaphysalis* and *Dermacentor* of which *Ixodes* is by far the most important.

As noted previously, the ixodids (Fig. 134) have a chitinous covering or scutum which extends over the whole dorsal surface of the male, but covers only a small area behind the head in the larva, nymph or female. The mouthparts carried on the capitulum are anterior and visible from the dorsal surface. Other distinguishing features are a series of grooves on the scutum and body, and in some species, a row of notches, called **festoons**, on the posterior border of the body. Chitinous plates are sometimes present on the ventral surface of the males. The genital opening is in the ventral mid-line and the anus is posterior. Some ticks have coloured enamel-like areas on the body and these are called '**ornate** ticks'. The adults have a pair of spiracles behind the fourth pair of legs. Eyes, when present, are situated on the outside margin of the scutum.

Like other ticks, the ixodids are temporary parasites and spend relatively short periods on the host. The number of hosts to which they attach during their parasitic life cycle varies from one to three, and, based on this, they are classified as **one-host ticks** where the entire parasitic development from larvae to adult takes place on the one host; **two-host ticks** where larvae and nymphs occur on one host and the adults on another; **three-host ticks**

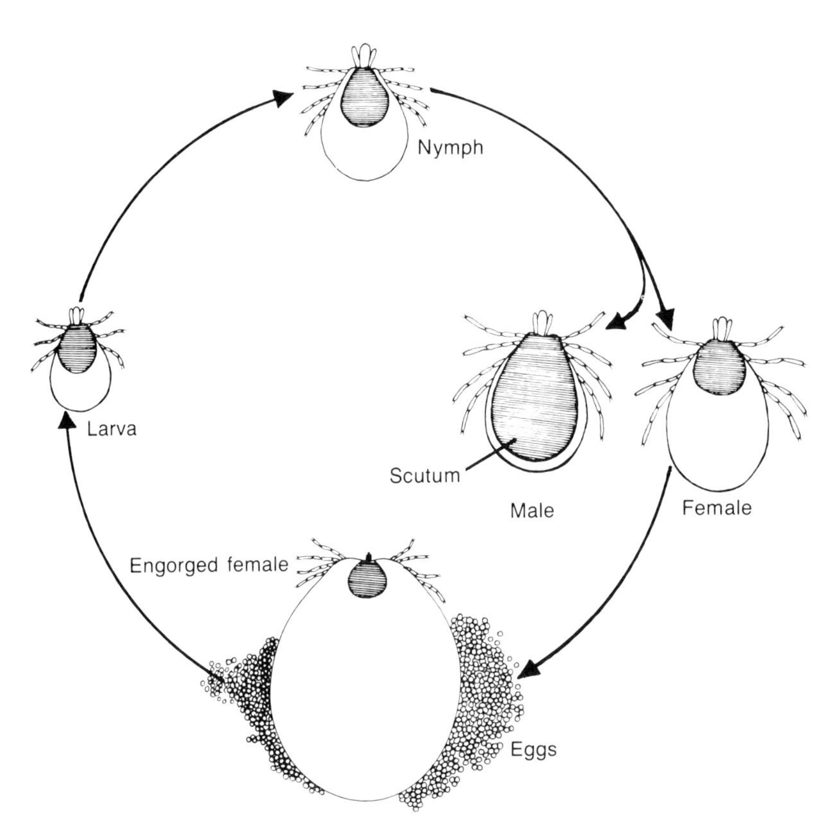

Fig 134 Structure and life cycle of an ixodid tick.

where each stage of development takes place on different hosts.

The three ixodid genera which are described first, *Ixodes*, *Haemaphysalis* and *Dermacentor* (Pl. VIII), are those which occur in western Europe as well as elsewhere in the world.

Ixodes

Hosts:
All mammals and birds. Of great veterinary significance in ruminants

Site:
All over the body but especially the axilla, inguinal region, face and ears

Species found in Europe:

Ixodes ricinus	castor bean tick
Ixodes canisuga	British dog tick
Ixodes hexagonus	hedgehog tick

Some other species:

Ixodes holocyclus	the paralysis tick of Australia
Ixodes rubicundus	the paralysis tick of South Africa
Ixodes scapularis	the shoulder tick of North America

GENUS IDENTIFICATION

These are inornate ticks, without festoons or eyes. The palps are long and the ventral surface of the male is almost entirely covered with a series of plates. An anal groove is present anterior to the anus.

Ixodes ricinus

DISTRIBUTION

Europe, North America, Australia, South Africa. In Britain it is more common in the western half of the country in areas of rough grazing and moorland. In mainland Europe, this species is most numerous in the north and central regions and largely confined to woodland.

IDENTIFICATION

The engorged female is light grey, up to 1.0 cm in length, bean shaped and has four pairs of legs. The males are only 2.0–3.0 mm long, and because of the small abdomen the four pairs of legs are readily visible.

Nymphs resemble the adults and also have four pairs of legs but are less than 2.0 mm in size while the larvae ('pepper ticks') are less than 1.0 mm, usually yellowish in colour and have only three pairs of legs.

In *I. ricinus*, as compared with *I. canisuga* and *I. hexagonus*, the tarsi are tapered and not humped and the posterior internal angle of the first coxa bears a spur which overlaps the second coxa (Fig. 135).

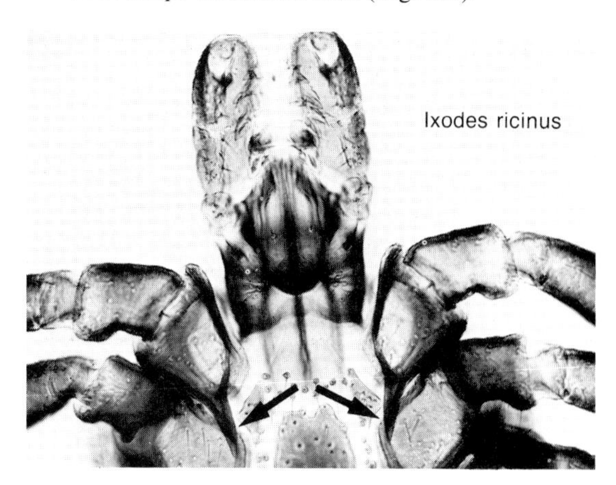

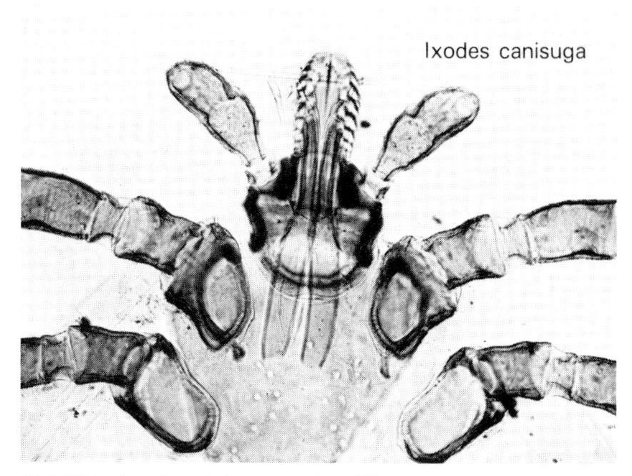

Fig 135 *Ixodes ricinus* may be differentiated from other *Ixodes* species by the presence of an overlapping spur (↑) on the posterior angle of the first coxa.

LIFE CYCLE AND EPIDEMIOLOGY

I. ricinus is a three-host tick and the life cycle requires three years. The tick feeds for only a few days each year, as a larva in the first year, a nymph in the second and an adult in the third (Fig. 134).

Mating takes place on the host. After attachment the female is inseminated once and subsequently completes her single large blood meal; in contrast, the males feed intermittently and mate repeatedly. During mating, the male crawls under the female and after manipulating the female genital opening with his mouthparts transfers the spermatophore, a sac containing the spermatozoa, into the opening presumably with the aid of his front legs. Once fertilised, the female subsequently feeds for about 14 days and then drops to the ground to lay several thousand eggs in sheltered spots, after which she dies. The larvae which hatch from the eggs will feed for about six days in the following year, then drop to the ground and moult to the nymphal stage. In the third year this stage feeds, drops off and becomes adult. Although the life cycle takes three years to complete, the larvae, nymphs and adults feed for a total of only 26–28 days. *I. ricinus* is therefore a temporary parasite.

There is a distinct seasonal activity of *I. ricinus* during which infestation of cattle, sheep or other hosts occurs and during this period ticks can be found on the tips of vegetation 'questing' or searching for a host which they locate with the aid of sensory bristles located on their legs. In the British Isles, two peak periods of activity occur, namely March–June and August–November, although some ticks may be acquired throughout most of the year.

It is thought that, by and large, these two distinct periods of activity reflect the occurrence of two physiologically distinct populations of ticks, one active in spring and the other in the autumn. In the border counties between Scotland and England only spring feeders occur whereas in the tick areas of South-West England, Wales, West Scotland and Ireland, autumn feeders also occur. The eggs of the spring feeders hatch in the autumn and the larvae and nymphs also moult at this time so that they overwinter in the unengorged or flat condition. In contrast, the eggs of the autumn feeders do not hatch until the following summer while the larvae and nymphs are in diapause over this whole period, so that they overwinter in the engorged state and do not moult until the summer.

The survival of *I. ricinus* and therefore its distribution is determined by its water requirements and it is unable to survive when the relative humidity is less than 90%. In Britain it is predominantly a tick of rough grazing where the annual death and subsequent decay of vegetation ensures the presence of a 'mat' with a high moisture-retaining capacity. In other areas where heather is prevalent there is little 'mat', but the damp open areas between the heather are covered in moisture-retaining mosses making this another suitable environment. In parts of mainland Europe ticks are mostly found in woodlands and in Ireland, where the water table is particularly high, they will survive on arable ground, particularly around hedgerows.

The cuticle of the tick is the key to its survival; this has an outer wax layer which is impermeable to water so that water loss is normally limited to tiny pores in the tegument. However, when the tick is active loss of water is accelerated due to the opening of the spiracles and during feeding by the secretion of saliva.

In Britain the movement of ticks from the vegetation mat to a questing position begins with the spring feeders when the mean day-night temperature exceeds 10 °C and ends with the autumn feeders when the temperature drops below 10 °C; in the summer there is no activity due to low humidity and high temperatures.

PATHOGENIC SIGNIFICANCE

The pathogenic effects are associated with the feeding mechanism of the parasite which is ideal for both penetrating the skin and transmitting micro-organisms. In the feeding process the scissor-like action of the digits at the end of the chelicerae is followed by a thrusting motion of the hypostome through the lacerated skin and the locking effect of its recurved teeth on the tissues. The salivary glands are thought to produce an hyaluronidase-like substance to assist in penetration, a cement-like material which assists locking and an anticoagulant. The dorsal groove in the hypostome provides a channel for the saliva to flow into the host and, subsequently, blood and lymph into the tick. As the female ticks engorge and swell they tend to bulge out from the site of attachment and the hind legs stiffen, the tick becoming tilted at right angles to the body. If the tick is forcibly removed during the locked phase of feeding there can be considerable reaction to the mouthparts, which often remain embedded in the tissues. During engorgement the bodyweight of a tick increases by about 200 times.

I. ricinus infestations are important for three reasons:

(i) **They suck blood** and in occasional heavy infestations can cause anaemia.
(ii) **The lesions** caused by the toothed mouthparts during feeding may become infected and **predispose to blowfly strike**. Also at slaughter the value of the hide or fleece may be reduced.
(iii) Most significant of all, this tick in W. Europe **transmits** *Babesia divergens*, the cause of redwater in cattle, and in sheep and cattle the virus of louping-ill and the rickettsia responsible for tick-borne fever. It is also associated with tick pyaemia, caused by *Staphylococcus aureus*, in lambs in Britain and Norway.

CLINICAL SIGNS

There are no obvious signs of tick infestation other than the presence of the parasites and the local skin reactions to their bites.

DIAGNOSIS

The adult ticks, particularly the engorged females, are easily seen on the skin, the predilection site being the face, ears, axilla and inguinal region. Usually small inflamed nodules are also seen in these areas, each a reaction to a previous tick bite.

CONTROL

This is discussed after a description of the other ixodid ticks.

Ixodes canisuga

This species, sometimes called the British dog tick, has been found in a variety of hosts, but is recognised as a problem in kennels where the tick is capable of survival in crevices and cracks in the floors and walls. Heavy infestations can cause pruritus, loss of hair and anaemia. It may be differentiated from *I. ricinus* by the presence of humped tarsi and the absence of a spur on the posterior internal angle of the first coxa (Fig 135).

Ixodes hexagonus

Usually called the hedgehog tick it also occurs in dogs, ferrets and weasels. It may be differentiated from *I. ricinus* by its humped tarsi and the fact that the spur on the first coxa does not overlap the second coxa.

Important Ixodes species outside Europe

Adult females of several *Ixodes* species, including *I. holocyclus* and *I. rubicundus* in Australia and South Africa respectively, produce a toxin responsible for tick paralysis which occurs in man and animals and is characterised by an acute ascending motor paralysis, occurring several days after attachment, which may terminate fatally unless the ticks are removed. The precise nature of the toxin is unknown, but current theory suggests it is produced in the salivary glands.

Haemaphysalis

These are inornate ricks with festoons present and eyes absent. The sensory palps are short and broad with the second segment extending beyond the basis capituli. The males have no ventral shields and the anal groove contours the anus posteriorly. They are three-host ticks.

Haemaphysalis punctata is widely distributed in Europe, including southern England and parts of Wales where it is responsible for the transmission of *Babesia major* and a non-pathogenic *Theileria* sp. in cattle, and in sheep *Babesia motasi* and the benign *Theileria ovis*.

In other countries it transmits *Babesia bigemina* in cattle, *B. motasi* in sheep and *Anaplasma marginale* and *A. centrale* in cattle; it has also been reported as causing tick paralysis.

There are many other species of *Haemaphysalis* to be found throughout the world. For example, *H. leachi*, the 'yellow dog tick', common in Africa, Asia and Australia, is responsible for the transmission of *B. canis* in dogs, and *H. longicornis*, primarily a cattle tick, has a wide distribution in the Far East and Australasia.

Dermacentor

These are ornate ticks with eyes and festoons present. The basis capituli is rectangular and the palps short. The males lack ventral plates and the fourth coxae are enlarged. They may be three-host or one-host ticks and are parasitic in many domestic and wild mammals.

Dermacentor reticulatus occurs in many parts of Europe including southern England and Wales and seems to prefer heavily wooded areas. It is a three-host tick which transmits *Babesia* infections in horses and dogs in countries where these parasites are present.

Several species of *Dermacentor* are important in North America. Thus, the three-host ticks *D. andersoni* and *D. variabilis* are responsible for the transmission of *Anaplasma marginale* in cattle and Rocky Mountain spotted fever in man caused by *Rickettsia rickettsii*. Both also cause tick paralysis. The one-host tick *D. nitens* is an important vector of equine babesiosis in the southern U.S.A., Central and South America and the Caribbean.

An abbreviated key to the important ticks of Western Europe is given in Table 5.

Table 5 Abbreviated key to the important adult ticks of domestic animals of western Europe

1.	Integument with scutum Capitulum visible dorsally or No scutum Capitulum ventral	Ixodidae Argasidae (very rare in Britain)	2
2.	Inornate or Ornate with eyes, short palps and rectangular basis capituli	 *Dermacentor reticulatus*	3
3.	Anal groove posterior to anus Festoons present or Anal groove anterior to anus Festoons absent	*Haemaphysalis punctata* *Ixodes*	 4
4.	Coxa 1 with a distinct posterior internal spur or Coxa 1 without a distinct posterior internal spur	 *Ixodes canisuga*	5

5. Tarsi tapering gradually to thin point. Spur overlaps Coxa 2 Scutum rounded posteriorly	*Ixodes ricinus*
or	
Tarsi humped. Spur does not overlap Coxa 2 Scutum sub-hexagonal Anal groove with parallel sides	*Ixodes hexagonus*

IXODID TICKS OCCURRING OUTSIDE WESTERN EUROPE

Amblyomma

These are large, usually ornate, ticks whose legs have bands of colour; eyes and festoons are present. The palps and hypostome are long and ventral plates are absent in the males. They are three-host ticks.

Important species are *Amblyomma variegatum* and *A. hebraeum*, the so-called 'bont ticks', ie. with patterns of colour on the back and legs. They are distributed mainly in Africa and transmit the important disease, heartwater, in cattle, caused by the rickettsia, *Cowdria ruminantium*.

In the southern U.S.A., *A. americanum*, the 'lone star tick', so called because of a single white spot on the scutum of the female, transmits Q fever, tularaemia and Rocky Mountain spotted fever; the 'Gulf Coast tick', *A. maculatum*, which occurs in the ears of cattle is a predisposing cause of screw-worm myiasis associated with *Cochliomyia* spp. *A. cajennense* is an important tick in South America, the bites caused by this genus being particularly painful, probably due to the long mouthparts.

Boophilus (including *Margaropus*)

Inornate ticks with eyes present and festoons absent. The palps and hypostome are short. The males have adanal or accessory ventral shields.

These, often known as 'blue ticks', are one-host ticks and the most important vectors of *Babesia* spp. and *Anaplasma marginale* in cattle in subtropical and tropical countries. The most important species are *B. microplus*, present in every continent except Europe, *B. annulatus* in Central and South America and Africa and *B. decoloratus* in East Africa.

Hyalomma

Usually inornate but with banded legs (the 'bont-legged tick'); eyes are present and festoons sometimes present. The palps and hypostome as in *Amblyomma* are long. The males have adanal shields. They are usually two-host ticks with the larvae and nymphs feeding on birds and small mammals and the adults on ruminants and equines.

Hyalomma spp. occur throughout Africa, Asia Minor and southern Europe and have been incriminated as vectors of several babesial, theilerial and rickettsial infections. *H. marginatum* and *H. detritum* are the important species in southern Europe and North Africa and *H. truncatum* throughout Africa.

This genus is mainly responsible for tick toxicosis, an entity distinct from tick paralysis, in parts of southern Africa and the Indian sub-continent. The 'toxin' produced by the adult tick causes a sweating sickness in ruminants and pigs characterised by a widespread hyperaemia of the mucous membranes and a profuse moist eczema.

Rhipicephalus

Usually inornate with eyes and festoons present. The palps and hypostome are short and the basis capituli hexagonal dorsally. The first coxa has two spurs. The males have adanal plates and accessory shields. The genus includes both two-host and three-host ticks.

Two important species are found exclusively in Africa south of the Sahara. The three-host tick, *Rhipicephalus appendiculatus*, the 'brown ear tick', is the most efficient vector of East Coast Fever of cattle caused by *Theileria parva* and also transmits *Babesia bigemina* and the virus of Nairobi sheep disease. The two-host tick *R. evertsi*, 'the red-legged tick', can also transmit theilerial infections and *Babesia bigemina* and *B. equi*. The common three-host species *R. sanguineus* has a more widespread distribution and is found throughout the southern hemisphere. It is primarily parasitic on dogs, is familiarly called 'the brown dog or kennel tick', and is responsible for the transmission of *Babesia canis* and *Ehrlichia canis* and can also cause tick paralysis in the dog. There seems little doubt that it can also transmit many protozoal, viral and rickettsial infections of animals and man.

No key for ticks outwith Europe is given because of the wide diversity of species in different areas, and advice should be sought from local experts.

EPIDEMIOLOGY OF IXODID TICKS IN TROPICAL AND SUBTROPICAL ENVIRONMENTS

The distribution of ticks in a temperate climate with frequent and non-seasonal rainfall is closely linked with the availability of a micro-environment with a high relative humidity such as occurs in the mat which forms under the surface of rough grazing. In contrast, in tropical grazing areas the grass cover on pastures is discontinuous and often interspersed with bare or eroded patches. Where suitable grass cover does exist it has been generally accepted, since temperatures are suitable for development throughout a large part of the year, that the distribution of ticks is mainly governed by rainfall, and with the exception of *Hyalomma* spp., a mean annual rainfall of more than 60 cm is required for survival.

However, recent studies in East Africa have shown that the factors underlying the maintenance of the necessary microclimate with a high relative humidity are rather more complex and depend on the transpiration of plant leaves. As long as this continues, adequate humidity is maintained in the microclimate despite the dryness of the ambient temperature. However, when the rate of evaporation increases beyond a certain level, the stomata on the leaves close, transpiration ceases and the low humidity created in the microclimate rapidly becomes lethal to the ticks.

In the field, of course, the stability of the microclimate is dependent on factors such as the quantity of herbage or plant debris and the grass species. The various genera of ticks have different thresholds of temperature and humidity within which they are active and feed and their distribution is governed by these thresholds. Generally, ticks are most active during the warm season provided there is sufficient rainfall, but in some species the larval and nymphal stages are also active in milder weather and this affects the duration and timing of control programmes.

CONTROL OF IXODID TICKS

The control of ixodid ticks is largely based on the use of chemical acaricides applied either by total immersion in a dipping bath or in the form of a spray or shower.

Where severely parasitised animals require individual treatment, special formulations of acaricides suspended in a greasy base may be applied to affected areas.

CONTROL IN WESTERN EUROPE

Cattle are not usually dipped for tick control and treatments are confined to hand-spraying when, on infrequent occasions, large populations are observed on cattle. It is also good policy to spray animals when they are being moved from endemic tick farms to those in which ticks are likely to be absent.

However, in many areas of Britain, such as the West Highlands and the Border counties of Scotland and England, sheep are routinely dipped with an effective acaricide each spring in order to control the tick population and to reduce the prevalence of tick-borne diseases. Although of value, the effectiveness of this single treatment seems to vary in individual flocks, possibly because the three-host tick, *I. ricinus*, spends such a short time on the host and its period of activity and feeding varies between years according to the climate.

When young lambs, or their ewes, are dipped it is important to ensure that they are properly 'mothered-up' before being released on to a wide area of hill. A common procedure for lambs is to immerse them individually in a small bath or drum containing the acaricide and subsequently allow them access to their mothers in a confined area.

CONTROL IN THE TROPICS

In these areas treatment is mainly directed towards cattle. Since ticks remain on the host for only a few days and often at defined times of the year, the timing and frequency of dipping is based on several factors. These include the seasonal activity and duration of feeding of the individual tick species, the significance of the diseases they transmit and the residual effect and toxicity to the host of the acaricide. A further consideration, which influences the regimen of control, is whether the tick is a one-host tick, in which all the instars feed and develop on the same host, or two- or three-host ticks which use two or three different hosts respectively. Clearly, the one-host tick is easier to control than the others.

An important point is that, unlike W. Europe, where tick activity only occurs in the spring or autumn, tropical ticks may be active throughout most of the year completing their life-cycle from egg to adult within a few months.

CONTROL OF ONE-HOST TICKS

The basis of successful control of one-host species such as *Boophilus*, prevalent in Australia, South Africa and Latin America, is to prevent the development of the engorged female ticks and so limit the deposition of large numbers of eggs. Since *Boophilus* has a parasitic life cycle which requires 20 days before adult females become fully engorged, an animal dipped with an acaricide which has a residual effect of 3–4 days should not harbour engorged females for at least 24 days (ie. 20 + 4). In theory, therefore, treatment every 21 days during the tick season should give good control, but since the nymphal stages appear to be less susceptible to most acaricides, a 12-day interval is often necessary between treatments at the beginning of the tick season.

CONTROL OF TWO- AND THREE-HOST TICKS

The control of the two- and three-host ticks prevalent in Africa and North America is similarly geared to the parasitic period required for the adult female stage to become fully engorged and this varies from 4–10 days according to the species. If an animal is treated with an acaricide which has a residual effect of, say, three days, it will be at least seven days before any fully engorged female reappears following dipping (ie. three days residual effect plus a minimum of four days for engorgement). Weekly dipping during the tick season should therefore kill the adult female ticks before they are engorged except in cases of very severe challenge when the dipping interval has to be reduced to four or five days. Dipping intervals of this latter frequency are also necessary for cattle infested with *R. appendiculatus* in areas where East Coast Fever is endemic so that the ticks are killed before the sporozoites of *T. parva* have time to develop to the infective stage in the salivary glands of the tick.

Theoretically, weekly dipping should also control the larvae and nymphs, but in several areas the peak infestations of larvae and nymphs occur at different seasons to the adult females and the duration of the dipping season has to be extended. Since many of the two- or three-host ticks occur on less accessible parts of the body, such as the anus, vulva, groin, scrotum, udder and ear, care must be exercised to ensure that the acaricide is properly applied.

There is considerable local variation in tick biology and times of dipping may vary widely within regions. Before embarking on any dipping control programme, local advice should be sought though the general principles described above should be observed.

ACARICIDES

Arsenic was the first compound to be widely used for tick control but due to problems of toxicity, lack of residual effect and resistance it was largely replaced by the organochlorines in the late 1940s. Consumer resistance to unacceptable levels of organochlorines in meat together with the onset of tick resistance to this group of insecticides led to their replacement in the 1960s by several organophosphorus compounds, the carbamate, butocarb, and more recently the formamidine, amitraz, and some synthetic pyrethroids. Ivermectin or closantel given by the parenteral route have also been shown to be a useful aid in the control against the one-host tick *Boophilus*.

OTHER CONTROL MEASURES

The development by ticks of resistance to most of the available acaricides poses such a threat to livestock production in the tropics that alternative methods of control are urgently being sought especially against the two- and three-host ticks which spend long periods off the host. Traditional methods such as burning of pastures are still used and are generally practised during a dry period before rains, when ticks are inactive. This technique is still a most useful one in extensive range conditions and provided it is used after seeding of the grasses has taken place, regeneration of the pastures will rapidly occur following the onset of rains. Cultivation of land, and in some areas, improved drainage help to reduce the prevalence of tick populations and can be used where more intensive systems of agriculture prevail. Pasture 'spelling' in which domestic livestock are removed from pastures for a period of time has been used in semi-extensive or extensive areas, but often has the disadvantage that ticks can still obtain blood from a wide variety of other hosts.

Other control systems, such as the use of the sterile male technique, tick vaccines and biological control by pathogens of ticks, are being investigated, but perhaps the most promising is the selection of cattle with a high natural resistance to tick infestations. Resistance of this type seems to be an heritable characteristic and is high in the humped breeds (*Bos indicus*) and low in the European breeds (*Bos taurus*). The potential of resistant humped breeds, possibly crossed with European cattle, is currently under investigation as a method of reducing tick populations and the diseases they transmit.

Family ARGASIDAE

The scutum is absent from these soft ticks and the mouthparts are not visible from the dorsal aspect (Fig. 136). They do not swell as much on engorgement as hard

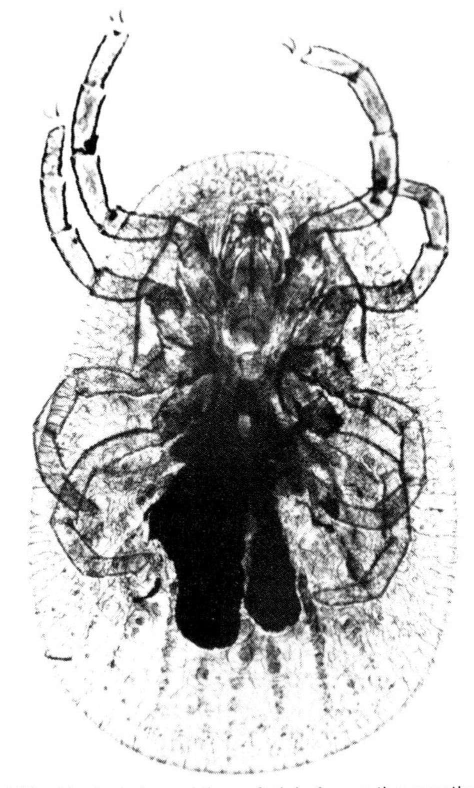

Fig 136 Ventral view of the soft tick *Argas*; the mouthparts are not visible from the dorsal aspect.

ticks since the females feed moderately and often. Mating takes place off the host and the eggs are laid in batches. These ticks, unlike the Ixodidae, are drought resistant and capable of living for several years. There are three genera of veterinary importance.

Argas (the fowl tick)

The common species, *Argas persicus*, is cosmopolitan in distribution especially on domestic poultry in the tropics; it has only been recorded from Britain on a few occasions. The life cycle involves one larval and at least two nymphal stages prior to the adult. These various stages live in cracks and crevices of the poultry house only approaching birds at night to suck blood about once per month. The adult stages live for several years, even in the absence of suitable hosts. These ticks cause sleeplessness, loss of productivity and anaemia, which can prove fatal. They transmit *Borrelia anserina*, the cause of fowl spirochaetosis, and *Aegyptianella pullorum*, a rickettsial infection.

Another species, *A. reflexus*, is a common parasite of pigeons.

Otobius

Otobius megnini, the spinose ear tick, is found in North and South America, India and southern Africa. Although these spiny ticks occur primarily in dogs' ears they can also infect many other hosts including man. Since the eggs are laid in crevices in walls and feeding troughs they are mainly a problem in housed stock. Only the larvae and nymphs are parasitic remaining on the same host for several months where they cause severe inflammation and a waxy exudate in the ear-canals, and in heavy infections, anaemia and loss of condition.

Ornithodoros

These soft ticks or sand tampans live in sandy soils, in primitive housing or in shaded areas around trees. *Ornithodoros moubata* and *O. savignyi* occur in Africa and the Middle East and the former at least is a reservoir host for the virus of African swine fever and for the spirochaete which causes relapsing fever in man. Only the nymphs and adults are parasitic and may be responsible for considerable irritation in man and animals and heavy infections can cause mortality of stock from blood-loss. Because of their location under the surface of sand and the short periods on the host, treatment and control may be difficult. Another species, *O. turicata*, also occurs in the U.S.A. and has been incriminated as a vector of Q fever in man and animals and also as a cause of tick paralysis.

The three important epidemiological characteristics of the soft ticks are, first, the fact that engorgement is completed rapidly and therefore allows advantage to be taken of the infrequent presence of suitable hosts; secondly, the great capacity of these soft ticks for survival in arid conditions; thirdly, the frequent feeding activity of the various stages allows many opportunities for the transmission of pathogens.

CONTROL OF ARGASID TICKS

Argasid ticks which exist in poultry houses or in animal shelters or enclosures can be controlled by application of an acaricide to their environment coupled with treatment of the population on the host. All niches and crevices in affected buildings should be sprayed with concentrations of acaricide containing up to 5% of the carbamate, carbaryl, or up to 2% gamma HCH. Nesting boxes and perches in poultry houses should also be painted with solutions containing these acaricides. At the same time as premises are treated, birds should either be dusted with a suitable acaricide, or in the case of larger animals, sprayed or dipped. Treatment should be repeated at monthly intervals.

In the case of the ear tick *Otobius*, which spends a long time on the host, control can be achieved by the topical use of creams containing gamma HCH and the spraying of premises with a solution of the same chemical.

For sand tampans the use of blocks of solid carbon dioxide to stimulate them to leave their secluded hiding places for the soil surface where they are exposed to acaricides is often recommended, but rather impractical. The recent introduction of ivermectin which has a residual effect against *Ornithodoros* offers a very promising method of control in domestic animals.

THE MITES

This group of acarines includes both parasitic and free-living forms, a few of the latter species being of interest to the veterinarian as intermediate hosts of anoplocephalid cestodes, including *Anoplocephala*, *Moniezia* and *Stilesia*.

The parasitic mites are small, most being less than 0.3 mm long, though a few blood-sucking species may attain 1.0 cm when fully engorged. With few exceptions they are in prolonged contact with the skin of the host, causing various forms of the condition generally known as **mange**.

Though, like the ticks, mites are obligate parasites, they differ from them in the important respect that most species spend their entire life cycles, from egg to adult, on the host so that transmission is mainly by contact. It will also be seen later that, unlike the ticks, once infection is established, pathogenic populations can build up on an animal without further acquisitions.

The mites have a complex taxonomy, occupying at least eight different families, and for veterinarians it is more useful to consider them according to their location on the host as **burrowing** and **non-burrowing** mites.

BURROWING MITES

With one exception, *Demodex*, to be considered later, the three important burrowing genera, *Sarcoptes*, *Notoedres* and *Knemidocoptes*, belong to a single family, the Sarcoptidae and have much in common.

Morphologically they have a general similarity, with circular bodies and very short legs which scarcely project beyond the body margin. Their generic characteristics are outlined in Fig. 137.

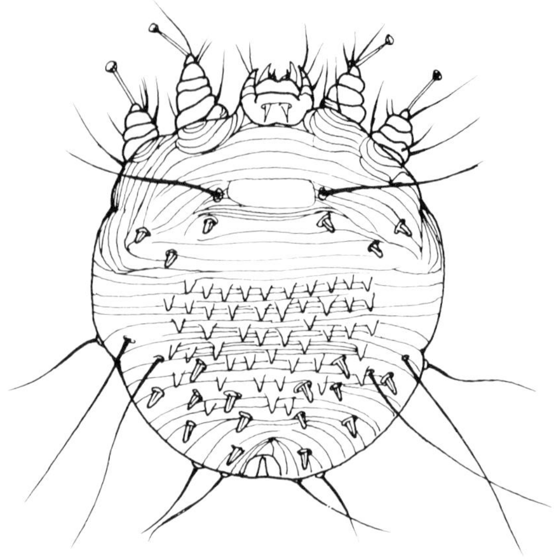

Fig 137
(a) Dorsal view of *Sarcoptes scabiei* showing transverse ridges and triangular scales.

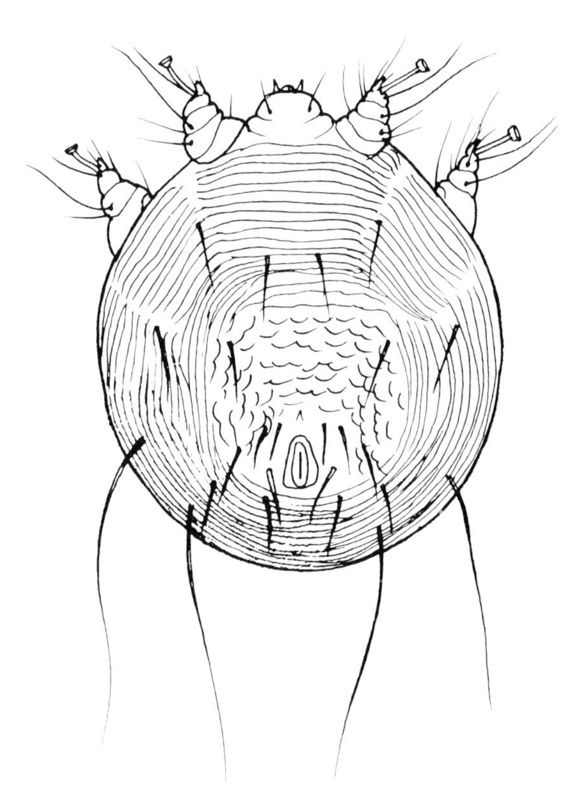

(b) Dorsal view of *Notoedres cati* showing concentric striations.

Family SARCOPTIDAE

Sarcoptes

The sole species of this mite occurs in a wide range of mammals, but by biological adaptation 'strains' have evolved which are largely host-specific. Thus, *Sarcoptes* is well known in both human and veterinary medicine as a cause of mange, the disease in man being generally known as scabies.

Hosts:
All domestic mammals and man

Species:
Sarcoptes scabiei

Distribution:
Worldwide

MORPHOLOGY

Sarcoptes is round in outline and up to 0.4 mm in diameter, with short legs which, like those of *Notoedres*, scarcely project beyond the body margin. Its most important recognition characters are the numerous transverse ridges and triangular scales on the dorsum, features possessed by no other mange mite of domestic mammals.

LIFE CYCLE

The fertilised female creates a winding burrow or tunnel in the upper layers of the epidermis, feeding on liquid oozing from the damaged tissues. The eggs are laid in these tunnels, hatch in 3–5 days, and the six-legged larvae crawl on to the skin surface. These larvae, in turn, burrow into the superficial layers of the skin to create small 'moulting pockets' in which the moults to nymph and adult are completed. The adult male then emerges and seeks a female either on the skin surface or in a moulting pocket. After fertilisation the females produce new tunnels, either *de novo* or by extension of the moulting pocket. The entire life cycle is completed in 17–21 days.

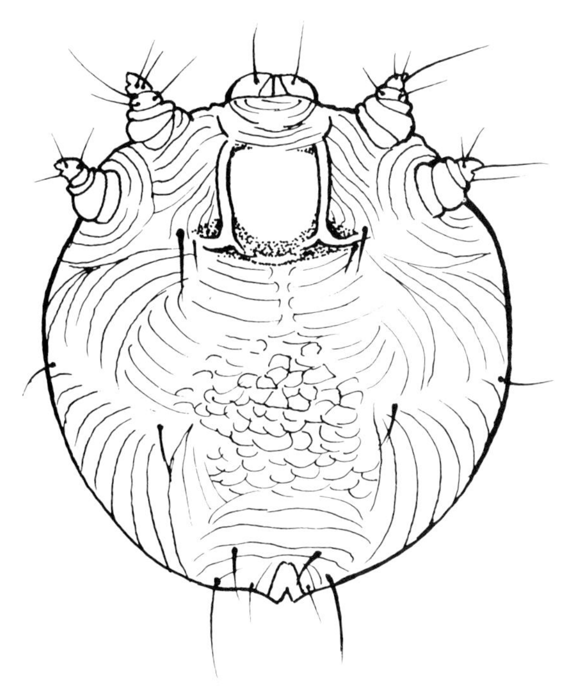

(c) Dorsal view of the poultry mite *Knemidocoptes.*

New hosts are infected by close contact, presumably from larvae which are commonly present on the skin surface.

SARCOPTIC MANGE OF DOGS

The predilection sites for the mites are areas such as the ears, muzzle, face, and elbows but, as in other manges, severe infestations may extend over the whole body (Pl. XI).

Visually, the condition begins as erythema, with papule formation, and this is followed by scale and crust formation and alopecia. It is a characteristic of this form of mange that there is intense pruritus, which often leads to self-inflicted trauma.

After a primary infection dogs begin to scratch within a week, often before lesions are visible. By analogy with infection in pig and man it seems likely that the degree of pruritus is exacerbated by the development of skin hypersensitivity to mite allergens.

In cases which are neglected for a number of months the whole skin surface may be involved, dogs becoming progressively weak and emaciated; a strong sour odour is a notable feature of this form of mange.

Useful diagnostic features of canine sarcoptic mange are:

(i) **The edges of the ears** are often first affected, and on rubbing a scratch reflex is readily elicited.

(ii) **There is always intense itching,** so that in cases of dermatitis where there is no itch, sarcoptic mange can be eliminated as a possibility.

(iii) **It is a highly contagious condition,** and single cases are rarely seen in groups of dogs kept in close contact.

Confirmatory diagnosis is by examination of skin scrapings for the presence of mites. However, since these are sometimes difficult to demonstrate, a negative finding should not preclude a tentative diagnosis of mange and initiation of treatment.

Based on the protected location of the parasites, the duration of the life cycle and the necessity of killing all mites, dogs should be bathed weekly with an acaricidal preparation for four weeks, or longer if necessary, until lesions have disappeared. Effective acaricides include the organochlorines, gamma HCH and bromocyclen, and the organophosphates such as ronnel. Many preparations are combined with a surfactant which aids contact with the mites, by removing skin scales and softening crusts and other debris.

Because this is a highly contagious mange, affected dogs should be isolated and it should be explained to owners that rapid cure cannot be expected. To ensure that an outbreak is contained, all dogs on the premises should be treated if possible.

In severely distressed dogs, oral or parenteral corticosteroids are valuable in reducing the pruritus and so preventing further excoriation.

SARCOPTIC MANGE OF CATS

Sarcoptic mange is rare in cats. In the few recorded cases the changes have been similar to those in *Notoedres* infection with progressive hair loss from the ears, face, and neck, extending to the abdomen. Treatment is as for *Notoedres.*

SARCOPTIC MANGE OF PIGS

The ears are the most common site, and are usually the primary focus from which the mite population spreads to other areas of the body (Pl. XI), especially the back, flanks, and abdomen. Many pigs harbour inapparent infections throughout their lives, and the main mode of transmission appears to be between carrier sows and their piglets during suckling. Signs may appear on the face and ears within three weeks of birth, later extending to other areas. Transmission may also occur during service, especially from an infected boar to gilts.

Affected pigs scratch continuously and may lose condition. The first lesions appear as small red papules or weals and general erythema about the eyes, around the snout, on the concave surface of the external ears, in the axillae and on the front of the hocks where the skin is thin. Scratching results in excoriation of these affected areas and the formation of brownish scabs on the damaged skin. Subsequently, the skin becomes wrinkled, covered with crusty lesions and thickened.

For confirmatory diagnosis the most reliable source of material for examination is wax from the ear.

A common control regimen is to treat the sow, the main reservoir of infection, before she goes into the farrowing crate or pen. This procedure will obviously be much more rewarding than having to treat partly grown pigs. Trials have shown that the economic benefit of such a regimen appears at the fattening stage, the offspring of treated sows showing better growth rates and shorter finishing periods than those of untreated sows. Boars should also be routinely treated at six-monthly intervals.

In the treatment of affected pigs, acaricide may be applied weekly, by wash or by spray, until the signs have regressed. Effective preparations which have been available for some time include amitraz, trichlorfon and bromocyclen. Newer and more convenient products with a better residual effect are the systemic organophosphate pour-on, phosmet, and injectable ivermectin. It is recommended that phosmet be applied to the back of the sow 3–7 days before farrowing, a small part of the dose being poured into the ears. Alternatively, ivermectin may be administered at 300 μg/kg.

SARCOPTIC MANGE OF CATTLE

Sarcoptic mange is potentially the most severe of the cattle manges, although many cases are mild. Nevertheless, it is being increasingly diagnosed in Britain and the highly relevant point should be made that in some areas including Canada and parts of the U.S.A., the disease is notifiable and the entry of cattle carrying *Sarcoptes*, whether clinically affected or not, is not permitted.

The mite has partial site preferences which have given it, in the U.S.A., the common name of 'neck and tail mange' (Pl. XI), but it may occur on any part of the body.

Mild infections show merely scaly skin with little hair-loss, but in severe cases the skin becomes thickened, there is marked loss of hair, and crusts form on the less well-haired parts of the body, such as the escutcheon of cows. As in all the sarcoptic manges, there is intense pruritus leading to loss of meat and milk production and to hides being down-graded because of damage by scratching and rubbing.

For confirmatory diagnosis, skin scrapings must be examined carefully since housed cattle, and those fed outdoors from high racks, often carry large populations of innocuous forage mites as well as the less harmful *Chorioptes*. Once encountered, *Sarcoptes*, with its circular outline, short, stump-like legs, and transverse striations, is easily differentiated from *Demodex* and from other cattle mites which have oval bodies and easily visible legs.

Until recently, treatment has largely depended on the use of repeated washes or sprays usually containing gamma HCH. Currently, a single injection of ivermectin is commonly used and has given very good results. Alternatively, the application of a pour-on organophosphate such as phosmet, on two occasions at an interval of 14 days, is also effective. Neither ivermectin nor phosmet are licensed for use in lactating animals whose milk is used for human consumption.

SARCOPTIC MANGE OF SHEEP

Sarcoptic mange has a wide geographic distribution in many sheep-raising areas of the world, such as the Middle East. In Africa it occurs in the local breeds of haired sheep and, because of hide damage, is of considerable economic importance, more than a million sheepskins being exported from the region annually.

Sarcoptic mange of sheep remains notifiable in Britain although it has not been encountered for more than 30 years.

The mite, unlike the non-burrowing genus *Psoroptes*, prefers regions without wool, such as the face, ears, axillae and groin, and has a slow spread. Affected areas are at first erythematous and scurfy. The intense pruritus characteristic of sarcoptic mange is present, and sheep scratch and rub the head, body and legs against trees, posts, and walls. Because of the itch, sheep are almost continuously restless and are unable to graze, so that there is progressive emaciation. In haired sheep the whole body may be affected.

Treatment and control is similar to that described later for the more common psoroptic mange of sheep.

SARCOPTIC MANGE OF GOATS

This form of mange in goats is worldwide in distribution, but is of greatest economic importance in areas where the goat is the basic domestic ruminant such as India and West Africa.

In goats the condition is often chronic, and may have been present simply as 'skin disease' for many months before definitive diagnosis has been made.

As in other sarcoptic infections the main signs are irritation with encrustations, loss of hair, and excoriation from rubbing and scratching. In long-standing cases the skin becomes thickened and nodules may develop on the less well-haired parts of the skin, including the muzzle, around the eyes, and inside the ears.

Repeated treatment is necessary, sometimes over several months in long-standing cases, the common acaricide used being gamma HCH.

Although not licensed for the treatment of goats, whose milk is consumed by man, a single injection of ivermectin is extremely effective.

Corticosteroid therapy has been reported to aid recovery in that it suppresses the pruritus.

SARCOPTIC MANGE OF HORSES

This mange is now uncommon, and though notifiable in Britain only two cases have been recorded since 1948. In both cases there was strong evidence that the infection had been acquired from other domestic species.

SARCOPTIC MANGE OF MAN

Though they have their own distinct 'strain' of *Sarcoptes*, humans readily become infected from domestic animals. Most cases of zoonotic scabies originate from dogs, but outbreaks have also occurred in stockmen closely involved with cattle or pigs.

The areas most often affected are those in direct contact with the animals, including the palms of the hands, the wrists, the arms, and the chest.

Primary scabies of animal origin is much less severe in man than infection with the human strain because the mites do not burrow into the skin and do not multiply. The condition appears as a reddish, papular eruption, with pruritus, which disappears in a few weeks. In people repeatedly exposed to infected animals a hypersensitive state, characterised by a transient rash, may appear within a few hours of animal contact.

Once affected animals have been successfully treated no further human cases occur.

Notoedres

This genus has somewhat similar behaviour and pathogenesis to *Sarcoptes*, but has a more restricted host range.

Hosts:
Cat; it is occasionally seen as a temporary parasite in the dog, and a variety or strain may cause severe mange of the face and head in rabbits

Species:
Notoedres cati

Distribution:
Worldwide

MORPHOLOGY

Notoedres closely resembles *Sarcoptes*, having a circular outline and short legs, but is distinguished by its concentric 'thumb print' striations and absence of spines (Fig. 137).

LIFE CYCLE

Similar to that of *Sarcoptes*, except that the females in the dermis do not occur singly, but are found in aggregations known as 'nests'.

EPIDEMIOLOGY AND PATHOGENESIS

Notoedric mange is highly contagious, but though it occurs in local, limited outbreaks, it is not a frequent cause of skin disease in cats. In Britain, for example, the disease is now rarely encountered.

The infection appears as dry, encrusted scaly lesions on the edges of the ears and on the face, the skin being thickened and somewhat leathery. The associated pruritus is often intense, and there may be severe excoriation of the head and neck from scratching. In typical cases the lesions appear first on the medial edge of the ear pinna, and then spread rapidly over the ears, face, eyelids and neck. It may be spread to the feet and tail by contact when the cat grooms and sleeps.

DIAGNOSIS

This is based on the host involved, the intense pruritus, the location of lesions, and the rapid spread to involve all kittens in a litter. Confirmation is by finding the mites in skin scrapings, which is more readily achieved than in sarcoptic mange, since a single 'nest' in a scraping yields many mites.

TREATMENT

Skin crusts should first be softened with liquid paraffin or soap solution before applying an acaricide. A 1% solution of selenium sulphide is specifically recommended for use in cats since certain compounds such as the organochlorines may prove toxic to cats. Treatment should be given at weekly intervals for 4–6 weeks, the prognosis being good.

Knemidocoptes (syn. *Cnemidocoptes*)

This is the only burrowing genus of domestic birds, and resembles *Sarcoptes* in many respects.

Hosts:
Poultry and cage birds

Distribution:
Worldwide

Species:
The conditions caused by the various species have acquired descriptive common names

Knemidocoptes mutans	poultry	'scaly leg'
K. gallinae	poultry	'depluming itch'
K. pilae	cage birds	'scaly face', 'tassel foot'

MORPHOLOGY

The circular body and short, stubby legs and the avian host are sufficient for generic diagnosis (Fig. 137).

LIFE CYCLE

Similar to *Sarcoptes*, the fertilised females burrowing into the dermis and laying eggs in tunnels.

EPIDEMIOLOGY AND CLINICAL SIGNS

In poultry, *K. mutans* affects the skin beneath the leg scales, causing the scales to loosen and rise, and giving a ragged appearance to the usually smooth limbs and toes (Fig. 138). Lameness and distortion of the feet and claws may be evident.

K. gallinae, responsible for 'depluming itch', burrows into the feather shafts, and the intense pain and irritation cause the bird to pull out body feathers.

K. pilae is most often seen in budgerigars because of their popularity, but other psittacines such as the parrot, parakeet and cockatiel, and finches, such as the canary, are equally susceptible. *K. pilae* attacks the bare and lightly-feathered areas, including the beak, cere, head, neck, inside of wings, legs and feet. The mites are deep in the skin, but unlike *Sarcoptes* cause little pruritus. Lesions develop slowly, over a number of months.

Infection may remain latent for a long time with a small static mite population until stress, such as chill or movement to a strange cage, occurs when the population increases.

On the head, the first change is a scaliness at the angle of the beak which spreads over the face ('scaly face') affecting the cere and horny tissue of the beak (Fig. 138). The beak may be distorted due to the mites burrowing in the matrix, and crossbeak may develop. If the matrix is destroyed there is no prospect of recovery. When the limbs are affected, an extreme form of scaly leg may develop and in severely affected birds toes may slough.

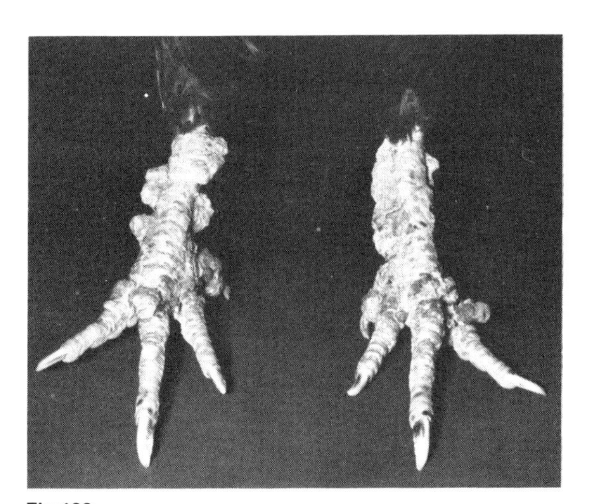

Fig 138
(a) 'Scaly leg', due to *K. mutans*.

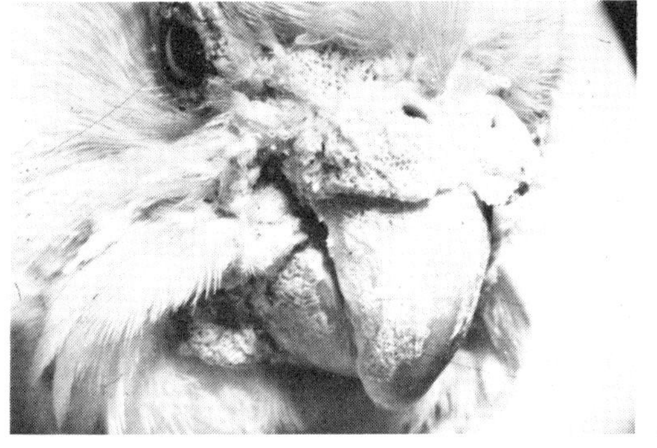

(b) 'Scaly face', due to *K. pilae*.

TREATMENT

For the treatment of mite infestation in poultry the acaricides most widely used are carbaryl or gamma HCH applied either as a dust or spray, paying particular attention to the undersides of the wings. For 'scaly leg' the legs should be dipped into the acaricide solution. The treatment should be repeated several times at 10 day intervals.

The poultry house should be thoroughly cleaned and the perches and nesting boxes sprayed with acaricide.

The treatment of cage birds is similar except that the acaricide is usually applied locally. Oil-based canine ear drops are useful, being easy to dispense and simple to apply.

Family DEMODICIDAE

Demodex

Mites of this genus belong to a separate family, the Demodicidae which, although burrowing mites, are quite different in form and behaviour from the Sarcoptidae.

Hosts:
All domestic mammals and man

Distribution:
Worldwide

Location:
Hair follicles and sebaceous glands

Species:
With the exceptions of *D. phylloides* of pigs, and *D. folliculorum* of man, all species of *Demodex* are named after their hosts, eg. *D. canis, D. bovis, D. equi*, etc.

MORPHOLOGY

Demodex has an elongate tapering body, up to 0.2 mm long, with four pairs of stumpy legs anteriorly (Fig. 139).

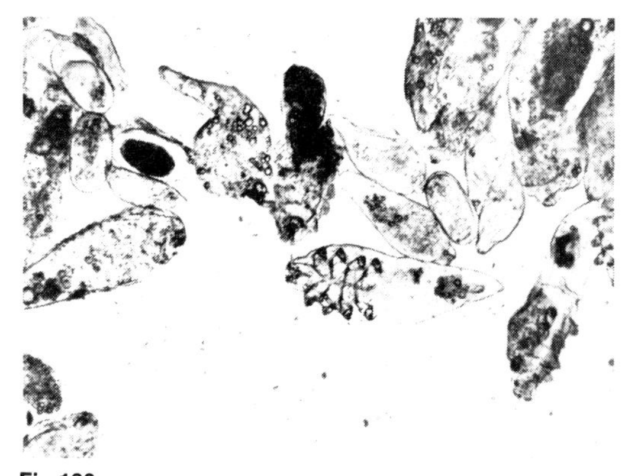

Fig 139

(a) Typical appearance of *Demodex* in skin scraping.

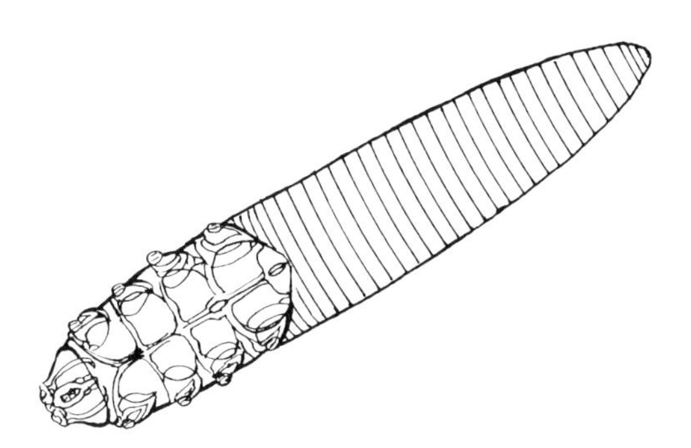

(b) Adult *Demodex*.

LIFE CYCLE

Demodex spp. live as commensals in the skin of most mammals, and are exceptional in being selective for particular skin sites, namely the hair follicles and sebaceous glands.

Most species spend the entire life cycle in the follicles or glands, in each of which they occur in large numbers in a characteristic head-downward posture. In the newborn and very young animal these sites are simple in structure, but later they become compound by outgrowths. The mites then move into the extended habitats, going much deeper into the dermis than the sarcoptids and hence being much less accessible to surface-acting acaricides.

This form of mange is best documented in the dog, but the pathogenesis and epidemiology in other animals suggests that their infections may have much in common with canine demodicosis.

DEMODECTIC MANGE OF DOGS

EPIDEMIOLOGY

Probably because of its location deep in the dermis, it is almost impossible to transmit *Demodex* between animals unless there is prolonged contact.

In nature such contact occurs only during suckling and it is thought that most infections are acquired in the early weeks of life, the commensal population in the skin of the bitch being transferred to the in-contact areas of the pup. Certainly, it is in these areas, the muzzle, face, periorbital region and forelimbs that lesions first appear (Pl. XII).

PATHOGENESIS

Early in infection there is a slight loss of hair on the face and forelimbs, followed by thickening of the skin, and the mange may progress no further than the in-contact areas; many of these localised mild infections resolve spontaneously without treatment. On the other hand, lesions may spread over the entire body, and this generalised demodicosis may take one of two forms:

(i) **Squamous** demodicosis is the less serious. It is a dry reaction, with little erythema, but widespread alopecia, desquamation and thickening of the skin. In some cases of this type only the face and paws are involved.

(ii) **Pustular** or follicular demodicosis is the severe form, and follows bacterial invasion of the lesions, often by staphylococci. The skin becomes wrinkled and thickened, with many small pustules from which serum, pus, and blood ooze, giving this form its common name of 'red mange' (Pl. XII); affected dogs have an offensive odour. Prolonged treatment is necessary, and survivors may be severely disfigured, so that early euthanasia is sometimes requested by owners, and especially by pedigree breeders.

A notable feature of all types of demodectic mange is the absence of pruritus.

The pathogenesis of *Demodex* is more complex than that of other mange mites because immune factors appear to play a large part in its occurrence and severity. It is thought that certain bitches carry a genetically transmitted factor which results in immunodeficiency in their offspring, making them more susceptible to invasion, and it has been observed that litter mates from such a bitch often develop the generalised form simultaneously, even though they have been reared separately. In addition, *Demodex* itself is thought to cause a cell-mediated immunodeficiency which suppresses the normal T-lymphocyte response; this defect disappears when the mites have been eradicated from the animal. Demodectic mange may erupt when dogs are given immunosuppressants for other conditions.

DIAGNOSIS

For confirmatory diagnosis, deep scrapings are necessary to reach the mites deep in the follicles and glands, and this is best achieved by taking a fold of skin, applying a drop of liquid paraffin, and scraping until capillary blood appears. Even in normal dogs a few commensal mites may be found in the material, but the presence of a high proportion of larvae and nymphs (Fig. 139) will indicate a rapidly increasing population, and hence an active infection. Skin biopsy, to detect mites in the follicles, has been used in severely affected dogs, but is rarely necessary.

In controlling the endemicity of demodicosis, it should be noted that since certain bitches are more prone than others to have susceptible offspring, it may be advisable to discard these from breeding establishments.

TREATMENT

With their deep location in the dermis the mites are not readily accessible to topically-applied acaricides, so that repeated treatment is necessary and rapid results should not be expected. In localised squamous mange recovery may be expected in 1–2 months, but in the generalised pustular form the prognosis should indicate that recovery will take at least three months, and should, even so, be guarded.

Before commencing specific treatment the dog should be clipped, washed with an anti-seborrhoeic shampoo and thoroughly dried. Of the available acaricides the most widely used are rotenone and the organophosphates, ronnel and cythioate. It is usually recommended that an alcoholic solution of rotenone is applied to half the affected area on one day and the other half the following day to minimise toxic effects. Treatment should be continued at weekly intervals for 4–5 weeks.

The application of ronnel to the entire body also requires care and should be limited to every four days. A common practice is to treat a different quarter of the body surface daily for four days. Treatment should be continued for a minimum of eight weeks. Sometimes, concurrent daily oral therapy with ronnel or cythioate may be recommended. Success of treatment should be monitored by regular skin scrapings and continued until three successive weekly samples are negative.

More recently, highly successful treatment has been reported using one or more applications of the formamidine compound, amitraz, at 14 day intervals.

Where the lesion is mild and localised it may be treated by local application of rotenone and where pyoderma is severe antibiotic therapy may be necessary.

DEMODECTIC MANGE OF CATS

Demodicosis is rare in cats. It takes a localised, self-limiting form, confined to the eyelids and periocular region, and is of the mild squamous type, with some alopecia. It may be treated by local application of rotenone.

DEMODECTIC MANGE OF CATTLE

The most important effect of bovine demodicosis is the formation of many pea-sized nodules, each containing caseous material and several thousand mites, which cause hide damage and economic loss. Though these nodules can be seen in smooth-coated animals they are often undetected in rough-coated cattle until the hide has been dressed.

In some parts of Australia 95% of hides are damaged, and surveys in U.S.A. have shown a quarter of the hides to be affected. In Britain 17% of hides have been found to have *Demodex* nodules.

As in the dog, transmission appears to occur during the early days of suckling, and the muzzle, neck, withers and back are common sites.

Control is rarely applied since there is little incentive for farmers to treat their animals, the cost of damage being borne by the hide merchant. If treatment is desired, 'pour-on' organophosphates or parenteral ivermectin may be tried.

DEMODECTIC MANGE OF SHEEP

This form of mange is rare in sheep and is of little economic importance, being confined to the face region and being mild in character.

DEMODECTIC MANGE OF GOATS

Demodicosis of goats is worldwide in distribution, and though formerly of greatest importance in warm countries, it is being increasingly diagnosed in Europe.

The disease is similar to that in cattle. The initial lesions on the face and neck extend to the chest and flanks and may eventually involve the whole body, with the formation of cutaneous nodules of up to 2.0 cm in diameter containing yellowish, caseous material with large numbers of mites.

This form of mange is rarely debilitating, and is of greatest importance as a cause of down-grading or condemnation of goat skins. Where treatment is attempted 'pour-on' organophosphates or parenteral ivermectin may be used.

DEMODECTIC MANGE OF PIGS

This mange is rare in pigs though sporadic incidences of up to 5% have been noted in eastern European countries. It is usually confined to the head, where there is pustule formation and thickening of the skin.

DEMODECTIC MANGE OF HORSES

In the horse demodectic mange is rare, but may occur either as the squamous or the pustular type, affecting initially the muzzle, forehead, and periocular area.

Family LAMINOSIOPTIDAE

Laminosioptes

Laminosioptes cysticola occurs in domestic poultry and pigeons in most parts of the world. Aggregations of these small, oval mites are found in yellow caseous nodules of several mm diameter in the subcutaneous muscle fascia and in deeper tissues in the lungs, peritoneum, muscle and abdominal viscera. The life cycle is unknown.

The subcutaneous nodules are often calcified and contain only dead mites, active mites occurring in the deep tissues.

Laminosioptes is never associated with clinical signs, and is only discovered at meat inspection, when infected carcasses are condemned partly on aesthetic grounds and partly because the infection appears somewhat similar to avian tuberculosis.

NON-BURROWING MITES

The mites in this category are diverse in biology and are classified in many different groups. Their common feature is that they do not burrow into the dermis, but feed superficially. Some feed solely on skin scales, but a few also suck tissue fluid from the skin, and several are bloodsuckers. In their behaviour they include species which live permanently on the skin surface, others live mainly in the hair or fur and a few genera which, like the ticks, visit the host only to feed. Others spend their adult existence as free-living acarines, but must feed on animals in their larval phase. The main diagnostic features of the important species are indicated in Fig. 140.

Family PSOROPTIDAE

Psoroptes

Hosts:
Sheep, cattle, equines

Species:

Psoroptes ovis	sheep and cattle
P. equi	equines
P. cuniculi	equines and rabbits

Distribution:
Worldwide

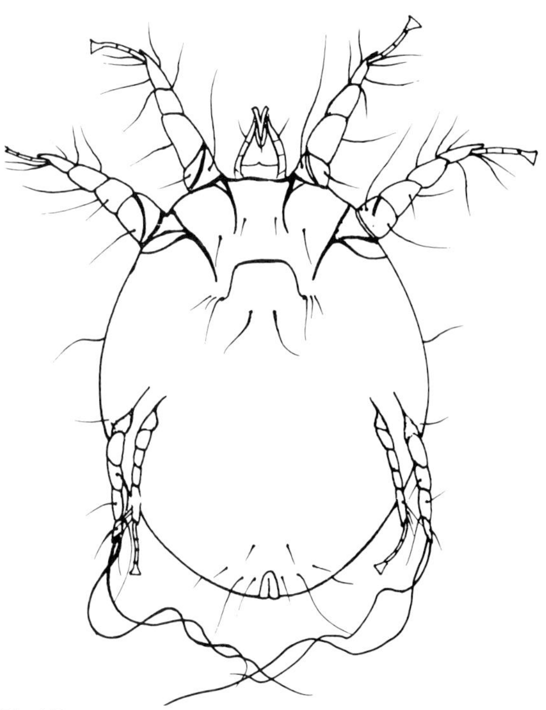

Fig 140
(a) Ventral view of female *Psoroptes* mite.

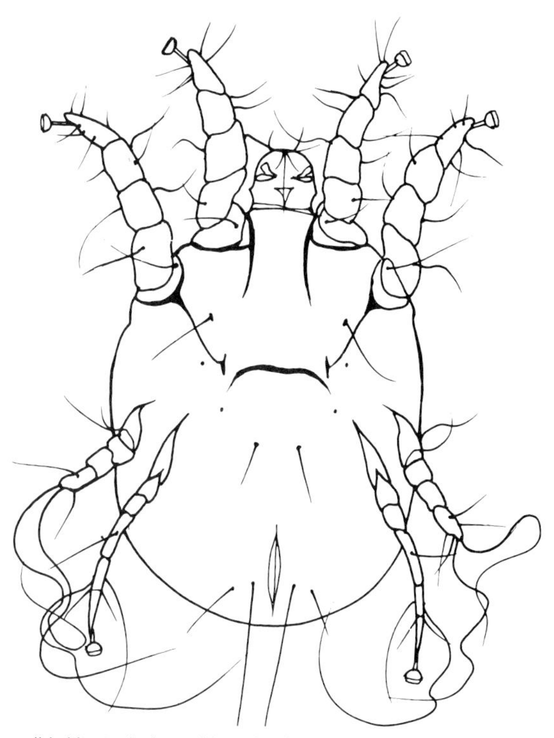

(b) Ventral view of female *Chorioptes* mite.

Fig 140 (continued)

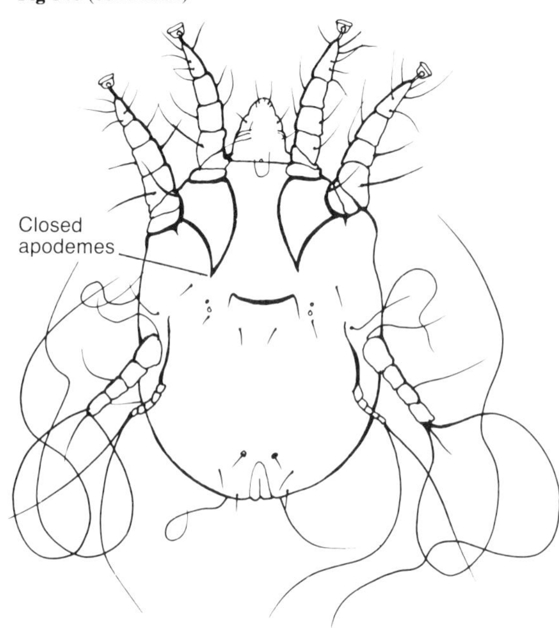

(c) Ventral view of female *Otodectes cynotis*.

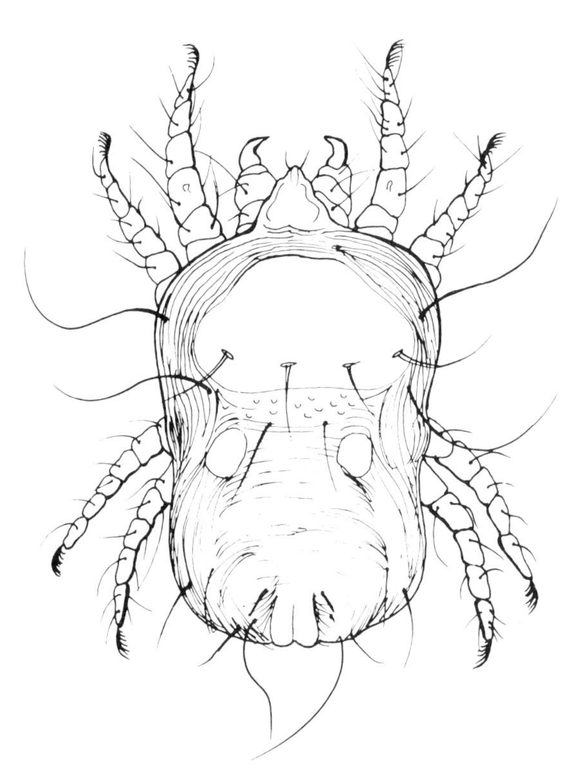

(d) Dorsal view of female *Cheyletiella*.

MORPHOLOGY

Psoroptes is a typical non-burrowing mite, up to 0.75 mm, oval in shape, and with all the legs projecting beyond the body margin. Its most important recognition features are the pointed mouthparts, the rounded abdominal tubercles of the male, and the 3-jointed pedicels bearing funnel-shaped suckers on most of the legs (Figs. 140 and 141).

LIFE CYCLE

In general, typical of the non-burrowing mites. The female lays about 90 eggs during her lifetime of 4–6 weeks, and development from the egg, through the larval and nymphal stages to mature adult, takes about 10 days. The greater pathogenicity of this mite is attributable to the fact that unlike most non-burrowing mites, it has piercing and chewing mouthparts which can severely damage the skin.

PSOROPTIC MANGE OF SHEEP (SHEEP SCAB)

EPIDEMIOLOGY AND PATHOGENESIS

The distribution of mites on the animals varies according to season, the infection being in a quiescent state in reservoir sites such as the axilla, groin, infra-orbital fossa, and inner surface of the pinna and auditory canal during spring, summer and early autumn, and spreading rapidly over the rest of the body in the colder months as the fleece thickens.

Though it is a non-burrowing mite, *Psoroptes* is very active in the keratin layer, and causes direct damage to the skin. The earliest phase of infection occurs as a zone of inflammation with small vesicles and serous exudate, but as the lesion spreads, the centre becomes dry and covered by a yellow crust while the borders, in which the mites are multiplying, are moist (Pl. XII). The first visible sign is usually a patch of lighter wool, but as the area of damage enlarges the sheep responds to the intense itching associated with mite activity by rubbing and scratching against fence posts and other objects, so that the wool becomes ragged and stained, and is shed from large areas (Fig. 141).

In addition to wool loss, the sheep are so restless and pre-occupied in scratching that they almost cease to feed, and in growing animals weight gains may be suppressed, while in adults there may be weight loss. In very severe infestations animals may even succumb, being so debilitated that they cannot compete for winter feed.

Though the majority of sheep become infected while the mites are active and multiplying, the quiescent phase is also very significant in the epidemiology of the disease, for apparently normal sheep carrying this phase in reservoir sites may be introduced to healthy flocks dur-

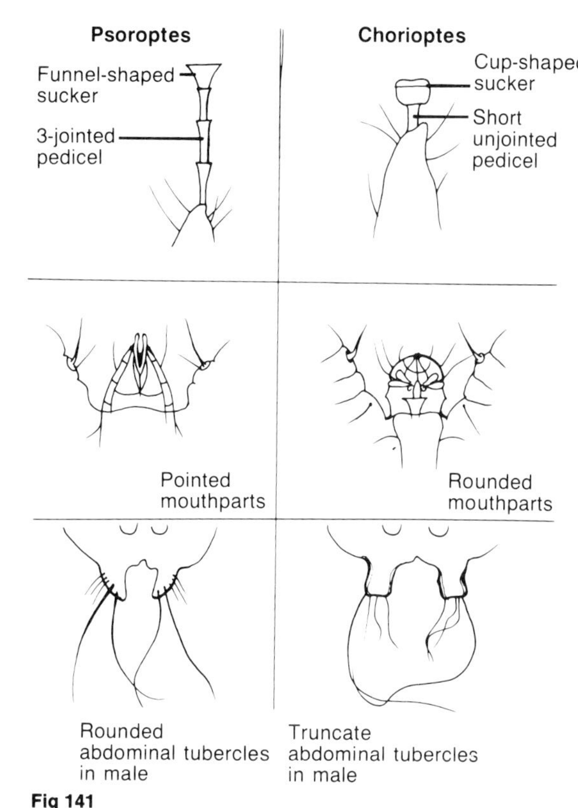

Fig 141

(a) Differential features of *Psoroptes* and *Chorioptes*.

(b) Pruritus and wool loss characteristic of sheep scab.

ing summer and autumn, and initiate outbreaks when the cold season arrives. There is a short-term risk of infection from pens and houses, but off the host, most mites die within a week, and premises may be presumed free of infection three weeks after removal of sheep.

DIAGNOSIS

Initial diagnosis is based on the season of occurrence and the signs of wet, discoloured wool, debility, and intense pruritus, with an easily elicited nibbling reflex.

Confirmatory diagnosis is made by identification of the mites. Material should be scraped from the edge of a lesion, placed in warm 10% potassium hydroxide, and examined microscopically.

As a non-burrowing mite *Psoroptes* has a generally oval outline, with all the legs projecting beyond the body margin. Another burrowing mite, *Chorioptes*, can be common in sheep, and it is essential that this relatively harmless mite should be differentiated from the pathogenic *Psoroptes*. The important differential features are shown in Fig. 141.

TREATMENT AND CONTROL

Because of its short population turnover period of 10 days there is very rapid spread, and it is this character which has led to legislative control in many countries since the economic consequences of uncontrolled sheep scab are serious. The disease was presumed to have been eradicated from the United Kingdom in 1952, there having been no notifications of outbreaks for a number of years; it reappeared in 1973, most probably having been introduced as the quiescent phase in imported sheep, and regulations were reimposed. It was eradicated from Australia and New Zealand many years ago, but remains notifiable in these countries. Legislation in support of control is based on inspection of flocks, limitation of movement of sheep in, and from, areas in which the infection has been diagnosed, and compulsory dipping of all sheep at prescribed times. In Britain, at present, two annual dips are mandatory, one in summer and one in autumn. In addition to compulsory dipping, British legislation requires any suspected outbreaks to be notified to the authorities. When a flock is found to be infected it is required that all sheep are dipped at 3-weekly intervals until no mites can be found in skin scrapings. Legislation also provides for the control of movement of sheep from surrounding farms. This is designated as an 'infected area' and it is forbidden to move sheep from there into 'clean' areas or to market unless they have been dipped within the previous four weeks or are going direct for slaughter.

Though several methods of applying acaricide, such as showering, have been tested, only plunge-dipping is generally recommended for sheep scab control, and in Britain definite directions are given for the process. Sheep must remain in the bath for at least one minute, and the head should be immersed at least twice. Sheep should be held in clean pens before dipping and it is customary to hold them in draining pens for a time afterwards to conserve dip. Modern acaricides have been developed which have an affinity for wool grease, so that as a succession of sheep go through the bath the acaricide is gradually 'stripped out', and manufacturers give directions for replenishment after a specified number of sheep have been dipped. It is usually recommended that the replenishing dip should have an acaricide concentration at least 1.5 times greater than the original, to replace the selective loss.

In most countries in which control is practised, only specified acaricides are permitted for use in dips. For many years only gamma BHC, now designated gamma HCH (hexachlorocyclohexane), was used, but has now been largely replaced by the organophosphates, diazinon and propetamphos, which in addition to giving the required persistence in the fleece, are rapidly detoxified and excreted from tissues, whereas gamma HCH leaves residues in the meat.

Though under test conditions injectable ivermectin has given complete clearance of *Psoroptes ovis*, it is not yet widely licensed for this purpose.

PSOROPTIC MANGE OF OTHER LIVESTOCK

Bovine psoroptic mange is of increasing importance in U.S.A. and parts of Europe, attributable mainly to more intensive husbandry methods.

The main effect is a pruritus caused by the biting and sucking activity of the mites which results in the formation of vesicles, the exudate drying on the skin to form a crust. Cattle are extremely restless, and in the U.S.A. it has been observed that the feed intake of infected animals may drop by 20%, so that up to three months' additional feeding may be necessary for them to reach their required weights. In Germany, extreme debility and some deaths have occurred in animals under a year old.

The skin areas most often affected in cows are the abdomen, tail-head and escutcheon, and in bulls the abdomen, tail-head and prepuce. These locations would suggest that one common mode of transfer of infection is when cattle mount each other.

Treatment is as for bovine sarcoptic mange using parenteral ivermectin, organophosphates or organochlorines. Ivermectin is most effective although treated animals remain contagious for 5 days after treatment. Pens should be disinfected and vacated for 2 weeks before restocking.

Though the species in cattle is *P. ovis*, the likelihood of transfer of infection from these animals to sheep would seem to be slight since in Britain, where twenty years of freedom from sheep scab was enjoyed from 1952, infected cattle were continuously present. It would appear that, as in some other manges, biological adaptation has resulted in restriction of the host range. Experimentally, infection has been transferred between sheep and cattle and vice versa, but in neither animal did the heterologous 'strain' persist for more than a few weeks.

Equine psoroptic mange due to *P. equi* remains notifiable in Britain, but it has not been officially recorded for man years. In the U.S.A. and Australia, *Psoroptes* has been found in association with a 'head-shaking' syndrome, in which infestation of the ears with what appears to be an equine strain of *P. cuniculi* causes acute irritation, so that horses adopt a lop-eared posture and resent being bridled. In Australia, where 'head-shaking' is relatively common, one survey has shown that 20% of horses harbour *P. cuniculi* in their ears.

Preparations containing gamma HCH applied daily for four days, and repeated in 10 days have been found effective.

In **rabbit** colonies, *P. cuniculi* also localises in the ears, where the mites are usually quiescent, but occasionally proliferate causing severe mange in which the auditory canal may be completely blocked with greyish debris; the infection may extend over the rest of the body with scabs, loss of hair, and excoriation from scratching.

Treatment is as for otodectic mange of cats and dogs.

Chorioptes

This genus and *Otodectes*, the next to be considered, feed only superficially. Unlike *Psoroptes* they have mouthparts which do not pierce the skin, but are adapted solely for chewing, feeding on shed scales and other skin debris.

Hosts:
Cattle, sheep, goats and equines

Distribution:
Worldwide

Species:
Although specific names have been given to *Chorioptes* found in cattle, sheep and equines (*C. bovis*, *C. ovis*, *C. equi*) they are now all considered to belong to the single species, *C. bovis*

MORPHOLOGY

See illustrations (Figs. 140 and 141), for comparison with *Psoroptes*. The mouthparts are distinctly rounder, the abdominal tubercles of the male are noticeably truncate and the pedicels are short and unjointed, with cup-shaped suckers.

LIFE CYCLE

Similar to that of *Psoroptes*, except that this mite feeds only on the skin surface.

EPIDEMIOLOGY AND PATHOGENESIS

In **cattle**, chorioptic mange occurs most often in housed animals, affecting mainly the neck, tail-head, udder and legs, and usually only a few animals in a group are clinically affected. It is a mild condition, and lesions tend to remain localised, with slow spread. Its importance is economic, the pruritus caused by the mites resulting in rubbing and scratching, with damage to the hide. The treatment is the same as for sarcoptic mange in cattle.

In **sheep**, the mites are found mainly on the legs, and though very common, little harm is caused. Lambs are thought to become infected by contact with the legs of the ewe. In some cases there may be spread from the limbs to the face and other regions, and in occasional severe cases pustular dermatitis, with wrinkling and thickening of the skin, may occur.

It has been noted in New Zealand that when the mange spreads to the scrotum the thickened and inflamed skin allows the scrotal temperature to remain high with, as a result, testicular atrophy and cessation of spermatogenesis. Infected rams have impaired reproductive ability or sterility, though their general health is not affected. The condition is not irreversible, and semen production and fertility return to normal after successful mange treatment.

Chorioptic mange in sheep is easily treated by dipping or by local treatment with an organophosphate acaricide.

Equine chorioptic mange occurs as crusty lesions with thickened skin on the legs below the knees and hocks, and is most prevalent in rough-legged animals and in those with heavy feather (Pl. XII). Though the mites are active only superficially their movement causes irritation and restlessness, especially at night when animals are housed and minor injuries may occur in the fetlock region from kicking against walls.

TREATMENT

A suitable acaricidal wash, such as bromocyclen, scrubbed on to the lesions on two occasions, 14 days apart, is effective.

Otodectes

This is the commonest mange mite of cats and dogs throughout the world.

Hosts:
Cat and dog; it occurs in a number of other small mammals including the ferret and red fox

Species:
Otodectes cynotis

MORPHOLOGY

Otodectes resembles *Psoroptes* and *Chorioptes* in general conformation, having an ovoid body and projecting legs. The most obvious distinguishing features, apart from the preferred location in the external ear of the host, are the closed apodemes adjacent to the first and second pairs of legs. The pedicels, like those of *Chorioptes*, are unjointed (Fig. 140).

LIFE CYCLE

Like *Chorioptes*, this mite feeds superficially, and a complete cycle occupies three weeks.

EPIDEMIOLOGY AND PATHOGENESIS

Cat

Most cats harbour this mite, and in adult animals it has almost a commensal association with the host, signs of irritation appearing only sporadically with the transient activity of the mites. It is supposed that the majority of infections are acquired by suckling kittens from their carrier dams and, being highly contagious, entire litters are affected.

Early in infections there is a brownish waxy exudate in the ear canal, which becomes crusty, the mites living deep in the crust, next to the skin. Secondary bacterial infection may result in purulent otitis.

The major signs are frequent shaking of the head and scratching of the ears from the pruritus, the presence of foetid waxy masses in the auditory canal, and on inspection of severe cases, otorrhea and ulceration of the auditory canal. Pus may be present, from secondary bacterial infection.

Scratching may cause excoriation of the posterior surface of the ear pinna, and this, with head-shaking, may result in a haematoma of the ear flap.

Dog

Otodectes is a common cause of otitis externa in dogs.

The changes and signs are similar to those in the cat, with brownish to black waxy deposits and exudate in the ear canal (Pl. XII), and intense pruritus. The resultant violent head-shaking and ear-scratching are a common cause of aural haematomata in this species. In long standing cases a severe purulent otitis is a common sequel.

DIAGNOSIS

Tentative diagnosis is based on the behaviour of the animal and the presence of dark, waxy deposits and exudate in the ear canal.

Confirmation depends on observing the mites either within the ear by an auroscope or, more simply, by removing some of the deposit and exudate and placing it on a dark surface where the mites will be seen by a hand lens as whitish moving specks.

TREATMENT

There are many effective preparations available commercially as ear drops, most of them including, in addition to an acaricide, antibiotics, fungicides, corticosteroids and local analgesics. The acaricide components include gamma HCH, piperonyl butoxide and rotenone.

The ear canal should first be thoroughly cleaned, and after the ear drops have been instilled, the base of the ear massaged to disperse the oily preparation.

Whatever preparation is used, treatment should be repeated in 10–14 days to kill any newly hatched mites.

In view of the ubiquity and high infectivity of the mite, all dogs or cats in the same household, or those in close contact in kennels and catteries, should be treated at the same time as clinically affected animals.

Family CHEYLETIDAE

Psorergates

This is the 'itch mite' of sheep.

Host:
Sheep

Distribution:
Australia, New Zealand, southern Africa, North and South America. It has not been reported in Europe

Species:
Psorergates ovis

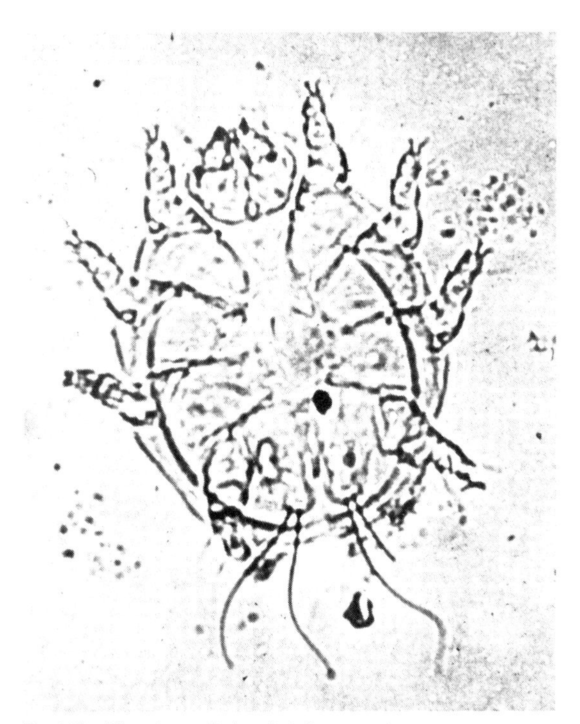

Fig 142 The sheep 'itch mite' *Psorergates*.

MORPHOLOGY

P. ovis is a small mite, roughly circular in form and less than 0.2 mm in diameter. The legs are very stout, with their bases adjacent, and are directed radially, giving the mite a crude star shape (Fig. 142).

LIFE CYCLE

Similar to that of *Psoroptes*, the mites feeding on the skin itself.

EPIDEMIOLOGY AND PATHOGENESIS

Infection is most common in fine wool breeds such as the Merino and Corriedale and is acquired by contact when the wool is short; as the fleece lengthens, it presents a barrier to the transfer of mites. The spread of the mite population is very slow, and infestation is rarely found in animals less than six months old; it may be three years or more before the whole fleece area is affected.

Though a 'non-burrowing' mite, *Psorergates* attacks the skin itself, living in the superficial layers and causing chronic irritation and skin thickening.

The earliest signs are small, pale areas of wool on the shoulders, body and flanks which gradually extend over the rest of the fleece, irritation increasing as the mite population grows. Sheep rub, bite and chew their wool, which becomes ragged, with loose strands trailing from the sides of the body. In long-standing cases large patches of wool may be lost.

The fleece itself contains much scurf and has a slightly yellowish hue while the staple is very dry and easily broken.

In severe cases the whole fleece, which is difficult to shear because of its matted consistency, must be discarded. In less severely affected sheep, and especially in older animals which have become tolerant of the itch because of their thickened, damaged skin, fleeces are down-graded.

DIAGNOSIS

To obtain mites it is necessary, having clipped away a patch of wool, to apply a drop of mineral oil and scrape the skin down to the blood capillary level. The mites themselves are easily identified. In endemic areas all stockowners are familiar with the appearance of sheep infested with 'itch mite'.

TREATMENT AND CONTROL

Psorergates is relatively insusceptible to most acaricides, although the formamidine, amitraz, has recently been shown to be of considerable value. Otherwise the older arsenic-sulphur preparations may be used. Sheep should be dipped soon after shearing.

Cheyletiella

This genus occurs mainly in dogs and cats, but is also a frequent parasite of rabbits. It has some public health importance as it will readily transfer from pets to humans.

Hosts:
Dog, cat and rabbit; man is a frequent erratic host

Species:

Cheyletiella yasguri	dog
C. blakei	cat
C. parasitivorax	rabbit

Distribution:
Worldwide

MORPHOLOGY

The body of the mite, up to 0.4 mm long, has a 'waist'. The palps are greatly enlarged, giving the appearance of an extra pair of legs, and each palp ends in a prehensile claw (Fig. 140).

There are six long hairs or setae on the body, one on each side of the terminal anus, and two on each side of the body between the second and third pairs of legs.

The legs terminate in 'combs' instead of claws or suckers.

LIFE CYCLE

The mites live in the hair and fur, only visiting the skin to feed. The life cycle itself is similar to those of *Psoroptes* and *Chorioptes* and is completed in about two weeks.

EPIDEMIOLOGY AND PATHOGENESIS

This highly contagious, though mild, mange can spread rapidly through catteries and kennels.

The mite is not highly pathogenic, and is often found in young kittens and pups in good physical condition. It is a characteristic of the dermatitis which it causes that many skin scales are shed into the hair (Pl. XII) or fur, giving it a powdery or mealy appearance, and the presence of moving mites among this debris has given it the common name of 'walking dandruff'. There is very little skin reaction or pruritus. In the rare severe case, involving much of the body surface, crusts are formed, but there is only slight hair loss.

Of all the mite infestations of domestic animals this is the most readily transferable to man, *C. blakei* being most often involved, probably because of the close physical relationship between owners and cats.

The mites can penetrate clothing and are easily transferred, even on short contact. It is often found that when a positive diagnosis has been made on a pet, there is a history of persistent skin rash in the owner's family. In contrast to the condition in its natural hosts, the infestation in humans causes severe irritation and intense pruritus. The early sign is an erythema which may progress to a vesicular and pustular eruption. Cases in humans invariably clear up spontaneously when the animal source has been treated.

DIAGNOSIS

In any case of excessive scurf or dandruff in a dog or cat *Cheyletiella* should be considered in the differential diagnosis.

On parting the coat along the back, and especially over the sacrum, scurf will be seen, and if this is combed out on to dark paper the movement of mites will be detected among the debris. Scraping is not necessary as the mites are always on the skin surface or in the coat.

TREATMENT AND CONTROL

Dogs should be treated with a shampoo containing an insecticide such as gamma HCH, while selenium sulphide shampoos are recommended for cats. The condition is usually cleared by three successive weekly treatments.

Where the potential for reinfection is high, as in kennels, catteries and pet shops, dichlorvos fly strips placed in the accommodation have been found useful.

Family DERMANYSSIDAE

Dermanyssus

The sole species of veterinary interest is the common 'red mite' of poultry.

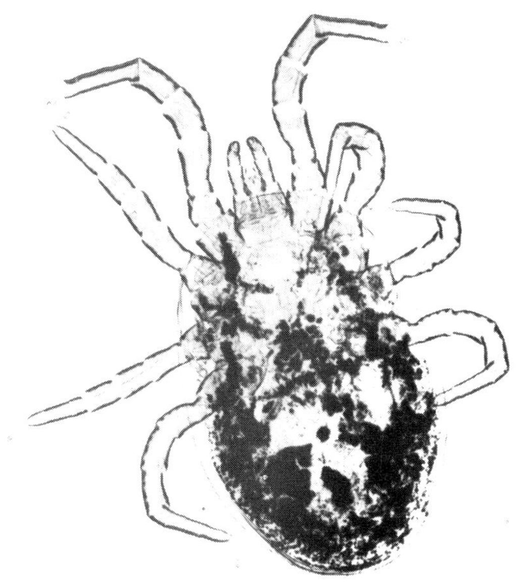

Fig 143 The poultry 'red mite' *Dermanyssus gallinae*.

Hosts:
Domestic poultry and wild birds; it is occasionally parasitic on mammals, including man

Species:
Dermanyssus gallinae

Distribution:
Worldwide

MORPHOLOGY

Dermanyssus is a small mite, with a body length of about 1.5 mm. It has long legs and is somewhat spider-like (Fig. 143).

The colour is white to greyish-black, becoming red when engorged with blood.

LIFE CYCLE, EPIDEMIOLOGY AND PATHOGENESIS

This mite spends much of its life cycle away from its host, the adult and nymph only visiting birds to feed, mainly at night. The favoured habitats are poultry houses, usually of timber construction, in the crevices of which the eggs are laid, the cycle being completed in a minimum of a week. The adult can survive for several months without feeding, so that a reservoir population can persist in unoccupied poultry houses and aviaries.

The mite is especially a threat to battery fowl housed in old buildings.

The feeding nymphs and adults cause irritation, restlessness and debility, and in heavy infections there may be severe, and occasionally fatal, anaemia.

In Australia it is a vector of *Borrelia anserina*, the cause of avian spirochaetosis.

Dermanyssus readily infects other animals, and can cause erythema and intense pruritus in cats which occupy old wooden poultry houses. Humans may develop skin lesions when mites enter rooms from wild birds' nests in the eaves of houses.

TREATMENT AND CONTROL

Treatment of birds is only palliative, and attention should be paid to the mite habitats in buildings. These should be cleaned, scalded with boiling water and treated with an acaricide such as carbaryl or synergised pyethroids. Where the mites are a problem in dwelling houses their ability to survive in nests, without feeding for several months, makes these important as reservoir sites, and all nests should be removed from eaves once the fledglings have departed; an alternative measure is to place dichlorvos-impregnated strips in the eaves.

Individual birds may be treated with an acaricide such as pyrethrum or carbaryl.

Ornithonyssus (syn. *Liponyssus*)

This genus is closely related to *Dermanyssus*. Its most important species, *Ornithonyssus sylvarium*, is the 'northern mite' of birds.

Hosts:
Poultry and wild birds

Species:
Ornithonyssus sylvarium
O. bursa

Distribution:
O. sylvarium is present in temperate areas throughout the world, and is replaced by *O. bursa* in the tropics

MORPHOLOGY

This rapidly moving, long-legged genus has an oval body, about 1.0 mm long, and the colour varies from white to reddish black depending on the amount of blood which it contains. The body carries many long setae and is much more hairy than *Dermanyssus*.

LIFE CYCLE

Unlike *Dermanyssus, Ornithonyssus* spends its entire life on the bird and can only survive for about 10 days away from a host.

EPIDEMIOLOGY AND PATHOGENESIS

Being almost a permanent parasite, infection is by contact or by placing birds in accommodation recently vacated by infected stock. This mite occurs not only on poultry, but in cage-bird colonies, and there is a permanent reservoir in wild birds. In heavy infections birds are restless and lose weight from irritation, egg production may be reduced, and there may be severe anaemia. Common signs, apart from debility, are thickened, crusty skin, and soiled feathers around the vent.

TREATMENT AND CONTROL

Dusts of carbaryl or organophosphorus preparations such as ronnel may be applied to the birds and their nesting boxes.

Partially parasitic mites

Mites of the family Trombiculidae are parasitic only at the larval stage, the nymphs and adults being free-living. There are two common genera, *Neotrombicula*, the 'har-

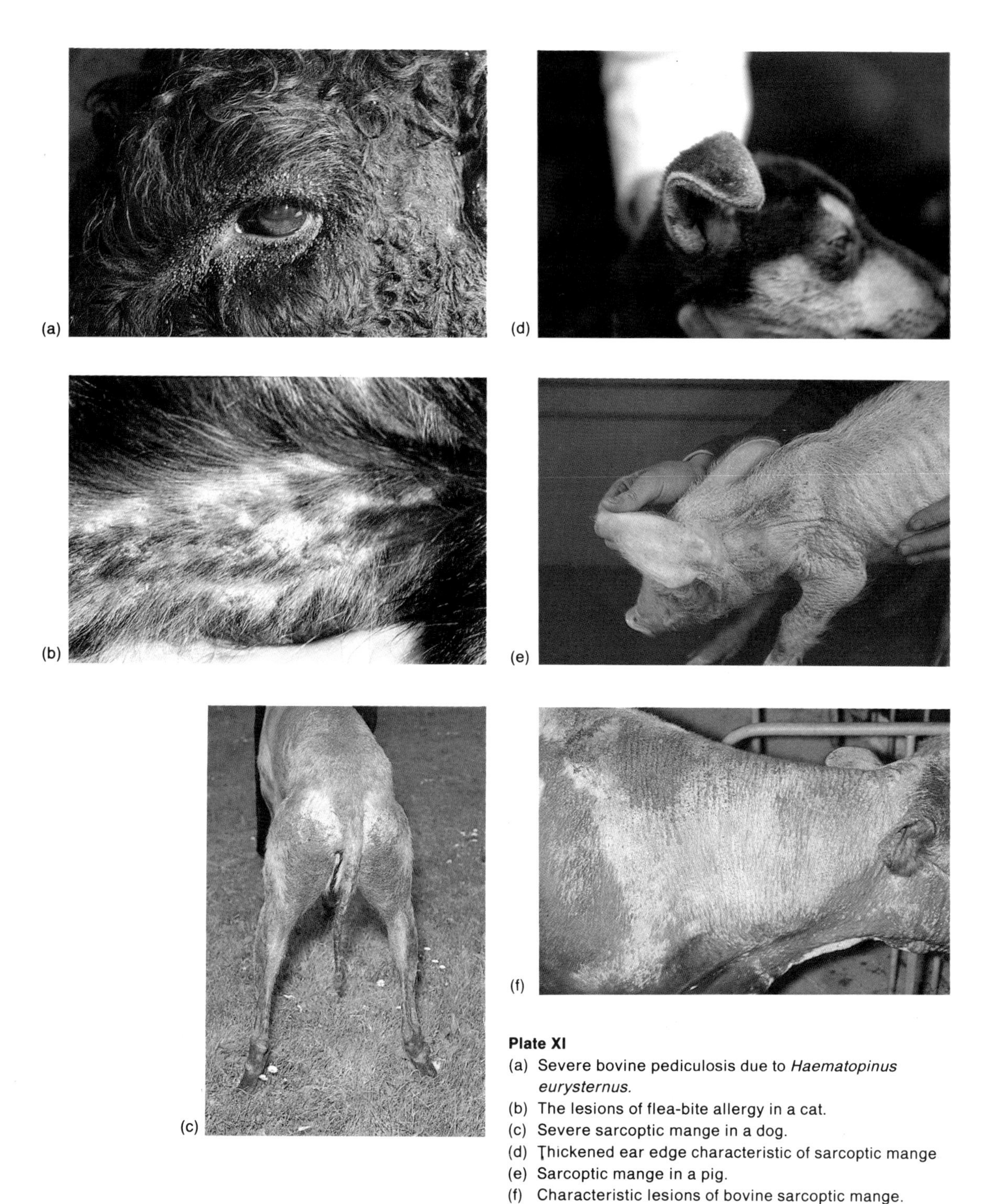

Plate XI

(a) Severe bovine pediculosis due to *Haematopinus eurysternus*.
(b) The lesions of flea-bite allergy in a cat.
(c) Severe sarcoptic mange in a dog.
(d) Thickened ear edge characteristic of sarcoptic mange
(e) Sarcoptic mange in a pig.
(f) Characteristic lesions of bovine sarcoptic mange.

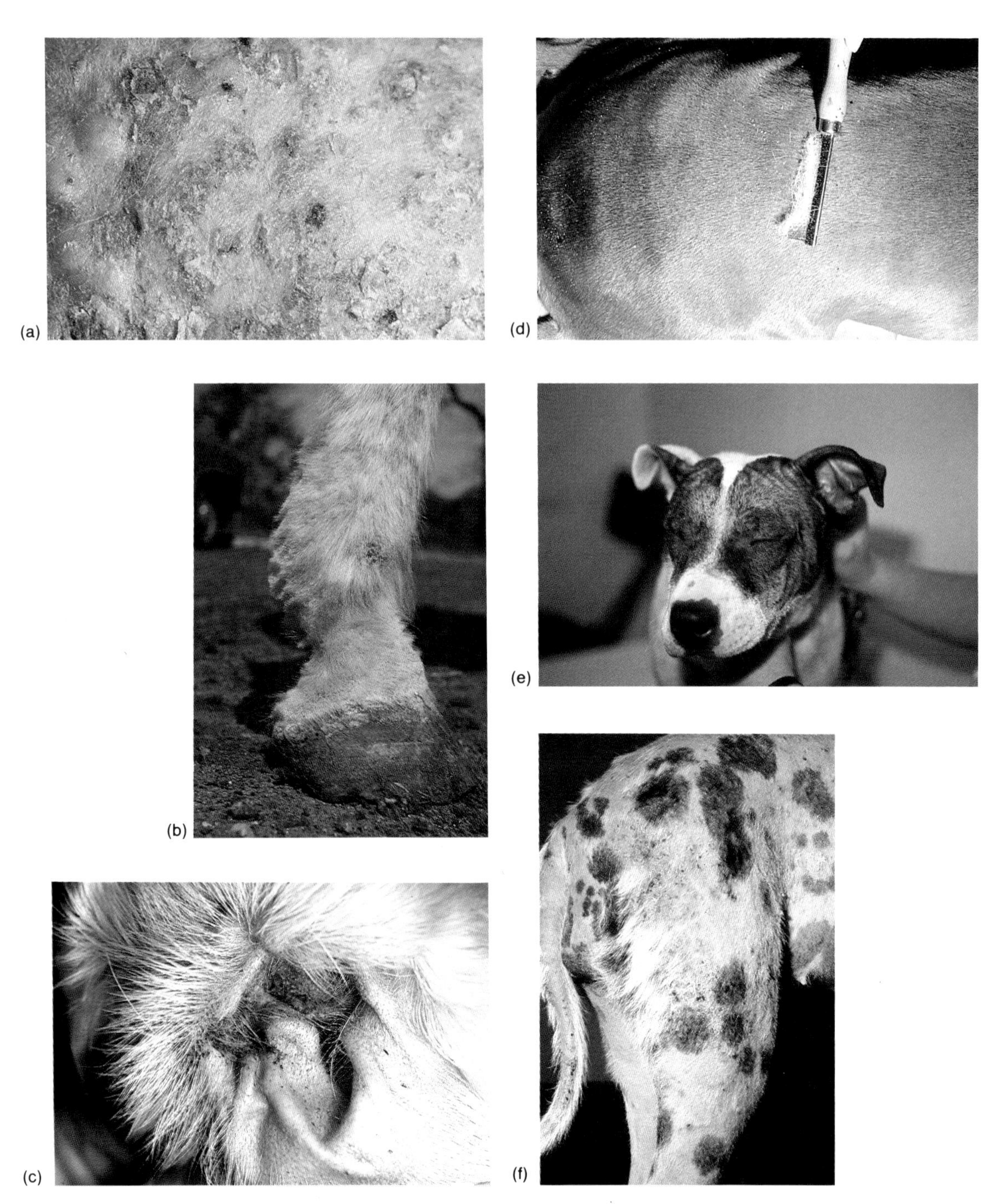

Plate XII

(a) Crusty lesions of sheep scab.
(b) Characteristic leg lesions of chorioptic mange in a horse.
(c) Dark waxy exudate caused by *Otodectes* infection in the dog.
(d) Marked 'dandruff' associated with *Cheyletiella* infection.
(e) Squamous demodectic mange.
(f) Pustular demodectic mange.

vest mite', which has a wide distribution in the Old World, and *Eutrombicula*, which occurs in North and South America, and whose larvae are known as 'chiggers'. Both of these genera will parasitise any animal, including man.

A third genus, with a more limited distribution, is *Leptotrombidium*, a vector of scrub typhus (tsutsugamushi fever) in the Far East.

The engorged larvae of trombiculids are ovoid and about 0.6 mm long, their most notable gross feature being their bright orange colour. Microscopically, the legs, which project considerably beyond the body margin, are seen to have plumose setae, unlike the 'hairs' of other parasitic mites.

The eggs are laid in soil and the hatched larvae crawl on to vegetation, where they can have contact with mammalian or avian hosts. In temperate areas larvae first appear in mid-summer, and populations are maximal in late summer, but in warmer regions they are present throughout the year.

The larvae insert their mouthparts into the skin, inject a cytolytic enzyme, and feed on the partly digested tissue. The feeding larvae cause irritation, which becomes more intense, with the formation of weals, papules and vesicles in successive attacks, due to the development of hypersensitivity to their secretions. Self-injury may result from rubbing and scratching in response to the irritation.

The sites of attachment are dictated by the height of the host, so that the mites occur on the head, ears and flanks of dogs and cats, and on the face and lower parts of the limbs of grazing animals.

Though both *Neotrombicula* and *Eutrombicula* have an apparent affinity for soft-fruit bushes they are common elsewhere, and occur in habitat types varying from semi-desert to swamp; in parts of the U.S.A., *Eutrombicula* is common on city lawns.

Control of these mites is almost impossible in cattle, sheep and equines, as the larval population is highest during summer when these animals are at pasture, but dogs and cats may be restricted from areas known to harbour large numbers of mites. The application of repellents has given only variable results.

Class PENTASTOMIDA

The adults of this strange class of aberrant arthropods are found in the respiratory passages of vertebrates and resemble annelid worms rather than arthropods.

Linguatula serrata is of some veterinary significance, the adult occurring in the nasal passages and sinuses of dogs, cats and foxes. It is up to 2.0 cm long, transversely striated, and shaped like an elongated tongue (Fig. 144) with a small mouth and tiny claws at the extremity of the thick anterior end. The eggs are expelled by coughing or sneezing or are swallowed and passed in the faeces.

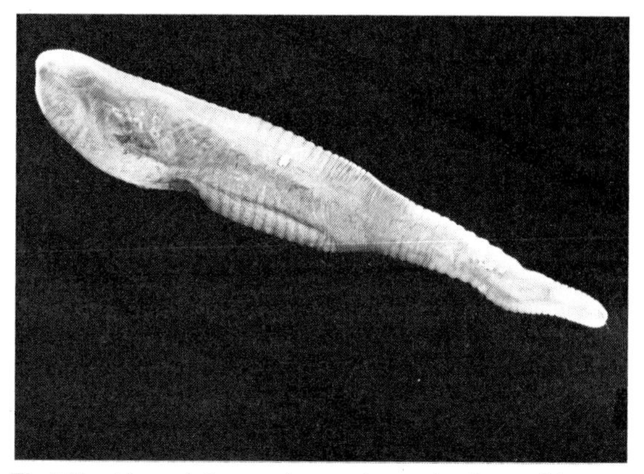

Fig 144 *Linguatula serrata*, an aberrant arthropod from the nasal passages of a dog.

When ingested by a herbivorous intermediate host, commonly sheep or cattle or rabbits, they hatch and the larvae locate in the mesenteric lymph glands. There they encyst to develop into the infective nymphal stages. The cysts, about 1.0 mm in diameter, may be visible in cut surfaces of mesenteric glands. The final host is infected by eating uncooked viscera.

Infrequently, heavy infections in dogs may cause sneezing, coughing and a nasal discharge. There is no specific treatment recommended, although organophosphate insecticides should be considered. In the absence of reinfection the parasites die after about one year.

VETERINARY PROTOZOOLOGY

Phylum PROTOZOA

The phylum Protozoa contains unicellular organisms which belong to the Animal Kingdom in that they obtain their energy by the intake of organic material, rather than, as in the Plant Kingdom, from the radiant energy of the sun by the process of photosynthesis in chloroplasts. Also, unlike plants, they do not possess a rigid cellulose wall exterior to the cell membrane. There are a few exceptions, and some free-living protozoa, for example, possess photosynthetic ability, but their consideration is irrelevant here.

Protozoa, like most organisms, are **eukaryotic** in that their genetic information is stored in chromosomes contained in a nuclear envelope. In this way they differ from bacteria which do not have a nucleus and whose single chromosome is coiled like a skein of wool in the cytoplasm. This primitive arrangement, found only in bacteria, rickettsia and certain algae, is called prokaryotic and such organisms may be regarded as neither animal nor plant, but as a separate kingdom of prokaryotic organisms, the Monera.

STRUCTURE AND FUNCTION OF PROTOZOA

Protozoa, like other eukaryotic cells, have a nucleus, an endoplasmic reticulum, mitochondria and a Golgi body and lysosomes. In addition, because they lead an independent existence they possess a variety of other subcellular structures or organelles with distinct organisational features and functions.

Thus locomotion, in, for example, the genus *Trypanosoma* (Fig. 145), is facilitated by a single **flagellum**, and in some other protozoa by several flagella. A flagellum is a contractile fibre, arising from a structure called a basal body, and in some species is attached to the body of the protozoan along its length, so that when the flagellum beats, the cell membrane (pellicle) is pulled up to form an **undulating membrane**. Sometimes, also, it projects beyond the protozoan body as a free flagellum. During

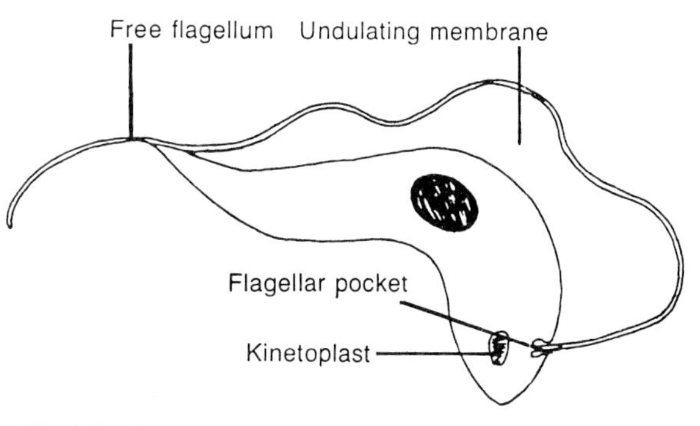

Fig 145 *Trypanosoma brucei* showing the flagellum and undulating membrane.

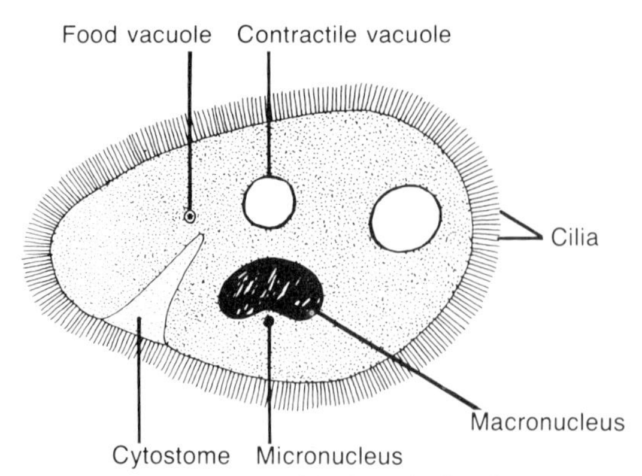

Fig 146 The morphology of the intestinal protozoan *Balantidium coli.*

movement the shape of these organisms is maintained by microtubules in the pellicle.

Other protozoa, such as *Balantidium* (Fig. 146), move by means of **cilia** which are fine, short hairs, each arising from a basal body; these cover much of the body surface and beat in unison to effect movement. In such species a mouth or **cytostome** is present and the ciliary movement is also used to waft food towards this opening.

A third means of locomotion, used by protozoa such as *Entamoeba* (Fig. 147) are **pseudopodia** which are prolongations of cytoplasm. Movement occurs as the rest of the cytoplasm flows into this prolongation. The pseudopodium also possesses a phagocytic capacity and can function as a cup which closes, enveloping particulate food material in a vacuole.

Finally some protozoa, such as the extra-cellular stages of the *Eimeria*, have no obvious means of locomotion, but are nevertheless capable of gliding movements.

The nutrition of parasitic protozoa usually occurs by pinocytosis or phagocytosis, depending on whether tiny droplets of fluid of small objects of macromolecular dimension are taken into the cell. In both cases, the process is the same, the cell membrane gradually enveloping the droplet or object which has become adherent to its outer surface. When this is complete, the particle is carried into the cell where fusion with lysosomes effects digestion. Finally, undigested material is extruded from the cell. As noted above, some ciliated protoza and also some stages of the organisms causing malaria obtain food through a cytostome. At the base of the cytostome the food enters a vacuole for digestion within the cell. Metabolic products are excreted by diffusion through the cell membrane.

The infective stage of some protozoa is called a **sporozoite** while the term **trophozoite** is applied to that stage of the protozoa in the host which feeds and grows until division commences. In most protozoa reproduction is asexual and is accomplished by binary fission or, in the case of *Babesia* within erythrocytes, by **budding**. Another form of asexual reproduction which occurs in the subphylum Sporozoa is **schizogony**. In the latter process, the trophozoite grows to a large size while the nucleus divides repeatedly. This structure is called a **schizont** and when mature, each nucleus has acquired a portion of the cytoplasm so that the schizont is filled with a large number of elongated separate organisms called **merozoites**. The schizont eventually ruptures liberating the individual merozoites.

Protozoa which only divide asexually generally have a short generation time, and since they cannot exchange genetic material, rely on mutants to provide the variants necessary for natural selection. However, most Sporozoa at certain stages in their life cycle also have a sexual phase of reproduction, called **gametogony** or **sporogony**. Sometimes, as in *Eimeria*, both asexual and sexual phases occur in the same host while in others such as *Plasmodium* the asexual phase occurs in the vertebrate host and the sexual phase in the arthropod vector.

Finally it should be noted that although this section deals with pathogenic protozoa of veterinary importance there are many other species, particularly in the rumen, which are purely commensal or even symbiotic in that they assist digestion of cellulose, and on being passed to the abomasum, act as a source of protein for the host.

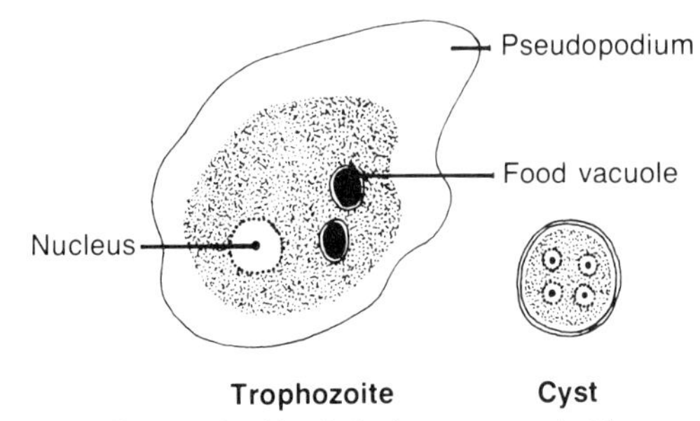

Fig 147 *Entamoeba histolytica* has an amoeboid trophozoite stage and a non-motile cystic stage with 4 nuclei.

CLASSIFICATION

Classification of the phylum Protozoa is extremely complex and the abridged version given below is simply intended to give an outline of the basic differences in the structure and life cycles of the main groups. To a large extent the common characteristics of each group are reflected by similarities in the diseases they cause.

There are four subphyla of protozoa of veterinary importance. These and the most important genera they contain are listed in Table 6.

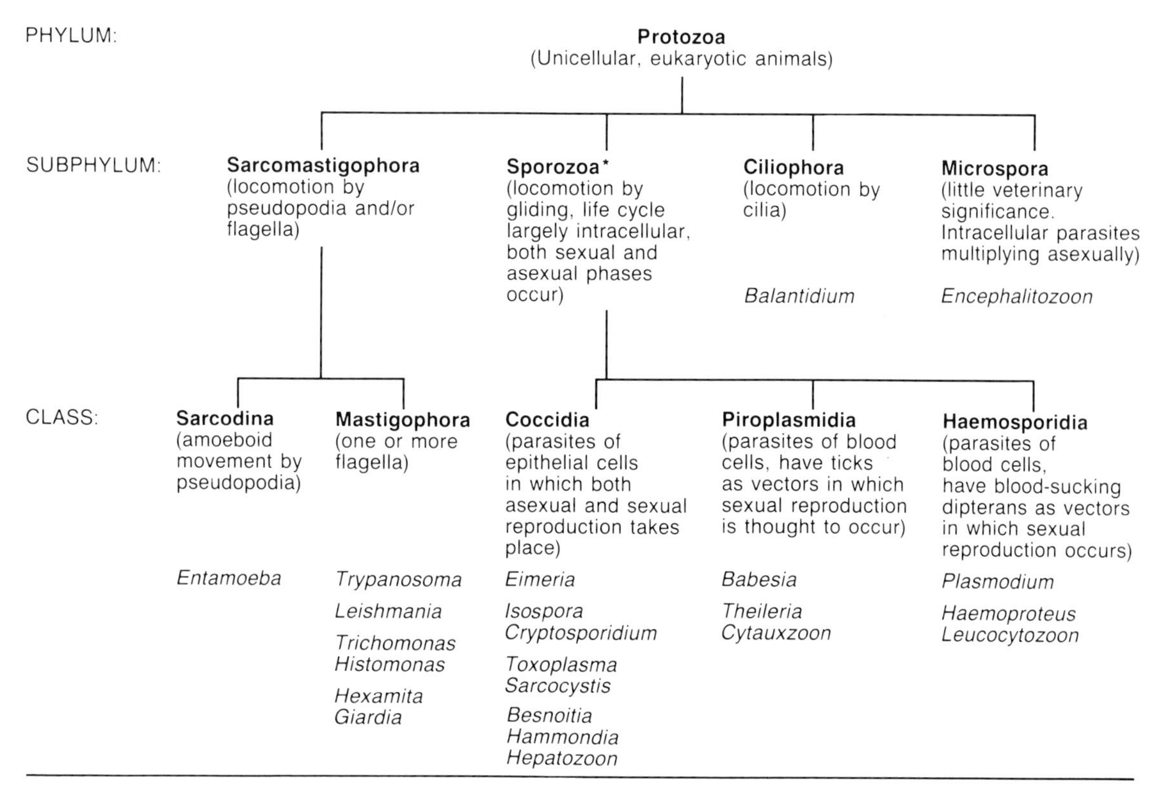

Table 6 Classification of the Protozoa

*Also called **Apicomplexa**. This alternative name refers to the group's possession of an 'apical complex', a structure which apparently assists penetration of the host cell. It is only visible with the electron microscope.

Subphylum SARCOMASTIGOPHORA

Class SARCODINA

Entamoeba histolytica

This pathogen is the cause of amoebic dysentery in man, a disease of worldwide distribution, although most common in the tropics. The amoeba-like trophozoites secrete proteolytic enzymes and produce characteristic flask-shaped ulcers in the mucosa of the large intestine. Their erosion may allow the parasites to enter the bloodstream when the most common sequel is the formation of amoebic abscesses in the liver.

Entamoeba multiplies by binary fission, but eventually encysts and is passed in the faeces. The cyst (Fig. 147), containing 4 nuclei, is relatively resistant and is the infective stage for the next host.

The veterinary significance of amoebiasis is that natural infections, usually without clinical signs, can occur occasionally in dogs from the human reservoir of active or carrier infections. Kittens are also susceptible to experimental infection, although they do not produce cysts. Monkeys have their own strains of *E. histolytica* and these can be infective to man.

Motile organisms and cysts of *E. histolytica* may be detected in smears from faeces, particularly if diarrhoeic, and provide a presumptive diagnosis; however their differentiation from less pathogenic species is a specialist task.

Dogs are not a significant reservoir of infection for man so that prophylaxis ultimately depends on personal and sanitary hygiene in the human population. Treatment, if required, relies on the combined use of metronidazole and diiodohydroxyquin.

Class MASTIGOPHORA

Trypanosoma

Members of this genus are found in the bloodstream and tissues of vertebrates throughout the world. However, a few species are of overwhelming importance as a serious cause of morbidity and mortality in animals and man in tropical regions. With one exception, all have an arthropod vector.

THE TRANSMISSION OF TRYPANOSOME INFECTION IN ANIMALS

Before discussing the separate species of the genus *Trypanosoma*, it is helpful to appreciate the various ways in which trypanosome infections are transmitted between animals.

With the single exception of *T. equiperdum* of equines which is a venereal disease, all have arthropod vectors in which transmission is either cyclical or non-cyclical.

In **cyclical transmission** the arthropod is a necessary intermediate host in which the trypanosomes multiply, undergoing a series of morphological transformations before forms infective for the next mammalian host are produced. When multiplication occurs in the digestive tract and proboscis, so that the new infection is transmitted when feeding, the process is known as **'anterior station development'** and the various species of trypanosomes which use this process are often considered as a group, the **Salivaria**. All are trypanosomes transmitted by tsetse flies, the main species being *T. congolense*, *T. vivax* and *T. brucei*.

In other trypanosomes, multiplication and transformaton occurs in the gut and the infective forms migrate to the rectum and are passed with the faeces; this is **'posterior station development'** and the trypanosome species are grouped together as the **Stercoraria**. In domestic animals these are all relatively non-pathogenic trypanosomes such as *T. theileri* and *T. melophagium* transmitted by tabanid flies and sheep keds respectively, but this is certainly not the case in man in which *T. cruzi*, the cause of the serious Chagas' disease in South America, is transmitted in the faeces of reduviid bugs.

Non-cyclical transmission is essentially mechanical transmission in which the trypanosomes are transferred from one mammalian host to another by the interrupted feeding of biting insects, notably tabanids and *Stomoxys*.

The trypanosomes in or on the contaminated proboscis do not multiply and die quickly so that cross-transmission is only possible for a few hours. *T. evansi*, widely distributed in livestock in Africa and Asia, is transmitted mechanically by biting flies. However, in Central and South America, *T. evansi* is also transmitted by the bites of vampire bats in which the parasites are capable of multiplying and surviving for a long period. Strictly speaking, this is more than mere mechanical transmission, since the bat is also a host, although it is certainly non-cyclical, since the multiplying trypanosomes in the bat's blood do not undergo any morphological transformation before they migrate into the buccal cavity.

It is important to note that the Salivarian trypanosomes, normally transmitted cyclically in tsetse flies, may on occasions be transmitted mechanically. Thus, in South America, *T. vivax* has established itself, presumably by the importation of infected cattle, and is thought to be transmitted mechanically by biting flies.

Finally, apart from classical cyclical and non-cyclical transmission, dogs, cats and wild carnivores may become infected by eating fresh carcasses or organs of animals which have died of trypanosomiasis, the parasites penetrating oral abrasions.

The important trypanosome infections of domestic animals differ considerably in many respects and are best treated separately. The African species responsible for the 'tsetse-transmitted trypanosomiases', ie. Salivaria, are generally considered to be the most significant and these are discussed first.

THE SALIVARIA

A number of species of *Trypanosoma*, found in domestic and wild animals, are all transmitted cyclically by *Glossina* in much of sub-Saharan Africa. The presence of trypanosomiasis precludes the rearing of livestock in many areas while in others, where the vectors are not so numerous, trypanosomiasis is often a serious problem, particularly in cattle. The disease, sometimes known as **nagana**, is characterised by lymphadenopathy and anaemia accompanied by progressive emaciation and, often, death.

Hosts:
All domestic livestock, but especially important in cattle. Also common in many wild animals such as the warthog, bush pig and various antelopes

Intermediate host:
Most species of *Glossina*, of which *G. morsitans* is perhaps the most widespread

Site:
All three species of trypanosome are characteristically present in the bloodstream. *T. brucei* is also found extravascularly in, for example, the myocardium, the central nervous system and the reproductive tract

Major species:
Trypanosoma brucei
T. congolense; the most common species
T. vivax

Minor species:
Probably the most important is *T. simiae* which is primarily a parasite of pigs and camels and morphologically resembles *T. congolense*

DISTRIBUTION

Approximately 10 million square kilometres of sub-Saharan Africa between latitudes 14°N and 29°S.

IDENTIFICATION

Elongated spindle-shaped protozoa ranging from

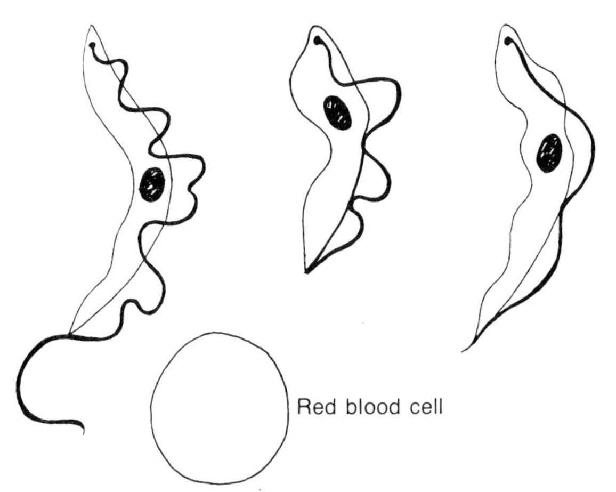

Fig 148 *Trypanosoma brucei* is pleomorphic showing long slender, short stumpy and intermediate forms.

8.0–39 μm long. All possess a flagellum which arises at the posterior end of the trypanosome from a basal body at the foot of a flagellar pocket (Fig. 145). The flagellum runs to the anterior end of the body and is attached along its length to the pellicle to form an undulating membrane. Thereafter the flagellum may continue forward as a free flagellum (Pl. XIII). Within a stained specimen a single centrally placed nucleus can be seen, and adjacent to the flagellar pocket, a small structure, the kinetoplast, which contains the DNA of the single mitochondrion.

IDENTIFICATION OF SPECIES

T. brucei is pleomorphic in form (Fig. 148) and ranges

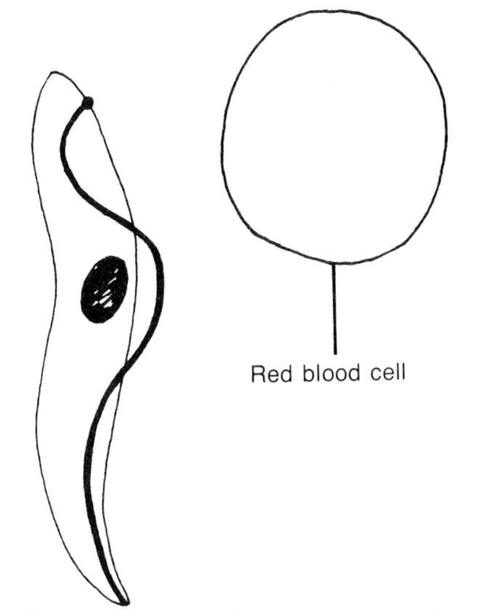

Fig 149 *Trypanosoma congolense* is monomorphic and possesses a marginal kinetoplast.

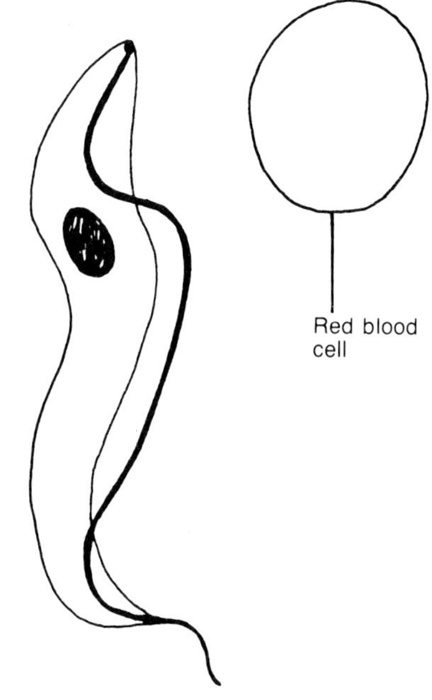

Fig 150 *Trypanosoma vivax* is monomorphic and has a short flagellum and terminal kinetoplast.

from long and slender (< 39 μm) to short and stumpy (< 18 μm), the two forms often being present in the same blood sample. The undulating membrane is conspicuous, the kinetoplast is small and sub-terminal and the posterior end is pointed. The slender form has a well-developed free flagellum while in the stumpy form it is either short or absent. In fresh unfixed blood films, *T. brucei* moves rapidly within small areas of the microscope field.

T. congolense is monomorphic in form (Fig. 149), ranging from 8.0–18 μm. The undulating membrane is inconspicuous, the medium-sized kinetoplast is marginal and the posterior end is blunt. There is no free flagellum.

In fresh blood films *T. congolense* moves sluggishly, often apparently attached to red cells.

T. vivax is monomorphic (Fig. 150) ranging from 20–27 μm. The undulating membrane is inconspicuous, the large kinetoplast is terminal and the posterior end is broad and rounded. A short free flagellum is present.

In fresh blood films *T. vivax* moves rapidly across the microscope field.

LIFE CYCLE

Tsetse flies ingest trypanosomes in the blood or lymph while feeding on an infected host. Thereafter the trypanosomes lose their glycoprotein surface coat, and in the case of *T. brucei* and *T. congolense*, become elongated and multiply in the midgut before migrating forward to the salivary glands (*T. brucei*) and the proboscis (*T. congolense*). There they undergo a transformation losing their typical trypanosome, or **trypomastigote**, form and acquire an **epimastigote** form characterised by the fact that the kinetoplast lies just in front of the nucleus. After further multiplication of the epimastigotes they transform again into small, typically trypomastigote forms with a glycoprotein surface coat. These are the infective forms for the next host and are called **metacyclic** trypanosomes. The entire process takes at least two to three weeks and the metacyclic trypanosomes are inoculated into the new host when the tsetse fly feeds.

With *T. vivax* a similar process of cyclical development takes place except that it occurs entirely within the proboscis.

At the site of inoculation the metacyclic forms multiply locally as the typical blood forms, producing within a few days a raised cutaneous inflammatory swelling called a **chancre**. Thereafter they enter the blood-stream, multiply, and a parasitaemia, detectable in the peripheral blood, usually becomes apparent 1–3 weeks later.

Subsequently, the parasitaemia may persist for many months although its level may wax and wane due to the immune response of the host.

PATHOGENESIS

With the exception of some strains of *T. vivax* which produce a hyperacute and fatal infection characterised by high parasitaemia, fever, severe anaemia and haemorrhages on the mucosal and serosal surfaces, the pathogenesis of trypanosomiasis may be considered under three headings:

(i) **Lymphoid enlargement** (Fig. 151) and **splenomegaly** develop associated with plasma cell hyperplasia and hypergammaglobulinaemia, which is primarily due to an increase in IgM. Concurrently there is a variable degree of suppression of immune responses to other antigens such as microbial pathogens or vaccines. Ultimately, in infections of long duration, the lymphoid organs and spleen become shrunken due to exhaustion of their cellular elements.

(ii) **Anaemia** is a cardinal feature of the disease, particularly in cattle, and initially is proportional to the degree of parasitaemia. It is haemolytic in that the red blood cells are removed from the circulation by the expanded mononuclear phagocytic system. Later, in infections of several months duration, when the parasitaemia often becomes low and intermittent, the anaemia may resolve to a variable degree. However, in some chronic cases it may persist despite chemotherapy.

(iii) **Cell degeneration** and **inflammatory infiltrates** occur in many organs such as the skeletal muscle and the central nervous system, but perhaps most significantly in the myocardium where there is separation and degeneration of the muscle fibres (Fig. 152). The mechanisms underlying these changes are still under study.

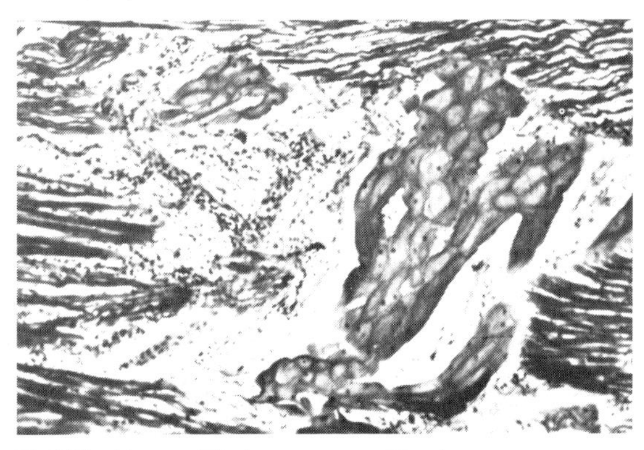

Fig 152 Myocarditis from a case of bovine trypanosomiasis showing myocardial degeneration, oedema and cellular infiltration.

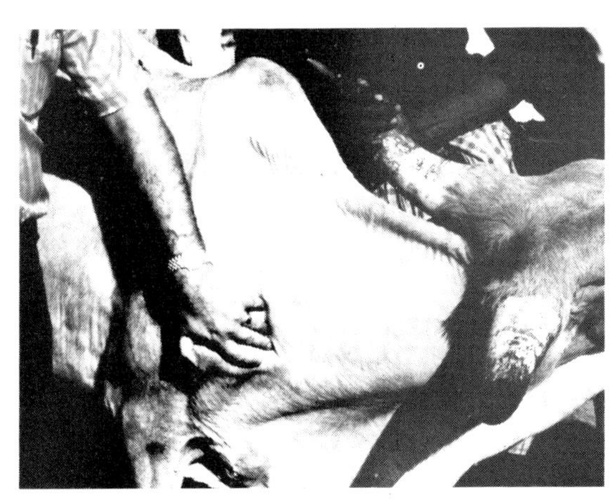

Fig 151 Enlarged prescapular lymph node of Zebu with trypanosomiasis.

Fig 153 Advanced case of bovine trypanosomiasis showing marked loss of bodily condition.

CLINICAL SIGNS

In **ruminants**, the major signs are anaemia, generalised enlargement of the superficial lymph glands, lethargy and progressive loss of bodily condition (Fig. 153). Fever and loss of appetite occur intermittently during parasitaemic peaks, the latter becoming marked in the terminal stages of the disease.

Typically, the disease is chronic extending over several months and usually terminates fatally if untreated.

As a herd phenomenon, the growth of young animals is stunted, while adults show decreased fertility, and if pregnant, may abort or give birth to weak offspring.

In the terminal stages animals become extremely weak, the lymph nodes are reduced in size and there is often a jugular pulse. Death is associated with congestive heart failure due to anaemia and myocarditis.

Occasionally, notably with some strains of *T. vivax*, the disease is acute, death occurring within 2–3 weeks of infection preceded by fever, anaemia and widespread haemorrhages.

In **horses**, *T. brucei* infections may be acute or chronic, often accompanied by oedema of the limbs and genitalia.

In the **pig**, *T. congolense* infections are usually mild and chronic in contrast to those associated with *T. simiae* where the disease is hyperacute, death occurring within a day or so of the onset of pyrexia.

The **dog** and **cat** are susceptible to *T. brucei* and *T. congolense*. The disease is usually acute, and apart from signs of fever, anaemia and myocarditis, corneal opacity is often a feature. There may also be neurological changes resulting in aggressive signs, ataxia or convulsions.

EPIDEMIOLOGY

The epidemiology depends on three factors, the distribution of the vectors, the virulence of the parasite and the response of the host.

The vectors

Of the three groups of *Glossina*, the savannah and riverine are the most important since they inhabit areas suitable for grazing and watering. Although the infection rate of *Glossina* with trypanosomes is usually low, ranging from 1–20% of the flies, each is infected for life, and their presence in any number makes the rearing of cattle, pigs and horses extremely difficult.

Biting flies may act as mechanical vectors, but their significance in Africa is still undefined. However in Central and South America, *T. vivax* is thought to be transmitted readily by such flies.

The parasites

Since parasitaemic animals commonly survive for prolonged periods, there are ample opportunities for fly transmission, especially of *T. brucei* and *T. congolense*. In contrast, some strains of *T. vivax* in cattle and *T. simiae* in domestic pigs kill their hosts within 1–2 weeks so that the chances of fly infection are more limited.

Perhaps the most important aspect of trypanosomiasis which accounts for the persistent parasitaemia, is the way in which the parasite evades the immune response of the host.

As noted previously, metacyclic and bloodstream trypanosomes possess a glycoprotein coat which is antigenic and provokes the formation of antibodies which cause opsonization and lysis of the trypanosomes. Unfortunately, by the time the antibody is produced, a proportion of the trypanosomes have altered the chemical composition of their glycoprotein coat and now, displaying a different antigenic surface, are unaffected by the antibody. Those trypanosomes possessing this new **variant antigen** multiply to produce a second wave of parasitaemia; the host produces a second antibody, but again the glycoprotein coat has altered in a number of trypanosomes so that a third wave of parasitaemia occurs. This process of **antigenic variation** associated with waves and remissions of parasitaemias, often at weekly intervals, may continue for months usually with a fatal outcome.

The repeated switching of the glycoprotein coat is now known to depend on a loosely ordered sequential expression of an undefined number of genes, each coding for a different glycoprotein coat. This, together with recent findings that metacyclic trypanosomes may be a mixture of antigenic types, each expressing a different genetic repertoire, explains why domestic animals, even if treated successfully, are often immediately susceptible to reinfection. The complexity of antigens potentially involved has also defeated attempts at vaccination.

The hosts

Trypanosomiasis is basically an infection of wildlife in which, by and large, it has achieved a *modus vivendi* in that the animal hosts are parasitaemic for prolonged periods, but generally remain in good health. This situation is known as **trypanotolerance**. In contrast, rearing of domestic livestock in endemic areas has always been associated with excessive morbidity and mortality although there is evidence that a degree of adaptation or selection has occurred in several breeds. Thus in West Africa small humpless cattle of the *Bos taurus* type, notably the N'dama (Fig. 154), survive and breed in areas of heavy trypanosome challenge despite the absence of control measures. However, their resistance is not absolute and trypanosomiasis exacts a heavy toll, particularly in productivity. In other areas of Africa, indigenous breeds of sheep and goats are known to be trypanotolerant, although this may be partly due to their being relatively unattractive hosts for *Glossina*.

Fig 154 The humpless N'dama breed of West Africa are trypanotolerant.

Precisely how trypanotolerant animals cope with antigenic variation is unknown. It is thought that the control and gradual elimination of their parasitaemias may depend on the possession of a particularly rapid and effective antibody response, although other factors may also be involved.

DIAGNOSIS

Confirmation of clinical diagnosis depends on the demonstration of trypanosomes in the blood and if a herd or flock is involved a representative number of blood samples should be examined since, in individual animals, the parasitaemia may be in remission or in long-standing cases may be extremely scanty.

Occasionally, when the parasitaemia is massive it is possible to detect motile trypanosomes in fresh films of blood. More usually, both thick and thin smears of blood are air-dried and examined later. Thick smears, dehaemoglobinised before staining with Giemsa or Leishman's stain, offer a better chance of finding trypanosomes while the stained thin smears are used for differentiation of the trypanosome species.

More sensitive techniques utilise centrifugation in a microhaematocrit tube followed by microscopic examination of the interface between the buffy coat and the plasma; alternatively, the tube may be snapped, the buffy coat expressed on to a slide, and the contents examined under dark-ground or phase-contrast microscopy for motile trypanosomes. With these techniques the packed red cell volume is also obtained which is of indirect value in diagnosis if one can eliminate other causes of anaemia, especially helminthiasis.

TREATMENT

This is a complex subject and only a general review is presented here.

In infected cattle, sheep or goats, the two drugs in common use are diminazene aceturate (Berenil) and homidium salts (Ethidium and Novidium). These are usually successful except where trypanosomes have developed resistance to the drug or in some very chronic cases. Treatment should be followed by surveillance since reinfection, followed by clinical signs and parasitaemia, may occur within a week or two. Alternatively, the animal may relapse after chemotherapy, due to a persisting focus of infection in its tissues or because the trypanosomes are drug-resistant.

Because of their arid habitat, camels are only rarely exposed to tsetse-transmitted trypanosomiasis although *T. brucei*, when it occurs, is particularly lethal. The treatment is suramin or the recently reintroduced quinapyramine sulphate (Trypacide, formerly called Antrycide), both of which are also used for the more common *T. evansi*. *T. simiae* also causes rapidly fatal infections in camels which offer little opportunity for treatment. It should be noted that diminazene is toxic to camels.

In pigs, *T. simiae* is the most important pathogen and the rapid onset of death again gives little chance of treatment.

Horses are particularly susceptible to *T. brucei* when suramin or quinapyramine are the drugs of choice. Diminazene is relatively toxic to horses.

In dogs and cats infected with *T. brucei*, suramin or quinapyramine should be used while *T. congolense* infections require diminazene. If cerebral trypanosomiasis, an occasional sequel of *T. brucei* infection, is suspected, the treatment should be followed by a course of the arsenical drug melarsoprol (Mel B).

CONTROL

This currently depends on the control of tsetse flies, discussed under *Glossina* (page 154), and on the use of drugs.

In cattle, and if necessary in sheep and goats, isometamidium (Samorin) is the drug of choice since it remains in the tissues and has a prophylactic effect for 2–6 months. Otherwise, diminazene or homidium salts may be used as cases arise, these being selected either by clinical examination or on the haematological detection of anaemic animals.

To reduce the possible development of drug resistance it may be advisable periodically to change from one trypanocidal drug to another. Surveillance is always necessary to ensure that the dosage of drug is adequate and to detect the early appearance of possible drug-resistance, when specialist advice should be sought.

Two important aspects of control are, first, the necessity to protect cattle from a tsetse-free zone while being trekked to market through an area of endemic trypanosomiasis and, secondly, an awareness of the dangers of stocking a tsetse-free ranch with cattle from areas where trypanosomiasis is present, as mechanical transmission may cause an outbreak of disease; in both cases treatment with a trypanocidal drug at an appropriate time is advisable.

An alternative approach, using trypanotolerant breeds of ruminants, perhaps combined with judicious drug therapy, may, in the future, offer a realistic solution in many areas where the disease is endemic and this aspect is currently under intensive study.

Chemoprophylactic regimens for other domestic species are occasionally practised and require to be adapted to the particular circumstances encountered.

[**Tsetse-transmitted trypanosomiases in man:** Two species of salivarian trypanosomes infect man. While both can cause 'sleeping sickness' by invasion of the central nervous system, *T. rhodesiense* usually causes an acute syndrome, while *T. gambiense* infections may be initially asymptomatic, although eventually the central nervous system is affected. *T. rhodesiense* occurs in East and Central Africa while *T. gambiense* is mainly found in West Africa.

Both species are morphologically identical to *T. brucei* and because of this and the limits on human experimentation it is difficult to delineate their precise relationship. However, in the light of present knowledge it appears that *T. brucei* is only infective to animals in contrast to *T. rhodesiense* and *T. gambiense* which may infect both man and animals. The latter include primates, wild ungulates and domestic animals and these may act as reservoirs of infection for man.]

THE STERCORARIA

Two relatively large trypanosomes, *Trypanosoma theileri* and *T. melophagium*, about 50 μm in length, are found in the blood of cattle and sheep respectively. *Trypanosoma theileri* is transmitted by tabanid flies and *T. melophagium* by the sheep ked, *Melophagus ovinus*, the worldwide distribution of each trypanosome corresponding to the range and prevalence of their intermediate hosts. Thus, both species occur in western Europe.

The metacyclic trypanosomes, present in the faeces of the vector, gain access to the blood of their mammalian host by penetrating abraded skin or following ingestion of the vector when the liberated trypanosomes penetrate the mucosa.

Both trypanosomes may produce transient parasitaemias, but infection is more usually asymptomatic and infection can only be diagnosed by incubating blood in culture medium suitable for the multiplication of trypanosomes. They are often referred to as 'the nonpathogenic trypanosomes'.

[***Trypanosoma cruzi*** **of man:** Although the stercorarian trypanosomes of ruminants are nonpathogenic the reverse is true of *T. cruzi*, the cause of Chagas' disease in man. The disease, found in Central and South America, is extremely serious both in its acute febrile parasitaemic stage and in its chronic stage characterised by myocarditis, megaoesophagus and megacolon. The vectors are haematophagous bugs of the Order Hemiptera, often called 'kissing bugs'. These are often present in primitive housing and infection of man results from rubbing faeces into skin abrasions such as those created by the bites of the bugs. Domestic animals, primarily the dog and cat, may be clinically affected or act as reservoir hosts. If the latter, they may, inadvertently, introduce the disease into hitherto clean areas such as, for example, parts of the United States.

Wild animals, notably the armadillo and opossum, also act as reservoir hosts].

MECHANICALLY TRANSMITTED TRYPANOSOMIASIS

Trypanosoma evansi

This species, although closely related to the salivarian trypanosome *T. brucei*, is mechanically transmitted by biting insects, and is the cause of a disease commonly called **surra**. It is widespread in North America and primarily affects horses and camels. The syndrome is similar to that caused by the tsetse-transmitted trypanosomes.

Trypanosoma evansi is identical in appearance to the slender forms of *T. brucei*. Species of Tabanidae and *Stomoxys* are vectors although in the Americas the vampire bat also acts as a vector as well as a reservoir host.

Depending on the virulence of the strain and the susceptibility of the individual host, the disease may be acute in horses, camels and dogs. Other domestic species such as cattle, buffalo and pigs are commonly infected, but overt disease is uncommon and their main significance is as reservoirs of infection.

Apart from the fever, anaemia and emaciation characteristic of trypanosomiasis, horses develop oedematous swellings ranging from cutaneous plaques to frank oedema of the ventral abdomen and genitalia. In more chronic cases progressive paralysis of the hindquarters is common.

In South America the disease is called **mal de caderas**, literally 'disease of the hip', and the trypanosome species is called *T. equinum*; it is identical to *T. evansi* except that it lacks a kinetoplast.

Suramin or quinapyramine (Trypacide) are the drugs of choice for treatment and also confer a short period of prophylaxis. For more prolonged protection a modified quinapyramine known as 'Trypacide Pro-Salt' is also available. Unfortunately, drug-resistance, at least to suramin, is not uncommon. Currently in camels, the use of isometamidium, administered intravenously because of local tissue reactions, is under study.

Trypanosoma equiperdum

This trypanosome, morphologically similar to *T. evansi*, causes a venereal disease of horses and donkeys called **dourine**. Previously widespread because of its unique method of transmission, it is now confined to parts of Africa, Asia and South and Central America. The clinical signs are those of genital and ventral abdominal oedema, transient urticarial plaques and progressive emaciation. In many cases, the C.N.S. is involved, causing an ascending motor paralysis which is ultimately fatal. Diagnosis generally relies on a complement fixation test because of the scarcity of trypanosomes in the blood.

Leishmania

Members of this genus are intracellular parasites of macrophages in man, the dog and a wide variety of wild animals and the disease they cause, leishmaniasis, has both cutaneous and visceral forms. Their vectors are blood-sucking sandflies in which the parasites undergo morphological transformation and multiplication. Leishmaniasis is of major importance as a disease of man.

Hosts:
Man, dog and a wide variety of wild animals

Intermediate hosts:
Blood-sucking phlebotomine sandflies

Site:
The protozoa multiply within macrophages, which are eventually destroyed, the liberated parasites entering other intact macrophages

SPECIES

Although *Leishmania* spp. are difficult to distinguish on morphological grounds, modern taxonomic methods have shown the existence of six main species which broadly conform to the geographical, epidemiological and clinical features of the disease they cause in man. Three of these species occur in dogs:

Leishmania tropica	causing cutaneous leishmaniasis or 'oriental sore', the lesions developing at the site of the insect bite
L. donovani	causing visceral leishmaniasis, or 'kala-azar', the infection being systemic
L. braziliensis	causing lesions similar to *L. tropica*

DISTRIBUTION

In the dog, *L. tropica* occurs in southern Europe, other Mediterranean countries, Africa and Asia; *L. donovani* is found in countries around the Mediterranean and in South America and *L. braziliensis* infection in parts of South America.

IDENTIFICATION AND LIFE CYCLE

Leishmania is related to the *Trypanosoma* insofar as the ovoid organism within the macrophage possesses a rod-shaped kinetoplast associated with a rudimentary flagellum which, however, does not extend beyond the cell margin. This leishmanial, or **amastigote** form, after ingestion by a sandfly, transforms into a **promastigote** form in the insect gut in which the kinetoplast is situated at the posterior end of the body (Fig. 155). These divide repeatedly by binary fission, migrate to the proboscis, and when the insect subsequently feeds, are inoculated into a new host. Once within a macrophage the promastigote reverts to the amastigote form and again starts to divide.

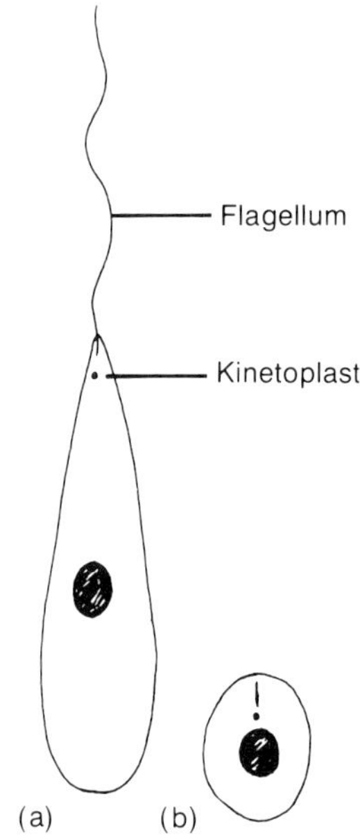

Fig 155 *Leishmania.*
(a) Promastigote form.
(b) Amastigote form.

PATHOGENESIS

The basic lesions are foci of activated proliferating macrophages infected with *Leishmania* organisms (Fig. 156). In some cases, these are ultimately surrounded by plasma cells and lymphocytes. The infected macrophages are then destroyed, the animal recovers and is immune to reinfection.

However, recovery seems to depend on the proper expression of cell-mediated immunity and if this does

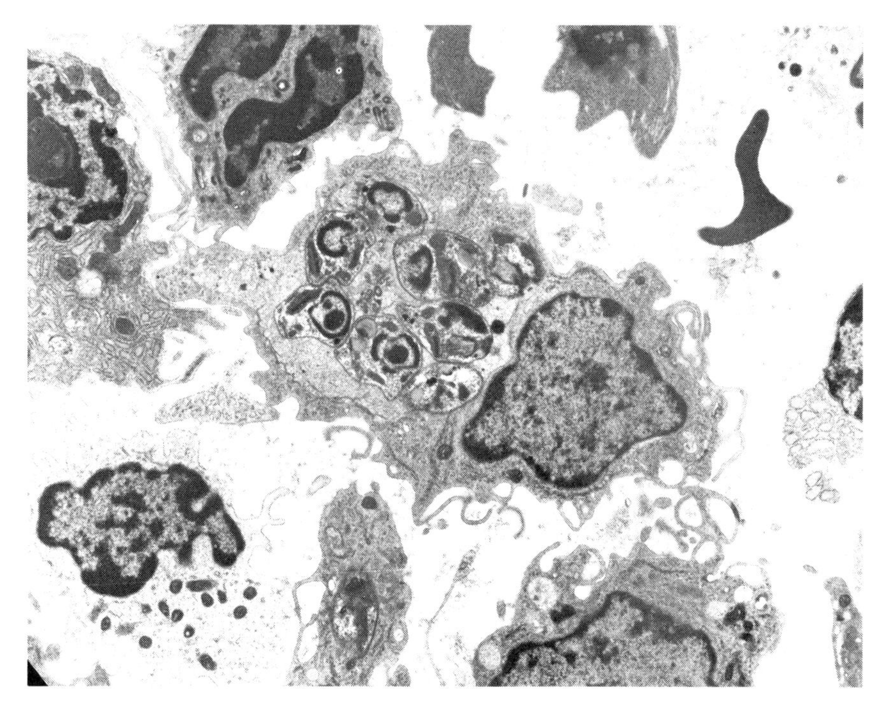

Fig 156 Electron micrograph showing cluster of *Leishmania* amastigote forms within cytoplasm of macrophage.

not occur, as in many cases of *L. donovani*, the active lesions persist leading to chronic enlargement of the spleen, liver and lymph nodes and persistent cutaneous lesions. Haemolytic anaemia is also a feature of *L. donovani* infection.

CLINICAL SIGNS

It may take many months for infected dogs to develop clinical signs, so that the disease may only become apparent long after dogs have left endemic areas.

In the cutaneous form, lesions are confined to shallow skin ulcers often on the lip or eyelid, from which recovery is often spontaneous. In the visceral form, which is more common, dogs initially develop 'spectacles' due to depilation of hair round the eyes and this is followed by generalised loss of body hair and eczema (Fig. 157), leishmanial organisms being present in large numbers in the infected skin. Intermittent fever, anaemia, cachexia and generalised lymphadenopathy are also typical signs.

Long periods of remission followed by the reappearance of clinical signs are not uncommon.

EPIDEMIOLOGY

Most *Leishmania* spp. primarily infect wild animals, especially rodents, and man only occasionally becomes infected by chance. Other *Leishmania* have a direct man to man transmission via the sandfly. The dog is a natural host and reservoir of infection for some strains of *L. donovani* and *L. tropica* which can infect man, especially children, in the Mediterranean basin.

DIAGNOSIS

This depends on the demonstration of the amastigote parasites in smears or scrapings from affected skin or from lymph-node or marrow biopsies.

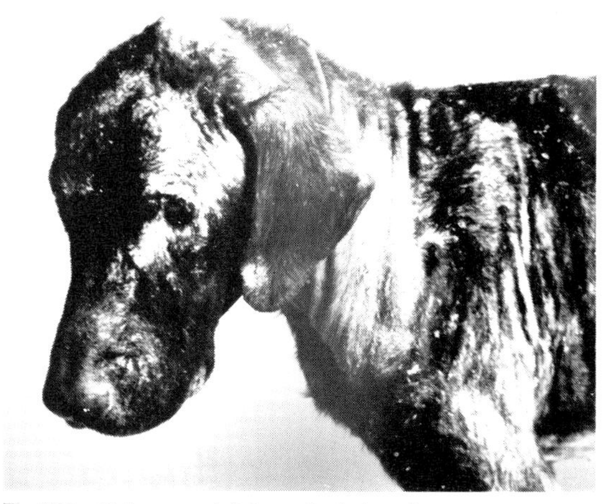

Fig 157 Cutaneous leishmaniasis in a dog.

TREATMENT

In man, antimonial compounds are commonly used, but these are not particularly effective in the canine disease and from the public health viewpoint infected dogs, particularly with visceral leishmaniasis, are best destroyed.

CONTROL

From the public health aspect, the destruction of infected dogs, and stray dogs generally, is desirable.

In many areas, the population of sandflies has been reduced as a result of mosquito control for malaria, and as a result the incidence of leishmaniasis has decreased.

Trichomonas

The most important pathogen in this genus is *Trichomonas foetus*, a venereally transmitted, multi-flagellated organism of the reproductive tract of cattle. In bulls, infection is inapparent, but in pregnant cows it produces early foetal death which is usually first recognised as an infertility problem.

Other species of *Trichomonas* occur in the digestive tract of animals, apparently as commensals, and only one species, *T. gallinae*, is clearly pathogenic in the oesophagus and crop of pigeons.

Trichomonas foetus

Hosts:
Cattle

Site:
In cows, the uterus and intermittently the vagina. In bulls, the preputial cavity

Distribution:
Worldwide. However, the prevalence has now decreased dramatically in areas where artificial insemination is widely practised and in Britain, for example, the disease is now probably extinct

IDENTIFICATION

The organism is pear-shaped, approximately 20 × 10 μm and has a single nucleus and four flagella, each arising from a basal body situated at the anterior rounded end (Fig. 158). Three of the flagella are free anteriorly, while the fourth extends backwards to form an undulating membrane along the length of the organism and then continues posteriorly as a free flagellum. The axostyle, a hyaline rod with a skeletal function, extends the length of the cell and usually projects posteriorly.

In fresh preparations, the organism is motile and progresses by rolling jerky movements, the flickering flagella and the movements of the undulating membrane being readily seen. Occasionally, rounded immobile forms are observed and these are possibly effete.

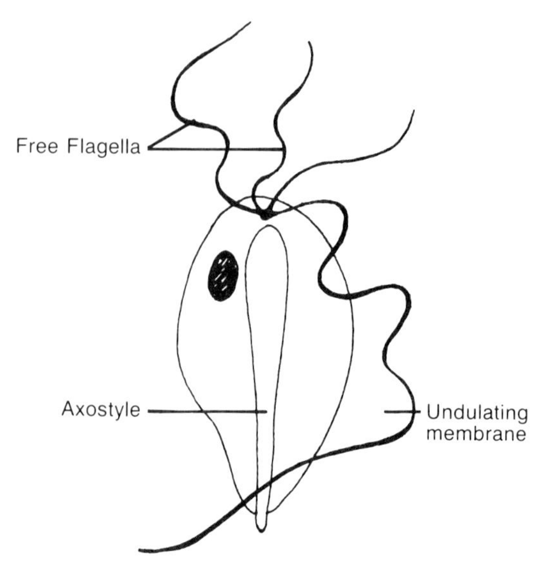

Fig 158 *Trichomonas foetus.*

LIFE CYCLE

Bulls, once infected, remain so permanently. The organisms inhabit the preputial cavity and transmission to the cow occurs during coitus. From the vagina, the trichomonads reach the uterus via the cervix to produce a low-grade endometritis. Intermittently, organisms are flushed into the vagina, often two or three days before oestrus.

Infection is usually followed by early abortion, the organisms being found in the amniotic and allantoic fluid. Subsequently cows appear to 'self-cure' and, in most cases, appear to develop a sterile immunity.

PATHOGENESIS

In the bull, a preputial discharge associated with small nodules on the preputial and penile membranes may develop shortly after infection. Thereafter there are no clinical signs or lesions.

In the cow, abortion before the fourth month of pregnancy is the commonest sequel and this is normally followed by recovery. Occasionally the developing foetal membranes are retained leading to a purulent endometritis, a persistent uterine discharge and anoestrus; infrequently the corpus luteum is retained and the cervical seal remains closed, when a massive pyometra develops which, visually, simulates the appearance of pregnancy.

CLINICAL SIGNS

In the bull, there are no clinical signs once the infection is established. In the cow, early abortion is a characteristic feature although this is often undetected because of the small size of the foetus and the case may present as one of an irregular oestrus cycle. Other clinical signs are those of purulent endometritis or a closed pyometra and, in these cases, the cow may become permanently sterile.

EPIDEMIOLOGY

Normally one might expect the overall prevalence of trichomoniasis to be high since it is venereally transmitted by bulls which show no clinical signs. In fact, the advent of supervised schemes of artificial insemination has largely eradicated the disease, and today it is limited to areas where there are many small farms each with their own bulls, or to countries where veterinary supervision is limited.

DIAGNOSIS

Apart from a problem of infertility, which usually follows the purchase of a mature bull, diagnosis depends on the demonstration of the organism.

Vaginal mucus collected from the anterior end of the vagina by suction into a sterile tube, or preputial washings from the bull, may be examined using a warm-stage microscope for the presence of organisms. However, since the organism is often only present intermittently, the examination may require to be repeated several times.

Alternatively, on a herd basis, samples of vaginal mucus may be examined in the laboratory for the presence of specific agglutinins against laboratory cultures of *T. foetus*.

TREATMENT

Since the disease is self-limiting in the female only symptomatic treatment and sexual rest for three months is normally necessary. In the bull, slaughter is the best policy, although dimetridazole orally or intravenously has been reported to be effective.

CONTROL

Artificial insemination from non-infected donors is the only entirely satisfactory method of control. If a return to natural service is contemplated, recovered cows should be disposed of since some may be carriers.

Trichomonas gallinae

This species causes yellow, necrotic lesions in the mouth, oesophagus and crop of pigeon squabs and is frequently fatal. Infection is acquired via regurgitated crop contents from adult birds, which, although immune, remain carriers. Turkeys and chickens may occasionally become infected.

Dimetridazole is recommended for treatment while control depends on preventing access of wild pigeons to drinking water.

[**Trichomoniasis in man:** *Trichomonas vaginalis* is a common and host-specific pathogen of the vagina and urethra. Like *T. foetus*, it is venereally transmitted and the clinical signs of inflammation are largely confined to females].

Histomonas meleagridis

Histomonas meleagridis is the cause of a disease in young turkeys known as infectious entero-hepatitis, histomoniasis or 'blackhead'. The characteristic necrotic lesions are confined to the caecum and liver and transmission between birds depends on the protozoan being carried in the egg of the poultry ascarid *Heterakis gallinarum*.

Hosts:
Turkeys, particularly poults: occasionally pathogenic in chickens and game birds

Transport host:
The adult and eggs of the ascarid worm, *Heterakis gallinarum*

Site:
The caecal mucosa and liver parenchyma.

Distribution:
Worldwide

IDENTIFICATION

A round or oval parasite, 6.0–20 μm in diameter, which in the lumen of the caecum bears a single flagellum (Fig. 159) although this appears to be lost when in the mucosal tissue or the liver. Both luminal and tissue stages exhibit pseudopodial movement.

LIFE CYCLE

Birds become infected by ingestion of the embryonated egg of the caecal worm, *Heterakis gallinarum*, the flagellate being carried in the unhatched larva. When the egg hatches, the histomonads are released from the larva and enter the caecal mucosa where they cause ulceration and necrosis. They reach the liver in the portal stream and colonise the liver parenchyma, producing circular necrotic foci which increase in size as the parasites multiply in the periphery of the lesion.

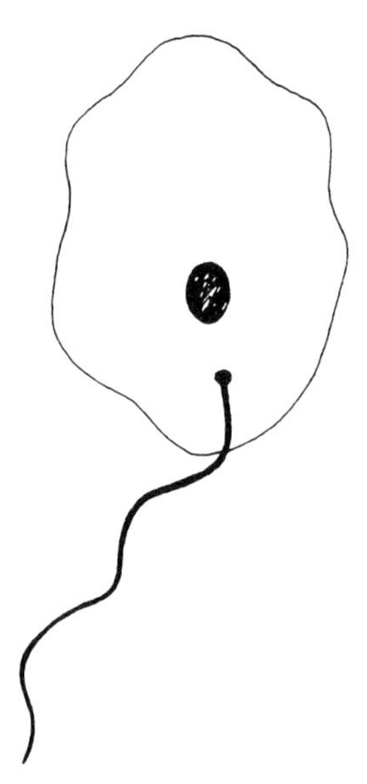

Fig 159 The flagellated caecal form of *Histomonas meleagridis*.

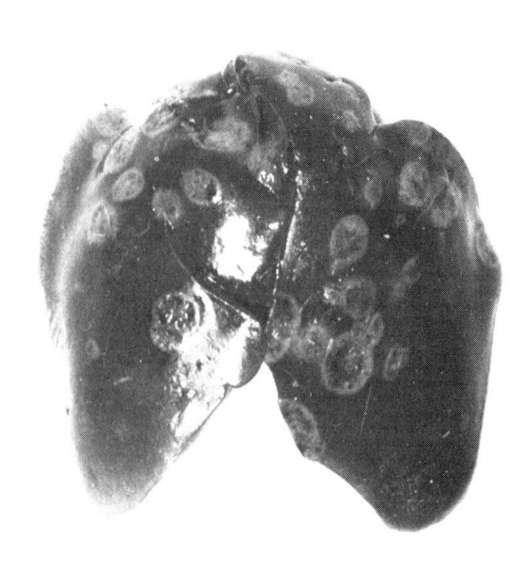

Fig 160 The characteristic circular necrotic liver lesions of histomoniasis.

The next phase of the life cycle is not clear, but it is presumed that the *Heterakis* worms become infected with the caecal histomonads, possibly by ingestion, and that these subsequently reach the ovary of the worm. It is certainly established that the histomonads become incorporated in a proportion of the *Heterakis* eggs and thus reach the exterior.

Infection of birds may also result from the ingestion of earthworms which are transport hosts for *Heterakis* eggs and larvae.

PATHOGENESIS

The disease is essentially one of young turkeys up to 14 weeks old and is characterised by necrotic lesions in the caecum and liver. The earliest lesions are small ulcers in the caeca, but these quickly enlarge and coalesce so that the entire mucosa becomes necrotic and detaches, forming, with the caecal contents, a caseous plug. The liver lesions are circular and up to 1.0 cm in diameter with yellow depressed centres; they are found both on the surface and in the substance of the liver (Fig. 160).

Mortality in poults may reach 100% and in birds which recover the caecum and liver may be permanently scarred.

CLINICAL SIGNS

Eight or more days after infection the poults become dull, the feathers are ruffled and the faeces become sulphur-yellow in colour. Unless treated, the birds usually die within one or two weeks.

In older turkeys the disease is more usually a chronic, wasting syndrome followed by recovery and subsequent immunity.

The name 'blackhead' was first coined to describe the disease when cyanosis of the head and wattles was thought to be a characteristic feature. However, this sign is not necessarily present, and anyway is not confined to histomoniasis.

EPIDEMIOLOGY

Although showing no signs of *Histomonas* infection the domestic chicken is commonly infected with *H. gallinarum*, whose eggs, if fed to turkeys, will regularly produce histomoniasis.

Typically, histomoniasis occurs when turkey poults are reared on ground shared, or recently vacated, by domestic chickens. However, since the organism may survive in embryonated *Heterakis* eggs in soil, or as larvae in earthworms, for over two years, outbreaks may arise on apparently clean ground.

Young turkeys may also become infected when reared by broody hens which are carriers.

DIAGNOSIS

This is based on history, clinical signs and necropsy findings. Although rarely necessary, histological sections of liver or caecum may be prepared for specialist examination.

TREATMENT

A number of drugs are effective including dimetridazole,

nitrothiazole compounds and nithiazide. In Britain, only dimetridazole is licensed and is available as a powder for inclusion in feed or, as a more readily administered soluble preparation, for medication of the drinking water. In both cases treatment should be given to all the turkeys, whether affected or not and should be continued for 12 days, after which a prophylactic regimen of the drug should be started.

CONTROL

Turkeys should be reared on ground not used by domestic chickens for at least two years, or on fresh litter or wire floors raised above the ground.

Continuous low level medication, usually in the feed rather than drinking water, with the drugs described above, is highly desirable where there is any risk of the disease.

Hexamita meleagridis

This protozoan rather resembles *Trichomonas foetus*, but is bilaterally symmetrical in that it possesses two nuclei, two sets of three anterior flagella and two flagella which pass through the body to emerge posteriorly. It is occasionally responsible for catarrhal enteritis in young turkeys.

Giardia lamblia

This organism is bilaterally symmetrical like *Hexamita* and also possesses eight flagella, six of which emerge as free flagella at intervals around the body (Fig. 161). It is unique in possessing a large adhesive disc on the flat ventral surface of the body which facilitates attachment to the epithelial cells of the intestinal mucosa. The organism is passed as multi-nucleated cysts in which the flagella may be visible, and occasionally as trophozoites in the faeces. Detection of these is the basis of laboratory diagnosis.

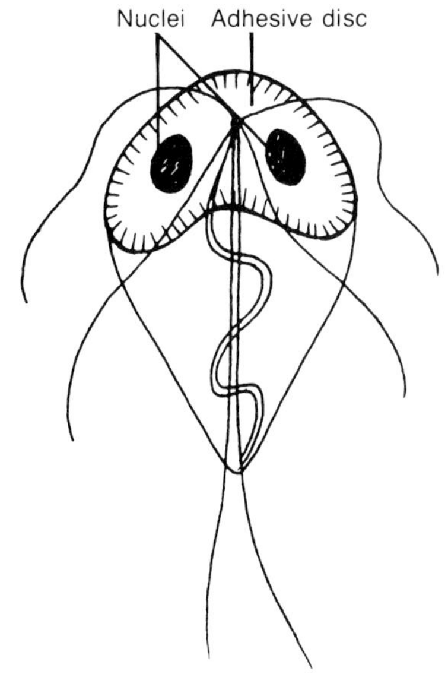

Fig 161 *Giardia lamblia.*

Giardia is a not uncommon cause of chronic diarrhoea in man often called lambliasis. In domestic animals several species have been described and reported to be responsible for diarrhoea, but the evidence is still inconclusive. However there is evidence from the U.S.A. that *Giardia* from man which gain access to municipal water reservoirs may successfully infect wild animals, especially beavers. These then act as a source of contamination of domestic water supplies.

Subphylum SPOROZOA

Protozoa within the subphylum Sporozoa are characterised by occurring intracellularly and having an apical complex at some stage of their development. The trophozoites have no cilia or flagella. Reproduction involves both asexual (schizogony) and sexual (gametogony) phases. Following gametogony, a zygote is formed which divides to produce spores (sporogony).

Of the three classes of veterinary significance the two most important are the Coccidia or alimentary sporozoa and the Piroplasmidia which are blood sporozoa.

Class COCCIDIA

The class Coccidia contains parasites which occur mainly in vertebrates. Since those of veterinary importance fall into two distinct family groups, the *Eimeriidae* and *Sarcocystidae*, it is proposed to discuss these separately.

Family EIMERIIDAE

These are mainly intracellular parasites of the intestinal epithelium. Schizogony and gametogony occur within the host and sporulation, or maturation of the fertilised zygote, usually takes place outwith the host. Although three genera, *Eimera*, *Isospora* and *Cryptosporidium*, are of considerable veterinary importance, the term coccidiosis is usually reserved for infections caused by *Eimeria* and *Isospora* spp. Since there are only minor differences in the pathogenesis and epidemiology of the different species it is proposed that, following a general account of these, the clinical signs, diagnosis, treatment and control will be considered in detail for each host.

Eimeria

Hosts:
Poultry, cattle, sheep, goats, pigs, horses and rabbits

Site:
Epithelial cells of the intestine and in two species the kidney and liver respectively

IMPORTANT SPECIES

Eimeria tenella, *E. necatrix*, *E. brunetti*, *E. maxima*, *E. mitis* and *E. acervulina* – chickens
E. meleagrimitis and *E. adenoeides* – turkeys
E. anseris, *E. nocens* and *E. truncata* (kidney) – geese
E. zuernii and *E. bovis* – cattle
E. crandallis, *E. ovinoidalis*, *E. ovina* – sheep
E. arloingi and *E. ninakohlyakimovae* – goats
E. debliecki – pigs
E. leuckarti – horses
E. flavescens, *E. intestinalis* and *E. stiedae* (liver) – rabbits

Distribution:
Worldwide

IDENTIFICATION

This can be made at microscopic level, either by examining the faeces for the presence of oocysts or by examination of scrapings or histological sections of affected tissues.

OOCYSTS

The oocysts may be identified according to shape and size. The most common shapes are spherical, ovoid or ellipsoidal and the size of the common species ranges from 15 to 50 μm. Oocysts have a refractile shell and some species possess a small pore at one end, the **micropyle**, which is often covered by a polar cap which may be prominent. The time taken for sporulation to occur under standard conditions can also be used as an aid to indentification.

Tissue stages

The mature schizonts may, in some cases, be identified histologically by their location, size and the number of

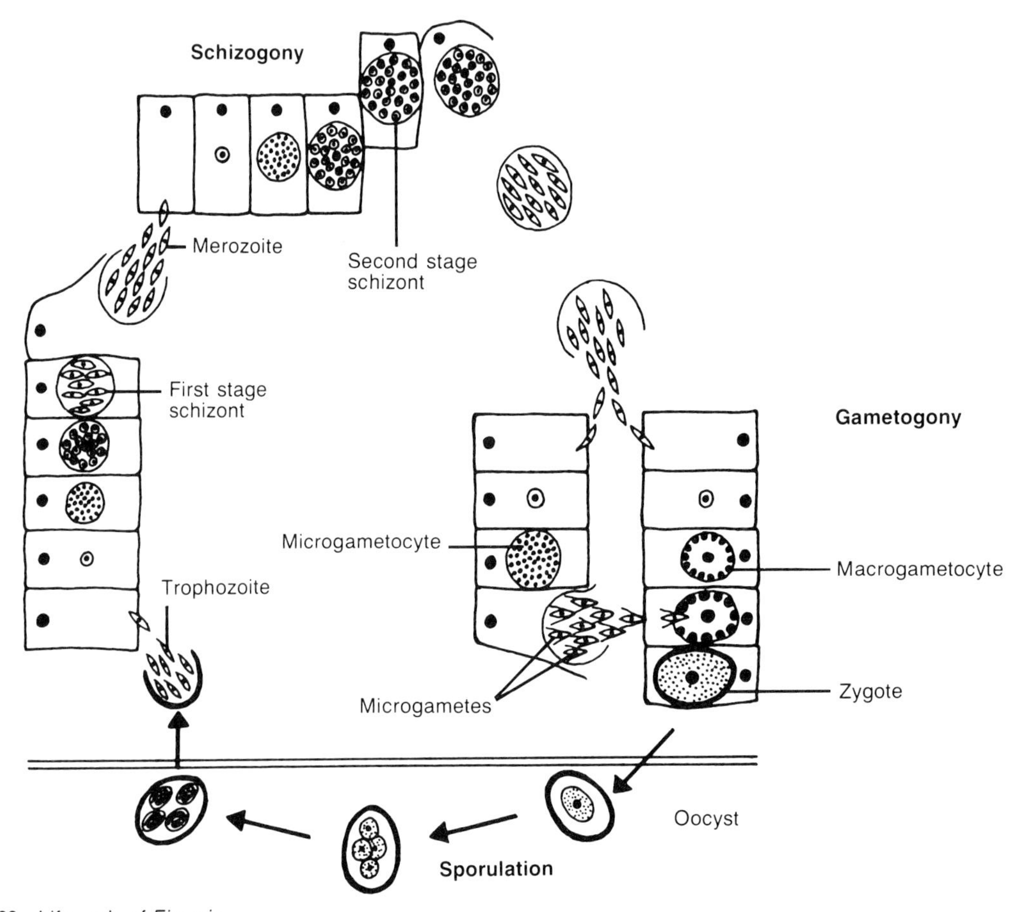

Fig 162 Life cycle of *Eimeria*.

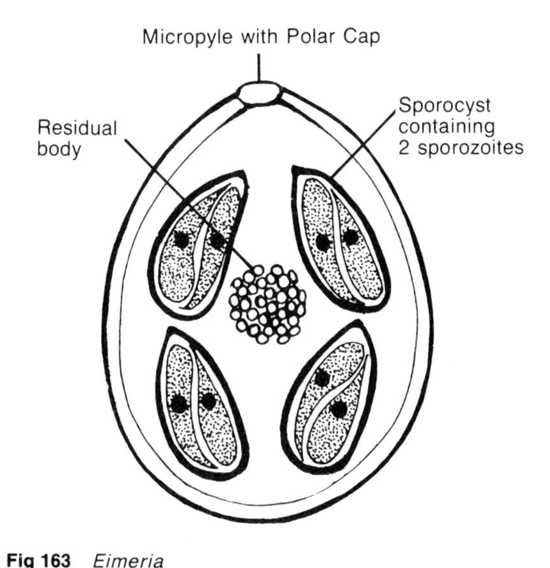

Fig 163 *Eimeria*

(a) Structure of sporulated oocyst.

merozoites they contain. The merozoites are arranged as a series of crescent shaped organisms (5.0–10 μm) rather akin to the appearance of a sliced onion. In contrast, in the mature microgametocyte the microgametes are arranged around the periphery of the cell and measure about 5.0 μm in length. The macrogametocyte has a large central nucleus with small granules arranged around the periphery of the cell. It is equivalent in size to the oocyst which will eventually develop from it.

LIFE CYCLE

This is divided into three phases: sporulation, infection and schizogony, and finally, gametogony and oocyst formation. (Fig. 162).

Sporulation

Unsporulated oocysts, consisting of a nucleated mass of protoplasm enclosed by a resistant wall, are passed to the exterior in the faeces. Under suitable conditions of oxygenation, high humidity and optimal temperatures

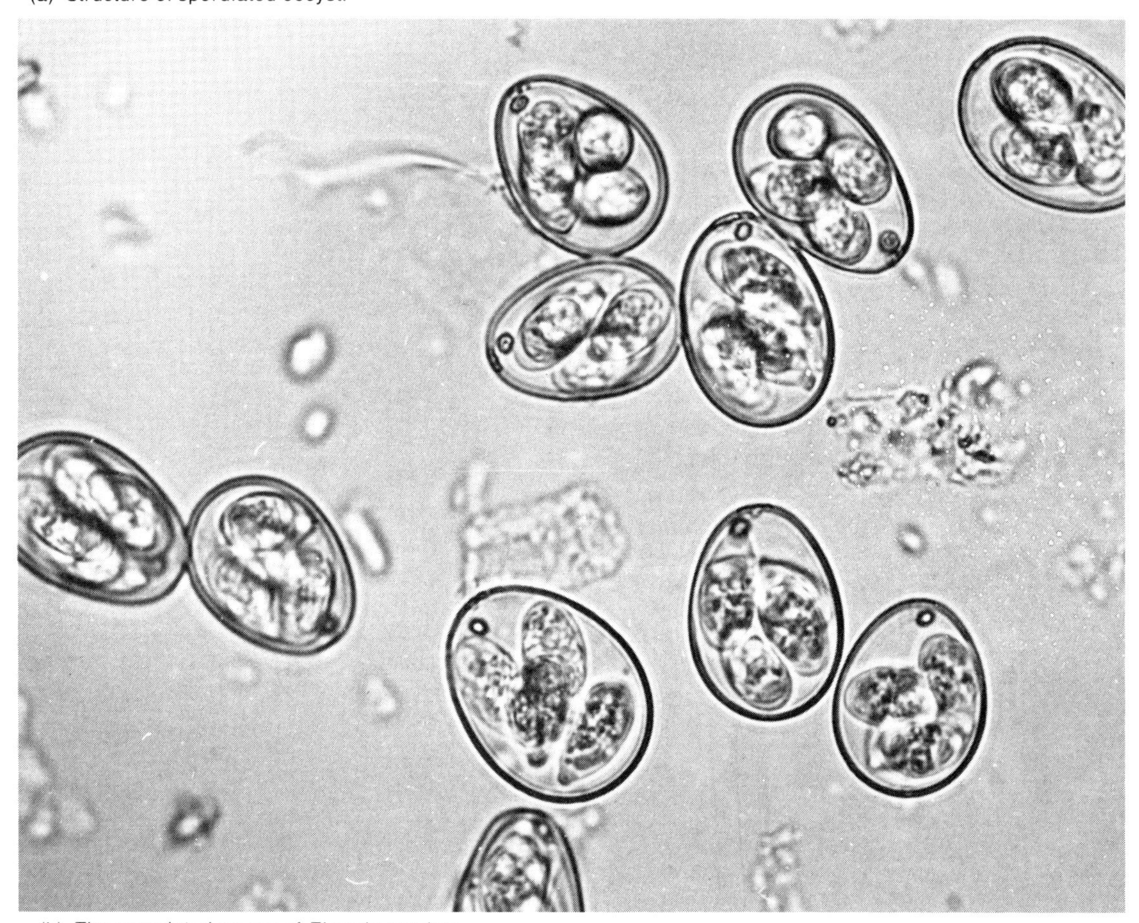

(b) The sporulated oocyst of *Eimeria maxima*.

of around 27 °C, the nucleus divides twice and the protoplasmic mass forms four conical bodies radiating from a central mass. Each of these nucleated cones becomes rounded to form a **sporoblast**, while in some species the remaining protoplasm forms the oocystic residual body. Each sporoblast secretes a wall of refractile material and becomes known as a **sporocyst**, while the protoplasm within divides into two banana-shaped **sporozoites**. In some species the remaining protoplasm within the sporocyst forms a sporocystic residual body.

The time taken for these changes varies according to temperature, but under optimal conditions usually requires 2–4 days. The oocyst, now consisting of an outer wall enclosing four sporocysts each containing two sporozoites, is referred to as a **sporulated oocyst** and is the infective stage (Fig. 163).

Infection and schizogony (asexual reproduction)

The host becomes infected by ingesting the sporulated oocyst. The sporocysts are then liberated either mechamically or by CO_2, and the sporozoites, activated by trypsin and bile, leave the sporocyst. In most species, each sporozoite penetrates an epithelial cell, rounds up, and is then known as a **trophozoite**. However, in a few species, for example *E. tenella*, the sporozoites penetrate the epithelium, are taken up by macrophages in the lamina propria of the villi and are transported to a position deep in the mucosa where they leave the macrophages and enter the epithelial cells to form trophozoites. After a few days each trophozoite has divided by multiple fission to form a **schizont** (Fig. 164), a structure consisting of a large number of elongated nucleated organisms known as **merozoites**. When division is complete and the schizont is mature, the host cell and the schizont rupture and the merozoites escape to invade neighbouring cells. Schizogony may be repeated, the number of schizont generations depending on the species.

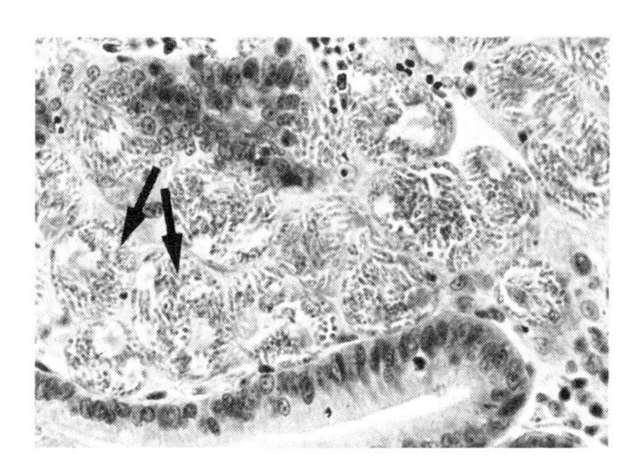

Fig 164 *Eimeria*
(a) Schizonts of *Eimeria necatrix* (↑).

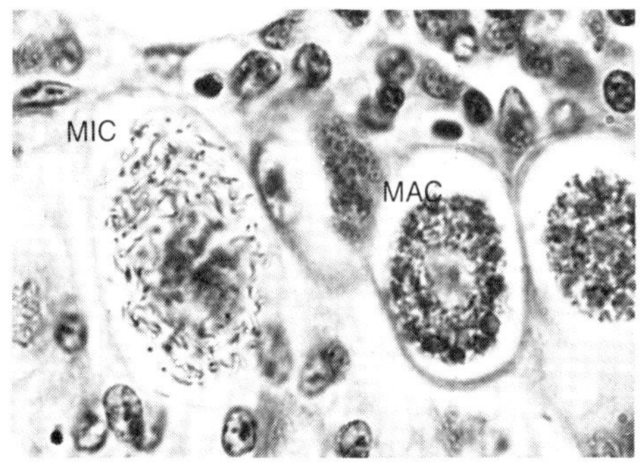

(b) Microgametocyte and macrogametocyte of *Eimeria maxima*.

Gametogony and oocyst formation (sexual reproduction)

Schizogony terminates when the merozoites give rise to male and female gametocytes (Fig. 164). The factors responsible for this switch to gametogony are not fully known. The **macrogametocytes** are female and remain unicellular, but increase in size to fill the parasitised cell. They may be distinguished from trophozoites or development schizonts by the fact that they have a single large nucleus. The male **microgametocytes** each undergo repeated division to form a large number of flagellated uninucleate organisms, the **microgametes**. It is only during this brief phase that coccidia have organs of locomotion. The microgametes are freed by rupture of the host cell, one penetrates a macrogamete, and fusion of the micro- and macrogamete nuclei then takes place. A cyst wall forms around the resulting **zygote** now known as an oocyst, and no further development usually takes place until this **unsporulated oocyst** is liberated from the body in the faeces.

The prepatent period varies considerably and may be as short as five days in poultry and up to 3–4 weeks in some ruminant species.

Isospora

The genus *Isospora* contains many species and like *Eimeria* parasitises a wide range of hosts.

The **important species** include *I. suis* in the pig, *I. canis* and *I. ohioensis* in the dog and *I. felis* and *I. rivolta* in the cat. The life cycle of *Isospora* species differs from *Eimeria* in three respects. First, that the sporulated oocyst contains two sporocysts each with four sporozoites. Secondly, that extraintestinal stages occurring in the spleen, liver and lymph nodes of the pig may reinvade the intestinal mucosa and cause clinical signs. Thirdly, that rodents may, by the ingestion of oocysts from the dog and cat, become infected with asexual stages and act as reservoirs.

GENERAL ASPECTS OF EIMERIA AND ISOSPORA

EPIDEMIOLOGY AND IMMUNITY

In both *Eimeria* and *Isospora*, certain types of management involving deep litter poultry houses, cattle yards or piggeries, offer optimal conditions of temperature and humidity for oocyst sporulation; with overcrowding, the risk of heavy infection is further increased. Although sporulation of oocysts can occur within two days of being passed in the faeces, on pasture this period may be much longer. Oocysts have a considerable longevity and can persist for several years.

There is also some evidence that the life cycle of certain coccidia of ruminants can be delayed or arrested in development at the schizogonous stage; resumption of development several months later with subsequent shedding of oocysts can play an important part in the epidemiology of bovine coccidiosis.

Immunity develops following infection, the immunogenic stages varying according to species, but generally being those involved in schizogony. The mechanism of the response is not fully understood, but is thought to be a combination of cellular and humoral factors. Both *Eimeria* spp. and *Isospora* spp. are highly host specific and immunity to any one species is only effective for that species.

PATHOLOGY

Both *Eimeria* and *Isospora* induce changes in the intestinal mucosa, the severity of which is related to parasite density and to the location of the parasites within the mucosa. Following rupture of the cells containing schizonts or gamonts, the tissue usually slowly recovers its basic morphology. In very heavy infections with species in which the developing schizonts are found deep in the mucosa or in the sub-mucosa, for example *E. tenella* in chickens, destruction is so severe that haemorrhage occurs. In lighter infections, the effect on the intestinal mucosa is to impair local absorption.

In species which develop more superficially, for example *E. acervulina* in chickens, the infection results in a change in villous architecture with a reduction in epithelial cell height and a diminution of the brush border giving the appearance of a 'flat mucosa'. These changes result in a reduction of the surface area available for absorption and consequently a reduced feed efficiency.

COCCIDIOSIS OF POULTRY

In domestic poultry this can conveniently be divided into caecal and intestinal coccidiosis.

CAECAL COCCIDIOSIS

E. tenella is the species primarily responsible for caecal coccidiosis although the gametogonous stages of *E. necatrix* also take place in the caecum and occasionally some stages of *E. brunetti*.

Coccidiosis due to *E. tenella* occurs principally in chickens of 3–7 weeks of age. As noted previously, the first stage schizonts of this species develop deep in the glands. The second stage schizonts are also unusual in that the epithelial cells in which they develop leave the mucosa and migrate into the lamina propria and submucosa. When these schizonts mature and rupture, about 72 hours after ingestion of oocysts, haemorrhage occurs, the mucosal surface is largely detached and clinical signs become apparent. The prepatent period is seven days and the ovoid oocysts sporulate in 2–3 days under normal conditions in poultry houses.

Clinical disease occurs when large numbers of oocysts are ingested over a short period and is characterised by the presence of soft faeces often containing blood. The chicks are dull and listless, with drooping feathers. In subclinical infections, there are poor weight gains and food conversion rates.

At post-mortem examination of chickens which had blood in their faeces, the caeca are found to be dilated and contain a mixture of clotted and unclotted blood (Pl. XIV). In longer-standing infections the caecal contents become caseous and adherent to the mucosa. As regeneration of the mucosa occurs these caecal plugs are detached and caseous material is shed in the faeces.

Although a good immunity to reinfection develops, it should be emphasised that recovered birds often continue to shed a few oocysts and so act as carriers.

COCCIDIOSIS OF THE SMALL INTESTINE

There are several important species of which *E. necatrix* is the most pathogenic. However, the prevalence of disease due to this species, and of caecal coccidiosis due to *E. tenella*, has declined since many of the anti-coccidial drugs in general use were developed specifically to control these two pathogenic species. As a result, the other small intestinal species have assumed a greater prevalence. These include *E. brunetti*, which is highly pathogenic, but more commonly *E. acervulina*, *E. maxima* and *E. mitis* which are moderately pathogenic and *E. praecox* which is only a minor pathogen. The prepatent periods vary from 4–7 days. As in caecal coccidiosis, clinical disease occurs about three days following the ingestion of large numbers of oocysts.

Generally, older chickens are affected by the species found in the small intestine, and clinical signs are similar to those of caecal coccidiosis with the exception that only certain species such as *E. necatrix* and *E. brunetti* cause sufficient damage for blood to appear in the faeces. Subclinical infections are more common than

Table 7 Differential characters of six important species of *Eimeria* in the domestic fowl

	E. tenella	*E. necatrix*	*E. brunetti*	*E. acervulina*	*E. maxima*	*E. mitis*
Region	Caeca	S.I.	Lower S.I.	Upper S.I.	Mid. S.I.	Lower S.I.
Intestinal lesions	Haemorrhage White spots	Haemorrhage Thickened wall White spots	Slight haemorrhage Coagulative necrosis	Watery exudate White transverse bands	Salmon pink exudate Thickened walls Haemorrhage with heavy infections	No visible lesions
Blood in faeces	+ +	+	±	−	−	−
Degree of pathogenicity	+ + + +	+ + + +	+ + +	+ +	+ +	+ +
Oocyst size μm	23 × 19	20 × 17	25 × 19	18 × 14	30 × 20	16 × 15
50% sporulation time in hours at 29 °C	21	20	38	12	38	19

overt disease and may be suspected when pullets have poor rates of growth and feed conversion, and the onset of egg laying is delayed.

At post-mortem examination, the site and severity of the lesions vary according to species and these are summarised in Table 7 together with the relevant oocyst morphology and sporulation times.

In turkeys, coccidiosis due to two species, *E. meleagrimitis* and *E. adenoeides*, occurs in the small intestine and in the small intestine and caeca, respectively. The prepatent period of both is five days. The oocysts of *E. meleagrimitis* are sub-spherical and measure approximately 19 × 16 μm, unlike those of *E. adenoeides* which are ellipsoidal and measure approximately 25 × 17 μm. Sporulation time for both at room temperature is 24 hours.

The pathogenesis of turkey coccidiosis is associated with schizogony and gametogony. *E. adenoeides* has two generations of schizonts and the clinical signs first appear four days after infection coincident with the rupture of the second stage schizonts. *E. meleagrimitis* has three generations of schizogony and disease occurs with the rupture of the third stage schizonts, also four days after infection.

Disease is seen in turkey poults 2–10 weeks of age and rarely in older birds because of acquired immunity. The affected poults are dull, listless, stand with ruffled feathers and have their heads tucked under their wings. Their droppings are white and mucoid and may contain blood, particularly in *E. adenoeides* infections.

There is comparatively little information on coccidiosis of ducks and geese. *E. anseris* and *E. nocens* have been reported as causing acute intestinal coccidiosis in goslings in Britain while another species *E. truncata*, found in the kidneys of geese, can cause an acute nephritis especially where domestic geese are reared intensively: outbreaks have also been recorded in geese in wildfowl sanctuaries.

DIAGNOSIS

Diagnosis is best based on post-mortem examination of a few affected birds. Although oocysts may be detected on faecal examination it would be wrong to diagnose solely on such evidence for two reasons. First, the major pathogenic effect usually occurs prior to oocyst production, and secondly, depending on the species involved, the presence of large numbers of oocysts is not necessarily correlated with severe pathological changes in the gut. At necropsy the location and type of lesions present provide a good guide to the species which can be confirmed by examination of the oocysts in the faeces and the schizonts and oocysts present in scrapings of the gut.

TREATMENT

This should be introduced as early as possible after a

diagnosis has been made. Sulphonamide drugs are the most widely used and it is recommended that these are given for two periods of three days in the drinking water, with an interval of two days between treatments. Sulphaquinoxaline, sometimes potentiated with diaveridine, or sulphamezathine are the drugs of choice. Where resistance has occurred to sulphonamides, mixtures of nitrofurazone and furazolidone, or amprolium and ethopabate have given good results.

In the successful treatment of an outbreak of coccidiosis the aim is to treat birds already affected and at the same time allow sufficient schizogonous development in the clinically unaffected birds to stimulate their resistance. For this reason, sulphonamides, which have their greatest effect on the second stage schizont without inhibiting development of the first stage are the best drugs in the treatment of poultry coccidiosis.

CONTROL

Prevention of avian coccidiosis is based on a combination of good management and the use of anticoccidial compounds in the feed or water. Thus, litter should always be kept dry and special attention given to litter near water fonts or feeding troughs. Fonts which prevent water reaching the litter should always be used and they should be placed on drip trays or over the droppings pit. Feeding and watering utensils should be of such a type and height that they cannot be contaminated by droppings. Good ventilation will also reduce the humidity in the house and help to keep litter dry. Preferably, clean litter should always be provided between batches of birds. If this is not possible, the litter should be heaped and left for 24 hours after it has reached a temperature of 50 °C; it should then be forked over again and the process repeated to ensure that all the oocysts in the litter have been destroyed.

The use of anti-coccidial agents depends on the type of management concerned.

Broiler chicks

These are on life-time medicated feed and the anticoccidials used are maintained at a level sufficient to prevent schizogony. The drugs most frequently used are monensin, dinitolmide, salinomycin and halofuginone. It is recommended that drugs are switched between batches of broilers, the so-called 'switch programme' or within the life span of each batch, the 'shuttle programme'. Most drugs have a minimum period for which they must be withdrawn before the birds can be slaughtered for human consumption. This is usually 5 to 7 days.

Replacement stock

Where replacement laying birds spend their whole life on wire floors, no medication is necessary; if they are reared on litter, for eventual production on wire, then a full level of coccidiostat is given as for broilers. If they are reared on litter, for production on litter, then a programme of anti-coccidials designed to stimulate immunity is used. Two preparations frequently used are dinitolmide and a mixture of amprolium, ethopabate and sulphaquinoxaline. The procedure is to administer these drugs in a decreasing level over the first 16 or 18 weeks of life. This may be done as a two-stage reduction, ie. between 0 and 8 weeks and 8 and 16 weeks, or, alternatively, as a three-stage reduction, from 0–6 weeks, 6–12 weeks and 12–18 weeks. Using this technique, complete protection against coccidial challenge is maintained in the very young birds and the reduced drug rate in older birds allows limited exposure to developing coccidia so that acquired immunity can develop.

When in-feed coccidiostats are used, there are two further factors to consider. First, outbreaks of coccidiosis may occur in birds on medicated feed either because the level of coccidiostat used is too low or because conditions in the house have changed to allow a massive sporulation of oocysts which, on ingestion, the level of drug can no longer control. Secondly, the influence of intercurrent infections in affecting appetite, and therefore uptake of coccidiostat, should also be considered.

In turkeys, sulphonamides are again the drugs of choice in treatment. For prophylaxis, coccidiostats are normally incorporated in the feed for the first 12–16 weeks of life. The drugs commonly used are sulphaquinoxaline, a mixture of amprolium, ethopabate and sulphaquinoxaline, or halofuginone at a low concentration of 3 p.p.m. Monensin is also used, although it is available only on prescription because of its greater toxicity for turkeys.

In the U.S.A. a 'vaccine' consisting of oocysts of eight species of coccidia is commercially available. Young chicks are given this live vaccine in the drinking water, and 10 days later a coccidiostat is introduced into the feed for a period of 3–4 weeks. Successful immunisation has also been achieved under experimental conditions with oocysts attenuated by irradiation. For success both techniques depend on subsequent exposure to oocysts to boost immunity and this may not occur unless litter is sufficiently moist to allow sporulation. There is considerable interest in developing more efficient vaccines in view of the increasing problem of drug resistance in coccidiosis.

COCCIDIOSIS OF CATTLE

Bovine coccidiosis occurs worldwide and usually affects cattle under one year old, but is occasionally seen in yearlings and adults. Of the 13 species recorded, the two principal pathogens are *E. zuernii* and *E. bovis*. The former is particularly pathogenic, attacking the caecum and colon and, in heavy infections, produces a severe bloodstained dysentery accompanied by tenesmus.

E. zuernii has a prepatent period of 17 days and produces small, spherical oocysts of 16 µm in diameter. *E. bovis* also affects the caecum and colon producing a severe enteritis and diarrhoea in heavy infections. Characteristically, schizonts may be found in the central lacteals of the villi (Fig. 165). The prepatent period is 18 days and the oocysts are large, egg-shaped and measure 28 × 20 µm. The disease is dependent on epidemiological conditions which precipitate a massive intake of oocysts, such as overcrowding in unhygienic yards or feed lots. It may also occur at pasture where livestock congregate around water troughs.

Diagnosis is based on history and clinical signs, and in patent infections, on the presence of oocysts of the pathogenic species in the faeces.

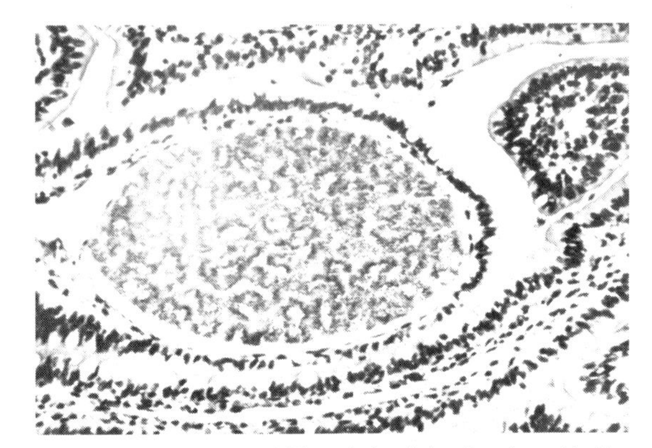

Fig 165 The schizont of *Eimeria bovis* is often found in the central lacteal of a villus.

Treatment with sulphamezathine, given parenterally and repeated at half the initial dose level on each of the next two days, is recommended. Alternatively, amprolium may be used as a drench or in the drinking water over a five day period.

Prevention is based on good management; in particular feed troughs and water containers should be moved regularly and bedding kept dry.

COCCIDIOSIS OF SHEEP AND GOATS

Clinical coccidiosis of sheep and goats occurs mainly in young lambs and kids and there appears to be an increasing prevalence under conditions of intensive husbandry. Although the majority of sheep, particularly those under one year old, carry coccidia only two of the eleven species are known to be highly pathogenic. These are *E. crandallis* and *E. ovinoidalis* both of which have a prepatent period of 15 days. The oocysts of *E. crandallis* are thick-shelled and sub-spherical while those of *E. ovinoidalis* are ellipsoidal, with a distinct inner shell; both have polar caps. Differentiation of the many species is a specialist task.

Heavy infections in lambs are responsible for severe diarrhoea which sometimes contains blood. The pathogenic lesions are mainly in the caecum and colon where gametogony of *E. crandallis* and second-stage schizogony and gametogony of *E. ovinoidalis* occur. The lesions cause local haemorrhage and oedema, and villous atrophy may be a sequel resulting in malabsorption.

Several species, including *E. ovinoidalis*, produce giant schizonts, up to 300 µm in diameter, which may be seen grossly as white spots in the lower small intestine. Also in this site papilloma-like lesions may occur, usually as a sequel to gametocyte formation by *E. ovina*, but these are not of great pathogenic significance (Pl. XIV).

Lambs are usually affected between four and seven weeks of age with a peak infection around six weeks. The outbreaks reported have occurred where ewes and lambs were housed in unhygienic conditions or grazed intensively. The feeding of concentrates in stationary troughs, around which has occurred heavy contamination with oocysts, can also be a precipitating factor. In the U.S.A. coccidiosis has occurred when older lambs are confined in feedlots after weaning.

In spring-lambing flocks in Western Europe, infection of lambs results both from oocysts which have survived the winter and from those produced by ewes during the periparturient period.

Diagnosis is based on the management history, the age of the lambs, post-mortem lesions and faecal examination for oocysts. The latter may be present in very large numbers in both healthy and diseased lambs so that a necropsy is always advisable.

Treatment is with sulphadimidine or amprolium as described for cattle or by a single parenteral injection of a long-acting sulphonamide.

Prevention is based on good management and regular moving of feed and water troughs, but in some intensive flocks in which the problem occurs annually, low levels of amprolium may be included in the concentrate feed.

Less is known about the problem of coccidiosis in goats, but oocysts are frequently recorded from the faeces and one species, *E. arloingi*, has been reported as causing severe pathology. It is not clear whether sheep species can affect goats, but this seems unlikely in view of the strong host-specificity expressed by *Eimeria* spp.

COCCIDIOSIS OF PIGS

Although some ten species of coccidia have been described from pigs their importance is not clear. *E. debliecki* has been described as causing clinical disease and severe pathology and it is only comparatively recently that another species, *Isospora suis*, has been incriminated as the cause of a naturally occurring severe enteritis in young piglets aged 1–2 weeks. *I. suis* has a short prepatent period of 4–6 days; the ellipsoidal oocysts measure 17 × 13 µm and when sporulated contain the two sporocysts each with four sporozoites characteristic of *Isospora*. The main clinical signs are diarrhoea, often biphasic, which varies in its severity from loose faeces to persistent fluid diarrhoea.

The source of infection appears to be oocysts produced by the sow during the periparturient period, the piglets becoming initially infected by coprophagia; the second phase of diarrhoea is initiated by reinvasion from tissue stages. Diagnosis of the condition is difficult unless post-mortem material is available since clinical signs occur prior to the shedding of oocysts and are very similar to those caused by other pathogens such as rotavirus.

Treatment with amprolium given orally to affected piglets is usually effective while prevention can be achieved by the in-feed administration of amprolium to sows during the periparturient period, that is, from one week prior to farrowing until three weeks post farrowing.

COCCIDIOSIS OF HORSES

Eimeria (syn. *Globidium*) *leuckarti* occurs in the small intestine of horses and donkeys and has been incriminated as the cause of an intermittent diarrhoea. The prepatent period is 15 days and the oval oocysts are very large, measuring 80 × 60 μm, with a thick dark shell and distinct micropyle.

The pathology includes marked inflammatory changes in the mucosa and a disruption of villous architecture.

Diagnosis is difficult, and because of the heavy nature of the oocysts, sedimentation techniques should be employed or, if flotation is used, a concentrated sugar solution is necessary. Little is known about treatment, but by analogy with other hosts sulphamezathine or amprolium should be tried.

COCCIDIOSIS OF RABBITS

There are three main pathogenic species in rabbits, namely *E. stiedae*, *E. flavescens* and *E. intestinalis*. The prepatent period of *E. stiedae* is 18 days and of the others 5–7 days.

The disease is commonest around weaning and clinical signs of *E. stiedae* infection include wasting, diarrhoea, ascites and polyuria. This species, which occurs in the bile ducts, reaches the liver via the portal vein and then locates in the epithelium of the bile ducts where it results in a severe cholangitis. Grossly the liver is enlarged and studded with white nodules. The intestinal species, *E. flavescens* and *E. intestinalis*, which are the more significant in commercial rabbit farms, cause the destruction of the crypts in the caecum resulting in diarrhoea and emaciation.

Diagnosis, as in other species, is best made by a post-mortem examination. However, in practice, the demonstration of many oocysts in the faeces is often used as an indication that rabbits require treatment. The oocysts of *E. stiedae* are ellipsoidal and 37 × 21 μm in size, those of *E. flavescens* are ovoidal and 31 × 21 μm with a micropyle at the broad end while those of *E. intestinalis* are pyriform and measure 27 × 18 μm.

Sulphamezathine or sulphaquinoxaline in the drinking water are used for treatment. Control of rabbit coccidiosis involves the daily cleaning of cages, hutches or pens and the provision of clean feeding troughs. In many large units control is achieved by rearing animals on wire floors, or alternatively, coccidiostats such as amprolium or robenidine are incorporated in the feed.

COCCIDIOSIS OF DOGS AND CATS

At one time it was thought that species of the genus *Isospora* were freely transmissible between dogs and cats, but it is now established that this is not the case.

In the dog, the common *Isospora* species are *I. canis* and *I. ohioensis*. The prepatent period of both is under 10 days and the oocyst of *I. canis* is the larger, measuring 38 × 30 μm, while that of *I. ohioensis* measures 25 × 20 μm. There is no real evidence that these species are pathogenic. The life cycles are normally direct although there is some evidence that a predator-prey relationship may be involved and that dogs can acquire infection from the tissues of rodents infected with asexual stages. It is important that the oocysts of these species be differentiated from those of *Sarcocystis* which are much smaller in size and sporulated when freshly shed in the faeces.

In the cat the common species are *I. felis* and *I. rivolta*, and, as in the dog, infection may be acquired directly or possibly by ingestion of infected small rodents. The prepatent periods are short being 7–8 days. The oocysts of *I. felis* measure 40 × 30 μm whereas those of *I. rivolta*

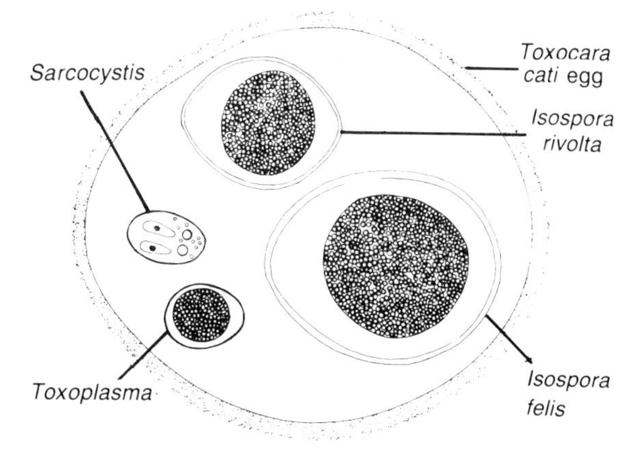

Fig 166 Diagram of oocysts of cat coccidia relative to *Toxocara cati* egg.

measure 25 × 20 μm. Their pathogenicity is generally thought to be low, although severe diarrhoea in young kittens has been associated with high oocyst counts.

Other oocysts found in cat faeces (Fig. 166) belong to members of the Family *Sarcocystidae* and it is important to differentiate the oocysts of *Isospora* from those of *Sarcocystis* spp. and *Toxoplasma gondii*, both of which can infect food animals and occasionally man. The former can be differentiated from *Isospora* by their smaller size (less than 15 × 11 μm) and the presence of sporocysts or sporulated oocysts in fresh faeces while *T. gondii* oocysts, although non-sporulated, measure only 12 × 10 μm.

Information on treatment in the dog and cat is scanty, although by analogy with other host species, sulpha drugs such as sulphamezathine or sulphadimidine should be tried.

Cryptosporidium

Cryptosporidium is thought to be a single species genus and has been found in a large number of domestic animals and man. It has been associated with outbreaks of diarrhoea in calves, lambs, kids, piglets, foals, young dogs and cats and turkey poults. It is remarkable in that, unlike other members of the Eimeriidae, it does not enter the cells of the host and lacks host specificity so that cross infection occurs between domestic and laboratory animals and man.

The life cycle of *Cryptosporidium* is basically similar to those of other intestinal coccidia although, like *Sarcocystis*, sporulation takes place within the host. The minute oocysts (4.0–4.5 μm), each with four sporozoites, are liberated in the faeces. Following ingestion, the sporozoites invade the microvillous brush border of the enterocytes and the trophozoites rapidly differentiate to form schizonts with 4–8 merozoites. Gametogony follows after one to two generations of schizonts and oocysts are produced in 72 hours.

Recent evidence also indicates that two types of oocysts are produced. The first, the majority, are thick-walled and are passed in the faeces. The remainder are thin-walled and release their sporozoites in the intestine, causing auto-infection.

The pathogenesis of infection with *Cryptosporidium* is not clear. The schizonts and gamonts develop in a parasitophorous envelope apparently derived from the microvilli (Fig. 167) and so the cell disruption seen in other coccidia does not apparently occur. However, mucosal changes are obvious in the ileum where there is stunting, swelling and eventually fusion of the villi; this has a marked effect on the activity of some of the membrane-bound enzymes.

Clinically the disease is characterised by anorexia and diarrhoea, often intermittent, which may result in poor growth rates. Vomiting and diarrhoea have been reported in young piglets with combined rotavirus and *Cryptosporidium* infections. Oocysts may be demonstrated by using Ziehl-Nielsen stained faecal smears in which the sporozoites appear as bright red granules (Pl. XIV). More accurate diagnosis is based on sophisticated staining techniques including immuno-fluorescence.

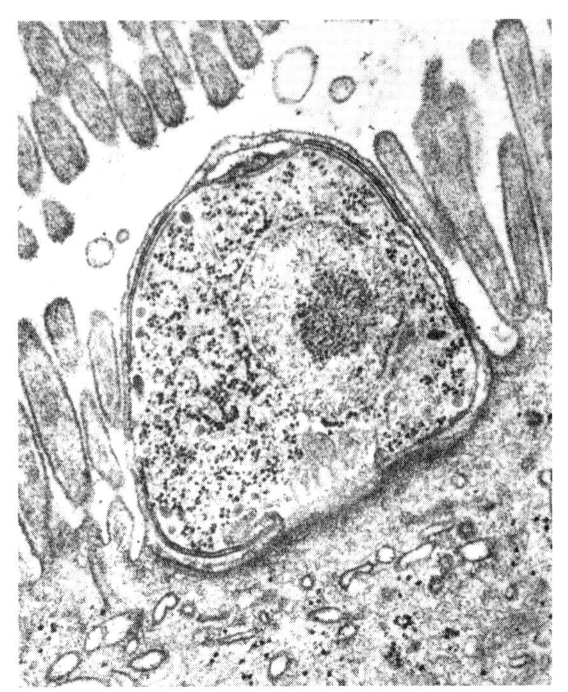

Fig 167 Electron micrograph showing *Cryptosporidium* macrogamont on the microvillous border of an epithelial cell.

There is no known treatment, although spiramycin may be of some value, and the infection is difficult to control since the oocysts are highly resistant to most disinfectants except formol saline and ammonia.

[**Cryptosporidiosis in man:** Immunosuppressed states, especially AIDS (acquired immunodeficiency syndrome), have been associated with severe cryptosporidiosis, presumably due to overwhelming auto-infection.]

Family SARCOCYSTIDAE

Four genera, *Toxoplasma*, *Sarcocystis*, *Besnoitia* and *Hammondia*, are of veterinary importance. Their life cycles are similar to *Eimeria* and *Isospora* except that the asexual and sexual stages occur in intermediate and final hosts respectively.

With the exception of the genus *Toxoplasma*, they are normally non-pathogenic to their final hosts and their significance is due to the cystic tissue stages in the intermediate hosts which include ruminants, pigs, horses and

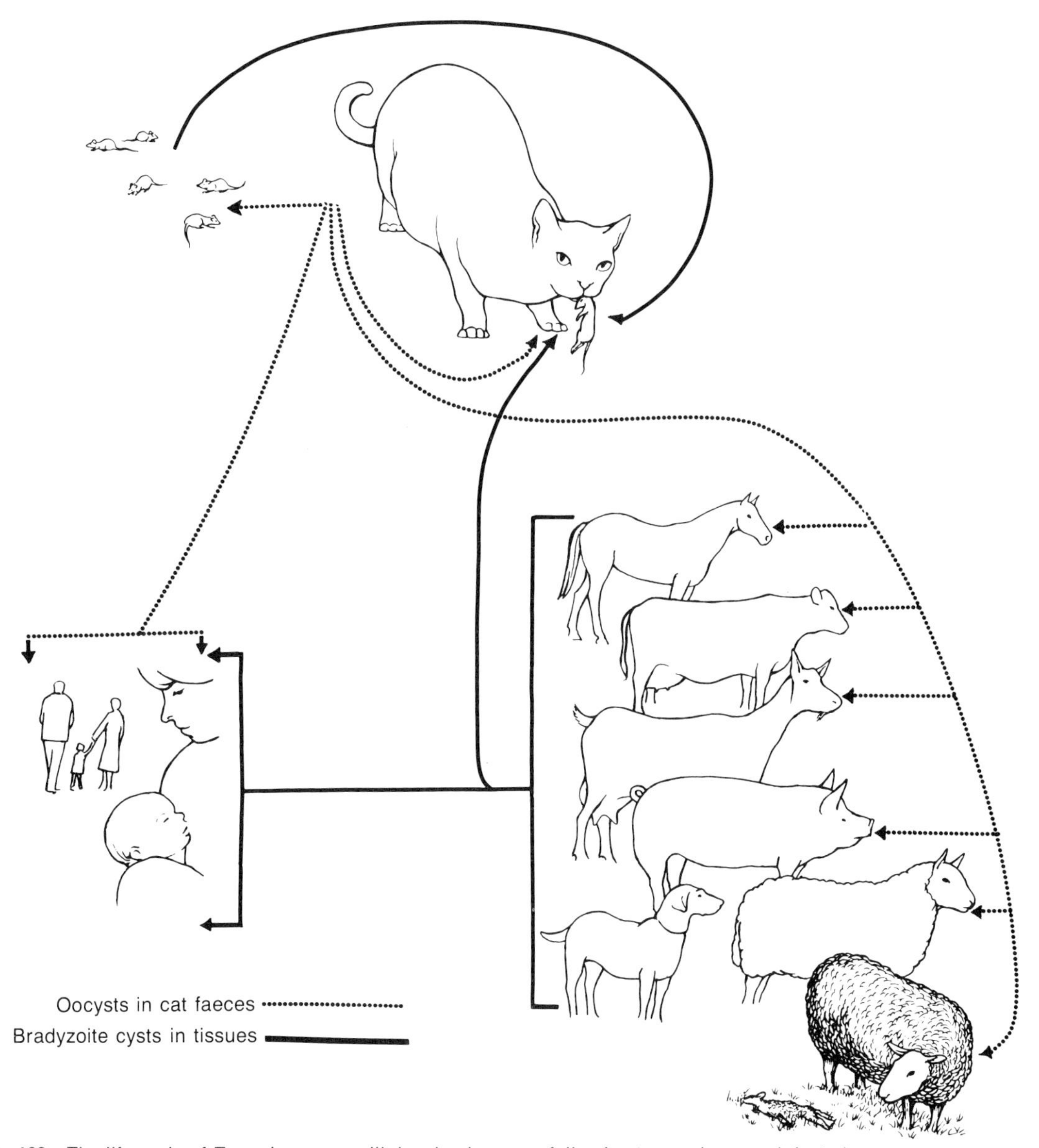

Fig 168 The life cycle of *Toxoplasma gondii* showing how non-feline hosts may become infected.
Note: The dog is an unlikely source of infection for man.

man. The tissue phase in the intermediate host is obligatory, except in *Toxoplasma* where it is facultative.

Toxoplasma

The genus *Toxoplasma* has a single species, *T. gondii*, which is an intestinal coccidian of cats. The life cycle includes a facultative systemic phase which is an important cause of abortion in sheep and may also cause a zoonosis. Human infections are particularly serious if they occur during pregnancy and may result in abortion or congenitally acquired disorders which primarily affect the central nervous system.

Species:
Toxoplasma gondii

Final hosts:
All felids. The domestic cat is the most important

Intermediate hosts:
Any mammal, including man, or birds

Note that the final host, the cat, may also be an intermediate host and harbour extra-intestinal stages

Sites in final host:
Schizonts and gamonts in the small intestine

Sites in intermediate host:
Tachyzoites and bradyzoites in extra-intestinal tissues including muscle, liver, lung and brain

Distribution:
Worldwide

IDENTIFICATION

Oocysts

These are found in the faeces of cats, are unsporulated and measure $12 \times 10\,\mu m$ (Fig. 166). When sporulated, which takes at least three days, the oocyst contains two sporocysts, each with four sporozoites.

Intestinal stages

Schizonts. These occur mainly in the jejunum and ileum, range in size from $4.0\,\mu m$ to $17\,\mu m$ in diameter and contain up to 32 merozoites.
Gamonts. These are most common in the ileum and measure approximately $10\,\mu m$ in diameter.

Extra-intestinal stages

Tachyzoites. These are found developing in vacuoles in many cell types, for example, fibroblasts, hepatocytes, reticular cells and myocardial cells. In any one cell there may be 8–16 organisms, each measuring 6.0–8.0 μm.
Bradyzoites. These are contained in cysts and occur mainly in the muscle, liver, lung and brain. The bradyzoites are lancet shaped and several thousand may be present in one cyst which can measure up to $100\,\mu m$ in diameter.

LIFE CYCLE

Final host

Most cats become infected by ingesting *Toxoplasma*-infected animals, usually rodents, whose tissues contain tachyzoites or bradyzoites, although direct transmission of oocysts between cats can also occur (Fig. 168). The ingestion of mature bradyzoites is the most important route and results in the shedding of higher numbers of oocysts than when infection is acquired from other stages.

Following infection, the cyst wall is digested in the cat's stomach, and in the intestinal epithelium the liberated bradyzoites initiate a cycle of schizogonous and gametogonous development culminating in the production of oocysts in 3–10 days. Oocysts are shed for only 1–2 weeks. During this cycle in the intestinal mucosa, the organisms may invade the extra-intestinal organs where the development of tachyzoites and bradyzoites proceeds as in intermediate hosts.

Intermediate hosts

This part of the cycle is extra-intestinal and results in the formation of tachyzoites and bradyzoites which are the only forms found in non-feline hosts. Infection of intermediate hosts may occur in two ways.

In the first, sporulated oocysts are ingested and the liberated sporozoites rapidly penetrate the intestinal wall and spread by the haematogenous route. This invasive and proliferative stage is called the **tachyzoite** (Fig. 169) and on entering a cell it multiplies asexually in a vacuole by a process of budding or **endodyogeny**, in which two individuals are formed within the mother cell, the pellicle of the latter being used by the daughter cells. When 8–16 tachyzoites have accumulated the cell ruptures and new cells are infected. This is the **acute phase of toxoplasmosis**.

In most instances, the host survives and antibody is produced which limits the invasiveness of the tachyzoites and results in the formation of cysts containing thousands of organisms which, because endodyogeny and growth are slow, are termed **bradyzoites** (Fig. 169).

Fig 169 Asexual stages of *Toxoplasma gondii*.
(a) Tachyzoites.

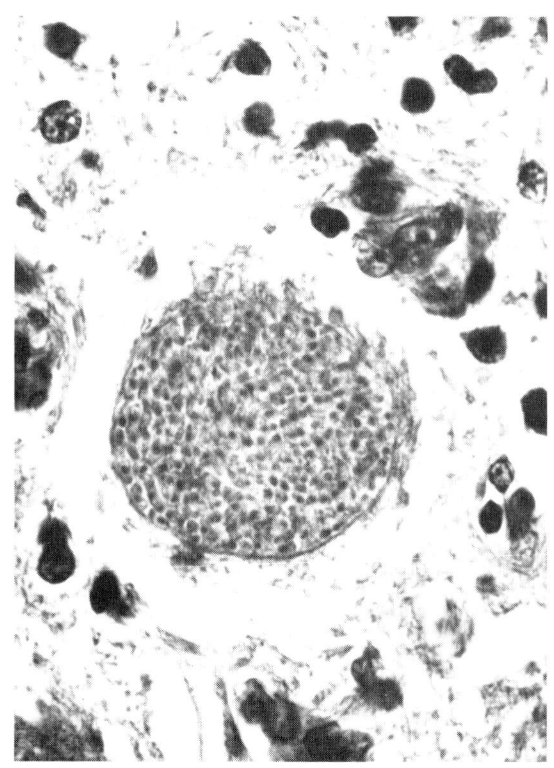

(b) Bradyzoite cyst.

The cyst containing the bradyzoites is the latent form, multiplication being held in check by the acquired immunity of the host. If this immunity wanes the cyst may rupture, releasing the bradyzoites which become active and resume the invasive characteristics of the tachyzoites.

Secondly, infection commonly occurs by the ingestion of bradyzoites and tachyzoites in the flesh of another intermediate host. Thus carnivores and humans can acquire infection by eating raw or underdone meat. The cycle of development following infection by tachyzoites or bradyzoites is similar to that following oocyst ingestion.

GENERAL PATHOLOGY

This is always related to the extra-intestinal phase of development.

Most infections are acquired via the digestive tract, and so organisms are disseminated by the lymphatics and portal system with subsequent invasion of various organs and tissues. In heavy infections, the multiplying tachyzoites may produce areas of necrosis in vital organs such as the myocardium, lungs, liver and brain and during this phase the host can become pyrexic and lymphadenopathy occurs. As the disease progresses bradyzoites are formed, this chronic phase being usually asymptomatic.

In pregnant animals or humans, exposed for the first time to *T. gondii* infection, congenital disease may occur. The predominant lesions are found in the central nervous system although other tissues may be affected. Thus, retinochoroiditis is a frequent lesion in congenital toxoplasmosis. The retina becomes inflamed and necrotic and the pigmented layer is disrupted by infiltration of inflammatory cells; eventually, granulation tissue forms and invades the vitreous humor.

Fortunately, the pathological changes described above are relatively uncommon and most *Toxoplasma* infections in animals and man are light and consequently asymptomatic.

SPECIAL PATHOLOGY AND CLINICAL SIGNS IN DIFFERENT HOSTS

Cats

Despite the fact that cats are frequently infected, clinical disease is rare although enteritis, enlarged mesenteric lymph nodes, pneumonia, degenerative changes in the central nervous system and encephalitis have been recorded in experimental infections. Congenital transmission, although uncommon, has occurred following activation of bradyzoite cysts during pregnancy.

Dogs

The onset of illness is marked by fever with lassitude, anorexia and diarrhoea. Pneumonia and neurological manifestations are common. Infection may occur in conjunction with distemper and has also been incriminated in distemper vaccination breakdowns.

At necropsy, bradyzoite cysts can be demonstrated in cells in the brain and the respiratory tract; the associated lymph nodes are enlarged.

Ruminants

There are only a few reports of clinical toxoplasmosis associated with fever, dyspnoea, nervous signs and abortion. At post-mortem, bradyzoites were demonstrable in the brain with focal necrosis in acute cases and glial nodules in chronic cases.

Undoubtedly the most important role of toxoplasmosis in ruminants is its association with **abortion in ewes** and **perinatal mortality in lambs**. If infection of the ewes occurs early in gestation (<55 days) there is death and expulsion of the small foetus, which is seldom observed. If infection occurs in mid-gestation abortion is more readily detected, the organisms being found in the typical white lesions, 2.0 mm in diameter, in the cotyledons of the placenta and in foetal tissues; alternatively the dead foetus may be retained, mummified, and expelled later. If the foetus survives *in utero*, the lamb may be still-born or, if alive, weak.

Ewes which abort due to *T. gondii* in one year usually lamb normally in subsequent years.

Other hosts

Toxoplasmosis has been occasionally reported in young pigs and poultry, while serological titres to *Toxoplasma* have been recorded in horses and wild rabbits.

Man

Infection of man may be acquired or congenital. Acquired infections occur in two ways. First, from the ingestion of oocysts shed in the faeces of cats. This may be directly from hands contaminated, for example, during the cleaning of litter trays or, more likely, indirectly from the ingestion of vegetables or food contaminated by cat faeces. Flies may also transfer oocysts on to food.

Secondly, an important source of infection is the ingestion of undercooked meat containing *Toxoplasma* cysts.

The majority of acquired infections are asymptomatic. Clinically apparent infections present as low grade fever and malaise with a general lymphadenopathy predominantly affecting the cervical nodes, symptoms which are similar to those of glandular fever. Involvement of vital organs is rare although myocarditis, encephalitis and retinochoroiditis have been recorded. Recrudescence of infection may occur in immunosuppressed patients.

Congenital infection, which occurs only when a woman is exposed to infection for the first time during pregnancy, can be serious, the tachyzoites crossing the placenta in the absence of maternal antibodies. While the majority of such congenital infections are asymptomatic, up to 10% result in abortions, stillbirths or damage to the central nervous system of the foetus. The frequency of disease is much higher when infection is acquired in the first trimster of pregnancy. Severely affected infants show retinochoroiditis and cerebral necrosis and there may be hepatosplenomegaly, liver failure, convulsions and hydrocephalus.

EPIDEMIOLOGY

The cat plays a central role in the epidemiology of toxoplasmosis and the disease is virtually absent from areas where cats do not occur.

Epidemiological investigations in the U.S.A and elsewhere indicate that 60% of cats are serologically positive to *Toxoplasma* antigen, the majority acquiring infection by predation. As might be expected infections are more prevalent in stray cats. Congenital infection is rare. Following infection, cats shed oocysts for only 1–2 weeks after which they are resistant to reinfection. However, a proportion remain as carriers, perhaps due to the persistence of some schizonts, and reactivation of infection with shedding of oocysts may occur in association with intercurrent disease, during the periparturient period in queens or following corticosteroid therapy. However, the oocysts appear to be very resistant and this compensates for the comparatively short period of oocyst excretion.

It is difficult to explain the widespread prevalence of toxoplasmosis in ruminants, particularly sheep, in view of the relatively low number of oocysts shed into the environment. It has been suggested that pregnant ewes are most commonly infected during periods of concentrate feeding prior to tupping or lambing, the stored food having been contaminated with cat faeces in which millions of oocysts may be present.

Further spread of oocysts may occur via coprophagous insects which can contaminate vegetables, meat and animal fodder. It has been suggested that venereal transmission can occur in sheep.

The prevalence of *Toxoplasma* infection in the human population, as estimated by serological titres, may be as high as 25% in some areas. The prevalence is higher in veterinarians, abattoir workers and those who handle cats.

DIAGNOSIS

Specific diagnosis is made by serological tests or by demonstration of the organisms in tissues of mice inoculated with suspect material.

Two of the most commonly used tests measure antibody, the Sabin-Feldman dye test and the indirect immunofluorescence test (IFA). The latter is preferred since it does not require live organisms. Whichever test is used it is important to employ paired samples taken at an interval of 1–2 weeks to determine if a rising titre, indicative of recent infection, is present.

More recently, an ELISA test has been developed which is capable of detecting a recent infection by the estimation of IgM, as compared to IgG, antibody.

The most convincing diagnosis is obtained by inoculating *Toxoplasma*-free mice by the intraperitoneal or intracerebral route with test material and the subsequent demonstration of tachyzoites or bradyzoites in smears of organs or serous cavities. It has the disadvantage that unless the strain of *Toxoplasma* is highly virulent, it requires three weeks before examination of the mice will yield recognisable *Toxoplasma* cysts.

TREATMENT

There is no completely satisfactory treatment A combination of the anti-malarial drug pyrimethamine and sulphadiazine has been reported to be effective against tachyzoites, but not bradyzoites, in humans but is rather toxic in cats.

Clindamycin is reported as being effective against murine toxoplasmosis, and like pyrimethamine, will reduce but not eliminate oocyst shedding in cats.

CONTROL

In domestic situations prevention of infection requires the daily cleaning of cat litter boxes and proper disposal of faeces. Hygienic precautions such as washing of hands prior to eating and the wearing of gloves when gardening should also be observed since flower and vegetable beds are favoured areas for cats to defaecate. Pregnant women should not undertake cleaning of cat litter boxes.

In addition, raw meat should not be fed to cats.

On farms, control is more difficult, but where possible animal feedstuffs should be covered to exclude access by cats and insects.

Sheep which abort following toxoplasmosis usually lamb normally in subsequent years. Current advice is that such sheep should be mixed with replacement stock some weeks before mating in the hope that these will become naturally infected and develop immunity before becoming pregnant. Presumably the value of this technique, if any, depends on the replacements being exposed to circumstances similar to those of the initial outbreak.

It is sometimes advised to mix replacement stock with ewes at the time of the outbreak of abortion in order to facilitate transmission of infection. This is extremely unwise since other causes of abortion, notably the agent of enzootic abortion of ewes, if also present, may affect the replacement stock and be responsible for abortion in subsequent years.

Sarcocystis

From a veterinary standpoint the important stages of the genus *Sarcocystis* are found in the intermediate hosts, both as schizonts in the endothelium of the blood vessels and as bradyzoite cysts in the skeletal and cardiac muscles.

Final hosts:
Dogs, cats, wild carnivores and man

Intermediate hosts:
Ruminants, pigs and horses

Site in final host:
Small intestine

Site in intermediate host:
Schizonts in endothelial cells of blood vessels; large cysts containing bradyzoites in muscles

Distribution:
Worldwide

SPECIES

The previously complex nomenclature for the large number of *Sarcocystis* spp. has largely been discarded by many workers in favour of a new system based on their biology. The new names generally incorporate those of the **intermediate** and **final hosts** in that order. Although unacceptable to systematists, this practice has the virtue of simplicity. At present the most important species recognised with the dog as a final host are:

Sarcocystis bovicanis (syn. *S. cruzi*)
S. ovicanis (syn. *S. tenella*)
S. capricanis
S. porcicanis (syn. *S. miescheriana*)
S. equicanis (syn. *S. bertrami*)
S. fayeri (horse/dog)

Those with the cat as the final host include:

S. bovifelis (syn. *S. hirsuta*)
S. ovifelis (syn. *S. tenella*)
S. porcifelis

Man is final host for two species, *S. bovihominis* and *S. porcihominis*, and these are reported as being responsible for anorexia, nausea and diarrhoea.

IDENTIFICATION

Oocysts: These, unlike *Isospora*, are sporulated when passed in the faeces and contain two sporocysts each with four sporozoites; usually the sporulated sporocyst is found free in the faeces (Fig. 170). In two common species, *S. bovicanis* and *S. ovicanis*, the sporulated sporocysts measure approximately $15 \times 10\,\mu m$ and $14 \times 9\,\mu m$ respectively.
Tissue stages: In the intermediate host the schizonts found in the endothelial cells are quite small measuring 2–8 μm in diameter (Fig. 171). In contrast the bradyzoite cysts (Fig. 172) can be very large and visible to the naked eye as whitish streaks running in the direction of the muscle fibres (Plate XIV). They have been reported as reaching several cm in length, but more commonly they range from 0.5 mm to 5.0 mm.

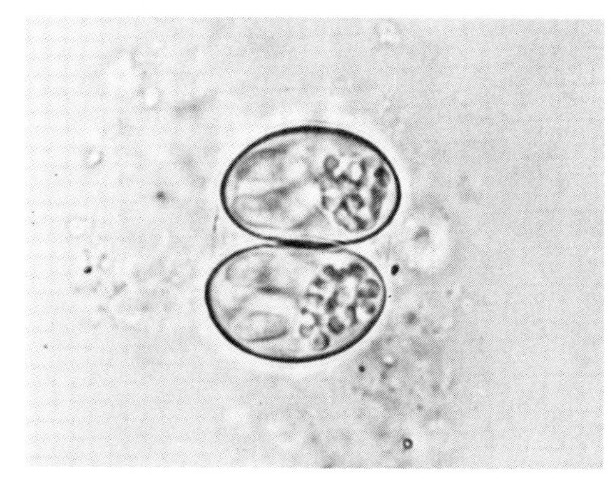

Fig 170 Sporulated sporocysts of *Sarcocystis* in smear of fresh faeces; the oocyst wall is just discernible.

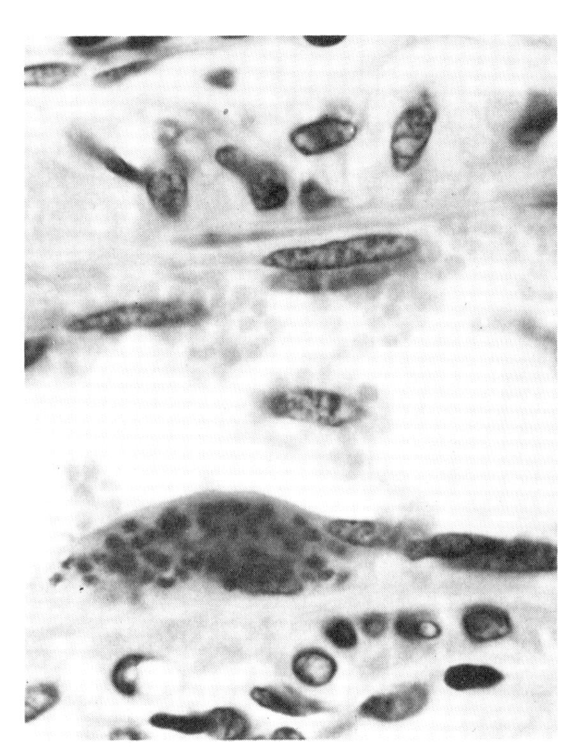

Fig 171 *Sarcocystis* schizont in endothelial cell.

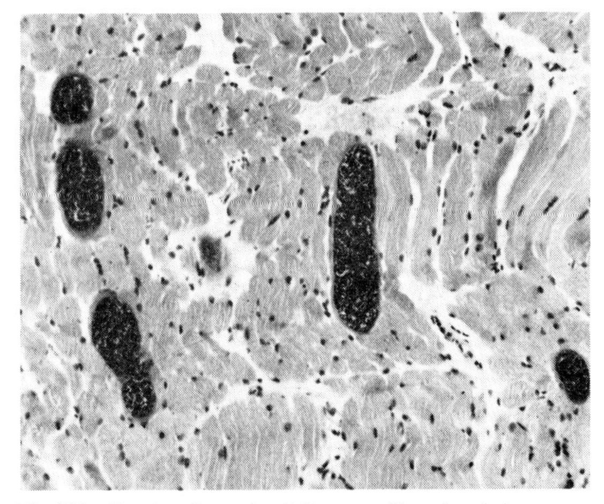

Fig 172 Bradyzoite cysts of *Sarcocystis ovicanis* in oesophagus.

LIFE CYCLE

Final host (gametogony): Infection is by ingestion of bradyzoite cysts in the muscles of the intermediate host. The bradyzoites are liberated in the intestine and the freed zoites pass to the sub-epithelial lamina propria and differentiate into micro- and macrogametocytes. Following conjugation of gametes, thin walled oocysts are formed which, unlike those of most other enteric sporozoans, sporulate within the body. Two sporocysts are formed, each containing four sporozoites. Usually the fragile oocyst wall ruptures and free sporocysts are found in the faeces.

Intermediate host (schizogony): Infection is by ingestion of the sporocysts and this is followed by at least three asexual generations. In the first, sporozoites, released from the sporocysts, invade the intestinal wall and enter the capillaries where they locate in endothelial cells and undergo two schizogonous cycles. A third asexual cycle occurs in the circulating lymphocytes, the resulting merozoites penetrating muscle cells. There they encyst and then divide by a process of budding or endodyogeny giving rise to broad banana-shaped bradyzoites contained within a cyst; this is the mature **Sarcocyst** and is the **infective stage** for the carnivorous final host. Although there are some variations according to species the time intervals in the life cycle are approximately as follows:

Prepatent period in carnivores	7–14 days
Patent period (period during which sporocysts are passed in faeces by carnivores)	1 week to several months
From ingestion of sporocysts to presence of infective bradyzoites in muscles of intermediate host	usually 2–3 months but may extend to 12 months in some species

PATHOGENESIS

Infection in the final host is normally non-pathogenic although mild diarrhoea has occasionally been reported.

In the intermediate host the principal pathogenic effect is attributable to the second stage of schizogony in the vascular endothelium. Heavy experimental infections of calves with *S. bovicanis* have resulted in mortality one month later, with, at necropsy, petechial haemorrhages in almost every organ including the heart, together with generalised lymphadenopathy. Experimental infection of adult cows has resulted in abortion.

A naturally occurring chronic disease of cattle, 'Dalmeny disease', has been recognised in Canada, U.S.A. and Britain. This is characterised by emaciation, submandibular oedema, recumbency and exophthalmia; at post-mortem examination, numerous schizonts are found in endothelial cells, and developing sarcocysts in areas of degenerative myositis.

S. ovicanis has been incriminated as the cause of abortion in ewes and severe myositis and encephalomyelitis in lambs in several countries.

Generally, however, clinical signs are rarely observed in this infection and the most significant effect is the presence of *Sarcocystis* in the muscles of food animals resulting in downgrading or condemnation of carcasses. While the dog-borne species are of primary importance in this context, there is increasing evidence that cat-borne species may also be responsible for lesions in meat.

In pigs there is a possibility of confusing the cysts of *Sarcocystis* with those of *Trichinella spiralis* and *Cysticercus cellulosae*. Differentiation requires the microscopic examination of squashed muscle preparations.

CLINICAL SIGNS

In heavy infections of the intermediate hosts there is anorexia, fever, anaemia, loss of weight, a disinclination to move and sometimes recumbency; in lambs a dog-sitting posture has been recorded. In cattle there is often a marked loss of hair at the end of the tail. These signs may be accompanied by submandibular oedema, exophthalmia and enlargement of lymph nodes. Abortions may occur in breeding stock.

EPIDEMIOLOGY

Little is known of the epidemiology, but from the high prevalence of symptomless infections observed in abattoirs it is clear that where dogs or cats are kept in close association with farm animals or their feed, then transmission is likely. For example, sheepdogs are known to play an important part in the transmission of *S. ovicanis* and care should be exercised that only cooked meat is fed to dogs. Acute outbreaks are probably most likely when livestock which have been reared without dog contact are subsequently exposed to large numbers of sporocysts from dog faeces. The longevity of the sporocysts shed in the faeces is not known.

DIAGNOSIS

Most cases of *Sarcocystis* infection are only revealed at meat inspection when the grossly visible sarcocysts in the muscle are discovered. However, in heavy infections of the intermediate hosts, diagnosis is based on the clinical signs and on histological demonstration of schizonts in the blood vessels of organs, such as kidney or heart and the presence of cysts in the muscles at necropsy or biopsy. An indirect haemagglutination test, using bradyzoites as antigen, is also a useful aid to diagnosis, but it should be remembered that the presence of a titre need not imply active lesions of *Sarcocystis*. Also, animals may die prior to a detectable humoral response.

In ruminants, the degenerative muscle changes closely resemble those of vitamin E-selenium deficiency,

Table 8 The major features of the life cycles of important Coccidia

	Eimeria	*Isospora*	*Cryptosporidium*	*Toxoplasma*	*Sarcocystis*
Life cycle	Direct	Direct	Direct	Indirect or direct Also between intermediate hosts	Always indirect
Infective stage for final host	Oocyst (2 sporocysts, each with 4 sporozoites)	Oocyst (4 sporocysts each with 2 sporozoites)	Very small oocyst with 4 sporozoites	Bradyzoite cysts Tachyzoites Small oocyst (2 sporocysts each with 4 sporozoites)	Bradyzoite cysts
Infective stage for intermediate host	—	—	—	Bradyzoite cyst Tachyzoites Oocyst	Sporocyst (4 sporozoites)
Asexual phase	Single host	Usually single host	Single host	Many hosts	Many hosts
Sexual phase				Cat	Dog and Cat

although the latter lacks an inflammatory cellular response.

Examination of faeces from cats or dogs on the farm for the presence of sporocysts may be helpful in the diagnosis.

TREATMENT

There is no effective treatment for infection, either in the final or in the intermediate host. Where an outbreak occurs in ruminants it has been suggested that the introduction of amprolium into the diet of the animals has a prophylactic effect.

CONTROL

The only control measures possible are those of simple hygiene. Farm dogs and cats should not be housed in, or allowed access to, fodder stores nor should they be allowed to defaecate in pens where livestock are housed. It is also important that they are not fed uncooked meat.

Since the differences in the life cycles of the various genera of the class Coccidia described up to this point are somewhat confusing, their major features are summarised in Table 8.

Besnoitia

The best known species is *Besnoitia besnoiti* which occurs worldwide, although especially important in Africa, and in which the final host is the cat and the intermediate hosts are cattle.

This genus differs from other members of the Sarcocystidae in that the cysts containing bradyzoites are found mainly in fibroblasts in or under the skin. The host cell enlarges and becomes multinucleate as the *Besnoitia* cyst grows within a parasitophorous vacuole, eventually reaching up to 0.6 mm in diameter.

Although infection of cattle is thought to be mainly by ingestion of sporulated oocysts from cat faeces, there is a suggestion that mechanical spread by biting flies feeding on skin lesions of cattle may be another route of transmission.

Following infection in cattle there is a systemic phase accompanied by lymphadenopathy and oedematous swellings in dependent parts of the body. Subsequently bradyzoites develop in fibroblasts in the dermis, subcutaneous tissues and fascia and in the nasal and laryngeal mucosa. The developing cysts in the skin result in a severe condition characterised by painful subcutaneous swellings and thickenings of the skin, loss of hair and necrosis. There is no known treatment.

Apart from the clinical manifestations which in severe cases can result in death, there can be considerable economic losses due to condemnation of hides at slaughter.

Hammondia

Only one species, *Hammondia hammondi*, is known. The final host is the cat and the intermediate hosts are small rodents. Unsporulated oocysts are produced in the faeces, and following infection of rodents, the multiplication of tachyzoites in the lamina propria of the intestinal wall is followed by the production of cysts containing bradyzoites in the skeletal muscle. It is not considered to be pathogenic to either host, but it is important to recognise that the oocysts of *Hammondia* closely resemble those of *Toxoplasma* and that their differentiation in cat faeces is a specialist task.

Hepatozoon

Hepatozoon canis occurs in the dog, and possibly the cat, in areas where the tick vector *Rhipicephalus sanguineus* is found, ie. Africa, Asia, S. Europe and the Texas Gulf Coast.

Syngamy occurs in the tick and since the sporozoites remain in the body cavity, the dog is apparently infected by ingesting the tick. Schizogony occurs in macrophages and endothelial cells in the skeletal muscle, heart and lungs followed by the production of large blue-staining gametocytes which parasitise the circulating neutrophil leucocytes. The cycle is completed when the tick ingests infected blood.

Infection may be asymptomatic and disease, when it occurs, often appears to be secondary to other pathogens. The clinical signs are those of recurrent fever, marked loss of condition and lumbar pain and may terminate fatally. Diagnosis depends on examination of stained blood smears and, if unsuccessful, muscle biopsy for the detection of schizonts.

Treatment is palliative using non-steroidal anti-inflammatory drugs and prophylaxis depends on regular tick control.

Class PIROPLASMIDIA

Babesia

Babesia are intraerythrocytic parasites of domestic animals and are the cause of anaemia and haemoglobinuria. They are transmitted by ticks in which the protozoan passes transovarially, via the egg, from one tick generation to the next. The disease, babesiosis, is particularly severe in naive animals introduced into endemic areas and is a considerable constraint on livestock development in many parts of the world.

Hosts:
All domestic animals

Intermediate hosts:
Hard ticks of the family Ixodidae in which **transovarian infection** ensures that *Babesia* are transmitted by stages

of the next generation of ticks. Depending on the species of *Babesia*, this may be by the larval, nymphal or adult stages or even all three

When infection persists from one stage to the next, in two- or three-host ticks feeding on different hosts, transmission is said to be **transtadial**

Site:
The organisms lie singly or in pairs inside the red blood cells

SPECIES

Babesia divergens, B. major *B. bigemina, B. bovis*	cattle
B. motasi, B. ovis	sheep and goats
B. caballi, B. equi	equines
B. perroncitoi, B. trautmanni	pigs
B. canis, B. gibsoni	dogs
B. felis	cats

Distribution:
The distribution of these many species throughout the world is discussed in the separate sections dealing with each final host

IDENTIFICATION AND MORPHOLOGY

Examination of stained blood films show the organisms to be within red cells, almost always singly or as pairs often arranged at a characteristic angle with their narrow ends opposed. Typically they are pyriform, but may be round, elongated or cigar-shaped (Fig. 173). Conventionally, the various species are grouped into the **small *Babesia*** whose pyriform bodies are 1.0–2.5 μm long, and **large *Babesia*** which are 2.5–5.0 μm long (Pl. XIII). With Romanowsky dyes the cytoplasm appears blue and the nucleus red.

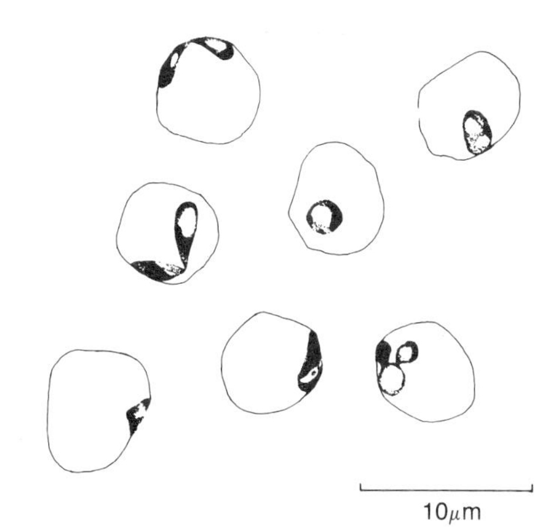

Fig 173 Diverse forms of *B. divergens* in bovine red cells.

Under the electron microscope the parasite is seen to possess at its blunt end an electron-dense 'apical complex' which is thought to be concerned with assisting penetration of the erythrocyte.

LIFE CYCLE

The organism divides asexually, by a process of budding, to form two, or sometimes four, individuals within the red cell. Eventually, the host cell ruptures and the organisms are liberated to penetrate new red cells.

The subsequent sequence of events, when the parasitaemic blood is ingested by the appropriate ixodid tick, usually the engorging adult female, is not clear, but it is now thought that a sexual phase occurs in the tick gut followed by schizogony which results in the production of elongated motile club-shaped bodies, called vermicules. These migrate to the tissues of the tick, especially the ovary, and undergo further multiplication to produce more vermicules. The entire process takes around seven days.

In the ovary of the tick the vermicules invade the eggs and, subsequently, continue to multiply in the tissues of the hatched larvae. When the larvae first feed, the vermicules enter the salivary acini and form, within a few days, the infective sporozoites, which are inoculated into the new host before feeding ceases.

When stage-to-stage transmission occurs, vermicules again reach the salivary glands of the next stage of the tick when feeding commences, and mature to become infective forms.

There is ample evidence that some species of *Babesia* may be transmitted through the ovary for two or more generations of female ticks; this is known as vertical transmission.

PATHOGENESIS

The rapidly dividing parasites in the red cells produce rapid destruction of the erythrocytes with accompanying haemoglobinaemia, haemoglobinuria and fever. This may be so acute as to cause death within a few days, during which the packed red cell volume falls below 20%. The parasitaemia, which is usually detectable once the clinical signs appear, may involve between 0.2% and 45% of the red cells, depending on the species of *Babesia*.

At necropsy, the carcass is pale and jaundiced, the bile is thick and granular and there may be sub-epicardial and sub-endocardial haemorrhages.

Milder forms of the disease, associated with less pathogenic species of *Babesia*, or with relatively resistant hosts, are characterised by fever, anorexia and perhaps slight jaundice for a period of several days.

In *B. bovis* and *B. canis* infections, clumping of erythrocytes may also occur in the capillaries of the brain, producing nervous signs of hyperexcitability and incoordination.

CLINICAL SIGNS

Typically the acute disease occurs 1–2 weeks after the tick commences to feed and is characterised by fever, and haemoglobinuria ('redwater'). The mucous membranes, at first congested, become jaundiced, the respiratory and pulse rates are increased, the heart beat is usually very audible, and in cattle ruminal movements cease and abortion may occur. If untreated, death commonly occurs in this phase. Otherwise convalescence is prolonged, there is loss of weight and milk production and diarrhoea followed by constipation is common.

In animals previously exposed to infection, or infected with a *Babesia* species of low pathogenicity, clinical signs may be mild or even inapparent.

EPIDEMIOLOGY

The following account is largely based on the epidemiology of the pathogenic bovine species which have been the most thoroughly studied, but the same principles almost certainly apply to all the *Babesia*. Essentially, the epidemiology depends on the interplay of a number of factors and these include:

(i) The virulence of the particular species of Babesia

For example, *B. divergens* in cattle and *B. canis* in dogs are relatively pathogenic while *B. major* in cattle and *B. ovis* in sheep usually produce only mild and transient anaemia.

(ii) The age of the host

It is frequently stated that there is an inverse age resistance to *Babesia* infection in that young animals are less susceptible to babesiosis than older animals. The reason for this is not known and indeed the assertion is now in some dispute.

(iii) The immune status of the host

In endemic areas, the young animal first acquires immunity passively, in the colostrum of the dam and, as a result, often suffers only transient infections with mild clinical signs. However, these infections are apparently sufficient to stimulate active immunity although recovery is followed by a long period during which they are carriers when, although showing no clinical signs, their blood remains infective to ticks for many months. It used to be thought that this active immunity was dependent on the persistence of the carrier state and the phenomenon was termed **premunity**. However, it seems unlikely that this is the case since it is now known that such animals may lose their infection either naturally or by chemotherapy, but still retain a solid immunity.

(iv) The level of tick challenge

In endemic areas, where there are many infected ticks, the immunity of the host is maintained at a high level through repeated challenge and overt disease is rare. In contrast, where there are few ticks or when they are confined to limited areas, the immune status of the population is low and the young animals receive little, if any, colostral protection. If, in these circumstances, the numbers of ticks suddenly increase due to favourable climatic conditions or to a reduction in dipping frequency, the incidence of clinical cases may rise sharply. This situation is known as **enzootic instability**.

(v) Stress

In endemic areas, the occasional outbreak of clinical disease, particularly in adult animals, is often associated with some form of stress, such as parturition or the presence of another disease, such as tick-borne fever.

These general aspects of epidemiology are dealt with in more detail in the sections dealing with babesiosis in the different host species.

DIAGNOSIS

The history and clinical signs are usually sufficient to justify a diagnosis of babesiosis. For confirmation, the examination of blood films, stained with Giemsa, will reveal the parasites in the red cells. However, once the acute febrile phase has subsided they are often impossible to find since they are rapidly removed from the circulation.

TREATMENT

This depends to some extent on the species of *Babesia* to be treated and the availability of particular drugs in individual countries. Quinuronium sulphate, pentamidine, amicarbilide, diminazene aceturate and imidocarb are the most commonly used drugs and their use is discussed in more detail in the sections dealing with babesiosis in the different animal hosts.

CONTROL

Specific control measures are not usually necessary for animals born of mothers in endemic areas since, as noted previously, their colostrally acquired immunity is gradually reinforced by repeated exposure to infection. Indeed, the veterinary importance of babesiosis is chiefly that it acts as a constraint to the introduction of improved livestock from other areas. Areas of enzootic instability also create problems when tick numbers suddenly increase or animals, for some reason, are forced to use an adjacent tick-infested area.

The numbers of ticks and therefore the quantum of *Babesia* infection may be reduced by regular spraying or dipping with insecticides. In addition, in cattle, the selection and breeding of cattle which acquire a high degree of resistance to ticks is practised, particularly in Australia.

In cattle, immunisation, using blood from carrier animals, has been practised for many years in tropical areas, and more recently in Australia, the use of rapidly passaged strains of *Babesia*, which are relatively non-pathogenic, have been widely utilised in live vaccines. It is possible that in the near future these will be superseded by adjuvanted vaccines prepared from soluble *Babesia* antigens. Otherwise the control of babesiosis in susceptible animals introduced into endemic areas depends on surveillance for the first few months after their arrival and, if necessary, treatment.

BABESIOSIS OF CATTLE

NORTHERN EUROPE

Two species occur in cattle, *B. divergens* and *B. major*. Of these, *B. divergens*, transmitted by *Ixodes ricinus*, is the most widespread and pathogenic, clinical cases occurring during the periods of tick activity, primarily in the spring and autumn. As with other *Babesia*, infection in the tick is transovarially transmitted and the larvae, nymphs and adults of the next generation are all able to transmit infection to cattle.

B. divergens is a 'small *Babesia*' and in blood films typically appear as paired, widely divergent organisms, $1.5 \times 0.4\,\mu m$, lying near the edge of the red cell, although other forms may be present. Normally no effort is made to control this infection in endemic areas although cattle recently introduced require surveillance for some months, since, on average, one in four will develop clinical disease and of these one in six will die if untreated. However, in some parts of mainland Europe, such as the Netherlands, where ticks are confined to rough vegetation on the edges of pastures and on roadsides it is often possible to take evasive measures. It is thought that the red and roe deer are not important reservoir hosts since only mild infections have been experimentally produced in splenectomised deer.

Quinuronium sulphate and amicarbalide are effective treatments and in some countries imidocarb is also licensed for use. The latter, due to its persistence in the tissues, has a prophylactic effect for several weeks. During the convalescent phase of the disease, blood transfusions may be valuable as are drugs designed to stimulate food and water intake.

Recently, long-acting preparations of oxytetracycline have been shown to have a prophylactic effect against *B. divergens* infection.

The second species, *B. major*, transmitted by the three-host tick *Haemaphysalis punctata*, is relatively uncommon and in Britain, for example, is confined to the south. *B. major* is a 'large Babesia', $3.2 \times 1.5\,\mu m$, being characteristically paired at an acute angle. It is only mildly pathogenic.

TROPICS AND SUBTROPICS

The other two important *Babesia* of cattle, *B. bovis* (syn. *B. argentina*) and *B. bigemina*, share the same primary vector, the one-host tick *Boophilus*, and co-exist over the same wide geographic area embracing Australia, Africa, Central and South America, Asia and southern Europe.

B. bovis, a small *Babesia*, whose piroplasms measure $2.0 \times 1.5\,\mu m$, is generally regarded as the most pathogenic of the bovine *Babesia*.

Although the classical signs of fever, anaemia and haemoglobinuria occur, the degree of anaemia is disproportional to the parasitaemia since haematocrit levels below 20% may be associated with infections of less than 1% of the red cells. The reason for this is unknown.

In addition, *B. bovis* infection is associated with sludging of the red cells in the small capillaries. In the cerebrum this causes blockage of the vessels by clumps of infected red cells leading to anoxia and tissue damage. The resulting clinical signs of aggression, incoordination or convulsions invariably terminate fatally.

Finally, recent work has indicated that some of the severity of *B. bovis* infections may be associated with the activation of certain plasma components, leading to circulatory stasis, shock and intravascular coagulation.

B. bigemina, large *Babesia* whose piroplasms measure $4.5 \times 2.0\,\mu m$, is of particular interest historically, since it was the first protozoan infection of man or animals demonstrated to have an arthropod intermediate host; this was shown in 1893 by Smith and Kilborne while investigating the cause of the locally known 'Texas Fever' in cattle in U.S.A. The disease has since been eradicated in that country.

Generally *B. bigemina* infections are not so virulent as those of *B. bovis*, despite the fact that the parasites may infect 40% of the red cells. Otherwise the disease is typically biphasic, the acute haemolytic crisis, if not fatal, being followed by a prolonged period of recovery.

Imidocarb and the diamidine derivatives such as diminazene aceturate, amicarbalide and phenamidine, are all effective against *B. bovis* and *B. bigemina*, especially if given early in the disease.

Vaccination of cattle against both *B. bovis* and *B. bigemina* infection is commonly practised in many countries by inoculating blood from donor animals. This is usually obtained from a recently recovered case, any untoward reactions in the vaccinates being controlled by babesicidal drugs. In Australia the procedure is more sophisticated in that the vaccine is produced from acute infections produced in splenectomised donors. For economy the blood is collected by exchange transfusion rather than by exsanguination. It is interesting that the rapid passage of the parasite by blood inoculation in splenectomised calves has fortuitously had the very desirable effect of decreasing the virulence of the infection in non-splenectomised calves to the extent that post-vaccination surveillance of cattle is frequently not performed.

The parasite count of the blood determines the dilution of the latter which is dispensed in plastic bags, packed in ice and despatched in insulated containers. Each dose of vaccine contains about 10 million parasites. Most of the vaccine is used in cattle under 12 months of age living in conditions of enzootic instability. The degree of protection induced is such that only 1% of vaccinated cattle subsequently develop clinical babesiosis from field challenge, compared to 18% of unvaccinated cattle.

The primary disadvantage of red cell vaccines is their lability and the fact that, unless their preparation is carefully supervised, they may spread diseases such as enzootic bovine leucosis. Currently, efforts are in progress to develop a soluble dead vaccine from *Babesia* cultured *in vitro*.

Recently, a regimen of four injections of long-acting oxytetracycline at weekly intervals, administered to naive cattle during their first month of grazing on tick-infested pastures, has been shown to confer prophylaxis against *B. bigemina* during this period, after which the cattle were immune to subsequent challenge.

BABESIOSIS OF SHEEP AND GOATS

Two species of *Babesia*, the smaller *B. ovis* and the larger *B. motasi*, are known to occur in sheep and goats in tropical and subtropical areas, including southern Europe.

Transmitted by various tick genera, such as *Rhipicephalus*, *Haemaphysalis*, *Dermacentor* and *Ixodes*, these infections are usually mild in indigenous sheep although severe clinical signs may occur in animals introduced from a non-endemic area. Diminazene aceturate is effective against *B. bovis* and *B. motasi*.

Recently, in Britain, the existence of two species of *Babesia* has been demonstrated in sheep. One of these is *B. motasi* transmitted by *Haemaphysalis punctata*, while the other is thought to be *B. capreoli* from Scottish red deer, apparently transmitted by *Ixodes ricinus*. Both species appear to be relatively non-pathogenic.

BABESIOSIS OF HORSES

Two species of *Babesia*, the small *B. equi* and the large *B. caballi*, may occur in horses and donkeys in the Americas, Africa, Asia and mainland Europe. Both are transmitted by a variety of tick species including *Dermacentor*, *Hyalomma* and *Rhipicephalus*.

The small *B. equi* is readily recognised in blood smears from acute cases, since apart from size, the piroplasms characteristically form a Maltese cross of four organisms. It is the more pathogenic species causing fever, anaemia, jaundice and haemoglobinuria. In contrast, *B. caballi* infections are usually characterised by fever and anaemia only.

Amicarbalide or imidocarb may be used for the treatment of either infection. Otherwise pentamidine is suitable for *B. equi* infections and quinuronium sulphate and diminazene aceturate for those of *B. caballi*. In each case the manufacturers' instructions should be carefully observed.

BABESIOSIS OF PIGS

Two species, the small *B. perroncitoi* and the large *B. trautmanni*, are found in southern Europe, Africa and Asia. Wild pigs may act as reservoirs of infection and the tick vectors include *Boophilus*, *Rhipicephalus* and *Dermacentor*. The clinical signs range from mild to severe.

BABESIOSIS OF DOGS

The most widespread and pathogenic species is the large *Babesia canis* found in mainland Europe, Africa, Asia and the Americas. *Rhipicephalus sanguineus* is the principal vector, in which transovarian followed by transtadial transmission occurs.

Although the commonest clinical signs are those of fever, anaemia, jaundice and haemoglobinuria, peracute cases may occur in dogs imported into an endemic area with collapse and profound anaemia leading to death in 1–2 days. Less commonly hyperexcitability, associated with cerebral babesiosis, may occur.

A variety of other signs and syndromes such as ascites, bronchitis, purpura haemorrhagica and severe muscular pains have been associated with *B. canis* infection, in that they respond to treatment with babesicidal drugs.

While specific diagnosis depends on the detection of the parasites in the red cells, the parasitaemia is often under 5% and may also be complicated by the concurrent presence of inclusions in the monocytes, of the rickettsial organism, *Ehrlichia canis*, also transmitted by *R. sanguineus*.

In every case, chemotherapy with pentamidine, phenamidine or diminazine aceturate is advisable immediately after clinical diagnosis, since death may occur rapidly.

Prophylaxis depends on regular treatment of dogs with a suitable acaricide, and since *R. sanguineus* may live in kennels, these should also be frequently treated with a suitable acaricide.

In addition, a degree of surveillance of dogs exposed to infection is advisable so that treatment can be administered as early as possible.

A second *Babesia*, *B. gibsoni*, occurs in dogs in the East. It is a small *Babesia* and produces a more chronic disease which is less susceptible to chemotherapy.

BABESIOSIS OF CATS

Babesia felis, a small *Babesia*, producing anaemia and icterus in cats, has been recorded in Africa.

BABESIOSIS IN MAN

Since 1957, several cases of fatal babesiosis due to *B. divergens* infection have occurred in man in Yugoslavia, U.S.S.R., Ireland and Scotland. In each case the individual had been splenectomised sometime previously or was currently undergoing immunosuppressive treatment.

More recently, about 100 cases of human babesiosis, only a few of which were fatal, have been diagnosed, mainly on the eastern seaboard of the U.S.A. and more recently in Wisconsin, apparently due to the rodent parasite *B. microti*. The majority of the affected individuals possessed their spleens.

In most of these cases the clinical syndrome and presence of intra-erythrocytic parasites had suggested an initial diagnosis of malaria. It is not known whether babesiosis in man is always associated with recent tick bites or may be due, on some occasions, to a flare-up of a latent *Babesia* infection. Some support for the latter theory is the recent demonstration of antibodies to *Babesia* in the sera of many healthy individuals in countries such as Mexico and Nigeria.

Theileria

The diseases caused by several species of *Theileria* are a serious constraint on livestock development in Africa, Asia and the Middle East. The parasites, which are tick-transmitted, undergo repeated schizogony in the lymphocytes, ultimately releasing small merozoites which invade the red cells to become piroplasms.

SPECIES, HOSTS AND DISTRIBUTION

Theileria are widely distributed in cattle and sheep in Africa, Asia, Europe and Australia, have a variety of tick vectors and are associated with infections which range from clinically inapparent to rapidly fatal.

Although the speciation of many *Theileria* is still controversial, largely because of their morphological similarity, there are three species of major veterinary importance. (See Table below.)

Minor and mildly pathogenic species infecting cattle include *T. mutans* and *T. taurotragi* in Africa and *T. sergenti* in Asia; the identities of the European and Australian bovine species are uncertain.

In sheep, the non-pathogenic *T. ovis* occurs in Europe and in Africa and Asia.

Theileria parva

This parasite is the cause of East Coast Fever in cattle in East and Central Africa. Because of the wide distribution of its tick vector, *Rhipicephalus*, and the fact that infection in cattle introduced into enzootic areas can be associated with a mortality of 100%, *T. parva* infection is an immense obstacle to livestock improvement.

IDENTIFICATION

In the erythrocytes the piroplasms (Pl. XIII) are predominantly rod-shaped and up to 2.0 μm long and 1.0 μm wide. Round, oval and ring-shaped forms also occur. With Giemsa stains, the cytoplasm of each is blue with a red chromatin dot at one end. Commonly, there is more than one parasite in each erythrocyte.

In the cytoplasm of the lymphocytes in the lymph nodes and spleen the schizonts, sometimes called Koch's blue bodies, are found (Pl. XIII). Under Giemsa stain, these are seen as two types; the **macroschizonts**, about 8.0 μm in diameter, are blue and contain up to eight nuclei while the next stage, the **microschizonts**, are similar in size, but contain up to 36 small nuclei; the latter are the developing micromerozoites which, on

Species	Host	Vector	Disease	Distribution
Theileria parva	Cattle	*Rhipicephalus*	East Coast Fever	East and Central Africa
T. annulata	Cattle	*Hyalomma*	Mediterranean or Tropical Theileriosis	North Africa, South Europe, Middle East, Asia
T. hirci	Sheep Goats	*Hyalomma*	Malignant Ovine (Caprine) Theileriosis	North Africa, South Europe, Middle East, Asia

rupture of the microschizont, invade the red cells to become piroplasms.

LIFE CYCLE

The sporozoites are inoculated into cattle by *Rhipicephalus appendiculatus*, the brown ear tick, and quickly enter lymphocytes in an associated lymph gland, usually the parotid. The parasitised lymphocyte transforms to a lymphoblast which divides rapidly as the macroschizont develops. This division is apparently stimulated by the parasite, which itself divides synchronously with the lymphoblast to produce two infected cells. The rate of proliferation is such that a ten-fold increase of infected cells may occur every three days.

About twelve days after infection a proportion of the macroschizonts develop into microschizonts and within a day or so these produce the micromerozoites which, liberated by rupture of the microschizonts, invade the red cells to become piroplasms. The piroplasms do not multiply in the red cells.

For completion of the life cycle the piroplasms require to be ingested by the larvae or nymphal stages of the three-host vector, *R. appendiculatus*. A sexual phase occurs in the tick gut followed by the formation of sporoblasts in the salivary glands. No further development occurs until the next stage of the tick starts to feed when the sporoblasts produce infective sporozoites from about four days onwards. Since female ticks feed continuously for about 10 days and males intermittently over a longer period this allows ample time for infection of the host.

Transmission is transtadial, ie. by the next stage of the tick, and transovarian transmission does not occur.

PATHOGENESIS

The sequence of events in a typical acute and fatal infection progresses through three phases each spanning about one week. The first is the incubation period of about one week when neither parasite nor lesions can be detected. This is followed during the second week by marked hyperplasia and expansion of the infected lymphoblast population, initially in the regional lymph node draining the site of the tick bite and ultimately throughout the body. During the third week there is a phase of lymphoid depletion and disorganisation associated with massive lymphocytolysis and depressed leucopoiesis. The cause of the lymphocytolysis is unknown, but is due perhaps to the activation of 'natural killer' cells, like macrophages.

Necropsy during the terminal phase shows atrophy of the cellular content of the lymph nodes and spleen, pulmonary oedema and emphysema, and petechial and ecchymotic haemorrhages on the gastrointestinal mucosa. Haemorrhages may also be present on the serosal and mucosal surfaces of many organs.

Occasionally nervous signs, the so-called 'turning sickness', have been reported and attributed to the presence of schizonts in cerebral capillaries.

CLINICAL SIGNS

About one week after infection, in a fully susceptible animal, the lymph node draining the area of tick-bite, usually the parotid, becomes enlarged and the animal becomes pyrexic. Within a few days there is generalised swelling of the superficial lymph nodes, the animal rapidly loses condition, becomes dyspnoeic and there is terminal diarrhoea, often blood-stained. Petechial haemorrhages may occur under the tongue and on the vulva. Recumbency and death almost invariably occurs, usually within three weeks of infection.

EPIDEMIOLOGY

Since the tick vector, *R. appendiculatus*, is most active following the onset of rain, outbreaks of East Coast Fever may be seasonal, or where rainfall is relatively constant, may occur at any time. Fortunately, indigenous cattle reared in endemic areas show a high degree of resistance and although transient mild infection occurs in early life, mortality is negligible. The mechanism of this resistance is unknown. However, such cattle may remain carriers and act as a reservoir of infection for ticks. Susceptible cattle introduced into such areas suffer high mortality, irrespective of age or breed, unless rigid precautions are observed.

In areas where the survival of the tick vector is marginal, challenge is low and indigenous cattle may have little immunity. Such areas, during a prolonged period of rain, may become ecologically suitable for the survival and proliferation of the ticks, ultimately resulting in disastrous outbreaks of East Coast Fever.

In some parts of East and Central Africa where populations of cattle and wild African buffalo overlap there is an additional epidemiological complication due to the presence of a strain of *T. parva*, known as *T. parva lawrenci*. This occurs naturally in African buffalo many of which remain as carriers. The tick vector is also *R. appendiculatus* and in cattle the disease causes high mortality. Since infected ticks may survive for nearly two years physical contact between buffalo and cattle need not be close.

DIAGNOSIS

East Coast Fever only occurs where *R. appendiculatus* is present, although occasionally outbreaks outwith such areas have been recorded due to the introduction of tick-infected cattle from an enzootic area.

In sick animals, macroschizonts are readily detected in biopsy smears of lymph nodes and in dead animals in impression smears of lymph nodes and spleen. In advanced cases, Giemsa-stained blood smears show piroplasms in the red cells, up to 80% of which may be parasitised.

The indirect fluorescent antibody test is of value in detecting cattle which have recovered from East Coast Fever.

TREATMENT

Although the tetracyclines have a therapeutic effect if given at the time of infection, they are of no value in the treatment of clinical cases. Recently however, with the introduction of parvaquone, a naphthaquinone compound, and halofuginone, both of which are highly effective, the successful treatment of clinical cases appears promising. However, more work is required to determine the ultimate productivity and carrier status of recovered cattle.

CONTROL

Traditionally, the control of East Coast Fever in areas where improved cattle are raised has relied on legislation to control the movement of cattle, on fencing to prevent access by nomadic cattle and buffalo and on repeated treatment of cattle with acaricides. In areas of high challenge, such treatments may require to be carried out twice weekly in order to kill the tick before the infective sporozoites develop in the salivary glands. This is not only expensive, but creates a population of fully susceptible cattle; if the acaricide fails, through human error or the acquisition of acaricide resistance by the ticks, the consequences can be disastrous.

Great efforts have been made to develop a suitable vaccine, but these have been thwarted by the complex immunological mechanisms involved in immunity to East Coast Fever and by the discovery of immunologically different strains of *T. parva* in the field. However, an 'infection and treatment' regime which involves the concurrent injection of a virulent stabilate of *T. parva* and long-acting tetracycline has been shown to be successful, although it has not been used on a large scale as yet. Apparently the tetracycline slows the rate of schizogony, giving the immune response time to develop.

Theileria annulata

The disease caused by *Theileria annulata* affects cattle and domestic buffalo and is similar in many respects to that caused by *T. parva*. However there are some features which differ and the most important of these are summarised below.

The disease is distributed in a wide tropical and subtropical belt embracing Portugal and Spain, the Balkans, those countries bordering the Mediterranean (hence the name 'Mediterranean theileriosis'), the Middle East and the Indian subcontinent and China.

T. annulata is transmitted transtadially by ticks of the genus *Hyalomma*. Like East Coast Fever, indigenous cattle in endemic areas are relatively resistant while improved cattle, particularly European breeds, are highly susceptible. However, unlike East Coast Fever, the disease in such cattle is not uniformly fatal, although the mortality rate may reach 70%.

The pathogenesis and clinical signs are initially similar to those of East Coast Fever with pyrexia and lymph node enlargement, but in the late stages there is a haemolytic anaemia and often icterus. Convalescence is protracted in those cases which recover.

Diagnosis depends on the detection of schizonts in both lymph node biopsies and, unlike *T. parva*, in blood smears. A low-grade piroplasm parasitaemia, in the absence of schizonts, is usually indicative of a recovered carrier animal.

In many areas, the prevention of *T. annulata* infection in imported dairy stock is based on permanent housing. However this is expensive and there is always the possibility that infected ticks may be brought in with the fodder to cause disease and colonise crevices in the cattle accommodation. In some countries immunisation with schizonts attenuated by prolonged *in vitro* culture has given excellent results. As with *T. parva* infection the value of naphthaquinone compounds and halofuginone in treatment is currently under study and appears promising.

Theileria hirci

This protozoan is the cause of an acute and highly fatal disease of sheep and goats in Eastern Europe, the Middle East, Asia and North Africa. The tick vector is unknown, although species of *Hyalomma* are suspected. The epidemiology and pathogenesis is similar to that of *T. annulata* with fever, enlarged lymph nodes, anaemia and icterus in susceptible animals.

Little specific information is available on tick control or treatment.

Cytauxzoon

Various species of *Cytauxzoon* occur as theileria-like piroplasms in the red cells of African wild ruminants. The genus differs from *Theileria* in that schizogony occurs in the reticuloendothelial cells rather than lymphocytes.

Recently, a fatal disease of domestic cats, characterised by fever, anaemia and icterus, has been described in the southern U.S.A. The presence of schizonts in the reticuloendothelial cells of the spleen, lymph nodes, liver and lungs has suggested that this may be a species of *Cytauxzoon*.

Class HAEMOSPORIDIA

Three separate genera in this class, *Plasmodium*, *Haemoproteus* and *Leucocytozoon*, are the causes of avian 'malaria' in domestic and wild birds, a disease most common in the tropics and transmitted by biting dipteran flies.

The vectors differ, in that avian species of *Plasmodium* are transmitted by mosquitoes, *Haemoproteus* by hippoboscid flies or midges and *Leucocytozoon* by *Simulium* spp.

Their life cycles are broadly similar to human malaria, schizogony in the reticuloendothelial system leading to the production of malaria-like gametocytes in the red cells and syngamy occurring in the insect vector with the production of sporozoites in the salivary glands. Depending on the species, chickens, ducks, turkeys, geese or pigeons may be affected, the clinical signs ranging from the inapparent to pyrexia, anaemia, paralysis and even sudden death.

Diagnosis depends on the recognition and differentiation of the parasites in the red cells of stained blood films.

Although anti-malarial drugs may be used in treatment, control of the insect vector is ultimately more important.

[**Malaria in man:** This is one of the most prevalent diseases of man in the world and is considered to cause 1.5 million deaths annually. It is caused by four species of *Plasmodium* of which *P. vivax* is the most common and *P. falciparum* the most pathogenic. The sporozoites are inoculated by the female anopheline mosquito and this is followed by schizogony in the parenchymal cells of the liver and then in the red blood cells. Ultimately micro- and macro-gametocytes are formed in the red cells, and, on ingestion by a mosquito, syngamy ocurs in the gut. The oocysts produce thousands of sporozoites which invade the salivary glands and are subsequently inoculated into another human host.

Control depends on the eradication of mosquitoes and, at least for temporary residents, the regular use of prophylactic antimalarial drugs. Since drug resistance to malaria is widespread in certain areas, informed medical advice should be obtained on the selection of drug].

Subphylum CILIOPHORA

Balantidium coli

Distribution:
Worldwide

Hosts:
The pig; occasionally other animals, including man and cattle

Site:
Large intestine

IDENTIFICATION

An actively motile organism, up to 300 μm, whose pellicle possesses rows of longitudinally arranged cilia (Fig. 146). At the anterior end there is a funnel-shaped depression, the peristome, which leads to the cytostome or mouth; from this, food particles are passed to vacuoles in the cytoplasm and digested. Internally there are two nuclei, a reniform macronucleus and adjacent micronucleus, and two contractile vacuoles which regulate osmotic pressure.

LIFE CYCLE

Balantidium coli probably exists as a commensal in the large intestine of most pigs. Reproduction is by binary fission. Conjugation, a temporary attachment of two individuals during which nuclear material is exchanged, also occurs, after which both cells separate. Eventually cysts are formed which are passed in the faeces; these have a thick yellowish wall, through which the parasite may be seen and are viable for two weeks at room temperature. Infection of a new host is by ingestion of the cysts.

PATHOGENESIS AND SIGNIFICANCE

Normally non-pathogenic, these protozoa may, for reasons unknown, occasionally cause ulceration of the mucosa and accompanying dysentery in the pig. Man may occasionally become clinically affected through contamination of foodstuffs or hands with pig faeces.

Tetracyclines are effective in treatment.

Subphylum MICROSPORA

Encephalitozoon cuniculi

This pathogen is one of several provisionally assigned to the phylum Protozoa in a subphylum, the Microspora. All are intracellular parasites with a schizogonous and sporogonous stage, the spore possessing a coiled polar filament. Most are parasites of fish.

Disease due to *E. cuniculi* has been reported on a few occasions in man, dog, blue fox, rodents and other mammals. The life cycle is obscure although infection can be by ingestion, the parasite subsequently undergoing schizogony and sporogony in the peritoneal macrophages, and in the chronic form, in the kidneys and brain. The infective spores are passed in the urine.

The clinical signs vary, range from mild to severe, and include abdominal distension, paralysis and nervous signs which may resemble rabies.

Protozoa of undetermined classification

Pneumocystis carinii

This organism is often tentatively regarded as a sporozoan of the toxoplasmid type. It is occasionally responsible for pneumonia in man, particularly in individuals who are very young, old, debilitated or immunosuppressed. The lesion is characterised by a massive plasma cell or histiocyte infiltration of the alveoli in which the organisms may be detected by a silver staining procedure.

The organism is apparently quite widely distributed in latent form in healthy individuals and in the dog, as well as a wide variety of other domestic and wild animals.

Order RICKETTSIALES

The organisms described in this section were initially thought to be protozoa, but are now known to be *Rickettsia*. However, because some may be readily confused with blood sporozoa, they have been traditionally included in veterinary parasitology.

Anaplasma

These organisms, found in the red cells of cattle, cause anaplasmosis, a disease characterised by fever, anaemia and jaundice. Infection is transmitted by ticks or mechanically by biting insects or even by contaminated hypodermic needles or surgical instruments.

Hosts:
Cattle. Wild ruminants, and perhaps sheep, may act as reservoirs of infection

Intermediate hosts:
Some 20 tick species, including the one-host *Boophilus*, have been shown to transmit infection experimentally. Since transovarial infection occurs in the majority of these species, they may be considered to be intermediate hosts, although there is little information on the development of the parasites in the tick

Site:
In red blood cells

Species:
Anaplasma marginale
A. centrale

Distribution:
Worldwide in tropics and subtropics, including southern Europe. It is also present in some temperate areas including parts of the U.S.A.

IDENTIFICATION

In Giemsa-stained blood films the organisms of *A. marginale* are seen as small, round, dark red 'inclusion bodies' within the red cell (Pl. XIII). Often there is only one organism in a red cell and characteristically this lies at the outer margin; however these two features are not constant.

The mildly pathogenic *A. centrale* is similar, except that the organisms are commonly found in the centre of the erythrocyte.

LIFE CYCLE

As described earlier, *Anaplasma* can be transmitted by ticks, and also mechanically by biting flies or contaminated surgical instruments.

Once in the blood, the organism enters the red cell by invaginating the cell membrane so that a vacuole is formed; thereafter it divides to form an inclusion body containing up to eight 'initial bodies' packed together. The inclusion bodies are most numerous during the acute phase of the infection, but some persist for years afterwards.

PATHOGENESIS

Typically, the changes are those of an acute febrile reaction accompanied by a severe haemolytic anaemia. After an incubation period of around four weeks, fever and parasitaemia appear, and as the latter develops, the anaemia becomes more severe so that within a week or so up to 70% of the erythrocytes are destroyed.

Necropsy at this time often reveals a jaundiced carcass, a grossly enlarged gallbladder and, on section, a liver suffused with bile. The spleen and lymph nodes are enlarged and congested and there are petechial haemorrhages in the heart muscle. The urine, unlike that in babesiosis, is normal in colour. In survivors recovery is prolonged.

CLINICAL SIGNS

The clinical signs are usually very mild in naive cattle under one year old. Thereafter, susceptibility increases so that cattle aged 2–3 years develop typical and often fatal anaplasmosis while in cattle over three years the disease is often peracute and frequently fatal.

The clinical features include pyrexia, anaemia and often jaundice, anorexia, laboured breathing and in cows a severe drop in milk yield or abortion. Occasionally peracute cases occur, which usually die within a day of the onset of clinical signs.

EPIDEMIOLOGY

Apart from the various modes of transmission described

above, little information is available. Reservoirs of infection are maintained in carrier cattle and perhaps in wild ruminants or sheep. Cattle, especially adults, introduced into endemic areas are particularly susceptible, the mortality rate being up to 80%. In contrast, cattle reared in endemic areas are much less susceptible, presumably due to previous exposure when young, although their acquired immunity usually co-exists with a carrier state. This balance may on occasions be disturbed and clinical anaplasmosis supervene when cattle are stressed by other diseases such as babesiosis.

DIAGNOSIS

The clinical signs supplemented, if possible, by a haematocrit estimation and the demonstration of *Anaplasma* inclusions in the red cells are usually sufficient for diagnosis. For the detection of immune carriers, the complement-fixation and capillary tube agglutination tests are available.

TREATMENT

Tetracycline compounds are effective in treatment if given early in the course of the disease and especially before the parasitaemia has reached its peak. More recently imidocarb has been shown to be effective and may also be used to sterilise carrier animals.

CONTROL

Vaccination of susceptible stock with small quantities of blood containing the mildly pathogenic *A. centrale* or a relatively avirulent strain of *A. marginale* is practised in several countries, any clinical signs in adults being controlled by drugs. In the U.S.A. a killed *A. marginale* vaccine containing erythrocyte stroma is also available. Although all are generally successful in the clinical sense, challenged cattle become carriers and so perpetuate transmission. The killed vaccine has the disadvantage that antibodies produced to the red cell stroma, if transferred in the colostrum, may produce isoerythrolysis in nursing calves. Improved inactivated vaccines are currently under development.

Otherwise, control at present depends largely on the reduction of ticks and biting flies.

Aegyptianella

Aegyptianella pullorum causes a disease in chickens, geese and ducks in South East Europe, Africa and Asia which is somewhat similar to anaplasmosis in cattle. *Anaplasma*-like bodies of various sizes are found in the cytoplasm of the red cells, the syndrome is characterised by anaemia, icterus and diarrhoea, and infection is tick transmitted by the soft tick, *Argas persicus*. Imported birds are especially susceptible and recovered birds are frequently carriers.

Tetracycline compounds are recommended for treatment.

Eperythrozoon and Haemobartonella

Typically, species of both of these genera, which are present on red cells, produce mild and clinically inapparent infections in a variety of domestic animals throughout the world. Their identification from staining artefacts requires good blood films and filtered Giemsa stain. Even then, differentiation between the two genera is difficult, since both appear as cocci or short rods on the surface of the erythrocytes, often completely surrounding the margin of the red cell. However, the organisms of *Eperythrozoon* are relatively loosely attached to the red cell surface (Pl. XIV) and are often found free in the plasma while those of *Haemobartonella* are tightly attached to the red cells and are rarely free in the plasma.

Among the *Eperythrozoon*, *E. suis* is the most pathogenic, producing icterus and anaemia in very young pigs, while the normally benign *E. ovis* of sheep and *E. wenyoni* of cattle are occasionally responsible for fever, anaemia and loss of weight.

Of the *Haemobartonella*, *H. felis* is the most significant as a cause of haemolytic anaemia in young cats. This may be acute or chronic with periodic recrudescence of clinical signs. Recovered cats may remain carriers.

Transmission of the disease probably depends on arthropods including lice, fleas, ticks and biting flies and also, at least in the case of *H. felis*, by ingestion of blood during fighting.

Tetracyclines are effective in treatment.

Ehrlichia

Members of this genus of Rickettsia are found in the blood leucocytes as intracytoplasmic inclusions and characteristically produce a short febrile illness associated with leucopaenia. The most important species are *E. phagocytophila*, the cause of tick-borne fever in sheep and cattle in the British Isles, Norway, Finland, the Netherlands and Austria, and *E. canis* which causes tropical pancytopaenia in dogs.

Ehrlichia (syn. *Cytoecetes*) *phagocytophila* is transmitted by *Ixodes ricinus* and in endemic areas the prevalence of infection in young hill lambs is virtually 100%.

Following an incubation period of seven days there is fever, dullness and inappetence which persists for around ten days. During this, although leucopaenia is marked, the characteristic 'morula' inclusions may be seen in a variable proportion of the polymorphonuclear leucocytes present (Pl. XIV). Recovery is usually uneventful, although such animals remain carriers for many months.

The veterinary significance of tick-borne fever is three-fold. First, although the disease in itself is transient, its occurrence in very young lambs on rough upland pastures may lead to death through inability to maintain contact with the dam. Secondly, the disease, possibly because of the associated leucopaenia, predisposes lambs to louping-ill, tick pyaemia (enzootic staphylococcosis) and pasteurellosis. Finally, the occurrence of the disease in adult sheep or cattle newly introduced into an endemic area may cause abortion or temporary sterility in males, possibly as consequences of the pyrexia.

Treatment of tick-borne fever is rarely called for and prophylaxis depends on tick control by dipping. When tick pyaemia in lambs is a problem this measure may be supplemented by one or two prophylactic injections of long-acting oxytetracycline, each of which protects against *Ehrlichia phagocytophila* infection for 2–3 weeks.

Ehrlichia canis, transmitted by *Rhipicephalus*, is the cause of the febrile disease canine pancytopaenia, which occurs in the tropics throughout the world. The inclusions are found in the monocytes, although leucopaenia and thrombocytopaenia are present. Death may occur due to secondary infections associated with the leucopaenia, or to mucosal and serosal haemorrhages due to platelet deficiencies. The tetracyclines are effective in treatment.

Two other febrile and haemorrhagic ehrlichial infections, presumably tick transmitted, which have been described in domestic animals are bovine petechial fever (Ondiri disease) in Kenya caused by *Cytoecetes ondiri* and equine ehrlichiosis, caused by *E. equi* in California. More recently, a syndrome, called Potomac horse fever, with fever, leucopaenia and diarrhoea in horses in the U.S.A. has been attributed to an ehrlichial organism in the blood leucocytes.

REVIEW TOPICS

THE EPIDEMIOLOGY OF PARASITIC DISEASES

Although the reasons for the occurrence of parasitic diseases are multiple and often interactive, the vast majority occur for one of four basic reasons (Table 9).

These are:

1. An increase in the numbers of infective stages.
2. An alteration in host susceptibility.
3. The introduction of susceptible stock.
4. The introduction of infection.

Each of these will be discussed in turn giving examples.

Table 9 Factors affecting the epidemiology of parasitic disease

1. An increase in the number of infective stages.	
(i) Contamination of the environment	Biotic potential Stock management Immune status Hypobiosis/ Diapause
(ii) Development and survival of infective stages	Microhabitat Seasonal development Stock management
2. An alteration in host susceptibility	
(i) Existing infections	Diet Pregnancy and Lactation Steroid therapy
(ii) New infections	Intercurrent infections Chemotherapy Hypersensitivity
3. The introduction of susceptible stock	
(i) Absence of acquired immunity	
(ii) Absence of age immunity	
(iii) Longevity of infective stages	
(iv) Genetic factors	between species between breeds
(v) Sex	
(vi) Strain of parasite	
4. The introduction of infection	
(i) Introduction of new stock	
(ii) Effluent	
(iii) Vectors	

AN INCREASE IN THE NUMBERS OF INFECTIVE STAGES

This category involves parasitic diseases which occur seasonally, and although more distinct in zones with a wide climatic variation, may also be observed in zones with minor variations in climates such as the humid tropics.

A multiplicity of causes are responsible for the seasonal fluctuations in the numbers and availability of infective stages, and these may conveniently be grouped as factors affecting contamination of the environment, and those controlling the development and survival of the free-living stages of the parasites and, where applicable, their intermediate hosts.

CONTAMINATION OF THE ENVIRONMENT

The level of contamination is influenced by several factors.

Biotic potential

This may be defined as the capacity of an organism for biological success as measured by its fecundity. Thus, some nematodes such as *Haemonchus contortus* and *Ascaris suum* produce many thousands of eggs daily, while others, like *Trichostrongylus*, produce only a few hundred. Egg production by some external parasites such as the blowfly, *Lucilia sericata*, or the tick, *Ixodes ricinus*, is also very high, whereas *Glossina* spp. produce relatively few offspring.

The biotic potential of parasites which multiply either within an intermediate or final host is also considerable. For example, the infection of *Lymnaea* with one miracidium of the trematode *Fasciola hepatica*, can give rise to several hundred cercariae. Within the final host, protozoal parasites such as *Eimeria*, because of schizogony and gametogony, also give rise to a rapid increase in the contamination of the environment.

Stock management

The density of stocking can influence the level of contamination and is particularly important in nematode and cestode infections in which no multiplication of the parasite takes place outside the final host. It has the greatest influence when climatic conditions are optimal for development of the contaminating eggs or larvae, such as in spring and summer in the northern hemisphere.

A high stocking density will also favour the spread of ectoparasitic conditions such as pediculosis and sarcoptic mange, where close contact between animals facilitates the spread of infection. This may occur under crowded conditions in cattle yards, or from mother to offspring where, for example, sows and their litters are in close contact.

In coccidiosis, where large numbers of oocysts are disseminated, management procedures which encourage the congregation of stock, such as the gathering of lambs around feeding troughs, may lead rapidly to heavy contamination.

In temperate countries, where livestock are stabled during the winter, the date of turning out to graze in spring will influence contamination of pasture with helminth eggs. Since many helminth infective stages, which have survived the winter, succumb during late spring, the withholding of stock until this time will minimise subsequent infection.

Immune status of the host

Clearly, the influence of stocking density will be greatest if all the stock are fully susceptible, or if the ratio of susceptible to immune stock is high, as in sheep flocks with a large percentage of twins or in multiple suckled beef herds.

However, even where the ratio of adults to juveniles is low it must be remembered that ewes, sows, female goats and to a lesser extent cows become more susceptible to many helminths during late pregnancy and early lactation due to the periparturient relaxation in immunity. In most areas of the world parturition in grazing animals, synchronised to occur with the climate most favourable to pasture growth, is also the time most suitable for development of the free-living stages of most helminths. Thus, the epidemiological significance of the periparturient relaxation of immunity is that it ensures increased contamination of the environment when the number of susceptible animals is increasing.

There is some evidence that resistance to intestinal protozoal infections such as coccidiosis and toxoplasmosis is also lowered during pregnancy and lactation, and so enhances spread of these important infections.

On the credit side, host immunity will limit the level of contamination by modifying the development of new infections either by their destruction or arrestment at the larval stages, while existing adult worm burdens are either expelled or their egg production severely curtailed.

Although immunity to ectoparasites is less well defined, in cattle it develops against most species of ticks, although in a herd this expression of resistance often inadvertently results in an overdispersed population of ticks with the susceptible young animals carrying most of the ticks.

In protozoal diseases, such as babesiosis or theileriosis, the presence of immune adults also limits the likelihood of ticks becoming infected; however, this effect is not absolute since such animals are often silent carriers of these protozoal infections.

Hypobiosis/diapause

These terms are used to describe an interruption in development of a parasite at a specific stage and for periods which may extend to several months.

Hypobiosis refers to the arrested development of nematode larvae within the host and occurs seasonally, usually at a time when conditions are adverse to the development and survival of the free-living stages. The epidemiological importance of hypobiosis is that the resumption of development of hypobiotic larvae usually occurs when conditions are optimal for free-living development and so results in an increased contamination of the environment. There are many examples of seasonal hypobiosis in nematodes including *Ostertagia* infections in ruminants, *Hyostrongylus rubidus* in pigs and *Trichonema* spp. in horses.

Diapause in arthropods, like hypobiosis in nematodes, is also considered to be an adaptation phenomenon whereby ectoparasites survive adverse conditions by a cessation of growth and metabolism at a particular stage. It is most common in temporary arthropod parasites in temperate climates. In these, feeding activity is restricted to the warmer months of the year and winter survival is often accomplished by a period of diapause. Depending on the extremity of the northern or southern latitudes, this may occur after one or several generations. For example, the head-fly *Hydrotoea irritans* has only one annual cycle and overwinters as a mature larva in diapause. Other insects, such as *Stomoxys calcitrans* or the British blowflies, have several generation cycles before entering diapause. Diapause occurs less in parasites which continuously infect the hosts such as mange mites or lice.

To date, similar phenomena have not been ascribed to protozoa, although there is one report of latent coccidiosis occurring in cattle for which a similar hypothesis has been proposed.

DEVELOPMENT AND SURVIVAL OF INFECTIVE STAGES

The factors which affect development and survival are mainly environmental, especially seasonal climatic change and certain management practices.

The microhabitat

Several environmental factors, which affect the microhabitats of free-living parasitic stages, are vital for development and survival. Thus moderate temperatures and high humidity favour development of most parasites, while cool temperatures prolong survival. The microclimate humidity depends, of course, not only on rainfall and temperature, but on other elements such as soil structure, vegetation type and drainage. Soil type influences the growth and species composition of the herbage and this, in turn, determines the degree to which a layer of 'mat' is formed between the soil and the herbage. The mat is abundant in older pastures and holds a permanent store of moisture in which the relative humidity remains high even after weeks of drought. The presence of this moisture and pockets of air trapped in the mat limit the rate of temperature change and these factors favour the development and survival of helminth larvae, ticks, larval stages of insects and coccidial oocysts.

In contrast, the use of rotational cropping of pastures reduces the influence of 'mat' and therefore parasite survival. In the arid tropics pasture growth is usually negligible causing a similar effect.

In the same way, a high ground water table is important for the development and survival of intermediate snail vectors of trematodes such as liver and rumen flukes.

The development and survival of helminth eggs or larvae within faeces are also dependent on temperature and moisture. The host species may also influence this situation since normal cattle faeces remains in its original form for a longer time than, say, sheep pellets. Thus the moisture content at the centre of a bovine faecal pat remains high for several weeks or even months and so provides shelter for developing larvae until the outside environment is suitable.

Dictyocaulus larvae may also be distributed with the spores of the fungus *Pilobolus* which grow in bovine faeces, while several species of nematode larvae including *Oesophagostomum* spp. of pigs are known to be spread mechanically by some dipteran flies.

Seasonal development

In temperate countries with distinct seasons of summer and winter there are a limited number of generations and the same is true of countries with distinct dry and wet seasons. For example, in Britain there is only one or, at the most, two parasitic generations of the common trichostrongyle infections of ruminants since larval development on the pasture occurs only from late spring through to early autumn, the peak levels of infective larvae being present from July until September. In contrast, in tropical climates there may be numerous generations per year, but even in this case there are times when conditions for the development and survival of the free-living stages are optimal.

The development of large numbers of infective stages of parasites within distinct seasons is usually followed by a high mortality rate within a few weeks. However, considerable numbers survive for much longer than is commonly realised. For example, in the helminths, significant numbers of metacercariae of *Fasciola hepatica* and infective larvae of trichostrongyles are capable of survival for at least nine months in Britain.

Dipteran fly populations also vary in the number of generations per year. Using the blowflies as an example, there are three or four generations, and therefore higher populations, in southern England whereas in Scotland there are only two, temperature being the limiting factor. In the humid tropical or subtropical countries the development of trichostrongyle larvae or fly populations proceeds throughout most of the year and although this may be slower at certain times there will be numerous generations per annum.

Although the permanent ectoparasites such as lice or mange mites live on or in the skin of animals and, therefore, in an apparently stable environment this is not really the case as the hair or wool alters in length due to seasonal factors or human intervention. In the northern hemisphere development of these parasites is optimal in the winter when the coat is long and the micro-environment humid and temperate.

Apart from the free-living stages of coccidian parasites which have seasonal requirements similar to those of the trichostrongyles the prevalence of other protozoan infections is related to the feeding activity of their arthropod vectors. For example, in Britain, babesiosis in cattle occurs at peak times of tick activity in the spring and autumn.

Stock management

The availability of helminth infective stages is also affected by certain management practices. Thus, a high density of stocking increases the level of contamination, and, by lowering the sward height, enhances the availability of the larval stages largely concentrated in the lower part of the herbage. Also, the scarcity of grass may induce animals to graze closer to faeces than otherwise. However, against this, the microclimate in a short sward is more susceptible to changes in temperature and humidity and so the free-living stages may, on adverse occasions, be particularly vulnerable. This may explain why the helminth burden of ruminants in close-cropped set-stocked pastures are often less than those in animals on rotated pastures.

Similarly, many pasture improvement schemes have direct or indirect effects on arthropod populations. Improved host nutrition results from pasture improvement and helps to maintain host resistance to parasitism. However, pasture improvement, particularly in the tropics, can increase the breeding success of ticks and of those dipteran flies which lay their eggs in faeces, by increasing the shelter available. Furthermore, the increased stocking rates on improved pastures may increase the chances of parasites finding a host.

The date of parturition in a flock or herd may also influence the likelihood of parasitic infection. Where livestock are born out of season the numbers of trichostrongyle infective stages are usually lower and the chance of infection postponed until the young animals are older and stronger.

AN ALTERATION IN SUSCEPTIBILITY TO INFECTION

This may refer to existing infections or to the acquisition of new infections.

ALTERED EFFECTS OF AN EXISTING INFECTION

This is observed principally in adolescent or adult stock which are harbouring parasite populations below the threshold usually associated with disease and may be explained by various dietary and host factors.

Diet

It is well known that adequately fed animals are better able to tolerate parasitism than animals on a low plane of nutrition.

Thus, ruminants affected with blood sucking helminths such as *Haemonchus contortus* or *Fasciola hepatica* may be able to maintain their haemoglobin levels as long as their iron intake is adequate. However, if iron reserves become low their haemopoeitic systems become exhausted and they may die. Similarly, cattle may grow at a reasonable rate with moderate trichostrongylid burdens even though some loss of protein is occurring through the alimentary mucosa. However, if there is a change in diet which reduces their protein intake they are unable to compensate for the loss of protein and lose weight. These deleterious effects of parasitism, without any change in the level of infection, are not uncommon in outwintered stock or, in the tropics, in animals during a period of drought.

Incidentally, the same effect is produced when food intake is not increased during pregnancy and lactation. Good examples of this are the accumulation of lice on poorly fed animals during the winter and the fact that the anaemia caused by ticks is greater in animals on poor nutrition.

Apart from protein and iron, dietary deficiencies in trace elements are also significant. Thus, trichostrongylosis in ruminants is known to impair the absorption of both calcium and phosphorus and where the dietary intake of these is sub-optimal osteoporosis can occur. Also, the deleterious effects of some abomasal parasites in sheep are greater where there is a cobalt deficiency and, in such animals, levels of parasitism generally considered to be non-pathogenic may be associated with severe diarrhoea and weight loss.

Pregnancy and lactation

The period of gestation in grazing livestock often coincides with that of inadequate nutrition and is geared to completion at a time when freshly growing pasture becomes available for their newborn progeny. In housed or outwintered livestock the cost of maintaining an adequate nutritional intake during pregnancy is often high and as a result the nutritional levels are often sub-optimal. If this occurs, quite low worm burdens can have a detrimental effect on the food conversion of the dam which in turn influences foetal growth and subsequently that of the neonate through poor milk production by the dam. This has been clearly illustrated in sows infected with moderate burdens of *Oesophagostomum dentatum* and in ewes infected with helminths such as *Haemonchus* or *Fasciola*.

Steroid therapy

Steroids are widely used in therapy of both man and animals and it is known that they may alter the susceptibility to parasitism. A good example of this is in the cat infected with *Toxoplasma gondii*; excretion of oocysts usually occurs for only about two weeks, but may reappear and be prolonged following the administration of steroids. Egg production by nematodes is also known to be increased following steroid treatment and so pasture contamination is increased.

ALTERED SUSCEPTIBILITY TO THE ACQUISITION OF NEW INFECTIONS

The role of intercurrent infections

The interaction of various parasites, or a parasite with another pathogen, resulting in an exaggerated clinical disease, has been reported on several occasions. For example: in lambs, the nematode *Nematodirus battus* and the protozoan *Eimeria*; in cattle, the trematode *Fasciola hepatica*, and the bacterium *Salmonella dublin*, and also *Fasciola hepatica* and the mange mite *Sarcoptes*; in pigs, the nematode, *Trichuris suis*, and the spirochaete *Treponema hyodysenteriae*.

The effect of chemotherapy

In certain instances, immunity to parasites appears to be dependent on the continuing presence of low threshold infections, commonly called premunity. If the balance between the host and the immunising infection is disturbed by therapy then re-infection of the host may occur, or in the case of helminths, an arrested larval population may develop to maturity from the reservoir of infection within the host. Thus, the use of anthelmitics, known to be effective against adult parasites, but not arrested nematode larvae, may precipitate development of the latter once the adults are removed; this is known to occur in infections with *Hyostrongylus rubidus* in the pig. Sometimes, also, the over zealous application of anthelmintics in grazing animals will result in the eventual establishment of higher numbers of trichostrongyles than were present prior to treatment. Excessive application of acaricides to control ticks may also lower herd immunity to babesial and theilerial infections, the so-called'enzootic instability'.

Hypersensitivity

In many instances, at least part of the immune response to parasites is associated with a marked IgE response and a hypersensitivity reaction. Where this occurs in the gut, as in intestinal nematode infections, the reaction is associated with an increased permeability of the gut to macromolecules such as protein, and this may be a significant factor in immune animals under heavy larval challenge. In sheep, for example, relatively poor growth rates and poor wool production may result.

A stunting effect has also been observed in tick resistant animals which are under constant challenge, while pet animals repeatedly exposed to mite infestations may have severely thickened, hyperaemic and sensitive skins, although only neglible numbers of mites are present.

PARASITISM RESULTING FROM THE MOVEMENT OF SUSCEPTIBLE STOCK INTO AN INFECTED ENVIROMENT

ABSENCE OF ACQUIRED IMMUNITY

The best examples of outbreaks of parasitic disease following the movement of calves into infected areas are provided by the common nematode diseases of ruminants. For example, in western Europe the cattle lungworm, *Dictyocaulus viviparus*, is endemic and the most severe outbreaks are seen in calves born in early spring and turned out in late summer to graze alongside older batches of calves which have grazed from early spring. Overwintered larval populations have cycled in these older calves and when the fresh populations of infective larvae which develop from these infections accrue on pasture, the younger calves, with no previous experience of infection, are extremely susceptible.

The occurrence of 'cysticercosis storms' in adult cattle, grazed on fields contaminated with eggs of the human tapeworm, *Taenia saginata*, or handled by infected stockmen, are occasionally reported in Europe and the U.S.A. This high degree of susceptibility is due to lack of previous exposure to infection. In contrast, in areas where cysticercosis is endemic, cattle are repeatedly infected and soon acquire a solid resistance to reinfection, only the cysts acquired in early life persisting in the muscles.

With protozoal diseases such as babesiosis, theileriosis, coccidiosis and toxoplasmosis, caution has to be exercised in introducing naive animals into infected areas. In the case of toxoplasmosis, the introduction of female sheep into a flock in which the disease is endemic has to be carefully controlled and these should be non-pregnant when purchased and allowed to graze with the flock for some months prior to mating.

ABSENCE OF AGE IMMUNITY

A significant age immunity develops against relatively few parasites, and adult stock not previously exposed to many helminth and protozoal infections are at risk if moved into an endemic area.

LONGEVITY OF INFECTIVE STAGES

Especially in temperate zones and in parts of the subtropics, the free-living stages of most parasites will survive in the environment or in intermediate hosts for periods sufficiently long to re-infect successive batches of young animals and may cause disease in these animals within a few weeks of exposure.

THE INFLUENCE OF GENETIC FACTORS

Between host species

Most parasites are host specific and this specificity has been utilised in integrated control programmes, such as mixed grazing of sheep and cattle, to control gastrointestinal nematodes. However, some economically important parasites are capable of infecting a wide range of hosts which vary in their susceptibility to the effects of the parasite. For example, cattle seem able to cope with liver fluke infestations which would cause death in sheep, and goats appear to be very much more susceptible than cattle or sheep to their common gastrointestinal trichostrongyles.

Between breeds

Evidence is accumulating that the susceptibility of various breeds of animals to parasites varies and is genetically determined. For example, some breeds of sheep are more susceptible to the abomasal nematode, *Haemonchus contortus*, than others; *Bos indicus* breeds of cattle are more resistant to ticks and other haematophagous insects than *Bos taurus* breeds. In Denmark, the Black Pied cattle are genetically deficient in their cellular immune responses and have proved more susceptible to liver fluke, while the N'dama breed of cattle in West Africa is known to be tolerant to trypanosomiasis.

Even within flocks or herds, individual responders and non-responders, in terms of their ability to develop resistance to internal and external parasites, are usually present and it is recommended by some experts that culling of the poorest responders should take place.

SEX

There is some evidence that entire male animals are more susceptible than females to some helminth infections. This could be of importance in countries where castration is not routinely practised, or where androgens are used to fatten castrates or cull cows.

STRAIN OF PARASITE

Although this aspect has received scant attention, except in protozoal infections, there is now evidence that strains of helminths occur which vary in infectivity and pathogenicity. The increasing prevalence of drug resistant strains of many parasites is another point which should be considered when disease outbreaks occur in herds, flocks or studs where control measures are routinely applied.

INTRODUCTION OF INFECTION INTO A CLEAN ENVIRONMENT

There are several ways in which a parasite may be introduced into an environment from which it has been eradicated or where it has never been found.

INTRODUCTION OF NEW STOCK

One of the current trends in the international livestock area is the movement of breeding stock from country to country. Quarantine restrictions and vaccination requirements are stringent in relation to epidemic diseases, but limited or non-existent for parasitic diseases. When infected animals are moved into an area previously free from any given parasite the infection may cycle, provided suitable conditions exist, and the consequences for the indigenous stock can be extremely serious. Examples of this category include the introduction of *Toxocara vitulorum* into Britain and Ireland, the source of infection being Charolais heifers from mainland Europe and transmission occurring via the dam's milk. The spread of *Parafilaria bovicola* in Sweden, presumably introduced with cattle, or by the muscid intermediate hosts inadvertently transported from southern Europe, is another example. In the United States and Australia the increased movement of human populations and their pets has seen the spread of heartworm infections in dogs to almost every state whereas it was previously limited to the more tropical areas; clearly, the mosquito vectors suitable for transmission were already present in the other states. Psoroptic mange in cattle, originally confined to southern Europe, is now endemic in Belgium and Germany due to trade in breeds of cattle. Protozoal diseases, such as toxoplasmosis, have been introduced into sheep flocks in countries where it was previously absent, by the importation of infected sheep. Babesiosis has also spread where animals carrying infected ticks have moved into non-endemic areas where the ticks were able to become established.

THE ROLE OF EFFLUENT

The transfer of infection from one farm to another via manure has also been reported. Thus outbreaks of ostertagiasis have occurred in farms following the application of cattle slurry as a fertiliser, while cysticercosis 'storms' due to *Cysticercus bovis* have occurred in cattle following the application of human sewage to pastures. Finally, the application of pig slurry containing ascarid eggs to pastures subsequently grazed by sheep has resulted in pneumonia due to migrating ascarid larvae.

THE ROLE OF INFECTED VECTORS

Several helminth infections are transmitted by winged insects, and these can serve to introduce infection into areas previously free of infection. Occasionally also, birds may mechanically transport infective stages of parasites to a new environment. This has occurred in the Netherlands where the ditches and dykes surrounding reclaimed land have become colonised by *Lymnaea* snails transported by wild birds. The introduction of livestock lightly infected with *Fasciola hepatica* resulted in the snails becoming infected, and subsequently, outbreaks of clinical fascioliasis.

RESISTANCE TO PARASITIC DISEASES

Broadly speaking, resistance to parasitic infections falls into two categories. The first of these includes species resistance, age resistance and breed resistance which, by and large, are not immunological in origin. The second category, acquired immunity, is completely dependent on antigenic stimulation and subsequent humoral and cellular responses. Although, for reasons explained below, there are very few vaccines yet available against parasitic diseases, the natural expression of acquired immunity plays a highly significant role in protecting animals against many infections and in modulating the epidemiology of many parasitic diseases.

SPECIES RESISTANCE

For a variety of parasitological, physiological and biochemical reasons, many parasites do not develop at all in other than their natural hosts; this is typified by, for example, the remarkable host specificity of the various species of *Eimeria*. In many instances however, a limited degree of development occurs, although this is not usually associated with clinical signs; thus, some larvae of the cattle parasite *Ostertagia ostertagi* undergo development in sheep, but very few reach the adult stage. However, in these unnatural or aberrant hosts, and especially with parasites which undergo tissue migration, there are occasionally serious consequences particularly if the migratory route becomes erratic. An example of this is *visceral larva migrans* in children, due to *Toxocara canis* which is associated with hepatomegaly and occasionally ocular and cerebral involvement.

Some parasites, of course, have a very wide host range, *Trichinella spiralis, Fasciola hepatica, Cryptosporidium* and the asexual stages of *Toxoplasma* being four examples.

AGE RESISTANCE

Many animals become more resistant to primary infections with some parasites as they reach maturity. For example, ascarid infections of animals are most likely to develop to patency if the hosts are a few months old. If infected at an older age the parasites either fail to develop, or are arrested as larval stages in the tissues; likewise, patent *Strongyloides* infections of ruminants and horses are most commonly seen in very young animals. Sheep of more than three months of age are relatively resistant to *Nematodirus battus* and in a similar fashion, dogs gradually develop resistance to infection with *Ancylostoma* over their first year of life.

The reasons underlying age resistance are unknown, although it has been suggested that the phenomenon is an indication that the host-parasite relationship has not yet fully evolved. Thus, while the parasite can develop in immature animals, it has not yet completely adapted to the adult.

On the other hand, where age resistance is encountered, most parasitic species seem to have developed an effective counter-mechanism. Thus *Ancylostoma caninum, Toxocara canis, Toxocara cati* and *Toxocara vitulorum* and *Strongyloides* spp. all survive as larval stages in the tissues of the host, only becoming activated during late pregnancy to infect the young *in utero* or by the transmammary route. In the case of *Nematodirus battus*, the critical hatching requirements for the egg, ie. prolonged chill followed by a temperature in excess of 10 °C, ensure the parasites' survival as a lamb-to-lamb infection.

Oddly enough, with *Babesia* and *Anaplasma* infection of cattle, there is generally thought to be an inverse age resistance in that young animals are more resistant than older naive animals.

BREED RESISTANCE

In recent years, there has been considerable practical interest in the fact that some breeds of domestic ruminants are more resistant to certain parasitic infection than others.

Probably the best example of this is the phenomenon of trypanotolerance displayed by West African humpless cattle, such as the N'dama, which survive in areas of heavy trypanosome challenge. The mechanism whereby these cattle control their parasitaemias is still unknown, although it is thought that immunological responses may play a role.

In helminth infections it has been shown that the Red Masai sheep, indigenous to East Africa, is more resistant to *Haemonchus contortus* infection than some imported breeds studied in that area, while in South Africa it has been reported that the Merino is less susceptible to trichostrongylosis than certain other breeds.

Within breeds, haemoglobin genotypes have been shown to reflect differences in susceptibility to *Haemonchus contortus* infection in that Merino, Scottish Blackface and Finn Dorset sheep which are homozygous for haemoglobin A, develop smaller worm burdens after infection than their haemoglobin B homozygous or heterozygous counterparts. Unfortunately, these genotypic differences in susceptibility often break down under heavy challenge.

Some work within a single breed has shown in Australia that individual Merino lambs may be divided into responders and non-responders on the basis of their immunological response to infection with *Trichostrongylus colubriformis* and that these differences are genetically transferred to the next generation.

The selection of resistant animals could be of great importance, especially in many developing areas of the world, but in practice would be most easily based on some easily recognisable feature such as 'coat colour' rather than be dependent on laboratory tests.

In Australia resistance to ticks, particularly *Boophilus*, has been shown to be influenced by genetics, being high in the humped, *Bos indicus*, Zebu breeds and low in the European, *Bos taurus*, breeds. However, where cattle are 50% Zebu, or greater, in genetic constitution, a high degree of resistance is still possible allowing a limited use of acaricides.

ACQUIRED IMMUNITY TO HELMINTH INFECTIONS

Immune responses to helminths are complex, possibly depending on antigenic stimulation by secretory or excretory products released during the development of the L_3 to the adult, and for this reason it has only been possible to develop one or two practical methods of artificial immunisation of which the radiation-attenuated vaccine against *Dictyocaulus viviparus* is perhaps the best example.

Despite this, there is no doubt that the success of many systems of grazing management depend on the gradual development by cattle and sheep of a naturally acquired degree of immunity to gastrointestinal nematodes. For example, experimental observations have shown that an immune adult sheep may ingest around 50,000 *Ostertagia* L_3 daily without showing any clinical signs of parasitic gastritis.

THE EFFECT OF THE IMMUNE RESPONSE

Dealing first with gastrointestinal and pulmonary nematodes the effects of the immune response may be grouped under 3 headings:

(i) The development of immunity after a primary infection is invariably associated with an **ability to expel the adult nematodes.**

(ii) Subsequently, the host can attempt to limit reinfection by **preventing the migration and establishment of larvae** or, sometimes, **by arresting their development at a larval stage.** This type of inhibition of development should not be confused with the more common hypobiosis triggered by environmental effects on infective larvae on pasture or, in the present state of knowledge, with the arrested larval development associated with age resistance in, for example, the ascarids.

(iii) Adults which do develop **are stunted in size** and their **fecundity is reduced.** The important practical aspect of this mechanism is perhaps not so much the reduced pathogenicity of such worms as the great reduction in pasture contamination with eggs and larvae, which in turn reduces the chance of subsequent reinfection.

Each of these mechanisms is exemplified in infections of the rat with the trichostrongyloid nematode *Nippostrongylus brasiliensis*, a much-studied laboratory model which has contributed greatly to our understanding of the mechanisms of host immunity in helminth infection. The infective stage of this parasite is normally a skin penetrator, but in the laboratory is usually injected subcutaneously for convenience. The larvae travel via the bloodstream to the lungs where, having moulted, they pass up the trachea and are swallowed. On reaching the small intestine they undergo a further moult and become adult, the time elapsing between infection and development to egg-laying adults being five to six days. The adult population remains static for about five more days. After this time the faecal worm egg output drops quickly, and the majority of the worms are rapidly expelled from the gut. This expulsion of adult worms, originally known as the 'self cure' phenomenon, has been shown to be due to an immune response.

If the rats are reinfected, a smaller proportion of the larval dose arrives in the intestine ie. their migration is stopped. The few adult worms which do develop in the gut remain stunted and are relatively infertile, and expulsion of worms from the gut starts earlier and proceeds at a faster rate.

Under natural grazing conditions larval infections of cattle and sheep are, of course, acquired over a period, but an approximately similar series of events occur. For example, calves exposed to *Dictyocaulus viviparus* quite rapidly acquire patent infections, readily recognisable by the clinical signs. After a period of a few weeks, immunity develops and the adult worm burdens are expelled. On subsequent exposure in succeeding years such animals are highly resistant to challenge, although if this is heavy, clinical signs associated with the reinfection syndrome ie. the immunological destruction of the invading larvae in the lungs, may be seen. With *Ostertagi* and *Trichostrongylus* infections, the pattern is the same with the build-up of an infestation of adult worms being followed by their expulsion and subsequent immunity; in later life only small, short-lived adult infections are established and eventually the infective larvae are expelled without any development at all. However, with gastrointestinal infections in ruminants, the ability to develop good immune responses is often delayed for some months because of neonatal immunological unresponsiveness.

The mechanism of this type of immunity to luminal parasites is still not understood despite a great deal of research. However it is generally agreed that such infections produce a state of gut hypersensitivity associated with an increase of mucosal mast cells in the lamina propria and the production of worm-specific IgE, much of which becomes bound to the surface of the mast cells. The reaction of worm antigen, from an existing infection or from a subsequent challenge, with these sensitised mast cells releases vasoactive amines which cause an increase in capillary and epithelial permeability and hyperproduction of mucus.

After this point there is some confusion. Some workers have concluded that these physiological changes simply affect the well-being of the worms by, for example, lowering the oxygen tension of their environment, so that they become detached from the mucosa and subsequently expelled.

Others have postulated that, in addition, the permeable mucosa allows the 'leakage' of IgG antiworm antibody from the plasma into the gut lumen, where it has access to the parasites.

Additional factors, such as the secretion of specific antiworm IgA on the mucosal surface and the significance of sensitised T cells which are known to promote the differentiation of mast cells, eosinophils and mucus-secreting cells are also currently under study.

With regard to tissue-invading helminths, the most closely studied have been the schistosomes. Recent work has shown that the schistosomulae of *Schistosoma mansoni* may be attacked by both eosinophils and macrophages, which attach to the antibody-coated parasite. Eosinophils, especially, attach closely to the parasites where their secretions damage the underlying parasite membrane.

Attempts to find if a similar mechanism exists against *Fasciola hepatica* have indicated that although eosinophils do attach to parts of the tegument of the young fluke, the latter seems able to shed its surface layer to evade damage.

EVASION OF THE HOST'S IMMUNE RESPONSE

Despite the evidence that animals are able to develop vigorous immune responses to many helminth infections, it is now clear that parasites, in the course of evolution, have capitalised on certain defects in this armoury. This aspect of parasitology is still in its infancy, but three examples of immune evasion are described below.

Neonatal immunological unresponsiveness

This is the inability of young animals to develop a proper immune response to some parasitic infections. For example, calves and lambs fail to develop any useful degree of immunity to reinfection with *Ostertagia* spp. until they have been exposed to constant reinfection for an entire grazing season. Similarly, lambs are repeatedly susceptible to *Haemonchus contortus* infection until they are between 6 months and 1 year old.

The cause of this unresponsiveness is unknown. However, while calves and lambs ultimately do develop a good immune response to *Ostertagia* infection, in the sheep/*Haemonchus contortus* system the neonatal unresponsiveness is apparently often succeeded by a long period of acquired immunological unresponsiveness eg. Merino sheep reared from birth in a *Haemonchus* endemic environment remain susceptible to reinfection throughout their entire lives.

Concomitant immunity

This term is used to describe an immunity which acts against invading larval stages, but not against an existing infection. Thus a host may be infected with adult parasites, but has a measure of immunity to further infection. Perhaps the best example is that found with schistosomes which are covered by a cytoplasmic syncytium which, unlike the chitinous-like cuticle of nematodes, would at first seem to be vulnerable to the action of antibody or cells. However, it has been found that adult schistosomes have the property of being able to incorporate host antigens, such as blood group antigens or host immunoglobulin, on their surface membrane to mask their own foreign antigens.

Concomitant immunity does not appear to operate with *Fasciola hepatica* in sheep, in that they are repeatedly susceptible to reinfection. On the other hand, cattle not only expel their primary adult burden of *Fasciola hepatica*, but also develop a marked resistance to reinfection.

Concomitant immunity also includes the situation where established larval cestodes may survive for years in the tissues of the host, although the latter is completely immune to reinfection. The mechanism is unknown, but it is thought that the established cyst may be 'masked' by host antigen or perhaps secrete an 'anticomplementary' substance which blocks the effect of an immune reaction.

Polyclonal stimulation of immunoglobulin

As well as stimulating the production of specific IgE antibody, helminths 'turn on' the production of large amounts of non-specific IgE. This may help the parasite in two ways. First, if mast cells are coated by non-specific IgE they are less likely to attract parasite-specific IgE and so will not degranulate when exposed to parasite antigen. Secondly, the fact that the host is producing immunoglobulin in a non-specific fashion means that specific antibody to the helminth is less likely to be produced in adequate quantity.

THE DEBIT SIDE OF THE IMMUNE RESPONSE

Sometimes immune responses are associated with lesions which are damaging to the host. For example the pathogenic effects of oesophogostomiasis are frequently attributable to the intestinal nodules of *Oe. columbianum*; similarly, the pathogenic effects of schistosomiasis are due to the egg granulomata, the result of cell-mediated reactions, in the liver and bladder.

ACQUIRED IMMUNITY TO PROTOZOAL INFECTIONS

As might be anticipated from their microscopic size and unicellular state, immunological responses against protozoa are similar to those directed against bacteria. The subject is, however, exceedingly complex and the following account is essentially a digest of current information on some of the more important pathogens. As with bacterial infections, immune responses are typically humoral or cell-mediated in type and occasionally both are involved.

Trypanosomiasis is a good example of a protozoal disease to which immunity is primarily humoral. Thus, *in vitro*, both IgG and IgM can be shown to lyse or agglutinate trypanosomes and *in vivo* even a small amount of immune serum will clear trypanosomes from the circulation, apparently by facilitating their uptake, through opsonisation, by phagocytic cells. Unfortunately, the phenomenon of antigenic variation, another method of immune evasion, prevents these infections being completely eliminated and typically allows the disease to run a characteristic course of continuous remissions and exacerbations of parasitaemia. It is likely, also, that the generalised immunosuppression induced by this disease, may, sooner or later, limit the responsiveness of the host.

It is also relevant to note that some of the important lesions of trypanosomiasis such as anaemia, myocarditis and lesions of the skeletel muscle are thought to be attributable to the deposition of trypanosome antigen or immune complexes on these cells leading to their subsequent destruction by macrophages or lymphocytes, a possible debit effect of the immune response.

Acquired immunity to babesiosis also appears to be mediated by antibody, perhaps acting as an opsonin, and facilitating the uptake of infected red cells by splenic macrophages. Antibody is also transferred in the colostrum of the mother to the new-born animal and confers a period of protection against infection.

Finally, in trichomoniasis, antibody, presumably produced largely by plasma cells in the lamina propria of the uterus and vagina, is present in the mucus secreted by these organs and to a lesser extent in the plasma. This, *in vitro*, kills or agglutinates the trichomonads and is presumably the major factor responsible for the self-limiting infections which typically occur in cows.

Of those infections against which immunity is primarily cell-mediated, leishmaniasis is of particular interest in that the amastigotes invade and proliferate in macrophages whose function, paradoxically, is the phagocytosis and destruction of foreign organisms. How they survive in macrophages is unknown, although it has been suggested that they may release substances which inhibit the enzyme activity of lysosomes or that the amastigote surface coat is refractory to lysosomal enzymes. The immunity which develops seems to be cell-mediated perhaps by cytotoxic T cells destroying infected macrophages or by the soluble products of sensitised T cells 'activating' macrophages to a point where they are able to destroy their intra-cellular parasites. Unfortunately in many cases the efficacy of the immune response and the consequent recovery is delayed or prevented by a variable degree of immunosuppression of uncertain etiology.

As noted above, sometimes both humoral and cell-mediated reactions are involved in immunity, and this seems to be the situation with coccidiosis, theileriosis and toxoplasmosis.

In coccidiosis, the protective antigens are associated with the developing asexual stages and the expression of immunity is dependent on T cell activity. It is thought that these function in two ways; first, as helper cells for the production of neutralising antibody against the extracellular sporozoites and merozoites and secondly, in a cell-mediated fashion, by releasing substances, such as lymphokines, which inhibit the multiplication of the intracellular stages. The net effect of these two immunological responses is manifested by a reduction in clinical signs and a decrease in oocyst production.

As described earlier, the proliferative stages of theilerial infections are the schizogonous stages which develop in lymphoblasts and divide synchronously with these cells to produce two infected daughter cells. During the course of infection, and provided it is not rapidly fatal, cell-mediated responses are stimulated in the form of cytotoxic T cells which target on the infected lymphoblasts by recognising two antigens on the host surface; one of these is derived from the *Theileria* parasite and the other is a histocompatibility antigen of the host cell. The role of antibodies in protection is less clear, although it has been recently demonstrated, using an *in vitro* test, that an antibody against the sporozoites inoculated by the tick may be highly effective in protection.

In toxoplasmosis also, both humoral and cell-mediated components appear to be involved in the immune response. However the relative importance of their roles remains to be ascertained, although it is generally believed that antibody formation by the host leads to a cessation in the production of tachyzoites and to the development of the latent bradyzoite cyst; also that recrudescence of tachyzoite activity may occur if the host becomes immunosuppressed as a consequence of therapy or some other disease.

ACQUIRED IMMUNITY TO ARTHROPOD INFECTIONS

It is known that animals exposed to repeated attacks by some insects gradually develop a degree of acquired immunity. For example, at least in man, over a period of time the skin reactions to the bites of *Culicoides* and mosquitoes usually decrease in severity. Likewise, sheep after several attacks of calliphorine myiasis become more resistant to further attack.

A similar sequence of events has been observed with many tick and mite infestations. The immune reaction to ticks, dependent on humoral and cell-mediated components to the oral secretions of the ticks, prevents proper engorgement of the parasites and has serious consequences on their subsequent fertility; dogs which have recovered from sarcoptic mange are usually immune to further infection.

Although these immune responses must moderate considerably the significance of many ectoparasitic infections, their primary importance to date is largely concerned with their debit side ie. the unfortunate consequences which often occur when an animal becomes sensitised to arthropod antigens. Three examples of this are flea dermatitis in dogs and cats, the pruritus and erythema associated with sarcoptic mange especially in the dog and pig, and 'sweet itch' of horses due to skin hypersensitivity to *Culicoides* bites.

ANTHELMINTICS

The control of parasitic helminths in domestic animals relies largely on the use of anthelmintic drugs. Although anthelmintics are used in all domestic species, the largest market is undoubtedly the ruminant market, especially cattle, where millions of pounds are spent annually in an effort to reduce the effects of parasitism.

It is not practical to give efficacy data and methods of application of the large number of drugs currently available against the vast range of helminths which parasitise domestic animals. Also, the number of compounds and their various formulations are continually changing and it is therefore more appropriate to discuss the use of anthelmintics in general terms, details of their use against individual species or groups of helminths having been described under the appropriate sections of the main text.

PROPERTIES OF ANTHELMINTIC COMPOUNDS

An ideal anthelmintic should possess the following properties:

1. **It should be efficient against all parasitic stages of a particular species.**

   It is also generally desirable that the spectrum of activity should include members of different genera, for example in dealing with the equine strongyles and *Parascaris equorum*. However in some circumstances, separate drugs have to be used at different times of year to control infections with unrelated helminths; the trichostrongyles responsible for ovine parasitic gastroenteritis and the liver fluke *Fasciola hepatica* are one such example.
2. It is important that any anthelmintic **should be non-toxic to the host**, or at least have a wide safety margin.

   This is especially important in the treatment of groups of animals such as a flock of sheep, where individual body weights cannot easily be obtained, rather than in the dosing of individual companion animals such as cats or dogs.
3. An anthelmintic **should be rapidly metabolised and excreted by the host**, otherwise there would require to be long withdrawal periods in meat and milk producing animals.
4. Anthelmintics **should be easily administered**, otherwise they will not be readily accepted by owners; different formulations are available for different domestic animal species. Oral and injectable products are widely used in ruminants, and pour-on preparations are available for cattle. Palatable in-feed and paste formulations are convenient for use in horses, while anthelmintics are usually available as tablets for dogs and cats.
5. **The cost of an anthelmintic should be reasonable.** This is of special importance in pigs and poultry where profit margins may be narrow.

USE OF ANTHELMINTICS

Anthelmintics are generally used in two ways, namely, therapeutically, to treat existing infections or clinical outbreaks, or prophylactically, in which the timing of treatment is based on a knowledge of the epidemiology. Clearly prophylactic use is preferable where administration of a drug at selected intervals or continuously over a period can prevent the occurrence of disease.

THERAPEUTIC USAGE

When used therapeutically, the following factors should be considered.

First, if the drug is not active against all stages it must be effective against the pathogenic stage of the parasite. For example, the larval stages of *F. hepatica* may cause acute disease, but some drugs are active only against adult flukes.

Secondly, use of the anthelmintic should, by successfully removing parasites, result in cessation of clinical signs of infection such as diarrhoea and respiratory distress; in other words, there should be a marked clinical improvement and rapid recovery after treatment.

PROPHYLACTIC USAGE

Several points should be considered where anthelmintics are used prophylactically.

First, the cost of prophylactic treatment should be justifiable economically, by increased production in food animals, or by preventing the occurrence of clinical or subclinical disease in, for example, horses with strongylosis or dogs with heartworm disease.

Secondly, the cost-benefit of anthelmintic prophylaxis should stand comparison with the control which can be achieved by other methods such as pasture management or, in the case of dictyocauliasis, by vaccination.

Thirdly, it is desirable that the use of anthelmintics should not interfere with the development of an acquired immunity, since there are reports of outbreaks of disease in older stock which have been overprotected by control measures during their earlier years.

ANTHELMINTICS AND THEIR MODE OF ACTION

The major groups of anthelmintics currently in use against nematodes, trematodes and cestodes are shown in Table 10.

The mode of action of many anthelmintics is not known in detail, but basically depends on interference with essential biochemical processes of the parasite, but not of the host.

PIPERAZINES

These drugs produce paralysis in helminths by an anticholinergic action at the neuromuscular junction. Piperazine salts are widely used against ascarids while diethylcarbamazine has been used against lungworms and filarial nematodes.

Table 10 The major groups of anthelmintics

Parasites	Chemical group	Drugs
Nematodes	Piperazines	Piperazine Salts, Diethylcarbamazine
	Imidazothiazoles/Tetrahydropyrimidines	Tetramisole, Levamisole: Morantel, Pyrantel
	Benzimidazoles/Pro-benzimidazoles	Thiabendazole, Mebendazole, Parbendazole, Fenbendazole, Oxfendazole, Albendazole, Oxibendazole, Cambendazole, Flubendazole, Febantel, Thiophanate, Netobimin
	Avermectins	Ivermectin
	Organophosphates	Dichlorvos, Haloxon Trichlorfon (Metriphonate)
	Salicylanilides/substituted Phenols	Nitroscanate, Closantel
Trematodes	Salicylanilides/substituted Phenols	Nitroxynil, Rafoxanide, Oxyclozanide, Brotianide, Diamphenethide, Niclofolan Closantel
	Others	Clorsulon
	Benzimidazoles	Triclabendazole, Albendazole
Cestodes	Salicylanilides/substituted Phenols	Niclosamide
	Others	Praziquantel Bunamidine Arecoline

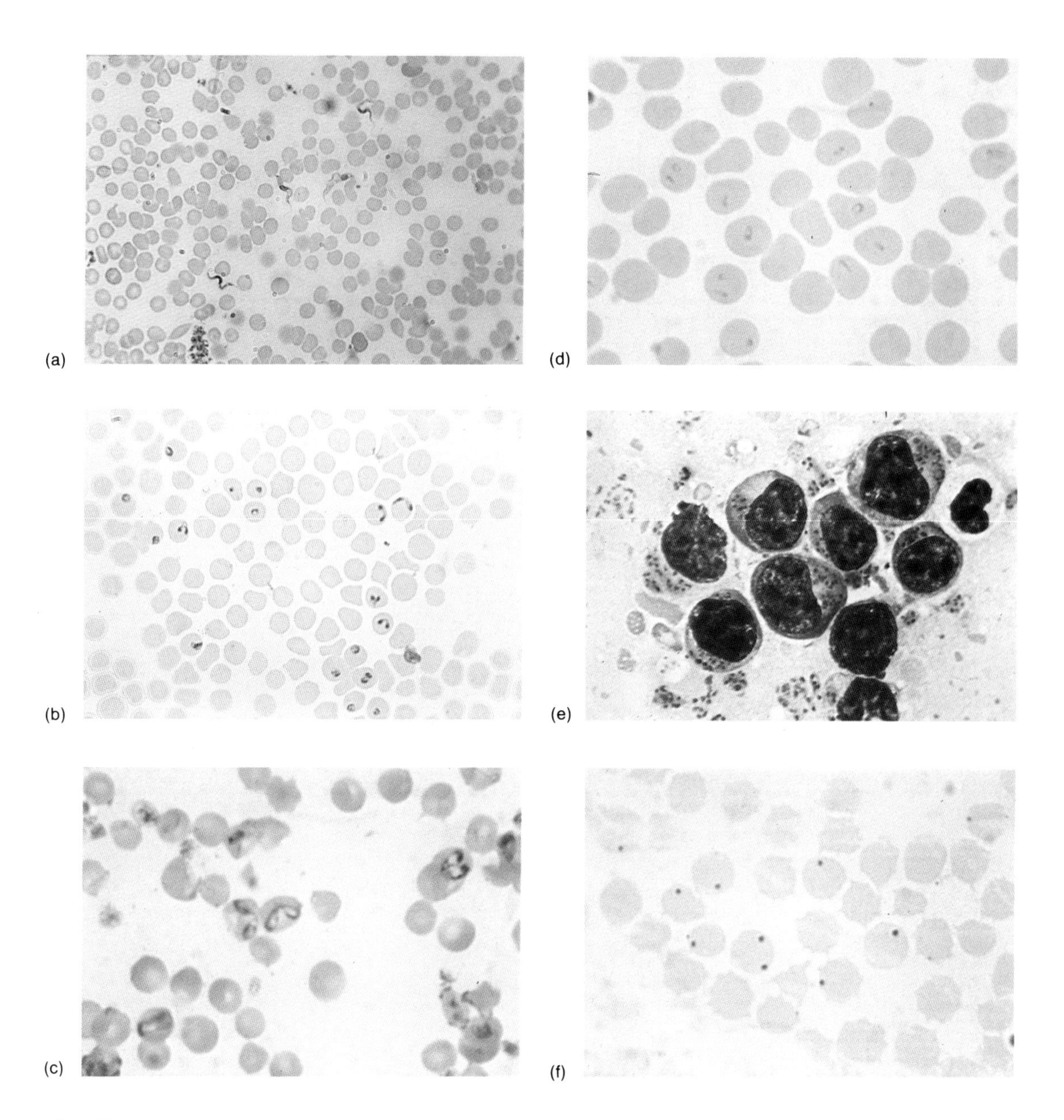

Plate XIII

(a) *Trypanosoma brucei* in blood film.
(b) The 'small' *Babesia divergens*.
(c) The 'large' *Babesia major*.
(d) Piroplasms of *Theileria parva*.
(e) Macroschizonts of *Theileria parva* in smear of lymph node.
(f) *Anaplasma marginale* in red blood cells.

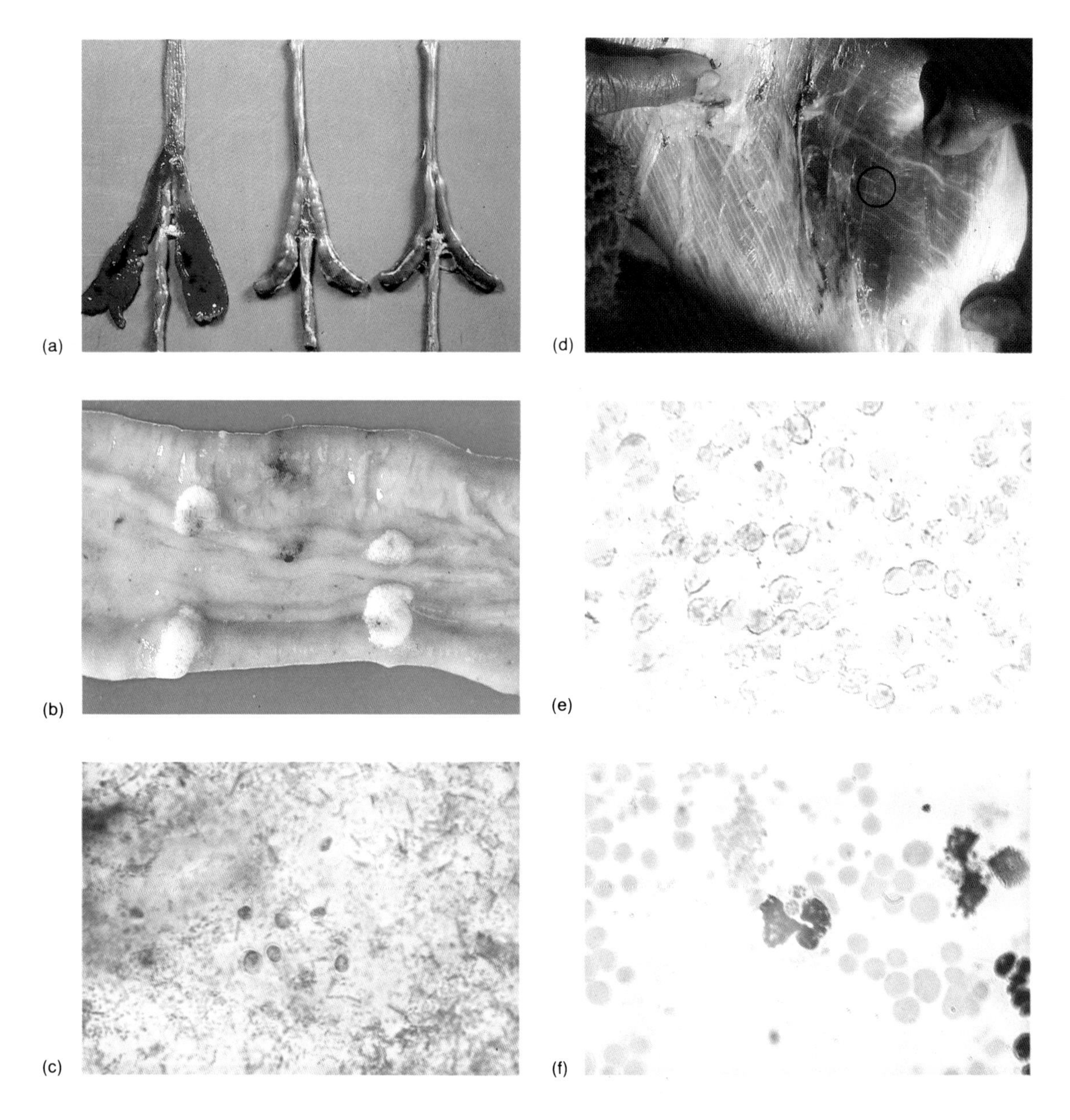

Plate XIV

(a) Caecal distension and haemorrhage due to *Eimeria tenella* infection of chickens.

(b) Papilloma-like lesions associated with *Eimeria ovina*.

(c) *Cryptosporidium* oocysts in bovine faecal smear stained with Ziehl–Nielsen.

(d) Two Bradyzoite cysts of *Sarcocystis* in bovine muscle.

(e) Smear showing *Eperythrozoon* on surface of red blood cells.

(f) Inclusions of *Ehrlichia phagocytophila* in polymorphonuclear leucocyte.

IMIDAZOTHIAZOLES/ TETRAHYDROPYRIMIDINES

Basically these compounds act as depolarising neuromuscular blocking agents in both nematodes and their hosts. In consequence the safety margin tends to be narrower than in some other groups. These drugs are active against a wide range of nematodes especially those in the gastrointestinal tract.

BENZIMIDAZOLES/PRO-BENZIMIDAZOLES

Drugs of this group generally act on the intestinal cells of helminths preventing glucose uptake and thus 'starving' the parasite. They are virtually without toxicity, in some cases even at over ten times the recommended dose rate. Parasite resistance to anthelmintics has most frequently been associated with repeated use of these drugs against nematodes of sheep and horses. Several compounds also have activity against tapeworms and flukes.

AVERMECTINS

These are a series of macrocyclic lactone derivatives which are fermentation products of the actinomycete *Streptomyces avermitilis*. Only one drug of this group, ivermectin, is generally available and it has been shown to have excellent activity, at very low dose rates, not only against a wide range of nematodes, but also against certain arthropod parasites. A further advantage is that this drug, if given parenterally, can remain active for at least two weeks after administration due to persistence in body fat. Ivermectin acts by potentiating the release and binding of gamma-aminobutyric acid (GABA) in certain nerve synapses. In nematodes, GABA acts as a neurotransmitter sending signals between interneurones and motor neurons; thus, when these signals are disrupted in the presence of ivermectin, paralysis of the nematode eventually ensues.

ORGANOPHOSPHATES

A few organophosphorus compounds are active against nematodes. They act by inhibiting cholinesterase resulting in a build-up of acetylcholine which leads to neuromuscular paralysis of nematodes and their expulsion. This group of drugs is relatively toxic and is used most frequently in horses, probably because of the additional insecticidal action against larvae of horse bots.

SALICYLANILIDES/SUBSTITUTED PHENOLS

Although details of the mode of action of drugs in these groups are not well understood, they appear to act by interfering with ATP production in parasites by uncoupling oxidative phosphorylation. They are most widely used against the liver fluke *Fasciola hepatica*, although nitroscanate is marketed for treatment of nematode and cestode infections of dogs, and niclosamide is widely used against tapeworms in many domestic species.

OTHER DRUGS

The mode of action of other drugs used to combat tapeworm infections is not well-known, but praziquantel apparently acts by causing spastic paralysis of muscle cells in the parasite.

METHODS OF ADMINISTRATION

Traditionally, anthelmintics have been administered orally or parenterally, usually by subcutaneous injection. Oral administration is common by drenching with liquids or suspensions, or by the incorporation of the drug in the feed or water for farm animals and by the administration of tablets to small animals. More recently, paste formulations have been introduced especially for horses and there are now a number of compounds which have systemic action when applied as pour-on or spot-on formulations to the skin. Also available are methods for injecting compounds directly into the rumen of cattle and sustained release devices, the latter being designed to remain in the rumen/reticulum and release anthelmintic over a period; this prevents the establishment of parasite populations and thus limits the contamination of pastures and the occurrence of disease. An apparatus for the delivery of anthelmintics into drinking water at daily or periodic intervals has also recently been developed.

A number of products are marketed for cattle and sheep consisting of a mixture of a roundworm anthelmintic and a fluke drug, but the timing of treatments for roundworms or flukes, whether curative or prophylactic, is often different and the necessity for such combination compounds is doubtful.

ANTHELMINTIC RESISTANCE

Helminth resistance to anthelmintics has been most frequently recorded in sheep and horses, and has mainly involved the benzimidazole group of compounds.

In sheep, resistance has occurred chiefly in geographical areas where *Haemonchus contortus* predominates and the annual number of cycles of infection and anthelmintic treatments are numerous. However, sometimes this resistance is incomplete and can be overcome by using higher dosage rates. Unfortunately, cross-resistance has been reported between different benzimidazoles, and, less frequently, multiple resistance embracing chemically unrelated compounds.

In western Europe anthelmintic resistance is less prevalent than in areas such as Australia and South Africa, although it has been reported with *Haemonchus* and to a lesser extent with *Ostertagia* infections of sheep and goats. In horses there is also evidence that extensive use of certain benzimidazoles has led to the selection of resistant strains of several species of small strongyles; cross-resistance between different benzimidazoles has also been reported.

Analysis of published reports on resistance suggests an association with frequent use of drugs with the same mode of action. Because of this Australian workers have recommended that the anthelmintics used should be rotated between the principal chemical groupings, for example, the benzimidazoles, the imidazothiazoles and the avermectins. Rotation should take place between generations of parasites. Thus, in western Europe, with only one or two complete generations of gastrointestinal nematodes annually, the compounds would be changed each year.

However, it is also possible that selection for resistance could be accelerated where livestock are moved on to helminthologically clean pastures immediately after anthelmintic treatment. Any contamination of these pastures would originate from helminths which had survived the treatment and so the pressure on selection for anthelmintic resistance might be increased.

ECTOPARASITICIDES (INSECTICIDES)

The control of the ectoparasites found on animals, including fleas, lice, ticks, mange mites, warbles and nuisance flies, is almost entirely based on the use of chemicals. There is a vast world market of approximately 300 million pounds in these chemicals, evenly divided between farm and companion animals.

ECTOPARASITICIDES AND THEIR MODE OF ACTION

Three main chemical groupings are used as the basis for the commonly used ectoparasiticides, namely, the organochlorines, the organophosphates and the synthetic pyrethroids. Other groups which are also used include the carbamates (primarily in poultry), the formamidines, the triazines, benzyl benzoate and natural plant products such as pyrethrin and rotenone. Recently, the avermectins have been shown to have a high activity against a range of ectoparasites and it is likely that this will form an important group in the future. The anthelmintic, closantel, is also used in some areas for the control of some ectoparasites.

ORGANOCHLORINES (OCs)

These include DDT, hexachlorocyclohexane (HCH), formerly called benzene hexachloride (BHC), the gamma isomer of which is the most potent (lindane); dieldrin and aldrin, now banned in many countries, and bromocyclen and toxaphene. They have the advantage that the effect of the drug persists for a longer time on the coat or fleece of the animal but the disadvantage, at least in food animals, that they persist in animal tissues. If toxicity occurs the signs are those of CNS stimulation with hypersensitivity, followed by increasing muscular spasm progressing to convulsions.

ORGANOPHOSPHATES (OPs)

These include a vast number of compounds of which chlorfenvinphos, coumaphos, crotoxyphos, crufomate, cythioate, diazinon, dichlofenthion, dichlorvos, fenthion, iodofenphos, malathion, phosmet, propetamphos, ronnel, tetrachlorvinphos and trichlorfon are among the most common. These can persist in the animals' coat or fleece for reasonable periods, but residues in animal tissues are short lived. Some have the ability to act systemically, given parenterally, orally or as a pour-on, but the effective blood levels of these are maintained for only 24 hours. The OPs are cholinesterase inhibitors and if toxicity occurs, the signs are salivation, dyspnoea, incoordination, muscle tremors and sometimes diarrhoea.

SYNTHETIC PYRETHROIDS (SPs)

The common synthetic pyrethroids in use include deltamethrin, permethrin, cypermethrin, fenvalerate and cyhalothrin. The main value of these compounds lies in their repellent effect and since they persist well on the coat or skin, but not in tissue, they are of particular value against parasites which feed on the skin surface such as lice, some mites and nuisance flies. Pyrethroids act as neurotoxins upon sensory and motor nerves of the neuroendocrine and central nervous system of insects. All the pyrethroids are lipophilic and this property helps them to act as contact insecticides. Some have the ability to repel and to 'knockdown' ie. affect flight and balance without causing complete paralysis. Because the synthetic pyrethroids have a strong affinity for sebum this property has been capitalised upon by incorporating the SPs into ear tags or tail bands. The SPs are fairly safe, but if toxicity does occur it is expressed in the peripheral nervous system as hypersensitivity and muscle tremors.

CARBAMATES

The most commonly used carbamates are butocarb, carbaryl and carbanolate. They are an important group in the control of poultry ectoparasites. The mode of action is similar to OPs and where toxicity occurs it is also similar.

AVERMECTINS

These are effective at very low dose levels against certain ectoparasites when given parenterally and by pour-on preparations, although the latter are not yet available commercially. They are particularly effective against ectoparasites with tissue stages such as warbles, bots and mites and have good activity against blood sucking parasites such as lice and one-host ticks. As in nematodes, they act by potentiating gamma aminobutyric acid (GABA), the only difference being that this occurs at the neuromuscular junction. They have a very wide safety margin.

The avermectins have a marked residual effect and a single treatment given parenterally is still effective against lice or mites hatching from eggs three to four weeks later.

METHODS OF APPLICATION AND USES: FARM ANIMALS

Traditionally, ectoparasiticides have been applied topically as dusts, sprays, foggers, washes, dips and occasionally used in baits to trap insects. However, the fairly recent advent of pour-on formulations with a systemic effect, the parenteral administration of drugs such as the avermectins and closantel, the use of impregnated ear tags, collars and tail-tags, and the mushrooming knowledge on sustained release technology is likely to change the methodology of control applications to animals.

TRADITIONAL METHODS

To be successful, the use of insecticides in dusts, sprays or washes usually requires two or more treatments, since even the most diligent applicant is unlikely to be successful in applying these formulations at the right concentration to all parts of the animal's body. The interval between treatments should be linked to the persistence of the chemical in the skin, hair or wool and to the life cycle of the parasite, further treatment being given prior to completion of another cycle.

Dip baths or spray races containing the necessary concentration of insecticide are used to control mites, lice and ticks and certain dipterans such as blowflies on sheep on a world wide basis and on cattle in tropical areas. This technique is more successful in sheep where the persistence of insecticide is greater in the wool fleece than in the hair coat found in cattle. It is important to remember that the concentration of insecticide in a dip bath is preferentially 'stripped' or removed as sheep or cattle are dipped, and so must be replenished at a higher than initial concentration, sufficient to maintain an adequate concentration of the active ingredient. Most dips are based on the organophosphate group with or without the addition of the organochlorine, gamma HCH.

Insect control in dairies or stables may be aided by the use of various resins strips incorporating the insecticide; dichlorvos and trichlorfon are often used for this purpose. Sometimes baits containing synthetic pheromones, sugars or hydrolysed yeasts, plus insecticide are spread around animals premises to attract and kill dipterans.

POUR-ON, SPOT-ON OR SPRAY-ON

Those available at present contain organophosphates with a systemic action such as fenthion or phosmet, or the synthetic pyrethroid, cypermethrin. They are recommended for the control of warbles and lice in cattle and lice and keds in sheep. A valuable development is that of pour-on phosmet for the control of sarcoptic mange in pigs and cattle. A single treatment in pigs gives very good results and if used in sows, prior to farrowing, prevents transmission to the litter; two treatments at an interval of 14 days are necessary in cattle. The synthetic pyrethroid cyhalothrin is recently available as a spray-on for the treatment of lice and the control of biting and nuisance flies in cattle.

EAR TAGS, COLLARS, LEG AND TAIL BANDS

These are based primarily on the synthetic pyrethroids and occasionally the organophosphates. They are recommended for the protection of cattle against nuisance flies. The tags are usually made of polyvinylchloride impregnated with the insecticide. When attached to an animal's ear the insecticide is released from the surface, dissolves in the sebum secreted by the skin and is then spread over the whole body by the normal grooming actions or ear flapping and tail swishing as well as by bodily contact between cattle. As the insecticide is rapidly bonded to the sebum on the animal's coat the treatment is rain-fast; also the tag or tail band continues to release a supply of chemical under all climatic conditions. Since the drugs are located in the sebum, they are not absorbed into the tissue so there is no need for a withdrawal period prior to slaughter nor is it necessary to discard milk. The SPs marketed for this purpose are cypermethrin, permethrin, fenvalerate and flucythrinate and one OP, tetrachlorvinphos. Under conditions of heavy fly challenge a tag should be inserted in each ear, possibly augmented by a tail band.

PARENTERAL TREATMENT

The avermectins and closantel may be given parenterally to control some ectoparasites. Ivermectin has good activity against warbles, lice, many mites and also the one-host tick *Boophilus*. Closantel is available in some tropical countries for use against one-host ticks and sucking lice.

METHODS OF APPLICATION AND USES: COMPANION OR PET ANIMALS

Ectoparasiticides are mainly used as dusting powders, aerosols, washes/shampoos and impregnated collars while two are available for oral use. They are mainly used for the control of fleas, lice and mange in dogs and cats and for lice, mange and nuisance flies in horses.

DUSTING POWDERS

The powders should be shaken well into the animal's fur or hair, and in the case of house pets, into the bedding. The powders commonly used are the organochlorine, bromocyclen, the natural pyrethroid, pybuthrin and the synthetic pyrethroid, permethrin. These are particularly useful for fleas and lice and repeat treatments are recommended every two to three weeks.

AEROSOLS

Although easy to use, some of the noisier sprays can upset pets. Over zealous spraying in confined spaces, such as in a cat basket, may produce toxic effects. Sprays available are based on bromocyclen, permethrin, a mixture of the organophosphates such as dichlorvos plus fenitrothion, or a mixture of the synergist piperonyl butoxide with OPs or pyrethroids. Depending on the spray, the aerosol container should be held at 15–30 cm from the animal and sprayed for up to five seconds for cats and a little longer for dogs. A repeat treatment is recommended in 7–14 days. The aerosol sprays are very effective for fleas and lice, but several treatments may be necessary for mange mites. The synthetic pyrethroid, permethrin is also available as a spot-on or swab and this is sometimes used in horses for the control of 'sweet-itch'.

An aerosol containing an insect growth regulator, methoprene, is also available for the control of larval populations of fleas in the environment (p. 174).

BATHS

These are available as shampoos, emulsifiable concentrates, wettable agents or creams for the control of fleas, lice and mange mites. Most preparations are for dogs and care is needed if they are used for cats. Common ingredients are the organochlorines, bromocyclen and gamma HCH, carbaryl and a range of OPs with a good persistent action such as iodofenphos and ronnel. Rotenone and benzyl benzoate are also available and the former is particularly useful for demodectic mange as is amitraz. The instructions for bathing should be carefully followed and, where necessary, care taken that the insecticide is properly rinsed from the coat. Organophosphate shampoos should not be used when dogs have insecticidal collars.

INSECTICIDAL COLLARS

These are used primarily for flea control and are entirely based on the organophosphates such as diazinon and dichlorvos. The period of protection is claimed to be three to four months, but the success of this method of application is variable. Occasional problems arise from contact dermatitis and care should be exercised that the animals do not receive other organophosphate treatments. Apart from collars, impregnated medallions are also available in some countries. Care should be taken with the use of collars in pedigreed long-haired cats and greyhound dogs due to individual susceptibility to OP poisoning.

ORAL PREPARATIONS

Two organophosphates, ronnel and cythioate, are marketed as oral preparations. These are specifically for the treatment of both demodectic mange and flea infestations in dogs and cats and the daily administration of tablets is recommended as a supplement to topical application.

OTHER PREPARATIONS

In horses, lice and areas of mange mite infestation can be treated topically, but the problem of nuisance or pasture flies remains. A recent suggestion is that ear tags impregnated with cypermethrin be attached to the saddle or mane as a possible means of incorporating the synthetic pyrethroid into the sebum.

POULTRY ECTOPARASITES

The carbamates and the organophosphate, malathion, are the most widely used. Individual birds are dusted and the insecticide applied in the poultry house, nesting boxes and litter. The possibility of using cypermethrin as a spot-on is currently under investigation.

RESISTANCE

As with other drugs, resistance has developed to most of the well-known ectoparasiticides. This is particularly true of the organochlorines and organophosphates used to control the ectoparasites of ruminants by dipping or spraying, but resistance to the synthetic pyrethroids used in impregnated ear tags has already been reported in Australia.

Resistance to insecticides is inherited and two types have been described for insects. The first is specific resistance, which is due to a single dominant gene or double recessive genes, and the other is polygenic or non-specific resistance in which resistance probably arises from the development in the insect of secondary physiological systems which bypass the primary system which is the target of the insecticide.

THE LABORATORY DIAGNOSIS OF PARASITISM

HELMINTH INFECTIONS

Although there is much current interest in the use of serology as an aid to the diagnosis of helminthiasis, particularly with the introduction of the enzyme linked immunosorbent assay (ELISA) test, faecal examination for the presence of worm eggs or larvae is the most common routine aid to diagnosis employed.

COLLECTION OF FAECES

Faecal samples should preferably be collected from the rectum and examined fresh. If it is difficult to take rectal samples, then fresh faeces can be collected from the field or floor. A plastic glove is suitable for collection, the glove being turned inside out to act as the receptacle. For small pets a thermometer or glass rod may be used.

Ideally, about 5 g of faeces should be collected, since this amount is required for some of the concentration methods of examination.

Since eggs embryonate rapidly the faeces should be stored in the refrigerator unless examination is carried out within a day. For samples sent through the post the addition of twice the faecal volume of 10% formalin to the faeces will minimise development and hatching.

METHODS OF EXAMINATION OF FAECES

Several methods are available for preparing faeces for microscopic examination to detect the presence of eggs or larvae. However, whatever method of preparation is used, the slides should first be examined under low power since most eggs can be detected at this magnification. If necessary, higher magnification can then be employed for measurement of the eggs or more detailed morphological differentiation. An eyepiece micrometer is very useful for sizing populations of eggs or larvae.

Direct smear method

A few drops of water plus an equivalent amount of faeces are mixed on a microscope slide. Tilting the slide then allows the lighter eggs to flow away from the heavier debris, a cover slip is placed on the fluid, and the preparation is then examined microscopically. It is possible to detect most eggs or larvae by this method, but due to the small amount of faeces used it may only detect relatively heavy infections.

Flotation methods

The basis of any flotation method is that when worm eggs are suspended in a liquid with a specific gravity higher than that of the eggs, the latter will float up to the surface. Nematode and cestode eggs float in a liquid with a specific gravity of between 1.10 and 1.20; trematode eggs, which are much heavier, require a specific gravity of 1.30–1.35.

The flotation solutions used for nematode and cestode ova are mainly based on sodium chloride or sometimes magnesium sulphate. A saturated solution of these is prepared and stored for a few days and the specific gravity checked prior to usage. In some laboratories a sugar solution of density 1.2 is preferred.

For trematode eggs, saturated solutions of zinc chloride or zinc sulphate are widely used. Some laboratories use the more expensive and toxic potassium mercury iodine solution.

Whatever solutions are employed the specific gravity should be checked regularly and examination of the solution containing the eggs or larvae made rapidly, otherwise distortion may take place.

Direct flotation

A small amount of fresh faeces, say 2.0 g, is added to 10 ml of the flotation solution and following thorough mixing the suspension is poured into a test tube and more flotation solution added to fill the tube to the top. A cover glass is then placed on top of the surface of the liquid and the tube and coverslip left standing for 10 to 15 minutes. The cover slip is then removed vertically and placed on a slide and examined under the microscope. If a centrifuge is available the flotation of the eggs in the flotation solution may be accelerated by centrifugation.

McMaster method

This quantitative technique is used where it is desirable to count the number of eggs or larvae per gram of faeces. The method is as follows:

1. Weigh 3.0 g of faeces or, if faeces are diarrhoeic, 3 teaspoonfuls.
2. Break up thoroughly in 42 ml of water in a plastic container.
   This can be done using a homogeniser if available or in a stoppered bottle containing glass beads.
3. Pour through a fine mesh sieve (aperture 250 microns, or 100 to 1 inch).

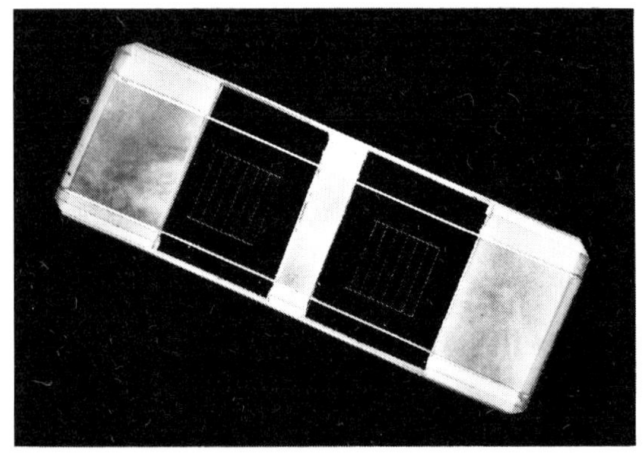

Fig 174 McMaster slide for estimating numbers of nematode eggs in faeces.

4. Collect filtrate, agitate, and fill a test tube, preferably 15 ml and flat bottomed.
5. Centrifuge at 2,000 r.p.m. for 2 minutes.
6. Pour off supernatant, agitate sediment and fill tube to previous level with flotation solution.
7. Invert tube 6 times and remove fluid with pipette to fill both chambers of McMaster slide (Fig. 174). Leave no fluid in pipette or else pipette rapidly, since the eggs will rise quickly in the flotation fluid.
8. Examine one chamber and multiply number of eggs or larvae under one etched area by 100, or two chambers and multiply by 50, to arrive at the number of eggs per gram of faeces (epg).

   If 3 g of faeces are dissolved in . . . . 42 ml,
   total volume is . . . . 45 ml,
   therefore 1 g . . . . 15 ml.
   The volume under etched area is . . . . 0.15 ml,
   therefore the number of eggs is multiplied by 100.
   If two chambers are examined, multiply by 50.

An abbreviated version of this technique is to homogenise the 3 g of faeces in 42 ml of salt solution, sieve, and pipette the filtrate directly into the McMaster slide. Although a faster process the slide contents are more difficult to 'read' because of their dark colour.

It is impossible to calculate from the epg the actual worm population of the host, since many factors influence egg production of worms and the number of eggs also varies with the species. Nevertheless, egg counts in excess of 1,000 are generally considered indicative of heavy infections and those over 500 of moderate infection. However, a low epg is not necessarily indicative of very low infections, since patency may just be newly established; alternatively, the epg may be affected by developing immunity. The eggs of some species, such as certain ascarids, *Strongyloides*, *Oxyuris*, *Trichuris* and *Capillaria*, can be easily recognised morphologically. However, with the exception of *Nematodirus* spp., the common trichostrongyle eggs require measurement for differentiation.

While this technique will detect the eggs and larvae of most nematodes, cestodes, and coccidia, it will not demonstrate trematode eggs which have a higher specific density. For these, a flotation fluid of higher specific gravity such as a saturated solution of zinc sulphate has to be used or a sedimentation method employed as described below.

Sedimentation methods

For trematode eggs. Homogenise 3 g of faeces with water and pass the suspension through a coarse mesh sieve (250 microns). Thoroughly wash the material retained on this screen using a fine water jet and discard the debris.

Transfer the filtrate to a conical flask and allow to stand for 2 minutes, remove the supernatant and transfer the remainder (approx. 12–15 ml) to a flat-bottomed tube.

After sedimentation for a further 2 minutes the supernatant is again drawn off, a few drops of 5% methylene blue added and the sediment screened using a low power stereomicroscope.

Any trematode eggs are readily visible against the pale blue background.

For lungworm larvae, the Baerman apparatus may be used. This consists of a glass funnel held in a retort stand. A rubber tube attached to the bottom of the funnel is constricted by a clip. A sieve (aperture 250 microns) is placed in the wide neck of the funnel, which has been partially filled with water, and a double layer of gauze is placed on top of the sieve. Faeces are placed on the gauze and the funnel is slowly filled with water until the faeces are immersed. Alternatively, faeces are spread on a filter paper which is then inverted and placed on the sieve (Fig. 175). The apparatus is left overnight at room temperature during which the larvae migrate out of the faeces and through the sieve to sediment in the neck of the funnel. The clip on the rubber is then removed and the water in the neck of the funnel collected in a small beaker for microscopic examination in a Petri dish.

Fig 175 The Baerman apparatus. Faeces are spread on a filter-paper which is then inverted and placed on the sieve.

A simple adaptation of the above method is to suspend the faeces enclosed in gauze in a urine glass filled with water and leave overnight. The larvae will leave the faeces, migrate through the gauze and settle at the bottom of the glass. After siphoning off the supernatant, the sediment is examined under the low power of the microscope as above.

CULTURE AND IDENTIFICATION OF LARVAE

Two techniques are widely used for the culture of infective larvae from nematode eggs.

In the first, faeces are placed in a jar with a lid and stored in the dark at a temperature of 21 to 24 °C. The lid should be lined with moist filter paper and should not be tightly attached. After 7 days incubation, the jar is filled with water and allowed to stand for 2 to 3 hours. The larvae will migrate into the water and the latter is poured into a cylinder for sedimentation. The larval suspension can be cleaned and concentrated by sedimentation in a Baerman apparatus as described above and then killed by adding a few drops of Lugol's iodine and examined microscopically.

An alternative method is to spread the faeces on the middle third of a filter paper placed in a moistened Petri dish. After storage at 21 to 24 °C for 7 to 10 days, the dish is flooded with water and the larvae harvested as before.

The identification of infective larvae is a specialist technique and for this, reference may be made to the publications listed at the end of this section.

RECOVERY OF ALIMENTARY NEMATODES

Details are given below of a technique for the collection, counting and identification of the alimentary nematodes of ruminants. The procedure is similar for other host species, information on identification being available in the text.

1. As soon as possible after removing the alimentary tract from the body cavity, the abomasal/duodenal junction should be ligatured to prevent transfer of parasites from one site to the other.
2. Separate the abomasum, small intestine and large intestine.
3. Open abomasum along the side of the greater curvature, wash contents into a bucket under running water and make the total volume up to 2 to 4 litres.
4. After thorough mixing transfer duplicate 200 ml samples to suitably labelled containers and preserve in 10% formalin.

5. Scrape off the abomasal mucosa and digest in a pepsin/HCl mixture at 42 °C for 6 hours; 200 g of mucosa will require 1 litre of mixture. Make digest up to a volume of 2 or 4 litres with cold water and again take duplicate 200 ml samples.

   Alternatively, the Williams technique may be used. In this, the washed abomasum is placed, mucosal surface down, in a bucket containing several litres of normal saline and maintained at 40 °C for 4 hours. Subsequently, the abomasum is gently rubbed in a second bucket of warm saline. The saline from both buckets is poured through a sieve (aperture 38 microns, about 600 to 1 inch) and the residue examined.
6. Open small intestine along its entire length and wash contents into a bucket. Treat as for the abomasal contents, but digestion of mucosal scrapings is unnecessary.
7. The contents of the large intestine are washed into a bucket, passed through a coarse mesh sieve (aperture 2–3 mm) and any parasites present collected and formalised.

Preparation of pepsin/HCl:
Dissolve 80 g of pepsin powder in 3 litres of cold water. Add 240 ml concentrated HCl slowly and stir well. Make final volume up to 8 litres. Store at 4°C.

Worm counting procedure

1. Add 2–3 ml of Iodine solution to one of the 200 ml samples.
2. After thorough mixing, transfer 4 ml of suspension to a petri dish, scored with lines to facilitate counting; add 2–3 ml sodium thiosulphate solution to decolourise debris.

   If necessary worms may be preserved after washing in saline in an aqueous solution of 10% formalin or 70% alcohol. To clear large worms for microscopic examination, immerse in lactophenol for a suitable period prior to examination.
3. Examine for the presence of parasites using a stereoscopic microscope (× 12 objective) and identify and count parasites as male, female and larval stages.

Iodine solution
Dissolve 907 g of potassium iodide in 650 ml boiling water. Add 510 g iodine crystals and make up to 1 litre.

Sodium thiosulphate solution
Dissolve 100 g of sodium thiosulphate in 5 litres of water.

A guide to the adult alimentary nematodes of sheep and cattle follows.

GUIDE TO ADULT ALIMENTARY NEMATODES

SHEEP

GROSS CHARACTERS

Abomasum

1. 2 cm long; bursa visible with naked eye: females have 'barber's pole' appearance; reddish when fresh	*Haemonchus*
2. 1 cm long; slender; reddish brown when fresh	*Ostertagia*
3. Less than 0.5 cm long; the smallest trichostrongyloid of ruminants; cannot be easily seen on abomasal wall or in contents; greyish when fresh	*Trichostrongylus axei*

Small intestine

1. 0.5 cm long; slender; greyish when fresh	*Trichostrongylus* or *Strongyloides*
2. 0.5 cm long; comma or watch-spring shape; slender; greyish	*Cooperia curticei*
3. 2 cm long; slender; much twisted, often tangled like cotton wool	*Nematodirus*
4. 2 cm long; stout white worms; head bent slightly	*Bunostomum*

Large intestine

1. Up to 8 cm long; whip-like, with long filamentous anterior part twice as long as posterior part	*Trichuris*
2. 1.5 to 2 cm long; large bell-shaped buccal capsule	*Chabertia*
3. Up to 2 cm long; buccal capsule tapered and not obvious as in *Chabertia*. Head end bent dorsally	*Oesophagostomum*

MICROSCOPIC CONFIRMATION

Abomasum

Haemonchus

Male: Dorsal ray of bursa asymmetric; spicules barbed near tips.

Female: Vulval flap, usually linguiform, present; gravid worm contains several hundred eggs; ovary coiled around intestine.

Ostertagia

Male: Spicules slender, rod-like (*O. circumcincta*) or stout with branch near middle (*O. trifurcata*).

Trichostrongylus axei

Both sexes: Excretory notch visible in oesophageal region.

Male: Spicules unequal in size and shape.

Female: Vulval flap absent; gravid worm contains 4 or 5 eggs, pole to pole.

Small intestine

Trichostrongylus

Both sexes: Excretory notch present in oesophageal region.

Male: Spicules leaf-shaped (*T. vitrinus*) or with 'step' near tip (*T. colubriformis*)

Female: Vulval flap absent; ovejectors present (cf *Strongyloides*, below).

Strongyloides
Only females present; long oesophagus; ovary and uterus show twisted thread appearance behind oesophagus; ovejectors absent.

Cooperia

Both sexes: Small cephalic vesicle present, giving anterior end a cylindrical appearance; prominent cuticular striations in oesophageal region.

Male: Spicules have 'wing' at middle region, bearing striations.

Nematodirus

Both sexes: Cephalic vesicle present.

Male: Spicules long, slender and fused, with expanded tip which is heart-shaped (*N. battus*); lanceloate (*N. filicollis*); bluntly rounded (*N. spathiger*). Bursa shows 2 sets of parallel rays (*N. battus*) or 4 sets (other species).

Female: Large eggs present; tip of tail is pointed (*N. battus*) or truncate with a small spine (other species).

Bunostomum
Large buccal capsule present.

Large intestine

Trichuris

Microscopic confirmation unnecessary, because of whip-like shape. Tail of female is bow-shaped and that of male spirally coiled with one spicule.

Chabertia

Large bell-shaped buccal capsule with no teeth and rudimentary leaf crowns.

Oesophagostomum

Relatively small buccal capsule; cervical vesicle with cervical groove behind it. Leaf crowns and cervical alae often present.

CATTLE

GROSS CHARACTERS

The gross characters are similar to those described for the nematodes of sheep.

MICROSCOPIC CONFIRMATION

Abomasum

Haemonchus

As in sheep, but vulval flap often bulb-shaped or vestigial.

Ostertagia

As in sheep, but male has stout, rod-like spicules with expanded tips (*O. ostertagi*) or very robust spicules, generally rectangular in outline (*O. lyrata*). Female has vulval flap of variable size, but usually skirt-like.

T. axei

As in sheep.

Small intestine

Trichostrongylus
As in sheep; *T. vitrinus* is very rare in cattle.

Cooperia

As in sheep, but the spicules of the common species, *C. oncophora*, have a stout, bow-like, appearance, with small terminal 'feet'.

Nematodirus

As in sheep; the spicules of the common bovine species, *N. helvetianus*, have a spear-shaped expansion at the tips.

Bunostomum

As in sheep.

Large intestine

As in sheep.

Based on the characters described above, the following key can be used to differentiate microscopically the genera of some common gastrointestinal nematodes of ruminants.

Body composed of a long filamentous anterior and a short broad posterior region	*Trichuris*
Body not so divided, oesophagus approximately one third of body length	*Strongyloides*
Short oesophagus and buccal capsule rudimentary	Tricho-strongyloidea (a)
Short oesophagus and buccal capsule well developed	Strongyloidea (b)
(a) **Trichostrongyloidea:**	
1. Distinct cephalic vesicle. Spicules very long uniting in a membrane at the tip	*Nematodirus*
Cephalic vesicle small. Spicules relatively short and unjoined posteriorly	*Cooperia*
2. No cephalic vesicle. Excretory notch present in both sexes.	*Trichostrongylus*
Absence of excretory notch	(3)
3. Dorsal lobe of bursa asymmetrical, barbed spicules. Large prominent vulval flap in female	*Haemonchus*
Dorsal lobe of bursa is symmetrical. Vulval flap small or absent	*Ostertagia*
(b) **Strongyloidea:**	
4. Buccal capsule small and cylindrical	*Oesophagostomum*
Buccal capsule well developed	(5)
5. Slight dorsal curvature of head and presence of teeth	*Bunostomum*
Absence of teeth, rudimentary leaf crowns present	*Chabertia*

RECOVERY OF LUNGWORMS

For *Dictyocaulus*, this is best done by opening the air passages starting from the trachea and cutting down to the small bronchi with fine blunt-pointed scissors. Visible worms are then removed from the opened lungs and transferred to glass beakers containing saline. The worms are best counted immediately, failing which they should be left overnight at 4 °C which will reduce clumping. Additional worms may be recovered if the opened lungs are soaked in warm saline overnight.

Another method is Inderbitzen's modification of the perfusion technique described by Wolff et al (1969) in which the lungs are perfused as follows:

The pericardial sac is incised and reflected to expose the pulmonary artery in which a 2 cm incision is made. Rubber tubing is introduced into the artery and fixed *in situ* by double ligatures. The remaining large blood vessels are tied off and water from a mains supply allowed to enter the pulmonary artery. The water ruptures the alveolar and bronchiolar walls, flushes out the bronchial lumina, and is expelled from the trachea. The fluid is collected and its contents concentrated by passing through a fine sieve (aperture 38 microns). As before, this is best examined immediately for the presence of adult worms and larvae.

The smaller genera of lungworms of small ruminants are difficult to recover and enumerate, although the Inderbitzen technique may be of value.

RECOVERY OF TREMATODE AND CESTODE PARASITES

For both *Fasciola* and *Dicrocoelium* the livers are removed and cut into slices approximately 1 cm thick. On squeezing the liver slices, any flukes seen grossly are removed and formalised and the slices immersed in warm water overnight. The gall bladder should also be opened and washed, and any flukes removed.

After soaking, the liver slices are again squeezed, rinsed in clean water and discarded. Both washings are passed through a fine sieve (aperture 100 microns) and the material retained formalised. In the case of intestinal paramphistomes, the first 4 metres of duodenum should be tied off, opened, washed and examined for adherent trematodes.

Counts are carried out microscopically, entire flukes plus the numbers of heads and tails being recorded. The highest number of either of the latter is added to the number of entire flukes to give the total count.

Cestodes are usually readily visible in the intestine or liver, but whenever possible these should be removed intact so that, if necessary, the head and the mature and gravid segments are all available for specialist examination. In the case of *Echinococcus* in canids, however, the worms are so small that the more detailed examination described in the text should be undertaken.

OTHER AIDS TO DIAGNOSIS

There are two other techniques which are useful aids in the diagnosis of trichostrongyle infections in ruminants. The first is the plasma pepsinogen test and the second the estimation of infective larvae on herbage.

Both of these techniques are usually beyond the scope of the general practitioner, but a short account is given here of the material required for these tests, the basis of the techniques and how the results may be interpreted.

THE PLASMA PEPSINOGEN TEST

The estimation of circulating pepsinogen is of value in the diagnosis of abomasal damage, and is especially elevated in cases of ostertagiasis. Elevations also occur with other gastric parasites such as *Trichostrongylus axei*, *Haemonchus contortus*, and in the pig *Hyostrongylus rubidus*.

The principle of the test, which is best carried out by a diagnostic laboratory, is that the sample of serum or plasma is acidified to pH 2.0, thus activating the inactive zymogen, pepsinogen, to the active proteolytic enzyme pepsin. This activated pepsin is then allowed to react with a protein substrate (usually bovine serum albumin) and the enzyme concentration calculated in international units (μ mols tryosine released per 100 ml serum per minute). The tyrosine liberated from the protein substrate by the pepsin is estimated by the blue colour which is formed when phenolic compounds react with Folin-Ciocalteu's reagent. The minimum requirements for the test, as carried out in most laboratories, is 1.5 ml serum or plasma. The anticoagulant used for plasma samples is either EDTA or heparin.

In parasitic gastritis of ruminants due to *Ostertagia* spp. and *T. axei* the levels of plasma pepsinogen become elevated. In parasite-free animals the level is less than 1.0 i.u. of tyrosine; in moderately infected animals, it is between 1.0 and 2.0 and in heavily infected animals it usually exceeds 3.0 reaching as high as 10.0 or more on occasion. Interpretation is simple in animals during their first 18 months, but thereafter becomes difficult as the level may become elevated when older and immune animals are under challenge (see page 15). In such cases the absence of the classical clinical signs of diarrhoea and weight loss indicates that there are few adult parasites present.

PASTURE LARVAL COUNTS

For this technique, samples of grass are plucked from the pasture and placed in a polythene bag which is then sealed and dispatched to a laboratory for processing. It is important to take a reasonable number of random samples, and one method is to traverse the pasture and remove four grass samples at intervals of about four paces until approximately 400 have been collected; another, primarily for lungworm larvae, is to collect a similar number of samples from the close proximity of faecal pats. At the laboratory, the grass is thoroughly soaked, washed and dried and the washings containing the larvae passed through a sieve (aperture 38 microns; 600 to 1 inch) to remove fine debris. The material retained in the sieve is then baermanised and the infective larvae identified and counted microscopically under the high power. The numbers present are expressed as L_3 per kg of dried herbage.

Where counts in excess of 1,000 L_3/kg of ruminant gastrointestinal trichostrongyles are recorded, the pasture can be regarded as moderately infective and values of over 5,000 L_3/kg can be expected to produce clinical disease in young cattle during their first season at grass.

Although this is a useful technique for detecting the level of gastrointestinal nematode L_3 on pastures, it is less valuable for detecting lungworm larvae because of the rapid fluctuations of these larvae on pastures. A more sophisticated technique, the Jorgensen method, which depends on migration of larvae through an agar medium containing bile, is used in some laboratories for estimating *Dictyocaulus* larval populations on pasture; since most lungworm larvae are concentrated close to faeces, herbage samples should be collected from around faecal deposits. In the present state of knowledge, the detection of any lungworm larvae in herbage samples should be regarded with suspicion and even a negative finding does not necessarily imply that the pasture is free of infection.

ECTOPARASITES

Arthropods of veterinary interest are divided into two major groups, the Insecta and the Arachnida. Most are temporary or permanent ectoparasites, found either in or on the skin with the exception of some flies whose larval stages may be found in the somatic tissues of the host. Parasitic insects include flies, lice and fleas, while the two groups of arachnids of veterinary importance are the ticks and mites. In all cases diagnosis of infection depends on the collection and identification of the parasite(s) concerned.

INSECTS

Adult dipteran flies visiting animals are usually caught either by netting or after being killed by insecticides, while larvae may be collected in areas where animals are housed or directly from animals where the larval stages are parasitic. Identification of the common flies of veterinary interest, at least to generic level, is fairly simple, the key characters being described in the main text whereas identification of larvae to generic and species level is rather more specialised and depends on examination of certain features such as the structure of the posterior spiracles. Publications dealing with this may be found at the end of the chapter.

LICE AND FLEAS

The detection of small ectoparasites such as lice and fleas depends on close examination; in the case of lice, the eggs, commonly known as 'nits', may also be found attached to the hair or feathers. Fleas may be more difficult to detect, but the finding of flea faeces in the coat, which appear as small dark pieces of grit and which on contact with moist cotton wool or tissue produce a red coloration due to ingested blood, allow confirmation of infection. Collection may be straightforward as in the case of many lice which may be brushed from the coat or removed by clipping hairs or feathers. Fleas may be removed by brushing or vacuum cleaning (see page 174). Alternatively, in the case of small animals, the parasites may be readily recovered if the host is placed on a sheet of paper or plastic before being sprayed with an insecticide. The gross characteristics of biting and sucking lice, and a key to the fleas which are commonly found on domestic animals are described in the text.

TICKS

Ticks are easily recognised on their hosts, especially when they are engorged, but care should be taken in their removal since their mouthparts are usually firmly embedded in the skin. The tick may be persuaded to withdraw its mouthparts if a piece of cotton wool, soaked in anaesthetic, is placed around it or, alternatively, if a lighted cigarette is held near its body.

One of the simplest methods used to recover ticks from pasture is to drag a blanket over the ground to which the unfed ticks become attached as they would to a host. Specific identification of the large variety of ticks which parasitise domestic animals is a specialised task. For those of western Europe, a key is given on page 179.

MITES

Some non-burrowing mites such as *Otodectes* and *Cheyletiella* can be found by close examination. For example *Otodectes* may be seen either on examination of the external auditory canal using an auroscope or on microscopic examination of ear wax removed by means of a swab; likewise, rigorous brushing of the coat and subsequent microscopic examination of this material will usually confirm infection with *Cheyletiella*. For the demonstration of some non-burrowing and burrowing mites it is often necessary to obtain a skin scraping which is subsequently examined microscopically. The area selected for scraping should be at the edge of a visible lesion and the hair over this area should be clipped away. A drop of lubricating oil such as liquid paraffin is placed on a microscope slide and a clean scalpel blade dipped in the oil before using it to scrape the surface of a fold of affected skin. Scraping should be continued until a slight amount of blood oozes from the skin surface and the material obtained then transferred to the oil on the slide. A coverslip should then be applied and the sample examined under low magnification (× 100). If during this initial examination no mites are detected a further sample may be heated on a slide with a drop of 10% caustic potash. After allowing this preparation to clear for 5–10 minutes it should be re-examined.

PRESERVATION

Most adult arthropods and their developing stages may be preserved satisfactorily in 70% alcohol in small glass or plastic tubes. A plug of cotton wool should be pushed down the tube to limit damage during transit and the tube firmly corked and labelled. Otherwise, the specimen may be pinned through the thorax on to the cork stopper of a specimen tube, but this is best left to the specialist for all but the largest flies.

PROTOZOAL INFECTIONS

The laboratory diagnosis of protozoal diseases is often relatively straightforward and well within the scope of the general practitioner, although on other occasions it may require specialised techniques and long experience. This section is concerned primarily with the former and supplements the information already given in the general text.

THE EXAMINATION OF FAECAL SAMPLES

The McMaster slide is the simplest technique for detecting the presence and estimating the number of coccidial oocysts in faeces. The technique is exactly the same as that described for helminthological diagnosis although the small size of the oocysts makes the microscopic examination more prolonged. If the animal has acute clinical signs of coccidiosis, such as blood-stained faeces, and many thousands of oocysts are present, one may reasonably consider that the diagnosis is confirmed. Unfortunately, with the more pathogenic species of coccidia, clinical signs may appear during the schizogonous phase or when oocyst production has just started, so that a negative or low oocyst count does not necessarily indicate that the clinical diagnosis was wrong. The oocyst count is also of little value in the less acute coccidial infections associated with production losses. In general, because of the limitations of the oocyst count, a postmortem examination, at least on poultry, is always advisable.

For the detection of intestinal protozoa such as *Entamoeba*, *Giardia* or *Balantidium*, a small amount of fresh faeces may be mixed with warm saline and examined under a warm stage microscope for the presence of trophozoites or cysts. However, their identification requires considerable experience and faecal samples preserved in formalin or polyvinyl alcohol may be sent to a specialist laboratory for confirmation.

The diagnosis of suspected *Cryptosporidium* infection depends on the examination of faecal smears stained by the Ziehl-Nielsen technique, the small thin-shelled oocysts appearing bright red.

THE EXAMINATION OF BLOOD AND LYMPH

Thin blood smears stained with Romanowsky dyes, such as Giemsa or Leishman, and examined under an oil immersion lens are commonly used for the detection of trypanosomes, babesial and theilerial piroplasms and rickettsial infections such as anaplasmosis, ehrlichiosis and eperythrozoonosis. On other occasions, needle biopsies of enlarged lymph nodes may be similarly stained for the detection of trypanosomes (especially *Trypanosoma brucei* or *T. vivax*) or theilerial schizonts.

In trypanosomiasis, the parasitaemia may be light and the chance of a positive diagnosis is increased if a thick blood film, dehaemoglobinised by immersing the slide in water before eosin staining, is used. For this a drop of fresh blood, with no added anti-coagulant, is gently stirred on a slide to cover an area of about 10 mm diameter and allowed to dry. Subsequently it may be stained by Field's technique as follows:

Field's Stain

Preparation of solutions:

Solution A	Methylene blue	0.4g
	Azure I	0.25g
	Solution B	250 ml
Solution B	$Na_2HPO_4 12H_2O$	25.2g
	KH_2PO_4	12.5g
	Distilled water	1000 ml
Solution C	Eosin	0.5g
	Solution B	250 ml

These solutions do not keep and should be freshly prepared each day.

1. Dip slide in solution 'A'	1 to 3 seconds
2. Rinse in solution 'B'	2 to 3 seconds
3. Dip slide in 'C'	1 to 3 seconds
4. Rinse in tap water	2 to 3 seconds
5. Stand upright to drain and dry	

This technique is commonly used in large scale survey work in the field.

A particularly efficient diagnostic technique for trypanosomiasis, described earlier in the text, is the examination, under darkground illumination, of the expressed buffy coat of a microhaematocrit tube for the detection of motile trypanosomes.

The inoculation of mice with fresh blood from suspected cases of *Trypanosoma congolense* or *T. brucei* infection is another common technique practised in the field. Three days later the tail blood of such mice should be examined and subsequently daily thereafter for about three to four weeks to establish if trypanosomes are present.

The detection of specific antibody in a specialist laboratory may also be useful in the diagnosis of several protozoal diseases such as theileriosis, trypanosomiasis, including *T. cruzi* infection, babesiosis, cryptosporidiosis and rickettsial infections such as anaplasmosis and ehrlichiosis. However a positive result does not necessarily imply the presence of a still active infection, but simply that the animal has at some time been exposed to the pathogen. An exception to this interpretation is the diagnosis of suspected toxoplasmosis in sheep, where rising antibody levels over a period of several weeks are reasonable evidence of recent and active infection.

EXAMINATION OF SKIN

Histological examination of skin biopsies or scrapings from the edges of skin ulcers, suspected to be due to leishmaniasis, may be used to demonstrate the amastigote parasites in the macrophages.

In dourine, caused by *Trypanosoma equiperdum*, fluid extracted from the cutaneous plaques usually offers a better chance of detecting trypanosomes than blood smears.

Finally, although not within the province of the general practitioner, the use of **xenodiagnosis** as a diagnostic technique should be noted. This is used to detect protozoal infections such as babesiosis, thieleriosis or *Trypanosoma cruzi* infection where the parasite cannot be found easily. It consists of allowing the correct intermediate host, such as a tick or a haematophagous bug, to feed on the animal. These arthropod vectors have, of course, to be reared in the laboratory so that they are free from infection. After feeding, the arthropod host is maintained for several weeks to allow any ingested organisms to multiply, after which it is killed and examined for evidence of infection. Although a valuable technique, especially for the detection of carrier states, the method has the disadvantage that the diagnosis may take several weeks.

REFERENCES

Edwards, K., Jepson, R. P. and Word, K. F. (1960). Value of plasma pepsinogen estimation, *British Medical Journal*, **1**, 30.

Jorgensen, R. (1975). Isolation of infective *Dictyocaulus* larvae from herbage, *Veterinary Parasitology*, **1**, 61.

Manual of Veterinary Parasitological Laboratory Techniques, Technical Bulletin No. 18 (1977). Ministry of Agriculture, Fisheries and Food. Her Majesty's Stationery Office, London.

Williams, J. C., Knox, J. W., Sheehan, D. and Fuselier, R. H. (1977). Efficacy of albendazole against inhibited early 4th stage larvae of *Ostertagia ostertagi*, *Veterinary Record*, **101**, 484.

Wolff, K., Ruosch, W. and Eckert, J. (1969). Perfusionstechnik zur Gewinnung von *Dicrocoelium dendriticum* aus Schaf-und Rinderlebern, *Zeitschrift für Parasitenkunde*, **33**, 85.

INDEX